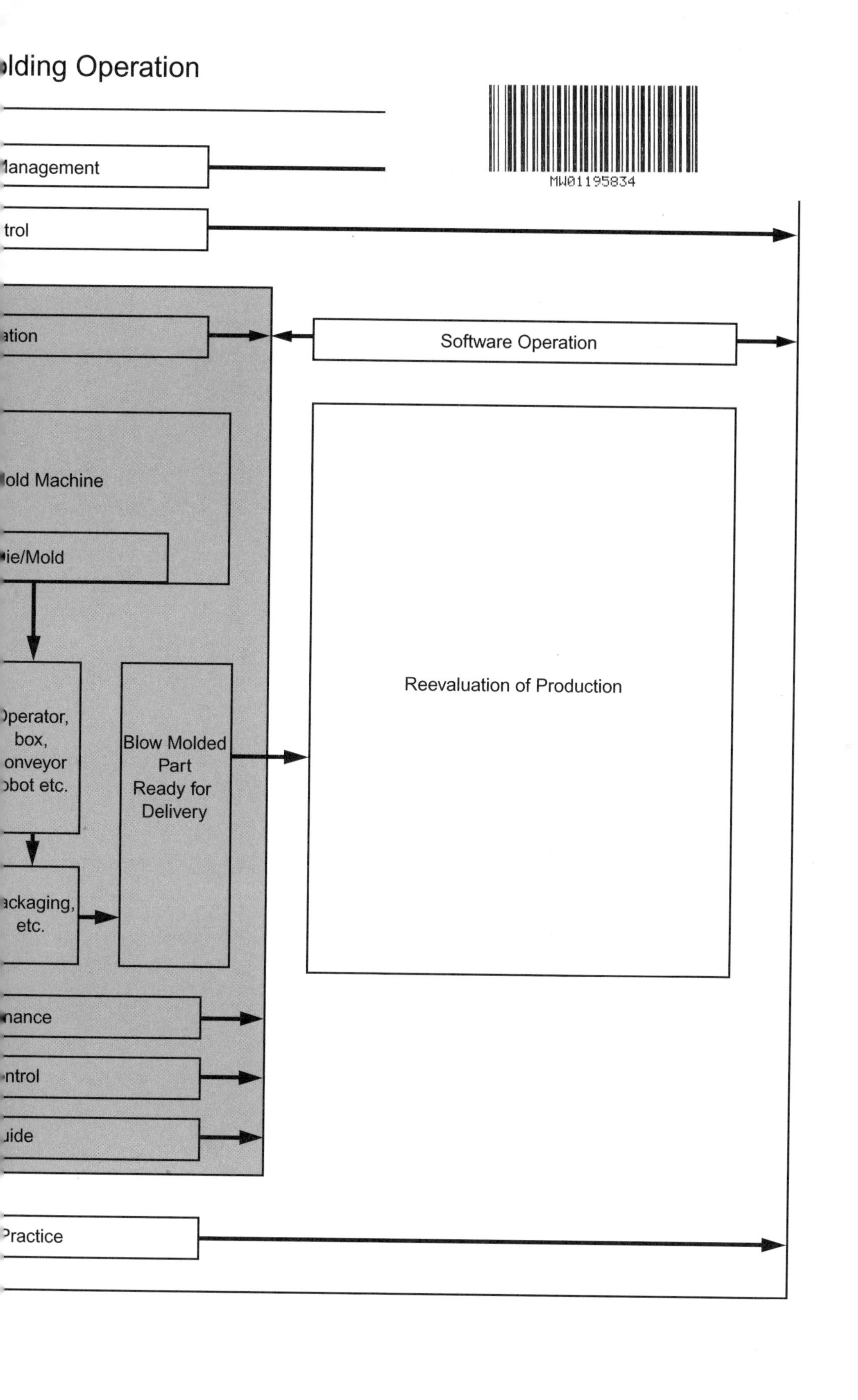
lding Operation
MW01195834
lanagement
trol
ation
Software Operation
lold Machine
ie/Mold
Reevaluation of Production
perator,
box,
onveyor
bot etc.
Blow Molded
Part
Ready for
Delivery
ackaging,
etc.
nance
ntrol
iide
Practice

Rosato / Rosato / Di Mattia
Blow Molding Handbook

Blow Molding Handbook

Technology, Performance, Markets, Economics
The Complete Blow Molding Operation

2nd revised Edition

by
Dominick V. Rosato, BSME, PE, PE
Andrew V. Rosato
David P. DiMattia

HANSER
Hanser Publishers, Munich
Hanser Gardner Publications, Inc., Cincinnati

The Authors:
Dominck V. Rosato, Andrew V. Rosato, David P. DiMattia, all: 215 Riverview Drive, Chatham, MA 02633, USA

Distributed in the USA and in Canada by
Hanser Gardner Publications, Inc.
6915 Valley Avenue, Cincinnati, Ohio 45244-3029, USA
Fax: (513) 527-8801
Phone: (513) 527-8977 or 1-800-950-8977
www.hansergardner.com

Distributed in all other countries by
Carl Hanser Verlag
Postfach 86 04 20, 81631 München, Germany
Fax: +49 (89) 98 48 09
www.hanser.de

Library of Congress Cataloging-in-Publication Data

Blow molding handbook : technology, performance, markets, economics :
the complete blow molding operation / [edited] by Dominick V. Rosato,
Andrew V. Rosato, David P. DiMattia.— 2nd ed.
p. cm.
Includes bibliographical references and index.
ISBN 1-56990-343-3
1. Plastics—Molding—Handbooks, manuals, etc. I. Rosato, Dominick V.
II. Rosato, Andrew V. III. Di Mattia, David P.
TP1150.B57 2003
668.4'12—dc21
2002155935

Bibliografische Information Der Deutschen Bibliothek
Die Deutsche Bibliothek verzeichnet diese Publikation in der Deutschen Nationalbibliografie;
detaillierte bibliografische Daten sind im Internet über <http://dnb.ddb.de> abrufbar.

ISBN 3-446-22017-8

Production Management: Oswald Immel
Typeset: Techset Composition, Salisbury, UK
Coverillustration: SIG Blowtec GmbH & Co. KG, Troisdorf, Germany
Coverdesign: MCP • Susanne Kraus GbR, Holzkirchen, Germany
Printed and bound by Druckhaus „Thomas Müntzer“, Bad Langensalza, Germany

Preface

This second edition has been written to update the subject of blow molding (BM) in the plastics industry. Many examples are provided of the BM processes using different plastics and relating them to critical factors, ranging from product designs to meeting performance requirements to reducing costs to zero defect targets. Detailed explanations and interpretations of individual topics are provided that include the operation of the basic extrusion and injection molding machines that are vital to successful BM products. Throughout the book there is comprehensive information on problems and solutions as well as extensive cross-referencing.

This book provides important details on fabricating the different types of BM products that includes the newer techniques. In addition it provides information and details (including operation and troubleshooting) on the basic extruders and injection molding machines that are required and important to successfully BM products.

In the manufacture of BM products there is always a challenge to utilize advanced techniques that includes, as reviewed in this book, understanding the different plastic melt flow behaviors, operational monitoring and control systems, testing and quality control, statistical analysis, and so on. However, these techniques are helpful only if the basic operations of molding are understood and characterized, to ensure the elimination or significant reduction of potential problems.

The book provides an understanding on the subject of BM for either the technical or nontechnical reader that is concise, practical, and comprehensive, starting with the plastic's melt flow behavior during processing. It is useful to different people including the fabricator, mold maker, designer, engineer, maintenance person, accountant, plant manager, testing and quality control individual, statistician, cost estimator, sales and marketing personnel, new venture individual, buyer, vendor, educator/trainer, workshop leader, librarian/information provider, lawyer, consultant, and others. People with different interests can focus on and interrelate across subjects with which they have limited or no familiarity. As explained throughout this book, understanding the BM process is required to be successful in the design, prototype, and manufacture of the many different, marketable BM products worldwide.

What makes this book unique is that the reader will have a useful reference of pertinent information readily available. As reviewers have commented, the information contained in the book is of value to even the most experienced fabricators, designers, and engineers; it also provides a firm basis for the beginner. The intent is to provide a complete review of the important aspects of the BM process that goes from the practical to the theoretical, and from the elementary to the advanced.

This book can provide people not familiar with BM with an understanding of how to fabricate products in order to obtain its benefits and advantages. It also provides information on the usual costly pitfalls or problems that can develop, resulting in poor product performances or failures. Accompanying the extensive problems listed are solutions. This information will enhance the intuitive skills of those people who are already working in plastics. The emphasis in this book is in providing a guide to understanding this worldwide technology and profitable business of BM products.

From a pragmatic standpoint, any theoretical aspect that is presented has been prepared so that it is understood and useful. The theorist, for example, will gain an insight into the limitations that exist relative to other materials such as steel, wood, and so on. Based on almost a century of worldwide production of all kinds of BM products, they can be processed successfully, meeting high quality, consistency, and profitability standards. One can apply the correct performance factors based on an intelligent understanding of the subject.

This book has been prepared with the awareness that its usefulness will depend on its simplicity and its ability to provide essential information. With the experience gained in working in the BM industry worldwide and in preparing the first edition as well as 23 other books on different aspects of plastics, we are able to provide a useful book. The book meets the criteria of providing a uniquely useful, practical reference work.

Note that the BM industry comprises about 10 wt% of all plastics. In turn the total plastic industry is ranked as the fourth largest industry in the United States. To a greater extent than with other materials, an opportunity will always exist to optimize the use of plastics, as new and useful developments in materials and processing continually are on the horizon. Examples of these developments are in this book, providing past to future trends in BM.

The limited plastics material properties information and data presented are provided as comparative guides; readers can obtain the latest information from material suppliers, industry software and/or as reviewed in this book's reference section. Our focus in the book is to present, interpret, analyze, and interrelate the basic elements of BM to processing plastic products. As explained in this book, even though there are over 35,000 plastic materials worldwide, selecting the right plastic requires applying certain factors. They include defining all product performance requirements, properly setting up or controlling the extrusion and injection molding processes to be used, and intelligently preparing a material specification purchase document and work order to produce the BM product.

New developments in plastic materials and equipment are always on the horizon, requiring updates. With the many varying properties of the different plastics, there are those that meet high performance requirements such as long time creep resistance, fatigue endurance, toughness, and so on. Conversely, the use of other plastics is driven by volume and low cost. Each of the different materials requires their specific BM operating procedures.

Patents or trademarks may cover information presented. No authorization to utilize these patents or trademarks is given or implied; they are discussed for information purposes only. The use of general descriptive names, proprietary names, trade names, commercial designations, or the like does not in any way imply that they may be used freely. While information presented represents useful information that can be studied or analyzed and is believed to be true and accurate, neither the authors nor the publisher can accept any legal responsibility for any errors, omissions, inaccuracies, or other factors.

In preparing this book to ensure its completeness and the correctness of the subjects reviewed, use was made of the authors' worldwide personal, industrial, and teaching experiences totaling over a century, as well as worldwide information from industry (personal contacts, material and equipment suppliers, conferences, books, articles, etc.) and trade associations.

Dominick V. Rosato
Drew V. Rosato
David P. DiMattia

CONTENTS

1 Introduction

1.1 Overview

Blow molding (BM) makes it possible to manufacture molded products economically, in unlimited quantities, with virtually no finishing required. The basic process of blow molding involves a softened thermoplastic hollow form which is inflated against the cooled surface of a closed mold. The expanded plastic form solidifies into a hollow product. The surfaces of the moldings are smooth and bright, or as grained and engraved as the surfaces of the mold cavity in which they are processed.

With plastics in BM products, to a greater extent than for other materials, an opportunity will always exist to optimize the use of BM, as new and useful materials and processes are developing continually and providing future trends in the plastics industry. Selecting the right plastic involves tasks such as setting up performance target requirements, choosing and adapting the process to be used, and intelligently preparing a specification purchase document and work order. Given the widely varying properties of the different plastics, some types meet high performance requirements such as long time creep resistance, fatigue endurance, toughness, and so on. Conversely, the use of some other plastics is volume and cost driven.

Blow molded components are eroding the market for traditional materials, particularly in liquid packaging applications. In the last few decades, since the introduction of polyethylene (PE) squeeze bottles for washing liquids, polyvinyl chloride (PVC) for cooking oils and fruit squash bottles, polyethylene terephthalate (PET) for carbonated beverage bottles, and others, there have been rapid advances, not only in process machinery, but also in the characteristics and range of materials available.

1.2 Industry Size

BM is the third largest plastic processing technique worldwide for manufacturing many different products in the plastics industry (Fig. 1.1), accounting for 10 wt% of

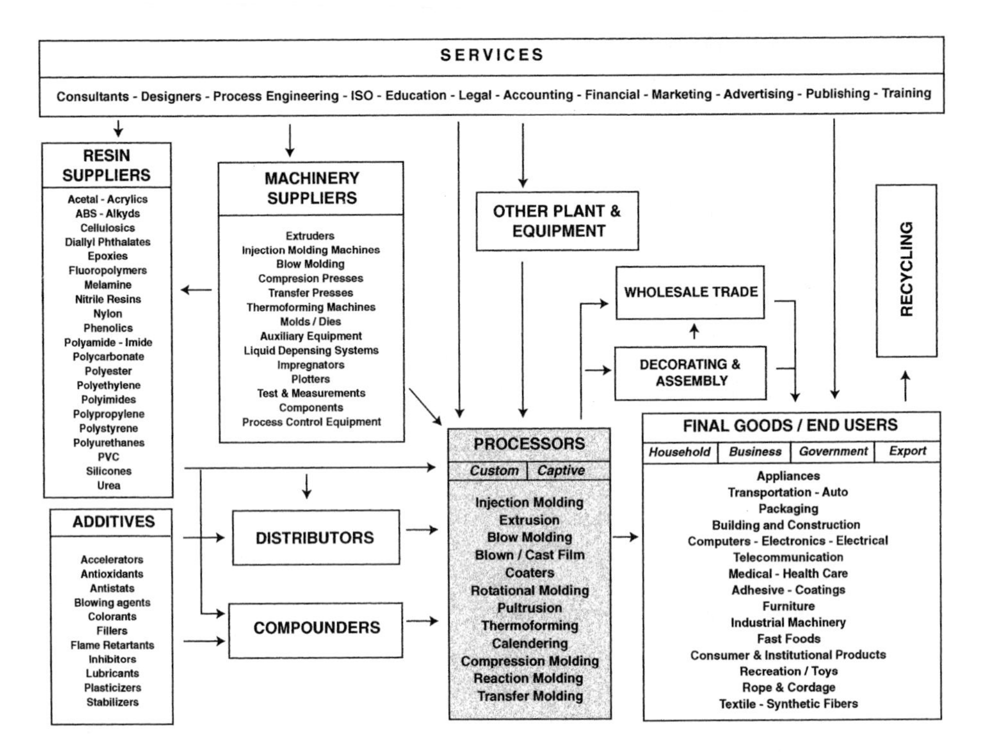

Fig. 1.1 *The plastics industry*

all plastics. Extrusion comprises 36%, injection molding 32%; calendering, 8%; coating, 5%; compression molding, 3%; and others, 3%. In the United States approximately 6,000 BM machines, 18,000 extruders, and 80,000 injection molding machines (IMMs) are in operation to process the many different types of plastics. The annual US sales of BM machines (BMMs) is about $350 million.

BM is one of the fastest growing industries worldwide. Demand for industrial and commercial molding of small to large containers, irregular hollow shaped industrial products, and bottles is predicted to continue growing. The market for these thermoplastic (TP) products consists of approximately 22 wt% food, 20% beverage, 15% household chemicals, 12% toiletries and cosmetics, 8% health, 7% industrial chemicals, 5% auto, and 11% others.

The Freedonia Group Inc. (Cleveland, OH) Plastics Processing Machinery Report predicted in mid-1999 that in the United States machinery sales demand will rise at 5.8% per year to $1.5 billion by the year 2003. IMM is the largest category, accounting for 51% of all the machinery sales. By 2003 sales of BM machines will grow the fastest, reaching $505 million; extrusion will reach $440 million; and thermoforming, $455 million. They also reported that there are now more than 350 machinery builders in the United States, with sales from five of these representing over 50% of the total.

1.3 Plastic Material

Annual US plastics utilization is now about 110 billion lb; 10% of this is used in BM. The type of plastics by weight is about 65% high-density polyethylene (HDPE), 22% PET, 6% PVC, 4% polypropylene (PP), 2% low-density polyethylene (LDPE), and 1% others. The basic consumer plastics such as PE, PET, PVC, and PP are increasingly being combined with special barrier layer materials to protect their contents from conditions such as oxygen and water vapor degradation; coatings are also successfully preserving flavor and extending product shelf life. New process technology brings a continuing wide-scale introduction of multilayer containers to our supermarket shelves, gasoline fuel tanks, and so forth. Figure 1.2 uses the image of a tree to show how BM fit into plastics product growth.

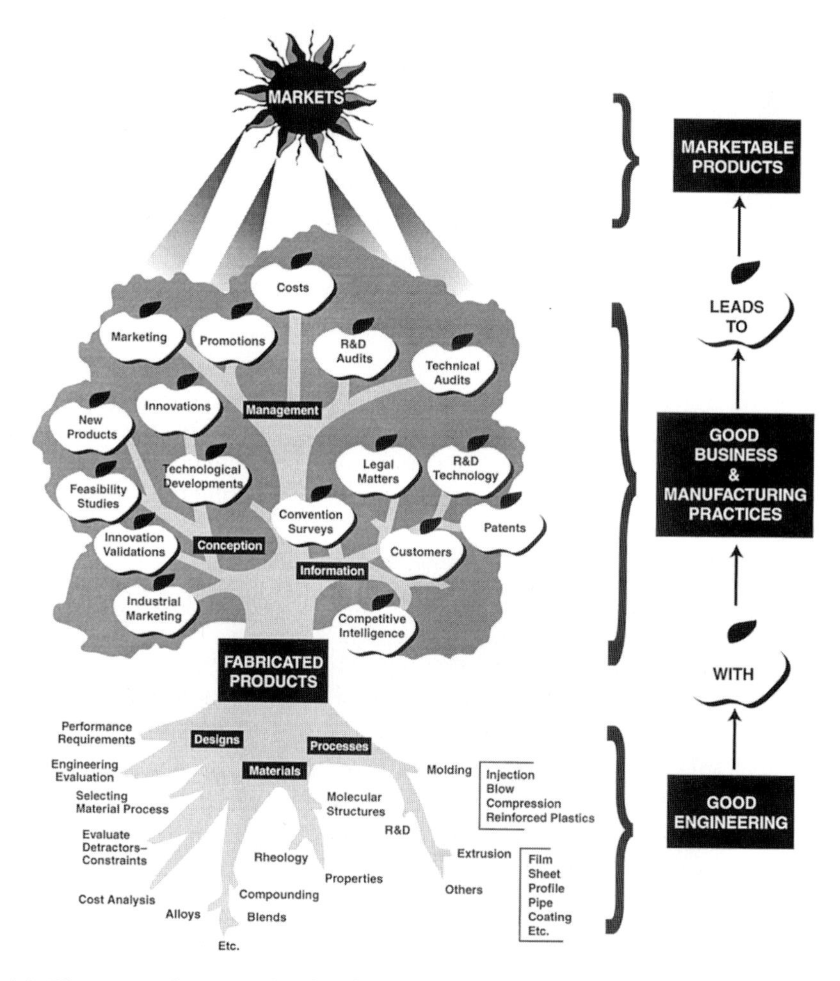

Fig. 1.2 *Plastics product growth related to tree growth*

When compared to other material, plastic provides a favorable balance of mechanical, thermal, chemical, and electrical properties. Often a single BM product can replace an assembled product made of plastics and/or other materials, substantially reducing costs. The process adapts to both low and high volume production runs with virtually no finishing required and is flexible in regard to product size and shape. It is principally a mass production method, however. Among the special manufacturing developments have been the complex three-dimensional (3D) products and the insertion of film in the mold to provide reinforcement, permitting reduction in BM material and eliminating the need for subsequent printing/decoration steps [2–4].

Different types of plastics are used to produce various BM products. Examples of BM products are shown in Fig. 1.3 and Table 1.1.

Figure 1.3a shows a cost-efficient BM machine that can produce 3- to 6-gallon (11.35- to 22.70-liter) polycarbonate bottles and containers. Optimized for PC processing, the new RS13500 P/C Uniloy Milacron has a top-fed head, fully packed smooth internal diameter (ID) neck, high-tonnage clamp system, and integrated trimming to the proven economies and reliability of the RSI reciprocating machine platform. The versatile workhorse machine produces high-strength bottles with or without handles to high consistent quality on aggressive cycle times. Top-fed, direct-flow design improves parison clarity and reduces the likelihood for material degradation and burning, while minimizing rejects and downtime for head dis-assembly and cleaning. A high-tonnage clamping system creates stronger parting line welds, reducing product loss to drop test failure.

Fig. 1.3 *(a) View of polycarbonate bottle with Uniloy Milacron BM machine (artistic assembly)*

Fig. 1.3 (b) BM products in all types of toys, tricycles, and games

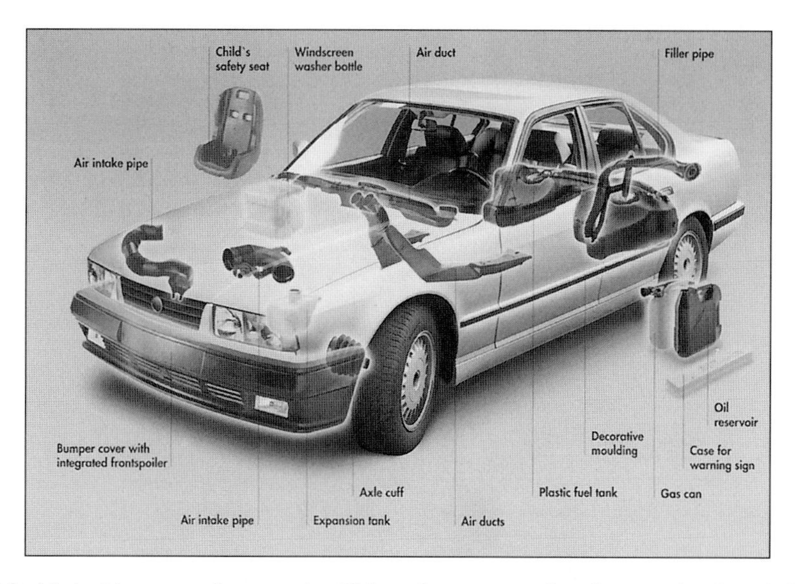

Fig. 1.3 (c) A wide range of automotive BM products are produced; examples from SIG Plastics International are shown

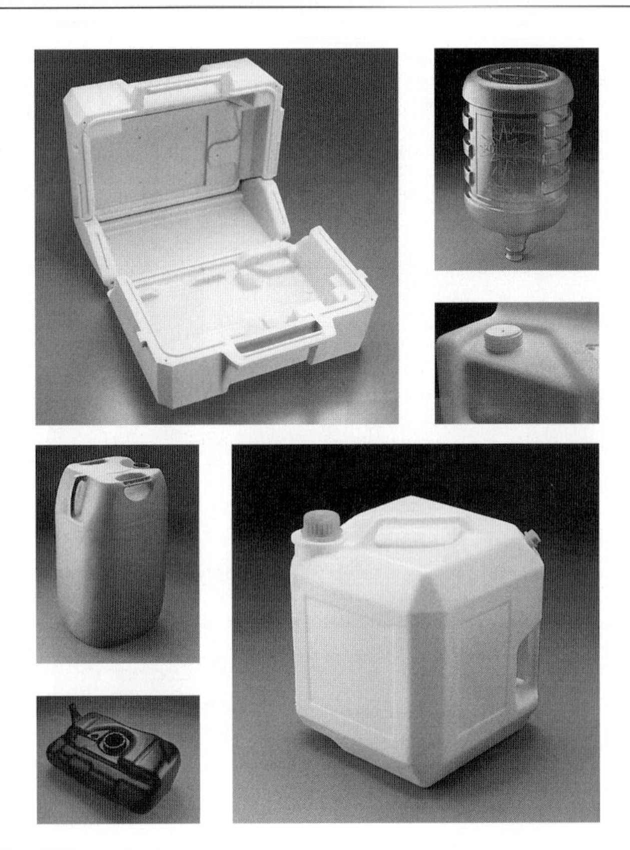

Fig. 1.3 *(d) Other BM products*

Fig. 1.3 *(e) Other BM products*

Table 1.1 Hollow and Structural BM Applications

Industry	Application	Required properties
Automotive	Spoilers	Low temperature, impact, cost
	Seat backs	Heat distortion, strength/weight
	Bumpers	Low temperature, impact, dimensional stability
	Underhood tubing	Chemical resistance, heat
Furniture	Work stations	Flame retardance, appearance
	Hospital furniture	Flame retardance, cleanability
	Office furniture	Flame retardance, cost
	Outdoor furniture	Weatherability, cost
Appliance	Air-handling equipment	Flame retardance, hollow
	Air-conditioning housings	Heat distortion, cost
Business machine	Housings	Flame retardance, cost
	Ductwork	Cost
Construction	Exterior panels	Weatherability, cost
Leisure	Flotation devices	Low temperature, impact strength, cost, weatherability
	Marine buoys	Low temperature, impact strength, cost, weatherability
	Sailboards	Low temperature, impact strength, cost, weatherability
	Toys	Low, temperature, impact strength, cost, weatherability
	Canoes/kayaks	Low temperature, impact strength, cost, weatherability
Industrial	Tool Boxes, ice chests	Low temperature, impact strength, cost
	Trash containers, drums	Low temperature, impact strength, cost
	Hot-water tanks	Low temperature, impact strength, cost

1.4 Basic Blow Molding Process

BM offers a number of technical and economic advantages in the manufacture of plastic items when compared to, for example, injection molding, including the possibilities of reentrant curves (irregular), low stresses, possibility of extremely variable wall thicknesses, and favorable cost factors. Reentrant curves are the most prominent features in BM items; this is true to the extent that it is difficult to find examples that do not incorporate them. No doubt this is due to the fact that such curves are included in the aesthetic design for merchandising reasons even when they are not absolutely necessary for the functional design of particular products.

The word most commonly used for this type of construction is hollow (note that Webster's Dictionary defines hollow as any concavity such as a bowl or a drinking glass). Such hollow shapes are injection molded in plastic on a large scale. When walls are designed so that they turn inward toward the center, they are "reentrant." BM has become the most prominent method for manufacture of plastic items with reentrant curves, providing technical, aesthetic, and cost advantages.

Since the mold equipment required for the Extrusion Blow Molding (EBM) process consists of female molds only, it is possible to vary the wall thickness and thus the weight of the finished product by simple changes of machine parts or extrusion conditions at the parison die. For an item where the exact thickness required in the finished product cannot be calculated with accuracy in advance, this is a great advantage in both time and costs. Thickness control of Injection Blow Molding (IBM) product is directly related to the preform shape produced; in conventional Injection Molding (IM), making such changes for each variable would be expensive. In vacuum forming, such changes would require availability of an infinite variety of sheet thicknesses.

Through vacuum forming, it is possible, using BM, to produce walls that are almost paper thin, which cannot be produced with IM. Both BM and IM can be successfully used for very thick walls. The final choice of process for a specific wall section would be strongly influenced by other factors such as reentrant curves, stress, and so forth.

BM can be used with some plastics, such as polyethylene (PE), with a much higher molecular weight than IM, lending greater toughness to the higher molecular weight plastics. This adds extra resistance to environmental stress cracking [89], which is necessary for plastic bottles used for detergents and for packages and products in contact with many industrial chemicals that can promote stress cracking.

The element of cost is so complex that it is difficult to generalize. However, the cost factors of BM are very pronounced in the capital equipment portion. Pound for pound of finished product, BM machine costs could average between one half and one third of injection machine costs for the same or similar items. Molds for BM have been known to cost as little as 10% that of injection molds for comparable items (Chapter 8).

With BM, the tight tolerances achieved with IM are not obtainable. However, to produce reentrant curved or irregular shaped IM products, different parts can be molded and in turn assembled (snap fit, solvent bond, ultrasonic bond, or others). BM the complete irregular/complex shape, even though tolerance cannot be equaled (most applications not required), reduces the cost of the blown container without the need for any secondary operations. Other advantages occur such as significantly reducing (if not eliminating) leaks, reducing cycle and production time, etc.

BM calls for an understanding of every element of the process, starting with the extruder or IMM where plastic is melted (plasticated). The basic components of the extruder and IMM are shown in Figs. 1.4 and 1.5. Plastics in different forms (pellets, flakes, etc.) are fed into the feed throat from the hopper. As the unmelted plastic exits the feed throat, it enters the plasticator to be melted. The plastic comes in contact with a rotating screw driven at a desired speed by an electric and/or hydraulic motor. The screw is contained within a heated barrel that is maintained at a higher temperature than the screw. Because plastic tends to stick to hotter surfaces, the unmelted material will stick to the barrel and slide on the screw. This allows the plastic to be forced forward by the flights of the rotating screw, and as it moves forward, it is heated by the barrel along with the frictional heat generated by the compounding and compression action of the screw. By the time the plastic leaves the screw and barrel (plasticator), it has been melted, compressed, and mixed into a rather homogeneous melt.

In some cases with EBM, a screen pack and breaker plate are positioned at the end of the barrel to filter out foreign contaminants that may have entered the hopper. The screen pack also increases backpressure in the barrel, contributing to a more homogeneous melt and steady flow. The breaker plate holds the screens, gives them support, and contributes to the backpressure [5].

There are basically two types of plasticators, the continuous type as seen in Fig. 1.4, and the reciprocating type shown in Fig. 1.5. The continuous extruder delivers a constant flow of plastic as the screw turns, while the reciprocating screw (basically an injection molding machine) moves not only in a circular motion, but also horizontally within the barrel. As the screw turns, it fills the forward end of the barrel with plastic, thus retracting the screw. When a predetermined distance is reached by the backward motion of the screw, a hydraulic or electrically operating cylinder is actuated, pushing the screw forward. This forward motion delivers a flow of material and the process is started once again [6].

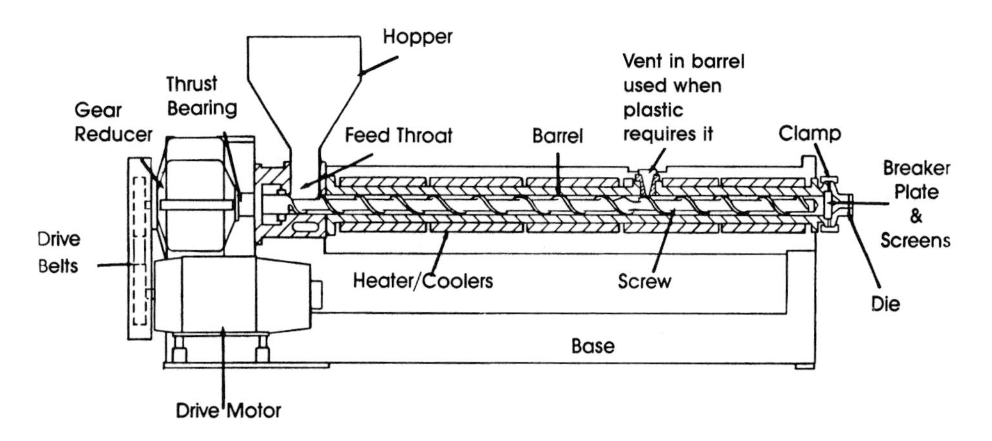

Fig. 1.4 Example of a single screw extruder with continuous melt flow

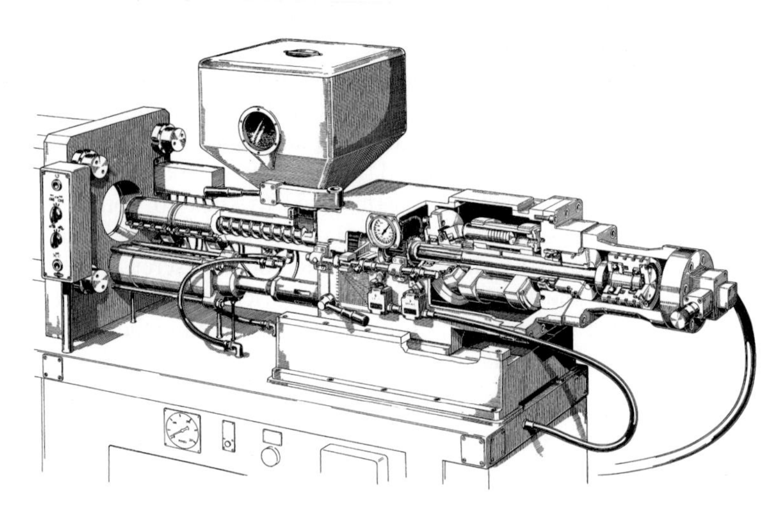

Fig. 1.5 Example of an injection molding plasticator that uses a reciprocating screw with a squash plate hydraulic plate

1.4.1 Blow Molding Type

BM can be divided into three major processing categories:

- Extrusion blow molding (EBM) with continuous or intermittent melt (parison) from an extruder.
- Injection blow molding (IBM) with noncontinuous melt from an injection molding machine that principally uses a preform supported by a metal core pin.
- Stretched/oriented EBM and IBM to obtain biaxially oriented products providing significantly improved performance-to-cost advantages.

In continuous EBM, an extruded tube or parison continuously exits the extruder die. The noncontinuous (intermittent) EBM machine has an accumulator in the die head. Parisons are extruded, which in turn form the blown product within a blow mold.

IBM, with its noncontinuous melt or preform from an IMM initially uses a preform mold supported by a metal core pin, which in turn forms the blown product within a blow mold.

Almost 75% of processes are EBM, almost 25% are IBM, and about 1% use other techniques such as dip BM. About 75% of all IBM products are stretched biaxially; there are also stretched EBM. With stretched BM, orientation takes place simultaneously in the hoop and longitudinal directions. These BM processes offer various advantages in producing different types of products based on the plastics to be used, performance requirements, production quantity, and costs. The typical BM processes used are summarized in Table 1.2. Typical plastics used to produce various BM products are summarized in Table 1.3.

Table 1.2 Examples of Blow Molding Processes

Basic process	Materials	Average production rate, parts per hour	Average container size	Approximate market penetration, percent
Continuous extrusion shuttle clamp	Polyethylene (PE), polypropylene (PP), polycarbonate (PC), polyvinyl chloride (PVC), polyethylene terephthalate glycol modified (PETG)	500–3,600	4 ounces– 7-$\frac{1}{2}$ gallons	50
Wheel clamp (6 to 24 clamps)	Polyethylene (PE), polyvinyl chloride (PVC), polypropylene (PP), coextrusions	2,000–5,000	32–300 ounces	10
Intermittent extrusion reciprocating screw	Polyethylene (PE) and polycarbonate (PC)	500–2,500	8–128 ounces	10
Accumulator head	Polyethylene (PE) and engineering plastics	50–500	5–2,000 gallons	5
Stretch blow molding	Polyvinyl chloride (PVC), polyethylene terephthalate (PET), polypropylene (PP), polyacrylonitrile (PAN)	1,200–4,000	16–48 ounces	20 (including reheat)
Injection blow molding	Polyethylene (PE), polypropylene (PP), polyvinyl chloride (PVC), polyethylene terephthalate (PET), polycarbonate (PC), polyacrylonitrile (PAN), polystyrene (PS)	500–3,000	1–16 ounces	5 (including dip molding)
Dip blow molding	Polyethylene (PE), polypropylene (PP), polyvinyl chloride (PVC), polyethylene terephthalate (PET), polyacrylonitrile (PAN), polystyrene (PS)	500–1,500	1–24 ounces	(see injection blow molding)
Reheat blow molding	Polyethylene terephthalate (PET)	240–15,000	16 ounces to 22 liters	(see stretch blow molding)
		1,200–4,000	16–48 ounces	
		1,200–4,000	8–48 ounces	

Table 1.3 Examples of Types of Plastics for Blow Molding

Term	Abbreviation	Term	Abbreviation
Acetals		Ultrahigh molecular weight polyethylene	UHMWPE
Acrylics			
Polymethylmethacrylate	PMMA	Polypropylene	PP
Polyacrylonitrile	PAN	Ionomer	
Alloys		Polybutylene	PB
Nitrile resins		Polyallomers	
Polyaryl sulfone		Styrenics	
Polyamide (nylon)	PA	Polystyrene	PS
Polycarbonate	PC	Acrylonitrile butadiene styrene	ABS
Polyester		Styrene acrylonitrile	SAN
Thermoplastic polyethylene		Styrene butadiene latricies	
Polybutylene terephthalate	PBT	Styrene base polymers	
Polytetramethylene terephthalate	PTMT	Thermoplastic elastomers	
		Vinyl	
Polyethylene terephthalate	PET	Polyvinyl chloride	PVC
Polyolefins	PO	Polyvinyl acetate	VAC
Low density polyethylene	LDPE	Polyvinylidene chloride	
Linear low density polyethylene	LLDPE	Polyvinyl alcohol	
High density polyethylene	HDPE	Ethylvinyl acetate	

BM multilayer plastics are used extensively. The coextrusion and coinjection technologies take advantage of using differing materials, including the use of plastic foams, recycled plastics, etc. that are systematically combined to meet cost to performance requirements.

1.4.2 Extrusion Blow Molding vs. Injection Blow Molding

The advantages of EBM compared to IBM include lower tooling costs and incorporation of blown handleware, etc. Disadvantages or limitations could be controlling parison swell, producing scrap, limited wall thickness control and plastic distribution, among others. If desired, blown solid handles can be molded during the BM process. Trimming can be accomplished in the mold for certain designed molds, or secondary trimming operations are included in the production lines. With 3D molding scrap is significantly reduced.

The main advantage of IBM is that no flash or scrap occurs during processing. It gives the best of all wall thicknesses and plastic distribution control and critical bottleneck finishes are easily molded to a higher accuracy. The initial IBM preforming cavities are designed to have the exact dimensions required after blowing the plastic melt as well as accounting for any shrinkage, etc. that may occur. Neck finishes, both internal and external, can be molded with an accuracy of at least ±4 mil (±0.10 mm). It also offers precise weight control in the finished product accurate to at least ±0.1 g.

Disadvantages of IBM could include its high tooling costs, incorporation of only solid handleware, and restriction to small products. (However, large and complex shaped products were fabricated once the market developed.) Start-up of the injection molding process takes longer and is more difficult than the extrusion blow molding process and the tooling costs are almost twice that of extrusion blow molding. Extrusion blow molding (EBM) normally produces larger containers on faster cycles than can be attained with IBM. Similar comparisons exist for biaxial orienting EBM or IBM. With respect to coextrusion, the two methods also have similar advantages and disadvantages, but mainly more advantages for both.

When compared to EBM, the IBM procedure permits the use of plastics that are suitable for EBM, but more importantly, permits the use of those unsuitable for EBM (other than when certain types are modified). Specifically, it is those with no controllable melt strength, such as the conventional PET, that is predominantly used in large quantities. An example is the stretch IBM method for carbonated beverage bottles (liter and other sizes).

1.5 Other Blow Molding Processes

In addition to the three major processing categories of EBM, IBM, and stretched/ oriented EBM and IBM, other BM techniques have been developed (Table 1.2).

1.5.1 Dip Injection Blow Molding

Dip injection BM (Fig. 1.6) was used when the BM industry started its market expansion. However, with extensive developments in EBM and IBM, the dip process has found only special limited use. At the beginning of BM hollow articles, the industry asked for a scrap-free energy-efficient process that allowed the use of low-cost tooling combined with short procurement times, and EBM therefore became the most common and major process to blow hollow articles. However, the inherent disadvantage in finishing the neck and bottom through mechanical shear action made the industry look for other methods.

IBM was used to overcome this latter drawback and further to improve the tolerances required for safety neck finishes. However, tooling costs were high because technically two molds are necessary to produce the hollow article: a preform mold and a blow mold. Both molds have to be constructed to very close tolerances [±0.005 in.

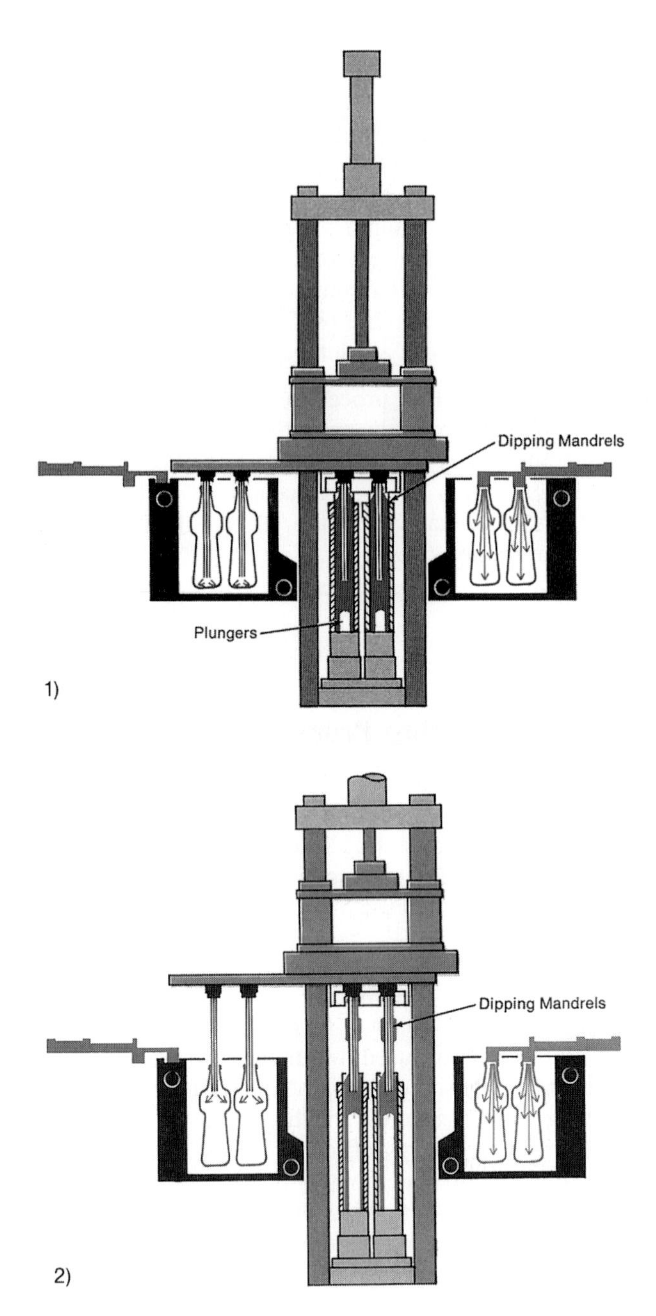

Fig. 1.6 Extrusion dip BM by Schloemann-Siemag AG

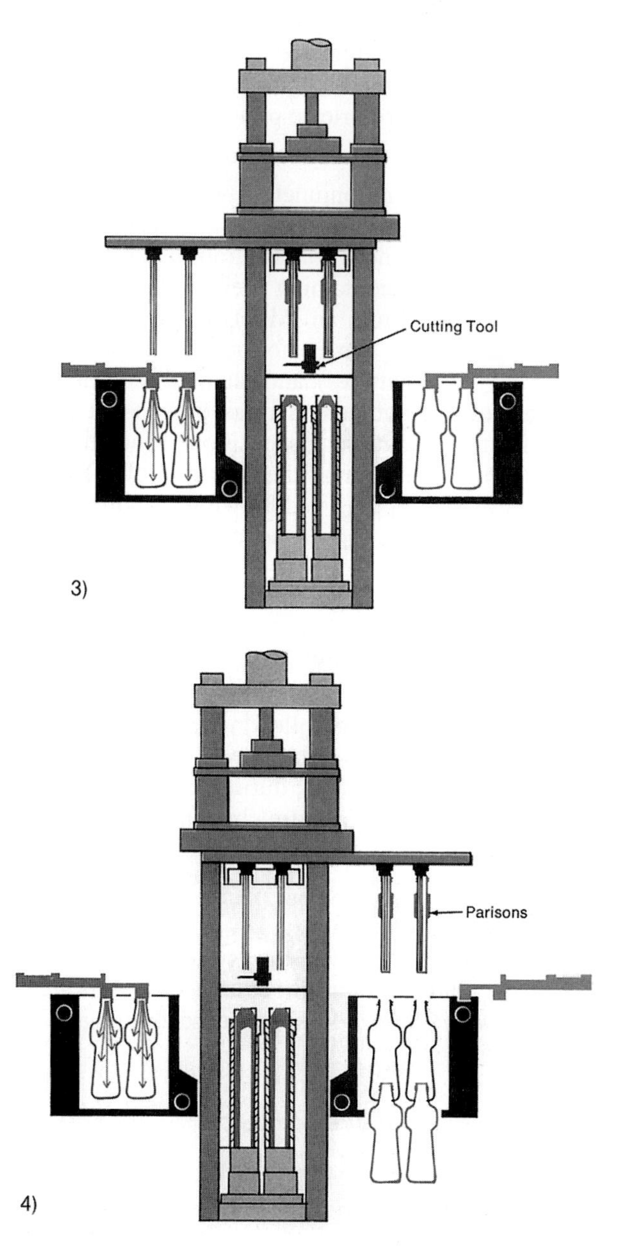

Fig. 1.6 Extrusion dip BM by Schloemann-Siemag AG (continued)

(0.013 cm)]. In addition, energy requirements are substantially higher than for EBM because of the need for hot water units to condition the preform, and so forth.

When dip BM entered the market, it appeared that it could combine the advantages of extrusion and injection BM. It allowed one to inject the neck finish precisely as in injection BM, but without requiring intricate preform molds that led to the name neck injection BM.

The dipping mandrel is working in conjunction with the plunger whose movement controls wall thickness. In one method, dip BM requires two sets of one or more dipping mandrels equipped with a neck forming tool that operate in tandem. One set is dipped into a feed cavity (pot) containing melted plastic that is injected into the preshaped cavities (melt injected from an IMM). The mandrel(s) is coated with the melt. At this stage the neck finish can be formed under pressure. Next the coated mandrel(s) is lifted from the cavity(ies) and shuttled sideways to descend into a cavity(ies) mold to be blown. Simultaneously as this coated mandrel(s) leaves the cavity(ies), the second set of mandrel(s) is positioned into the cavity(ies) that had been refilled with melt. In the meantime the blown bottle(s) is cooled and is also being lifted to be released. The dipping and blowing mandrels can be positioned on the same support plate so that the simultaneous actions occur following repeated cycles.

1.5.2 Compression–Stretch Blow Molding

Compression stretch BM consists of the following stages: (1) starts with an extruded sheet; (2) circular blanks are cut from the sheet; (3) the blanks are compression molded into the desired preliminary shape; during the compression action, the blank can be simultaneously stretched; (4) blow stretching can take place after compression molding; and (5) any trimming that may be required is the final step.

Compression–stretch blow molding (CSBM) patents include: (1) those held by Valyi Institute for Plastic Forming (VIPF) located at the University of Massachusetts-Lowell; (2) Dynaplast S.A. has the Co-Blow system; (3) American Can's OMNI container; (4) Petainer's cold forming process; (5) Dow Chemical's coforming (COFO) (Fig. 1.7); (6) Dow Chemical's solid phase forming (SPF) (Fig. 1.8); and (7) others.

1.5.3 Molding with Rotation

Injection molding with rotation, also called injection spin molding or injection stretched molding, is an example of processing at lower temperatures, pressure, and so forth. This BM process called molding with rotation (MWR) combines injection molding and injection BM with melt orientation (Dow Chemical patent). The equipment used is what is commercially available for IM except the mold is modified so that either the core pin or outside cavity rotates. The MWR process is a

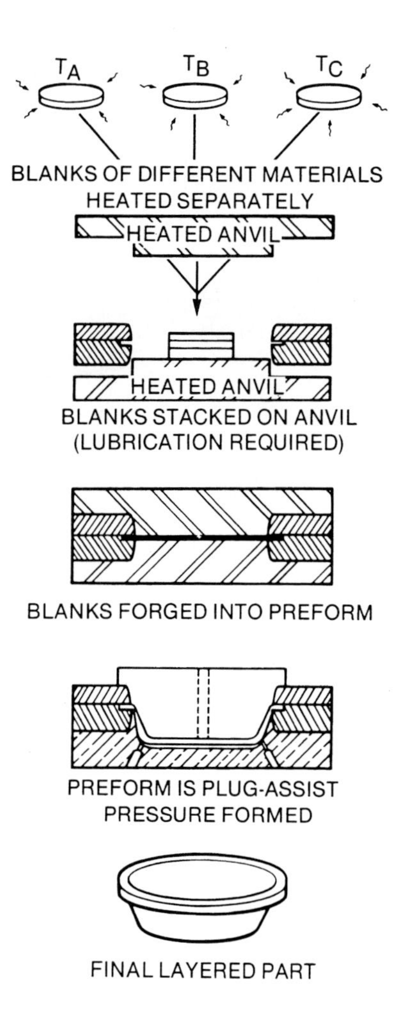

Fig. 1.7 *Multilayer formed technique starts with plastic blanks*

fabrication method using a rotating mold element in the IMM (Fig. 1.9). The end product can come directly from the IMM mold or be a result of two-stage fabrication: making a parison and BM the parison (Fig. 1.10).

This technology is most effective when employed with articles which have: (1) a polar axis of symmetry; (2) reasonably uniform wall thickness; and (3) dimensional specifications and part-to-part trueness that are important to market acceptance. Note that within these requirements are many products having variable surface and wall geometries.

The MWR process requires no sacrifice of either cycle time or surface finish. Both laboratory and early commercial runs identify good potentials for reducing cycle time; for either reducing the amount of plastic required or improving properties with

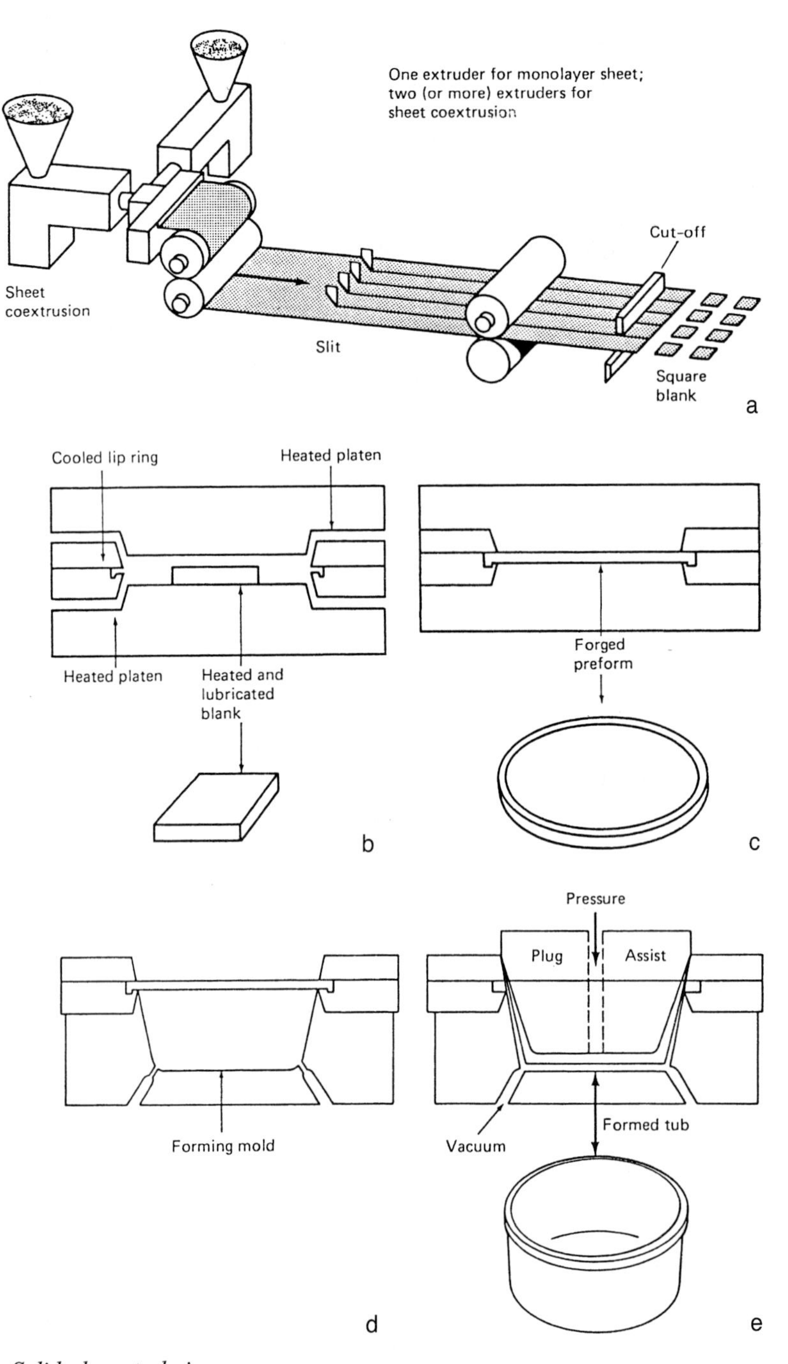

Fig. 1.8 *Solid-phase technique*

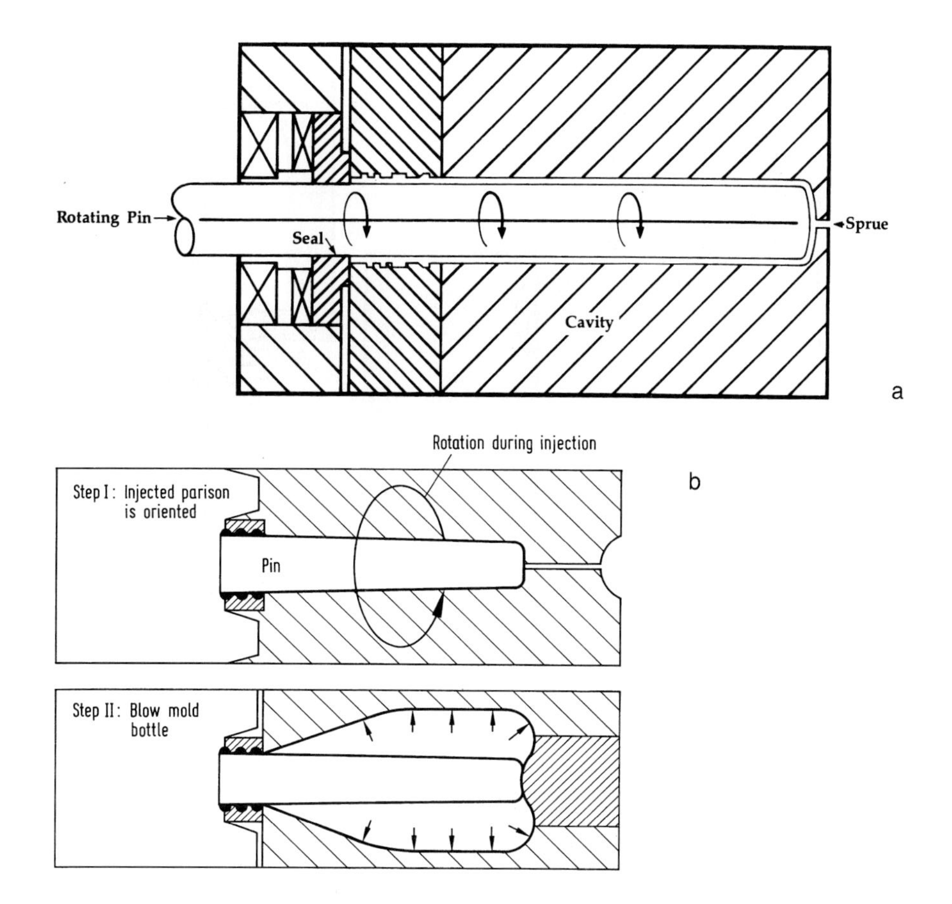

Fig. 1.9 Mold schematic for MWR. (a) Pin rotating and (b) cavity rotating

the same amount of plastic, or both; and for substituting less expensive plastics while achieving adequate properties in the fabricated product.

The MWR process permits a balancing of plastic temperature, plastic rheology, pressures, time, and either mold core or mold surface rotation to achieve a carefully controlled degree of multiaxial orientation within the thermoplastic mass. During fabrication using the MWR process, two forces act on the plastic: injection (longitudinal) and rotation (hoop). The targeted balanced orientation is a result of those forces. As the product wall cools, additional high-magnitude, cross-laminated orientation is developed frozen-in and throughout the wall thickness. Orientation on molecular planes occurs as each layer cools after injection.

This orientation can change direction and magnitude as a function of wall thickness. The result is analogous to that for plywood or reinforced plastics [7, 10, 11] and the strength improvements are as dramatic. In the MWR process, there is an infinite number of layers each with its own controlled direction of orientation. By appropriate

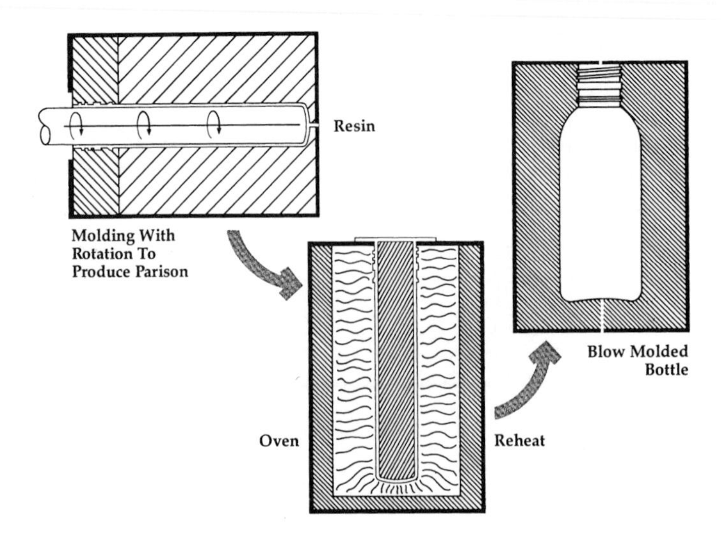

Fig. 1.10 *Two-stage MWR blow molding procedure*

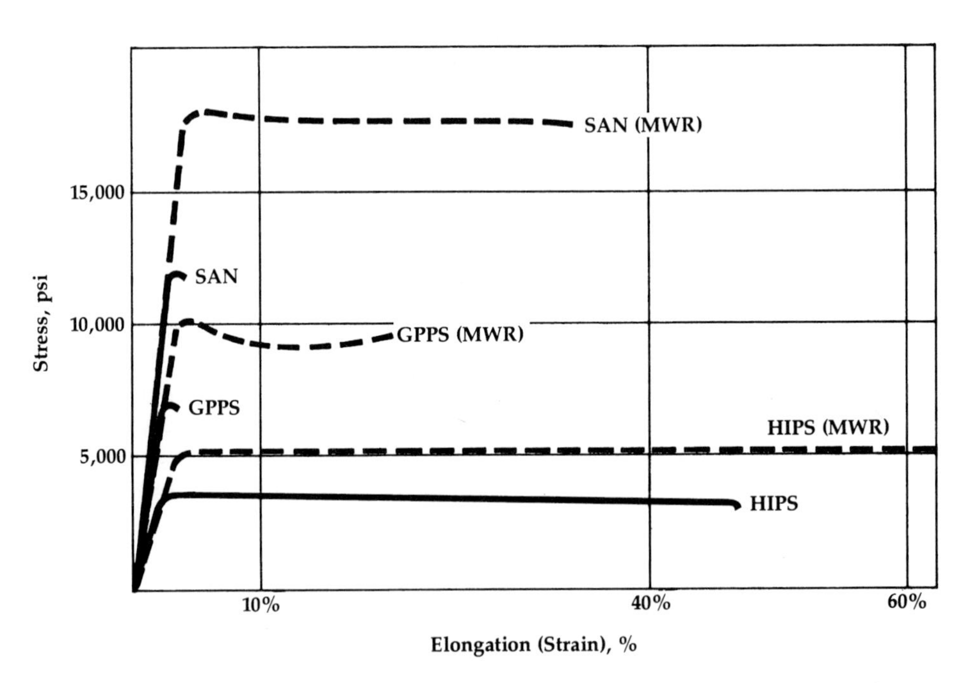

Fig. 1.11 *Tensile stress–strain diagrams of MWR vs. unoriented (non-MWR) BM products*

processing conditions, both the magnitude and direction of the orientation and strength properties can be varied and controlled throughout the wall thickness (Fig. 1.11).

The effect of cycle time on IM economics is great. With MWR, one potential for reducing cycle time relates to the ability to obtain satisfactory end product performance with less plastic because less plastic, heat, and cooling time shortens the cycle. Another potential for reduced cycle time occurs because IM with MWR is most effectively done when plastic melt occurs at a much lower temperature than would be used for IM without MWR.

As is common with conventional injection molding, MWR also results in products having excellent dimensional properties. In addition, MWR permits products with high length-to-diameter ratios to be molded without problems of core deflection and consequently thinner–thicker sections in the product wall. With core rotation during MWR, the pin self-centers and product wall uniformity is excellent. The final molded product therefore is uniformly strong about its circumference, which has particular value if the molded product is a parison. Parisons fabricated with MWR can be reheated and blown without problems caused by wall eccentricity.

1.5.4 Other Molding Processes

There are different processing methods that to some degree are a takeoff of BM in which forming occurs against only one mold surface. One such method is the thermoforming BM process. Figure 1.12 is an example of the National Can's process that combines thermoforming with BM. After the liner materials are extruded with the desired properties and thickness (a), they are laminated in a process called Corona bonding (a) and this operation is said to facilitate the separation of the web layers after forming for reuse (b), as it does not require the use of adhesive layers. After heating (c), the bonded layers are thermoformed to a deep, test tube configuration to be compatible with the BM process (d). Trimming of surplus material takes place in the

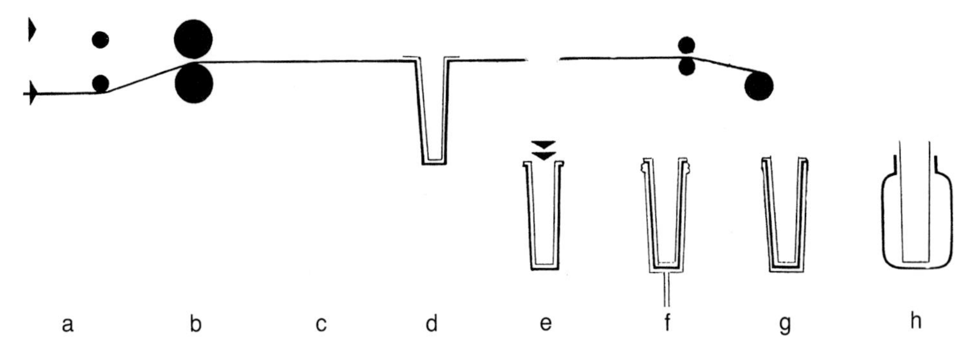

Fig. 1.12 Thermoforming BM process

same operation as the liner layers are released (e). The remains of the laminated web are separated and recycled. Liners are automatically fed onto cores, where the structural material is then injection molded over them to form lined parison (f). Lastly, after reaching the proper temperature, the lined parisons (g) are BM, cooled, and ejected (h). Horizontal, rotary, and IBM are all compatible with this process. Among the food applications for these bottles are sugar substitutes, coffee and other powdered or granulated products.

1.6 Process Control

As BM became more complex, molders required greater accuracy and increased variations in the types of cycles that could be adapted to their machines. Different types of machine process controls (PCs) can be used to meet the requirements based on the molders' operating needs. Knowledge of the equipment and its operating needs is a prerequisite before an intelligent process control program can be developed and used (Chapter 9).

1.7 Product Design

Product design starts by incorporating and combining the factors pertaining to functional and appearance requirements, the properties of the material, and the feasibility of processing in arriving at an acceptable solution in terms of performance and cost (Chapter 8).

1.8 Fabricating Basics

BM processing techniques range from the unsophisticated (high labor costs with low capital costs) to sophisticated (zero or almost zero labor costs with very high capital costs). Production quantity, material being processed, available equipment, and total cost govern decisions on the appropriate technique. Small quantities are usually produced with an unsophisticated approach.

A selection from the many pieces of equipment that are available needs to address the economics of the product being fabricated, the degree of precision necessary, and the production rate required. The plastic compounding mix can be manually proportioned using a balancing and weighing scale [2]. Proportioning pumps partition according to a desired ratio, usually by weight but also by volume. Diaphragm pumps may be used with adjustable displacement to draw the plastic components from reservoirs and to meter the correct proportions into a mixing chamber. Gear pumps with drive ratios can be used to deliver the correct portion, but only when the mix is relatively free of abrasive fillers or additives.

As summarized in Fig. 1.2, fabricating is an important part of the overall project to manufacture acceptable plastic products. It highlights the flow pattern for the fabricator (manufacturer) to be successful and profitable.

Factors such as good engineering, process control, etc. are very important but represent only part of the overall requirements to ensure profitable products. In addition to good engineering and manufacturing, proper business and sales marketing play an equally important contribution to a business. This diagram is essentially a philosophical approach to the overall industry as it provides examples that touch all aspects of the technology and business ranging from local to global competition.

To understand potential problems and solutions of design and fabrication, it is helpful to consider the relationships of machine capabilities, plastics processing variables, and product performance (Fig. 1.13). In turn, as an example, a distinction has to be

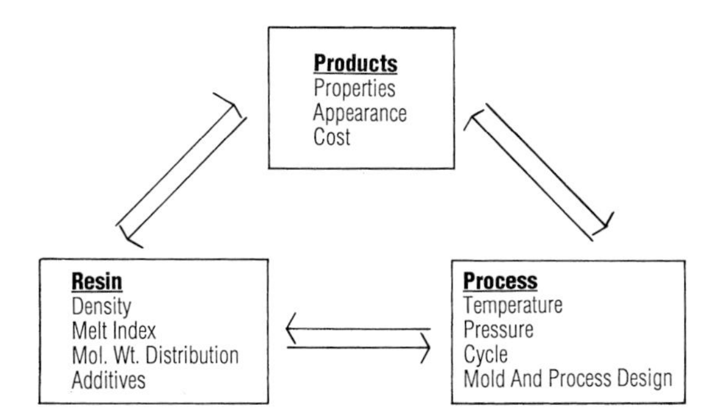

Fig. 1.13 *Interrelating product–plastic–process performance*

made here between machine conditions and processing variables. Machine conditions include the operating temperature and pressure, mold and die temperature, machine output rate, and so on, whereas processing variables are more specific, such as the melt condition in the mold or die, the flow rate vs. temperature, and so on.

It is the processing variables, properly defined and measured, not necessarily the machine setting, that can be correlated with product performance. Design and processing are now much more interrelated so that products can be consistently successful.

Those familiar with processing can detect and correct visible problems or readily measure factors such as color, surface conditions, and dimensions. However, less apparent property changes may not show up until the products are in service, unless extensive testing and quality control are used.

1.9 Processing Fundamentals

Understanding and measuring factors such as melt flow or heat behavior of plastics during processing is important for two reasons. First, it provides a means for determining whether a plastic can be fabricated into a useful product such as a usable extruded extrudate, completely fill a mold cavity, provide mixing action in a screw, meet product thickness requirements, and so forth. Second, the flow is an indication of whether its final properties will be consistent with those required by the product. The target is to provide the necessary homogeneous uniformly heated melt during processing so the melt operation is completely stable and in equilibrium.

In practice, this perfectly stable situation is never achieved as two variables affect the quality and output rate: (1) the variables of the machine's design and manufacture and (2) the operating or dynamic variables that control how the machine is operating.

An important factor is obtaining the best processing temperature for the plastics used, with past experience and/or the input of the resin supplier. Plastics with or without extensive additives, fillers, and/or reinforcements influence the temperature setting. Crystalline and amorphous thermoplastics have different melt temperature requirements. This type of information on initial startup of the fabricating equipment is important but only provides a guide. Recognize that if the same plastic is used with a different machine (with identical operating specifications) new temperature settings may be required because, like the material, machines have variables.

One method of defining the melt behavior and property performance of plastics is to use information concerning their molecular weight (MW). MW refers to the average weight of plastics composed of different weight molecules. These differences are important to the processor, who uses the molecular weight distribution (MWD) to evaluate materials. A narrow MWD enhances the performance of plastic products whereas wide MWD permits easier processing. The melt flow rate is dependent on the MWD (Chapter 7).

1.10 Variabilities

Existing material and equipment variabilities are continually reduced owing to improvement in their manufacturing and process control capabilities. To ensure control of material and machine, setting up controls via different tests and setting limits is important, although inferior products may still result. For example, the material specification from a supplier will provide an available minimum to maximum value such as MWD. It is determined that when all the material arrives on the maximum side it produces acceptable products, but when it arrives on the minimum value, the process control may need to be changed to produce acceptable results.

To judge performance capabilities within the controlled variabilities, a reference to measure performance against is required. As an example, the IBM preform mold cavity pressure profile is easily influenced by variations in the materials. Related to this parameter are four groups of variables that when put together influence the profile: (1) melt viscosity and fill rate, (2) boost time, (3) pack and hold pressures, and (4) recovery of plasticator. Thus material variations may be directly related to the cavity pressure variation. Similar relationships of melt viscosity, pressure, and time to the extrudate parison can be used with EBM.

1.10.1 Material

Even though equipment operations have understandable but controllable variables that influence processing, the most uncontrollable variable in the process is usually the plastic material. A specific plastic will have a range performance. More significant, however, is the degree of properly compounding or blending by the plastic manufacturer, converter, or by the in-house fabricator. Most additives, fillers, and/or reinforcements when not properly compounded will significantly influence processability and molded product performances.

A very important factor that should not be overlooked by a designer, processor, analyst, statistician, or others is that most conventional and commercial tabulated material data and plots, such as tensile strength, are mean values. They would imply a 50% survival rate when the material value below the mean processes unacceptable products. The target is to obtain some level of reliability that will account for material and other variations that can occur during the product design to processing the plastics.

1.10.2 Process

In addition to material variables, a number of equipment hardware and control factors can cause processing variabilities. They include accuracy of machined component equipment parts, method and degree of accuracy during the assembly of component parts, and temperature/pressure control capability, particularly when interrelated with time and heat transfer uniformity in metal components such as those used in molds and dies. Location of sensors (temperature, pressure, etc.) has a significant effect on the range of the variables.

These variables are controllable within limits to produce useful products. Improvements in equipment (as in materials) have made exceptional strides in significantly reducing operating variabilities or limitations that will continue into the future as there is a rather endless improvement in performance of steels and other materials and control methods such as fuzzy logic. Applications in fuzzy logic, introduced in 1981 and based on mimicking the control actions of the human operator (Chapter 8), are growing.

1.10.3 Product and Tool Design

The designer matches the end use requirements with the properties of the selected material using a practical or engineering technique. The target is to achieve the basic three general requirements of design success: (1) economical, (2) functional, and (3) attractive appearance. In turn the functional aspect relates to the product's three environmental conditions of load, temperature, and time. The production method to be used will often set limitations on designs and vice versa. The way in which a product is manufactured has a profound influence on its design [2].

1.10.4 Plastic–Process Interaction

Many different processing factors could influence the repeatability of meeting performance requirements. Some products may require only the compliance of one or two processing factors, but others will require many. Computer programs have been developed to provide the capability of integrating all the applicable factors, thus replacing traditional trial-and-error methods (Chapters 9 and 10).

1.11 Product Performance

It is important to understand what happens to plastics during the manufacture of products and how it fits into the overall manufacturing operation. With proper fabricating controls, products are reproduced that meet performance requirements at a minimum cost.

In this section different processing behaviors are presented that affect the properties of the different plastic products. Examples include processing parameters such as melt temperature with flow rate or fill rate, die or mold temperature, and packing pressure in injection molding. Excessively high processing temperatures can increase the rate of thermal decomposition of plastics or produce deleterious effects. Too low a processing temperature can cause melt viscosity of these materials to be too high, thereby requiring excessive costly pressure to fill the mold or extrude the parison. High melt viscosity and high processing pressure can lead to high shearing stresses in the material and may impart increased molecular orientation to the final product. Molecular orientation makes the product anisotropic; its properties are not the same when measured in different directions.

Higher pressures make it possible to develop tighter dimensional tolerances with higher mechanical performance, but there is also a tendency to develop undesirable stresses (orientations) if the processes are not properly understood or controlled.

Most products are designed to fit processes of proven reliability and consistent production. Various options may exist for processing different shapes, sizes, and weights (Table 1.4). Parameters that will help one to select the right options are (1) setting up specific performance requirements; (2) evaluating the requirements and processing capabilities of materials; (3) designing products on the basis of material and processing characteristics, considering product complexity and size as well as a product and process cost comparison (Fig. 1.14); (4) designing and manufacturing tools (molds, dies, etc.) to permit ease of processing; (5) setting up the complete line, including auxiliary equipment, testing, and providing quality control, from delivery of the plastics through production to the product; and (6) interfacing all these parameters by using logic and experience or obtaining a required update on technology.

The EBM process has fewer problems with thermoplastics than does IBM but has greater problems in dimensional control and shape. The dimensional control and the die shape required to achieve the desired product shape are usually solved by trial-and-error settings. The more experience one has in the specific plastic and equipment to be used, the less trial time that is needed.

The choice of molder and fabricator should place no limits on a design. There is a way to make a product if the projected values justify the price; any job can be done "at a price." The real limiting factors are tool design considerations, material shrinkage, finishing operations, dimensional tolerances, allowances, undercuts, insert inclusions, parting lines, fragile sections, production rate or cycle time, and selling price.

Applying the following principles will aid the designer in specifying products that can be produced at minimum cost:

- maintaining simplicity;
- using standard materials and components;
- specifying liberal tolerances;
- employing the most processable plastics;
- collaborating with manufacturing people;
- avoiding secondary operations;
- designing what is appropriate to the expected level of production;

Table 1.4 *Examples of Competitive Processes vs. Different Products*

	Injection molding	Extrusion	Blow molding	Thermoforming	Reaction injection molding	Rotational molding	Compression and transfer molding	Matched spray
Bottles, necked containers, etc.	2, A	—	1	2, A	—	2	—	2
Cups, trays, open containers, etc.	1	—	—	1	1	—	1	2
Tanks, drums, large hollow shapes, etc.		—	1	2, A	—	1	—	2
Caps, covers, closures, etc.	1	—	—	2	2	—	1	—
Hoods, housings, auto parts, etc.	1	—	2	2	2	—	1	1
Complex shapes, thickness changes, etc.	1	—	—	—	—	—	1	2
linear shapes, pipe, profiles, etc.	2, B	1	—	—	—	—	2, B	—
Sheets, panels, laminates, etc.	—	1, C	—	—	—	—	2	2

1, Primary process; 2, Secondary process
A, Combine two or more parts with ultrasonics, adhesives, etc.; B, short sections can be molded; C, also calendering process

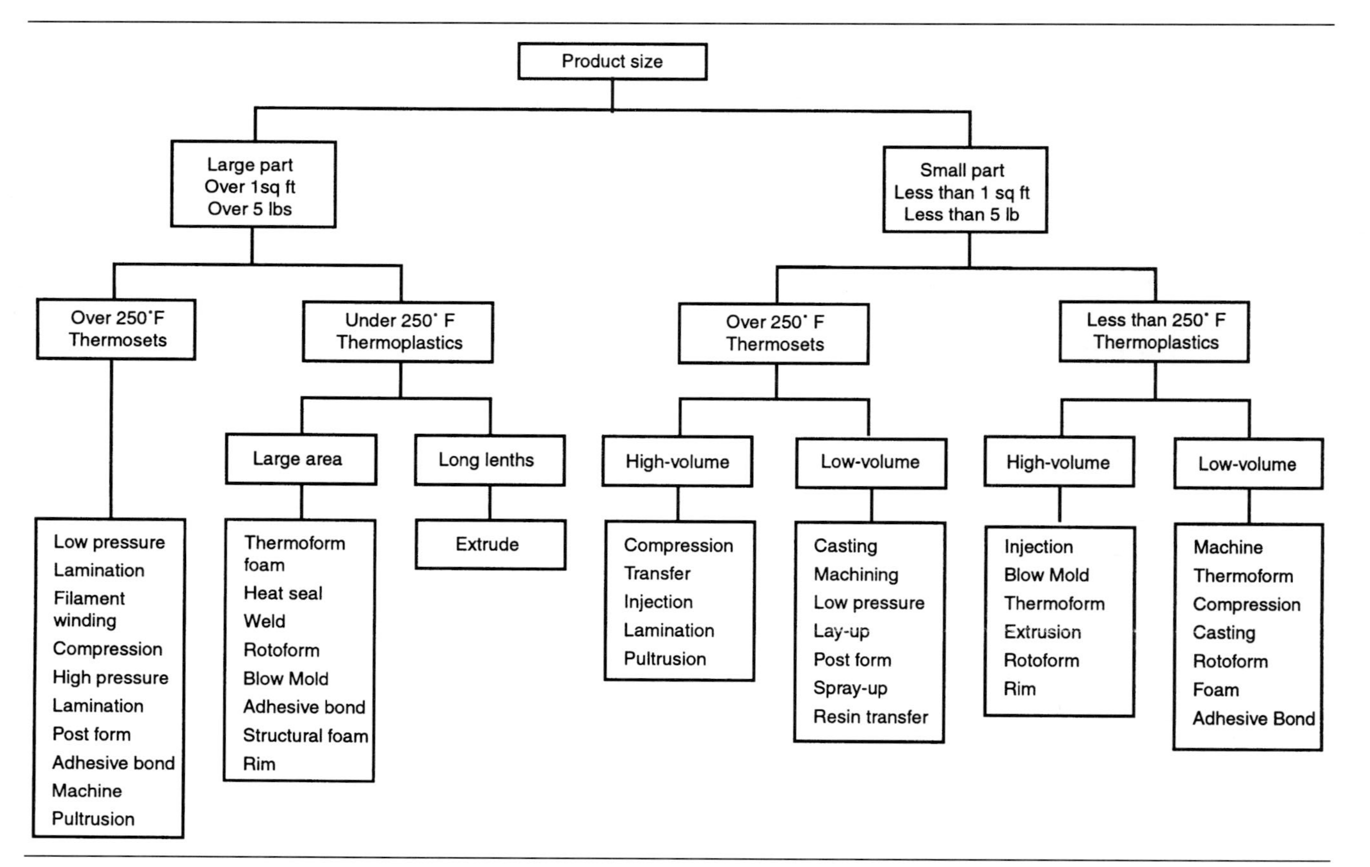

Figure 1.14 *Detailed Guide to Product Size*

- utilizing special process characteristics; and
- avoiding processing restrictions.

Several processing treatments can embrittle the surface of a plastic product, including contamination with certain chemical agents, oxidative degradation or crosslinking of the plastic molecules by heat aging or exposure to ultraviolet light; coating with brittle paints; and plating. Under certain circumstances cracks starting in this brittle surface layer can propagate into the underlying ductile plastic at sufficiently high velocity to induce a brittle failure of the entire product. Surface embrittlement can have a significant effect on the observed stress/strain, fatigue, and impact behavior of the plastic.

In light of the many types of behavior plastics can manifest and the considerable effect this behavior can have on the performance of the finished product, designers need to become familiar with specific behavior characteristics of each plastic considered for an application and to recognize potential problems. A major cause of problems is not poor product design but that the processes operated outside of their required operating processing window.

1.12 Coextrusion/Coinjection

With the coextruded and coinjection multilayer technology now established and the hardware in position, the variety of achievable properties is readily extendable through the correct combination of different materials. The potential for BM packaged products has been much more than simple bottles today. Now the know-how and economics are such that many previously futuristic ideas are much closer to reality (Chapters 2 and 5).

1.13 Processing Rule

In all design and manufacturing applications, one should always be aware that a change in processing (as well as in material, design, etc.) may create a problem.

Rules to remember:

- Processing is a marriage of machine, die/mold, material, process control, and operator, all working together

- Heat always goes from hot to cold through any substance at a controlled rate
- Hydraulic fluids or electric drives are pushed, not pulled, at a rate that depends on pressure and melt flow
- The fastest cycle or rate of output that produces the most products uses:
  - Minimal melt temperature for fast cooling,
  - Minimal pressure for lowest stress in products,
  - Minimal production time
- All problems have a logical cause (understand the problem, solve it, and then allow the machine to equalize its production to adjust to the change).

1.14 FALLO Approach

What has made the millions of plastic products successful worldwide, including BM products, was that there were those who understood the behavior of plastics and how to properly apply this knowledge. They did not have the "tools" that make it easier for us to now design and fabricate products. Although expanding knowledge has made it easier, a design plan conceived in the human mind and intended for subsequent fabricating execution by the proper method is still needed.

To create products at the lowest cost designers and processors have unconsciously used the FALLO approach, which emphasizes that many steps are involved to achieve success, all of which must be coordinated and interrelated. It starts with the design involves specifying the plastic and specifying the manufacturing process. The specific process (injection, extrusion, blow molding, thermoforming, and so forth) is an important part of the overall scheme and should not be problematic. Basically the FALLO approach consists of: (1) designing a product to meet performance and manufacturing requirements at the lowest cost; (2) specifying the proper plastic material that meet product performance requirements after being processed; (3) specifying the complete equipment line by (a) designing the tool (die, mold) "around" the product, (b) putting the "proper performing" fabricating process "around" the tool, (c) setting up auxiliary equipment (up-stream to down-stream) to "match" the operation of the complete line, and (d) setting up the required "complete controls" (such as testing, quality control, troubleshooting, maintenance, data recording, etc.) to produce "zero defects"; and (4) purchasing and properly warehousing plastic materials and maintaining equipment.

Using this type of approach leads to maximizing the product's profitability. Following the product design is producing a tool (die, mold, and so forth) around the product. Next is putting the proper fabricating process around the tool, which requires setting up the necessary auxiliary and/or secondary equipment to interface in the complete fabricating line, and finally setting up completely integrated equipment controls to target for the goal of zero defects. The final step (being repeated in order to emphasize its importance) in the FALLO process is that of purchasing equipment as well as materials, then properly warehousing the material and maintaining equipment. If processing is to be contracted ensure that the proper equipment is available and used. This interrelationship is different from that for most other materials, where the designer is usually limited to using specific prefabricated forms that are bonded, welded, bent, and so on.

2 Plasticator Melting Operation

The EBM and IBM machines use plasticators to melt plastics. They are a very important component in a melting process with its usual barrel and screw. If the proper screw design or barrel heat profile is not used, products may not maximize their performance or meet their low cost requirements.

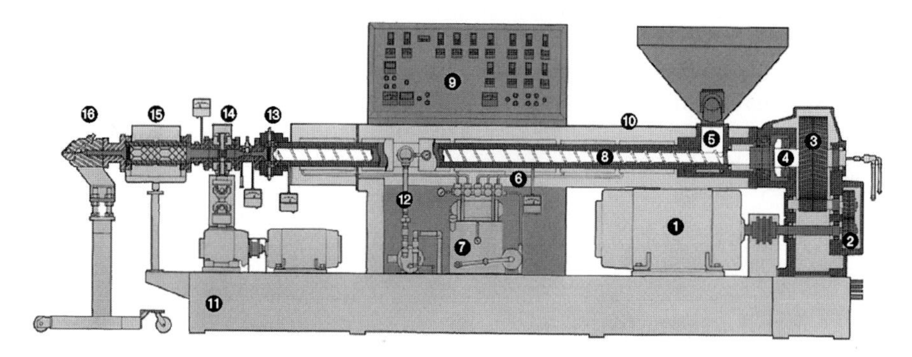

Fig. 2.1 *Example of a Welex extruder schematic*

Figure 2.1 shows the barrel/screw features of the extruder plasticator. This schematic shows: (1) drive motor (from 20- to 2,000-hp infinitely variable speed drive directly coupled to reducer for maximum efficiency all designed to save floor space); (2) high-efficiency gears to process all plastics; (3) heavy duty, heat treated helical or herringbone gears equipped with shaft-driven oil pumps and oil cooler; (4) long-lasting thrust bearing (with life expectant well in excess of 30 years of continuous operation); (5) large rectangular standard feed opening (round with lining, optional, for use with crammer feeders); (6) cast-in heater/cooler elements (heat quickly and last a long time); (7) cast-in stainless steel cooling tubes run parallel with heating elements (closed-loop, nonferrous distilled water is automatically adjusted via microprocessor-based temperature controller providing uniform, efficient cooling); (8) high-performance screws with bimetallic lined cylinder designed for the plastic requirement (long life at all temperatures and cored for cooling); (9) prepiped and prewired (read, for single power drop installation); (10) guards fully insulated (one-piece hinged, no loose parts, no disassembly needed for access); (11) heave fabricated steel single-unit base (preassembled so all parts are in place ready to be used); (12)

when required, patented two-stage venting which eliminate predrying for most plastics (vents can be plugged in minutes); (13) screen changer (optional) for continuous operation without shut down (hinged swing-bolt gate standard); (14) gear pump (optional) to ensure absolute volumetric output stability; (15) static mixer (optional) to provide thermal and viscosity homogeneity; and (16) die (optional) (designed to produce single or multilayer sheet without modification also strand dies, etc.).

Generally, the material being fed flows by gravity (usually controlled weightwise) from the feed hopper down into the throat of the plasticator barrel in the extruder or injection molding operations. As the screw turns in the heated barrel, plastic falls down into its channel. Frictional forces develop in the plastic during plasticizing so that the melt moves forward toward the die or mold.

In the initial action the plastic is in a solid state at a temperature below its melting point. As the temperature of the plastic rises above its melting point and the plasticizing action starts, a melt film will form on the barrel surface. This changeover can occur at any location downstream in the plasticator, although generally not prior to the start of the compression (transition) zone. The actual location depends on the rheological behavior of the plastic, machine geometry that is generally fixed, and operating conditions. The solid state of the plastic declines as it moves downstream to a complete melt by the end of the transition zone.

2.1 Feed Operation

Plastic resin and additives are fed to the plasticator operation. A hopper stores the premix and feeds the appropriate quantity to the feed throat of the plasticator. The hopper shape and size, material (recycled, flakes, pellets, etc), and auxiliary devices all affect the final material flow consistency and quality.

2.1.1 The Hopper

Plastic usage for a given process should be measured so as to determine how much plastic should be loaded into the hopper. As a guide, the hopper should hold enough plastic for possibly $\frac{1}{2}$ to 1 h of production. This action is taken so as to prevent storage in the hopper for any length of time. Care should be taken to ensure that hygroscopic plastics are in an unheated hopper for no more than a $\frac{1}{2}$ to 1 h, or as determined from experience and/or specified by the material supplier.

The hopper can be fitted with devices to perform different functions, for example, a hinged or tightly fitted sliding cover and a magnetic screen for protection against moisture pickup and metal ingress, respectively. It is usually advisable to install a hopper drier, especially when processing certain materials such as regrind, colors, and hygroscopic plastics. It can be of value in limiting material handling, as well as removing moisture.

In addition to designing the hopper for minimal flow problems, mechanical and vibrational devices can be used to promote the continuous flow of material to the extruder. Horizontal, flat, grooved discs are installed at the bottom of a hopper feeding a plasticator to control the feed rate by varying the discs' speed of rotation and/or varying the clearance between discs. A scraper is used to remove plastic material from the discs. Likewise, vibrating mechanisms can be placed against the hopper walls to stimulate particle movement and shift static/bridging material.

2.1.2 Feed Rate

To maintain the maximum and most consistent feeding, it is necessary to exercise care when changing hopper dimensions, feed throat openings, or adding any intermediate sections (side feeders, magnet packs, adapters, etc.).

When considering reengineering the solid's delivery system consider the following:

- Minimum taper for hoppers is 60° included angle for general purpose use with some plastics requiring smaller angle/steeper sides.
- Be sure the system is streamlined with no ledges, projections, or rough surfaces.
- Avoid, as much as possible, changes in shape, such as round to square, because each change of shape causes a restriction to flow.
- The absolute minimum cross section in any solid's flow channel should be the cross section of the barrel bore with a preference of about $1\frac{1}{2}$ times the cross section of the barrel bore.

In these respects, the flow of solids is much like liquid flow. That is, anything that has entry problems is not streamlined, has shape changes, or restrictive flow area that will result in excessive pressure drops. Unless there is a minimum pressure of the solid's mass at the entry to the screw, the screw channel will not fully fill. This occurs particularly as the screw speed increases and obviously varies with the characteristics of the solids.

2.1.3 The Hopper Feeder

Auxiliary equipment provides controlled flow of materials (powders, pellets, etc.) to or from a processing operation. Different equipment designs are used for dispensing, metering, and mixing that use volume and the popular, useful and cost saving gravimeter/weight blenders located over the feed hopper. Motor-driven augers as well as air-driven valves can be included to process materials such as flakes, powders, granular material, liquids, and pellets.

A gravimetric feeder/blender has the ability to pinpoint material accuracy despite variations in bulk density. It consists of the feeder (including discharge device), scale, and control unit, with a separate feeder system used for blending. There are both batch and continuous type units.

The volumetric feeder has an enclosed chamber that meters a specific volume of materials. Bulk handling and particle size as well as moisture content usually influence uniformity of the metering capacity.

Vibratory devices convey dry materials from storage hoppers to processing equipment, comprising a tray vibrated by mechanical or electrical pulses. The frequency and/or amplitude of the vibrations control the rate of material flow.

Hopper-mounted coloring feeder/blenders combine virgin plastics, regrinds, one or more colorants, and/or additives (such as slip agents, inhibitors, etc.). Materials are mixed by tumbling and the mixture is dropped by gravity into the process hopper. Some coloring loaders allow use of dry, powdered colorants; color concentrates; or liquid colorants through the same unit without major equipment alterations. They are self-loading and mount directly over processing machines so there is no need to manually handle component materials and risk contamination and waste. When powdered dry colorants are used, they can be placed in a canister in a separate color room and the filled canister is then mounted on the coloring loader (Chapter 11).

2.2 The Barrel

The barrel is also called a cylinder or plasticator barrel. It contains a screw together with which it provides the bearing surface where shear is imparted to the plastic material. Heating and cooling media are housed around it to keep the melt at the desired temperature profile.

There are many options for barrel construction, with most designed to withstand up to at least 69 MPa (10,000 psi) internal pressures for extruders; higher pressure units (210 MPa (30,000 psi)) are manufactured for the injection molding processes. The SPI Machinery Division Bulletin entitled Recommended Guidelines for Single Barrels provides working tolerances that produce effective performance with economy of manufacture. The need for corrosion and/or wear protection, cost, repair, or their combinations may determine the choice of materials. Barrels can be made from a solid piece of metal. The most common material is carbon steel with a nitride interior that gives a very thin hardened layer of only about 0.3 mm (0.01 in.) which results in an extended life span, especially when processing abrasive filled compounds. For large extruders repair can be achieved by inlaying weld material. Smaller machines can be repaired by machining out the opening diameter and inserting a liner.

Solid barrels can also be fabricated from stainless steel with a hardened interior. However, the hardening process often lessens the protection against corrosion afforded by stainless steel. Other negative aspects to solid stainless steel are poor heat transfer and higher cost. Therefore, this is usually an option for smaller machines having inside diameters of less than 60 mm (2.3 in.). Applying castings to the interior in place of hardening results in a superior wear and/or corrosion protection but at a higher cost.

Lined barrels are used where repair or high-performance alloys are considerations. The liners are thin walled and inserted into an opening in the larger load-bearing barrel. Liners can be made from hardened carbon steel, stainless steel, nickel based alloys, and so forth and can be used with a bimetallic coating. Heat transfer can become a problem if air gaps between liner and barrel materials exist.

Bimetallic coatings are used if wear and corrosion are of concern. They can be applied by traditional welding techniques for large barrels. Centrifugal casting is the most common technique for both single and twin bore barrels. A wide variety of bimetallic coatings are used. The choice of coating depends on the plastic to be processed. For noncorrosive processes with up to about 30 to 40 wt% of filler, the iron-based alloy is usually considered. Highly corrosive systems using materials such as PVC and fluoropolymers can benefit from the Ni/Co family of alloys. Processes compounding high levels of abrasive fillers should use the tungsten family of alloys.

2.2.1 Barrel Length-to-Diameter Ratio

The barrel's inside diameter together with its length classifies sizes. It is common practice to refer to the L/D ratio, which is the barrel length (L) to diameter ratio (D). As reviewed later, there is also a screw length-to-diameter ratio. For low-output 40 to 60 mm (1.6 to 2.3 in.) diameter extruders are normally used whereas for higher output 120 and 150 mm (4.7 and 5.8 in.) diameter screws are more common. Injection molding barrel diameters are approximately the same, with the smaller and larger diameters providing the smaller and larger melt shot sizes, respectively.

The trend for extruders has been toward the use of longer barrel lengths. In the past, L/D ratios of 24 : 1 were the rule. Now 30 : 1 and longer are frequently specified. Usual injection moldings L/D are 20 : 1 and up to at least 24 : 1. These longer barrels permit more melt inventory time, and consequently a higher output rate is possible without sacrificing the plasticization rate or reducing the homogenization of the melt. To take full advantage of the increased output rates of longer barrels, a more powerful drive unit should be used; actually power requirement is based on the plastic-melt torque requirement. Typically for extruders, 150-hp motors are used on 120 mm (3.2 in.), 24 : 1 L/D extruders. For a 30 : 1 L/D, 200 hp are frequently specified. Smaller 60 mm (1.6 in.) machines have drive motors of up to 50 hp.

2.2.2 Barrel and Feed Unit

When selecting a plasticator barrel the size and shape of the feed throat are very important. Feed throat size and shape can have a significant effect on output and its stability. The feed throat and feed hopper units are important in ensuring that plastics are properly plasticized. The feed throat is the section in a barrel where plastic is directed into the screw channel. It is usually fitted around the first few flights of the screw. Some machines do not have a separate feed throat, instead the feed throat is an integral part of the barrel.

In general, the smaller the hole, the more adverse the effect. Sometimes smaller feed holes can be compensated for by screw designs, but more often the feed hole geometry must be modified. Output rates have been observed to vary as much as 25%, with the only variable being the feed throat geometry. Round feed throats are sufficient for 100% pellet feed but when 20% or more regrind is added with the virgin feed the rate is reduced; a rectangular or oblong opening will improve the feeding characteristics. An elongated opening also helps in reducing (possibly eliminating) potential bridging problems in the throat.

Some materials, such as flakes, regrind, and others, have poor flow properties that lead to problems. The feed throat and feed hopper units are important in ensuring that plastics are properly plasticized.

2.2.3 Feed Cooling Unit Operation

The feed throat casting is generally water cooled to prevent an early temperature rise of the plastics. A good starting point is to have the temperature about 110 to 120 °F (43 to 49 °C), or “warm to the touch” to help ensuring that a stable feed is developed. If the temperature rises too high, it may cause the plastic to stick to the surface of the feed opening, causing a material conveying problem to the screw. The overheated plastic solidifies at the base of the hopper or above the barrel bore, causing bridging whereby material can no longer enter the screw. The problem can also develop on the screw, with plastic sticking to it, restricting forward movement of material.

Overcooling the feed throat can have a negative effect on performance because of the heat sink effect that pulls most of the heat from the feed zone of the barrel. The idea of block cooling is primarily to prevent sticking or bridging in that area. Thus, the temperature should not be colder than necessary. Always control water flow in the throat cooling systems from the outlet side to prevent steam flashing and to minimize air pockets.

2.2.4 Barrel Grooves

In the feed section, grooves can be placed in the internal barrel surface to permit considerably more friction between the solid plastic particles and the barrel surface. This results in higher temperatures, increased output, and/or improved process stability. The grooves are typically placed in the feed section for a length of several screw diameters (Fig. 2.2). The grooves usually run along the length (in the axial direction) of the barrel. To prevent the material from melting in the grooves, sufficient cooling channels must be provided along the length of the barrel. However, changing to a grooved barrel will affect the behavior of the feed and the conveying capabilities of the operation. A special screw design will be required for the grooved feed extruders or the anticipated benefits will not materialize.

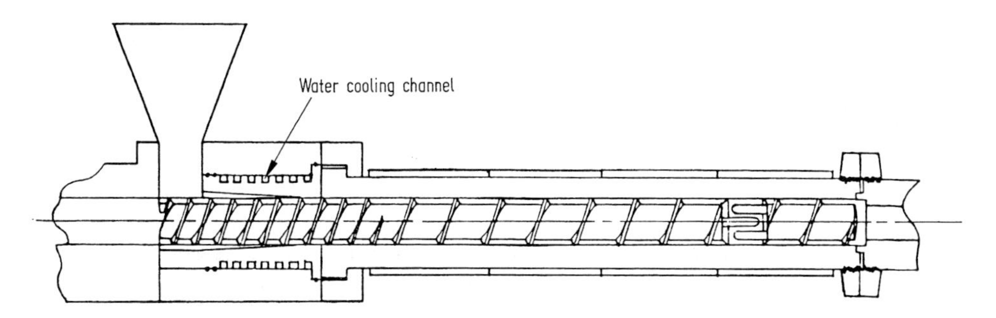

Fig. 2.2 Grooved barrel extruder with zero compression ratio screw

2.2.5 Barrel Check

On receiving or replacing machines with barrels, it is best to have them measured so that you can determine if any wear or damage will occur after they are put into operation.

Machines are designed to process certain quantities of different plastics at certain rates. If too little is being used, problems could develop with excessive residence time, particularly with engineered plastics. Instead of replacing the entire machine to remedy the problem, the plasticator is changed (best to start with the properly sized plasticator). Downsizing, usually to a limited amount with plasticators, requires a smaller screw, smaller heaters, modification of the barrel shroud, and so forth. Upsizing, which is rare, to increase output will reduce the available pressure, set up torque limitations on the larger screw, and so forth.

2.2.6 Alignment

It is important that alignment at installation and routine maintenance checks are made to ensure the barrel and screw are properly aligned. (Alignment also involves other attachments to the plasticator such as the extruder die or IM mold as well as with inline equipment.) Without proper machine installation, the precision alignment built into the machines (plasticator, etc.) is lost when not properly supported on all its mounting points. Installation involves factors such as ground support stability, precise alignment of equipment, uniform support, and effective control of vibration. A good foundation provides a level bed for the machine and does not move or settle during operation. The contractor laying the foundation should be familiar with the underlying soil conditions and meet the machinery recommended requirements. Instructions should be in the manufacturer's manual; if not do not buy the equipment.

Depending on the soil and other ground conditions, a bed of reinforced concrete is put down with sufficient thickness and strength so that it can distribute both static and dynamic loads to the ground without settling. To ensure proper load distribution, a steel sole plate is usually set in the concrete. The machine and in-line equipment is attached to their respective foundations using anchor bolts, jack screws, shims, and grout. These mounts are carefully adjusted to permit accurate leveling and absorption of any shock. The anchor bolts are part of the foundation. A washer or other device attached at the bottom is needed to prevent the bolts from unscrewing during tightening of the supports.

2.2.6.1 Plasticator Wobble

If the barrel of a plasticator tends to wobble when the screw is rotating, it is a symptom of a misalignment. It could be an alignment deficiency between the transmission, feed block, and/or barrel. This should be investigated immediately as it can result in a high wear situation, screw breakage, heater creep, and/or shutdown of the machine.

2.2.7 Barrel Borescoping

Barrel borescoping is the alignment of a barrel with the screw. The clearance can range from 0.05 to 0.20 mm (small- to large-diameter screw) on any side of the screw. Borescoping is not a guarantee that this will correct all problems. With an alignment scope one can tell what the internal shape of a barrel is at any point, whether it is a straight, curved, or even S-shaped that may be due to machining or after use due to wear. Other areas are also to be examined. Regardless, aligning with a scope provides a much better opportunity to produce better products with less downtime, less scrap, and extension of the life of the barrel/screw. Most machines can be adjusted in a day at very little cost. The machine will have at least a 25% life extension. As an example, if a barrel costs $12,000 and a screw $9,000 for a total of $21,000, extending their life only 25% saves over $5,250 and increases production.

2.2.8 Heating and Cooling Barrels

The extruder barrel is composed of smaller barrel units, which have independent capabilities for both heating and cooling. Each barrel unit represents a zone along the length of the barrel. Typical barrels can be as short as one or two zones. Longer barrels will have more zones permitting individual controlled heating zones facilitating the required melt temperature profile as the plastic travels through the barrel.

For barrel heating, electrical resistance or induction heater zones (sections) are mounted on or around the barrel in different locations along its axis.

The cooling of certain thermoplastics during extrusion usually requires cooling control in addition to heaters to ensure that overheating of the melt does not occur. Different methods are used such as liquid cooling channels, or coils around the barrel, and/or forced air around the barrel some of which may have fins thus providing more cooling surfaces.

Injection molding machines (IMM) for processing thermoset plastics and rubbers (thermosets) control the barrel temperature most of the time indirectly with an external heat exchanger. It is operated by a liquid heat-transfer medium such as oil or brine. Figure 2.3 shows heating by fluid circulation using two independent zones. In this example, by removing the manifold indicated by 1, the barrel can be converted to a three-zone heated unit. Curing of the plastic or rubber occurs in the mold cavity(ies) by the application of higher heat than the temperature in the barrel melt. A chemical crosslinking action occurs with the additional heat resulting in the solidification of the thermoset materials.

Depending on the IMM operation capability as well as on the type of plastic being processed, the melt passing from the plasticator through the nozzle may require heat

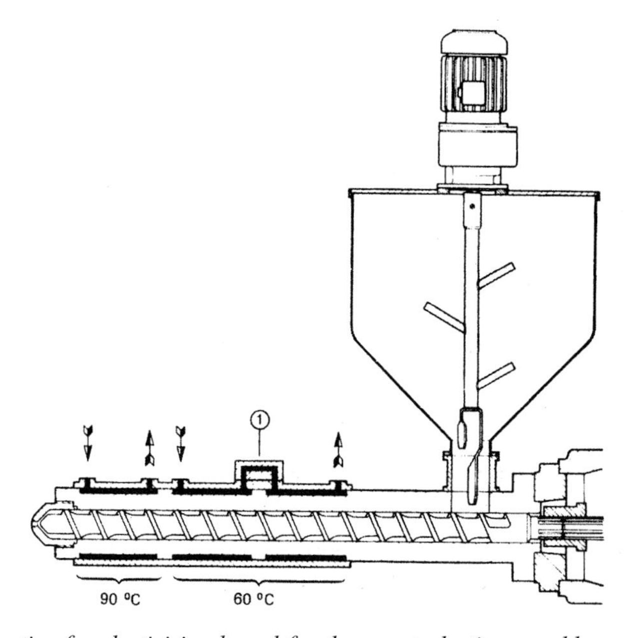

Fig. 2.3 *Schematic of a plasticizing barrel for thermoset plastics or rubbers*

control. This action is usually necessary when processing certain heat-sensitive plastics and/or if a long nozzle is used.

2.2.8.1 Barrel Temperature Measurement

Temperatures of barrels (and BM mold cavities) can be measured with thermocouples (TCs) and/or resistance temperature detectors (RTDs). They are mostly equipped with a spring-loaded bayonet fitting (or equivalent), fitting its tip closely to the bottom in the barrel well. Good contact is required or falsified readings will occur affecting thermal conduction performance.

2.2.8.2 Heating Methods

Heat to soften the plastic is supplied in two ways: by external barrel heating and internal frictional forces brought about on the plastic due to the action of the metal screw in the metal barrel. The amount of such frictional heat supplied in the plasticator is appreciable; in many extrusion operations it represents most of the total heat supplied to the plastic.

There are three principal methods of barrel heating: electrical, fluid, and steam. Electrical heating can cover a much larger temperature range and is clean, relatively inexpensive, and efficient. The heaters are generally placed around and along the barrel, grouped in zones. Each zone is usually controlled independently, so the desired temperature profile can be maintained along the barrel. Electrical heating is generally preferred because it is the most convenient, responds rapidly, is the easiest to adjust, easy to clean, requires a minimum of maintenance, covers a much larger temperature range, and is generally the least expensive in terms of initial investments.

Many different types and shapes of heaters are used. There are electrical resistance or induction heaters (ceramic, mica, etc.) in the form of a cuff encircling the barrel, providing the necessary heat. Common with extruder barrels are cast aluminum heaters/coolers or calrod elements in grooved aluminum elements.

Fluid heating, such as the use of heated oil, allows an even temperature over the entire heat transfer area, avoiding local overheating. If the same heat transfer fluid is used for cooling, an even reduction in temperature can be achieved. The maximum operating temperature of most fluids is relatively low for processing thermoplastics, generally below 250 °C (482 °F). With its even temperature, the required fluid heating is desirable with thermoset plastics so that no accidental overheating occurs to chemically react and solidify in the barrel.

Although steam heating was used in the past, particularly when processing natural rubber, it is rarely used now. Steam is a good heat transfer fluid because of its high specific heat capacity, but it is difficult to get steam to the temperatures required for thermoplastic processing of 200 °C (392 °F) and greater.

2.2.8.3 Cooling Methods

The cooling of barrels is an important aspect. The target is to minimize any cooling and, where practical, to eliminate it because, in a sense, cooling is a waste of money. Any amount of cooling reduces the energy efficiency of the process, because cooling directly translates into lost energy; it contributes to the machine's power requirement. If an extruder requires a substantial amount of cooling, when compared to other machines, it is usually a strong indication of improper process control, improper screw design, excessive L/D, and/or incorrect choice of extruder (single vs. twin screw).

Cooling is usually accomplished with forced air blowers mounted underneath the barrel. The external surface of the heaters or the spacers between the heaters are often made with cooling ribs to increase the heat transfer area (ribbed surfaces will have a larger area than flat surface), which significantly increases cooling efficiency. Forced air is not required with small-diameter extruders because their barrel surface area is rather large compared to the channel/rib volume, providing a relatively large amount of radiant heat losses.

Fluid cooling is used when substantial or intensive cooling is required. Air cooling is rather gentle because its heat transfer rates are rather small compared to water cooling. But it does have the advantage in that, when the air-cooling is turned on, the change in temperature occurs gradually. Water cooling produces rapid and steep change, which is a requirement in certain operations. This faster action requires much more accurate control and is more difficult to handle without proper control equipment.

The larger barrels are often liquid cooled, using cored channels to circulate the cooling medium because they require intense cooling action (feed throats also use water cooling). However, if not properly controlled, problems could develop. If the water temperature exceeds its boiling point, evaporation can occur, resulting in poor cooling control. The water system is an effective way to extract heat, but can cause a sudden increase in cooling rate resulting in a nonlinear control problem, which makes it more difficult to regulate temperature. Nevertheless, the water cooling approach is used very successfully with adequate installation in the extruder and adequate control and startup procedures.

2.2.9 Metal Detection

Ensuring no metal contamination exists in virgin plastics and particularly recycled plastics is important to eliminate problems during processing and/or product use. Magnets of different designs are used to nest and fit in hoppers (also in blenders, granulators, etc.).

Metal detection and separation technology offers a number of options ranging from simple magnetic removal of ferrous (iron containing) metals to electrostatic and other techniques capable of removing all manner of metals, including copper, aluminum, and stainless steel. The sensitivity of the equipment is measured by the size of the metal particles it can detect. When different types of equipment are combined, the result can be output with metal contamination as low as 5 ppm.

Most systems for detecting both ferrous and nonferrous metals are based on the use of conventional or rare earth magnets to create magnetic eddy currents. They are available in different forms. An example of a separation system is an electrostatic sensor under the plastics being fed over a conveyor system, where an air jet propels unwanted particles out of the plastic material stream.

2.2.10 Barrel–Die Coupling

The coupling between an extruder barrel and die can be carried out in various ways using bolts or locking devices including:

- Flange fitting with a clamp ring on the barrel and a fixed flange on the die
- Flanges on the barrel and die with tapered links and two bolted half-clamps, or a ring clamp hinged at one side and bolted to the other side
- Swing-bolt flange connection between the barrel flange and a die flange. In most extruders, a breaker plate with screens is used between the barrel and die assembly.

2.2.10.1 Bell End

The bell end is a flange at the discharge end of the barrel that provides added strength to withstand internal pressure.

2.2.11 Barrel Safety

Barrels contain some type of pressure safety devices such as fail-safe rupture discs/bolts. If barrel pressure exceeds its rated burst pressure, disc(s) and/or bolt(s) (adapter, etc.) rupture to relieve pressure. These safety devices are to be carefully handled during maintenance of the barrels.

When using bimetallic liners in the barrels any exposed edge of the liner can be easily damaged when inserting the screw. Protect it with a ring made to fit the end.

2.3 The Screw

Basically, a screw is a geometric helical flighted hard steel shaft that rotates within a plasticizing barrel to mechanically process and advance the plastic being prepared for

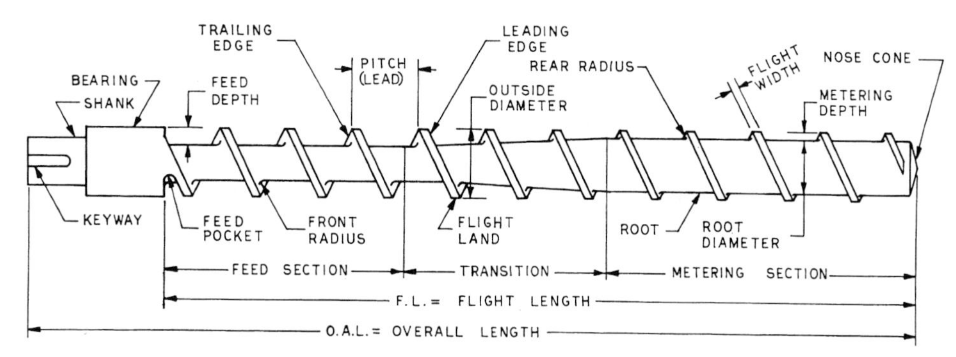

Fig. 2.4 Typical metering EBM screw with the usual 17.8° flight helical angle [5]

different processes that principally involve EBM and IBM (Figs. 2.4 and 2.5). It has a wide operating range to meet different performance requirements of all the different plastics processed. Its rotating drive system can be via a hydraulic or an electrical motor. Electrical motors tend to increase melt processing efficiency which in turn increases production rate. It has a wide operating range to meet different performance requirements of all the different plastics processed. It is important to obtain maximum throughput with as close to a perfect melt quality. It is an endless target owing to the limits and/or variabilities of the plastics, machines, and controls. Since the start of using screw plasticators definite improvements have been occurring in the melt quality and will continue because of growing advancements in applying advanced screw designs and the changing melting characteristics of plastic materials.

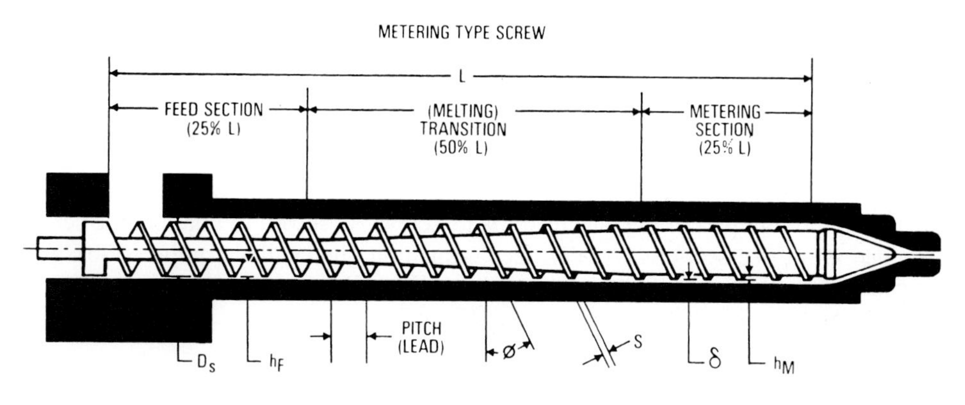

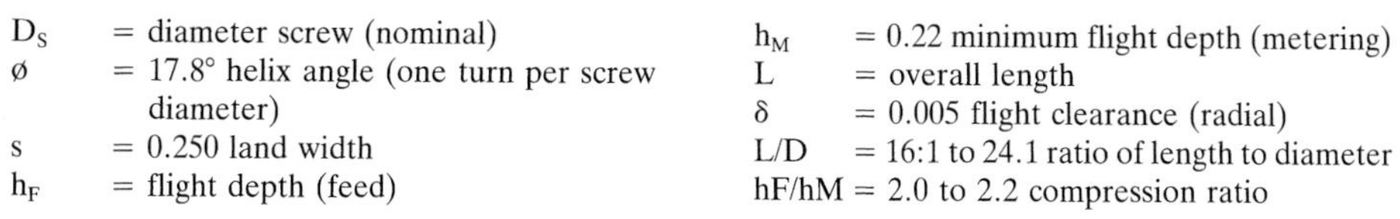

D_S = diameter screw (nominal)

ø = 17.8° helix angle (one turn per screw diameter)

s = 0.250 land width

h_F = flight depth (feed)

h_M = 0.22 minimum flight depth (metering)

L = overall length

δ = 0.005 flight clearance (radial)

L/D = 16:1 to 24.1 ratio of length to diameter

hF/hM = 2.0 to 2.2 compression ratio

Fig. 2.5 Typical metering IBM screw with smear head and the usual 17.8° flight helical angle [6]

The interplay among the many variables characterizes the screw requirements. Examples are the screw diameter with its geometry as well as the screw drive that must be carefully selected based on melt volume and rate of travel. Many different screw designs are available to achieve the desired performance for the different plastics being processed.

The features common to all screw plasticators are screw(s) with matching barrel(s) that have at least one hopper/feeder intake entrance for plastics, and one discharge port/exit for the melt. The essential factor in their "pumping" process is the interaction between the rotating flights of the screw and the stationary barrel wall. If the plastic is to be mixed and conveyed at all, its friction must be low at the screw surface but high at the barrel wall. If this basic criterion is not met the material may rotate with the screw without moving at all in the axial direction and out through the die. The clearance between the screw and barrel is usually extremely small.

As practically all plastics processed are thermoplastics, most reviews on screw design concerns processing thermoplastics. When extruding thermoplastics the screw is usually limited in design, with a compression ratio of 1 to eliminate possible overheating of the thermosets. The thermosets cannot be permitted to overheat in the barrel, or it will solidify. If it does, the screw must be removed from the barrel and the solidified thermosets removed from the screw.

In the output zone, both screw and barrel surfaces are usually covered with the melt, and external forces between the melt and the screw channel walls have no influence except when processing extremely high viscosity materials such as rigid PVC and UHMWPE. The flow of the melt in the output section is affected by the coefficient of internal friction (viscosity), particularly when the die offers a high resistance to the flow of the melt.

2.3.1 Screw Sections

The most important part of the plasticator is its screw. As shown in Fig. 2.6, the screw is divided into three sections (zones): feed, transition (also called compression), and metering. This schematic illustrates a screw for IBM. Removing the cone (nonreturn valve) in front would make it a typical EBM screw.

2.3.1.1 Feed Section

The feed section is the portion of the screw that picks up the plastic at the feed opening (throat) plus an additional portion downstream. Many screws, particularly for extruders, have an initial constant lead and depth section, all of which is considered the feed section. This section can be welded onto the barrel or a separate part bolted onto the upstream end of the barrel. It is usually jacked for fluid heating and/or cooling.

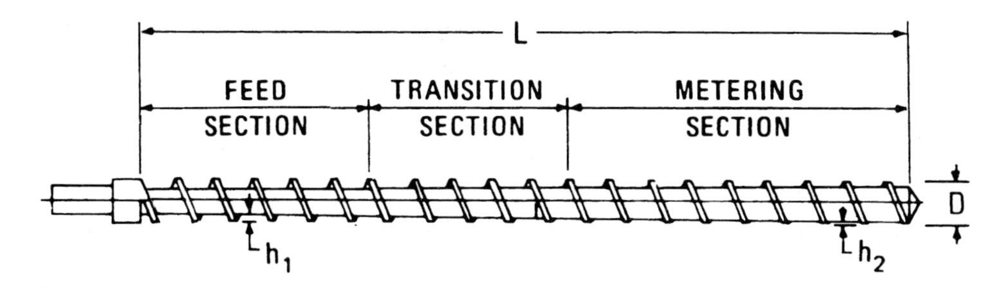

Screw diameter (D) inches, (mm)	Feed zone depth (h_1) inches, (mm)	Metering zone depth (h_2) inches, (mm)
1.5 (38)	0.250 (6.35)	0.080 (2.03)
2.0 (51)	0.320 (8.13)	0.100 (2.54)
2.5 (63)	0.380 (9.65)	0.120 (2.79)
3.5 (89)	0.400 (10.16)	0.125 (3.17)

For injection molding machines, the following general configurations are suggested:

	Upper Range (%)	Lower Range (%)	Recommended (%)
Feed section	60	33⅓	50
Transition section	33⅓	20	25
Metering section	33⅓	20	25

Fig. 2.6 *Examples of gradual transition screws for EBM (top chart) and IBM*

2.3.1.2 Transition Section

The transition section, also called the compression section, is the section of a screw between the feed zone and metering zone in which the flight depth decreases in the direction of discharge. In this zone the plastic starts in both a solid and a molten state with the target of having only molten plastic on leaving this zone.

2.3.1.3 Metering Section

This section is a relatively shallow portion of the screw at the discharge end with a constant depth and lead, usually having the melt move three or four runs of the flight length.

2.3.2 Screw Action

The constantly turning screw augers the plastic through the heated barrel where it is heated to a proper temperature profile and blended into a homogeneous melt (Fig.

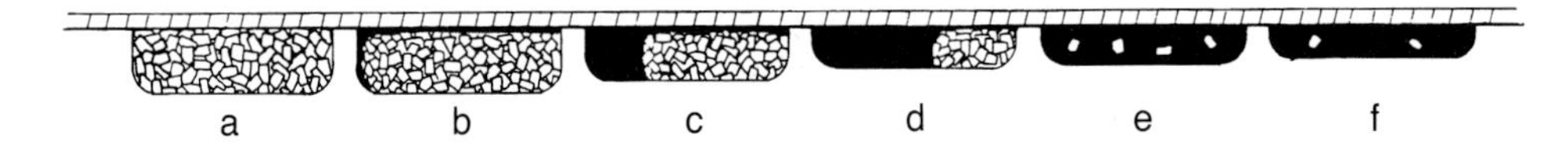

Fig. 2.7 Melt model for a standard screw

2.7). The rotating geometrically-flighted screw mechanically plasticizes and advances the material through the barrel. With the help of heat and pressure at a controlled flow rate, the material in the screw channel is subject to thermal and mechanical changes during the operation (Fig. 2.8). The shearing action of the screw is the major contributor to heating the plastic once the initial barrel heat startup occurs. The screw action shown in Figure 2.6 can be described as follows:

- The feed section initiates solids conveying. Sliding with low friction on the screw and high friction on the barrel enhances this. In this section, there is also some material compacting and a little heating of the plastic.
- At the beginning of the transition, the plastic is further heated and more compression occurs. The solid plastic is forced against the barrel, causing a sliding action. This frictional heat creates a thin film of melted plastic on the inner barrel surface.
- As the plastic proceeds down the transition, there is more melting, usually in the transition zone, and more compression. Here the plastic is divided into three parts: a compacted solid bed, a melt film along the barrel surface, and a melt pool. The melt pool is formed as the melt film is collected by the advancing flight. Most of the melting continues to be the result of sliding friction of the solids bed against the heated barrel. This is a rapid and efficient melting action similar to melting an ice cube by pushing it against a hot grinding wheel.
- The channel depth continues to decrease as plastic progresses down the transition zone. Melting continues and the width/volume of the solids bed decreases, while the width/volume of the melt pool increases. Unfortunately, as the channel gets shallower, shear rate increases. Now the already melted plastic continues to be heated. With too high a heating, undesirable conditions can develop such as degradation of the plastic.
- Continuing downstream through the plasticator, the solids bed breaks up; the unmelted plastics are distributed throughout the channel like ice cubes in water. The efficient melting by friction of the solids bed against the barrel tends to stop and now only less efficient melting occurs where overheating in the melt continues in the shallow metering zone. The complete melting action should occur within this zone.
- Plastic continues down the shallow metering section to its exit from the plasticator. There is a possibility that unmelted plastics or the melt has nonuniform temperature and viscosities. This situation of a nonuniform melts usually results in poor product performances, color mixing, and so on. Improved mixing can be obtained by reducing the screw's channel depth; however, overheating and less

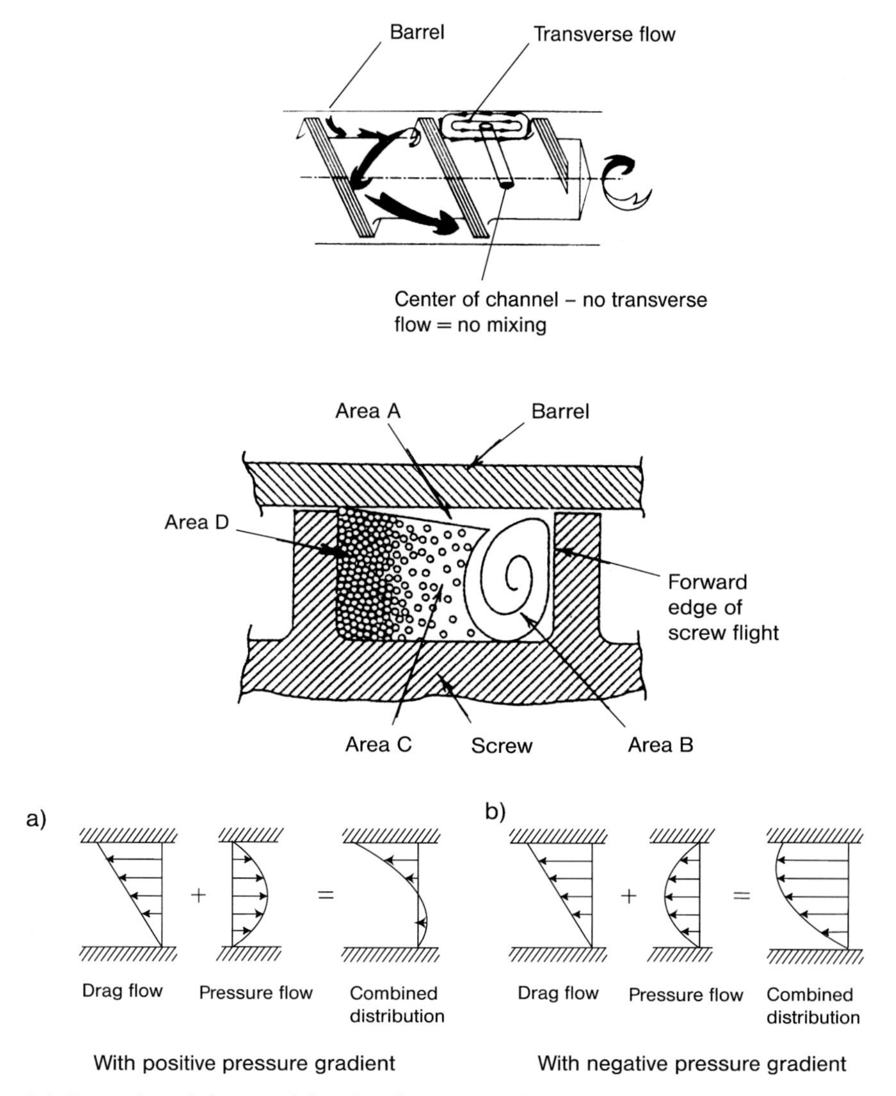

Fig. 2.8 *Examples of plastic solid and melt action in the screw channels*

output occurs. Feeding plastics from the hopper can help (or resolve) this situation. The constant depth metering section is not considered a good mixer when improved mixing is required in this section because smooth laminar flow patterns are established, causing the different portions of melt to continue to move in a fairly constant pattern so that no mixing action occurs to eliminate the problem. Screw design plays an important part in eliminating the problem. Material properties as well as process variations will change the optimal processing lengths for the screw sections. Table 2.1 shows some general screw design configurations for various plastics.

Table 2.1 *Hypothetical Screw Designs for Various Plastics*

Dimension	Rigid PVC	Impact polystyrene	Low-density polyethylene	High-density polyethylene	Nylon	Cellulose acet/butyrate
Diameter	$4\frac{1}{2}$	$4\frac{1}{2}$	$4\frac{1}{2}$	$4\frac{1}{2}$	$4\frac{1}{2}$	$4\frac{1}{2}$
Total length	90	90	90	90	90	90
Feed zone (F)	$13\frac{1}{2}$	27	$22\frac{1}{2}$	36	$67\frac{1}{2}$	0
Compression zone	$76\frac{1}{2}$	18	45	18	$4\frac{1}{2}$	90
Metering zone (M)	0	45	$22\frac{1}{2}$	36	18	0
Depth in (M)	0.200	0.140	0.125	0.155	0.125	0.125
Depth in (F)	0.600	0.600	0600	0.650	0.650	0.600

2.3.3 Screw Design

Even in today's highly technological world, screw design is still dominated by experienced trial and error approaches providing the exact capabilities of the screws for a particular plastic operating under specific conditions. Each operation of the screw subjects the plastic to different thermal and shear situations. Consequently, the plasticizing process becomes rather complex. However, it is controllable and repeatable within the limits of the equipment and material capabilities. A fixed screw speed, screw pitch, and channel depth determines the output. A deep-channel screw is much more sensitive to pressure changes than a shallow screw. In the lower pressure range, a deep channel will provide more output; however, the reverse is true at high pressures. Shallower channels tend to give better mixing and flow patterns.

In addition, computer models (based on proper data input and, very important, experience with a setup similar to the one being studied) play a very important role. When new materials are developed or improvements in existing materials are required, one must go to the laboratory to obtain rheological and thermal properties before computer modeling can be performed effectively. New screws improve one or more of the basic screw functions of melt quality, mixing efficiency, melting performance along the screw, melt heat level, output rate, output stability, and power usage or energy efficiency.

2.3.3.1 Screw Mixing

A general theory of mixing usually considers a nonrandom or segregated mass of at least two components and their deformation by a laminar or shearing deformation process. The object of the shearing is to mix the mass in such a way that samples taken from the mass exhibit minimal variations, ultimately tending to be zero. The basic principles are:

- Interfacial area between different components must be greatly increased to decrease striation thickness.
- Elements of the interface must be distributed uniformly.
- The ratio of mix within any unit or the whole is the same. The mixing theory suffers from limitations in this regard.

- It is limited to purely viscous (Newtonian) materials.
- No van der Waals or other forces between particles are assumed.
- Mixing process is assumed to be isothermal yet heat generated usually is sufficiently high that the energy dissipation during mixing is considerable. Even though a major rheological gap between theory and practice exists, qualitative and semiquantitative concepts are obtained.

Different screw designs are used to meet the different mixing requirements based on the plastic's melt requirements. As an example, the Spirex Pulsar mixing screw is used where low shear action is required; the Spirex Z-mixer is for the higher shear melts [41].

Mixing can be distributive and/or dispersive. Distributive and dispersive types of mixing are not physically separated. In dispersive mixing, there will always be distributive mixing, but the reverse is not always true. In distributive mixing, there can be dispersive mixing only if there is a component exhibiting a yield stress and if the stresses acting on this component exceeds the yield stress. For a dispersive mixing device to be efficient, it should have the following characteristics:

- The mixing section should have a region where the plastic is subjected to high stresses.
- High stress region should be designed so that exposure to high stresses occurs only for a short time.
- All fluid elements should experience the same high stress level to accomplish uniform mixing. In addition, they should follow the general rules for mixing: minimum pressure drop in the mixing section, streamline flow, and complete barrel surface wiping action.

Many of these dynamic barrier mixers are used extensively worldwide to improve screw performance. Static mixers are sometimes also inserted at the end of the plasticator. Each type of mixer offers its own advantages and limitations. These mixing elements are usually installed as near as possible to the end of the metering zone. Where practical, they should be located in a region where the melt viscosity is low, or where there is sufficient back pressure to push the material through. Because some of these installations may have to operate at a lower speed to avoid problems such as surging, independently driven mixers can be used so machines can operate at optimum speed [60]. Other benefits of independently driven mixers involve feeding capability and performance. For example, metering pumps can inject liquid additives with precision directly into the mixer.

2.3.3.2 Melting Plastic

Although melting technology is considered to be basically empirical, scientific approaches to screw designs based on an analytical melting model can be used. The production rate of an acceptable melt from a screw, which is its most important function, is often limited by its melting capacity. The melting capacity of the screw depends on the plastic properties, the processing conditions, and the particular geometry of the screw. Once the melting capacity is predicted, the screw can be designed to match the melting capacity.

Different plastic materials have different melt requirements. As an example standard, single-stage metering PE screws generally have the following features:

- A length of 20 to 30 times the diameter of the screw (L/D 20 to 30 : 1)
- A ratio of 3 to 4 : 1 for the depth or volume of the flights in the feed section to the depth or volume of the flights in the metering section (compression ratio)
- A constant distance between flights (pitch) equal to the diameter (square pitch)
- A mixing device at the end of the metering section to promote homogeneity.

The design of the screw is important for obtaining the desired mixing and melt properties as well as output rate and temperature tolerance on melt. Generally, machines use a single constant-pitch, metering-type screw for handling the majority of plastic materials. A straight compression-type screw or metering screws with special tips (heads) are used to process heat-sensitive thermoplastics, and so forth. In injection molding special valving is required on the screw tip to properly melt and in turn inject certain amounts of melt into the preform mold [6].

A general heat profile for an extruder running high density polyethylene (HDPE) is as follows: zone 1 (rear)= 300 °F, zone 2= 325 °F, zone 3= 350 °F, zone 4= 350 °F, and zone 5 (front)= 375 °F. Actual temperature profiles may vary, depending on grade of plastic used, extruder output rate (Table 2.2), overall cycle, and so forth.

2.3.3.3 Compression Ratio

The compression ratio (CR) relates to the compression that occurs on the plastic basically in the transition (or compression) section; it is the ratio of the volume at the start of the feed section divided by the volume in the metering section (determined by dividing the feed depth by the metering depth). The CR should be high enough to compress the low bulk unmelted plastic into a solid melt without air pockets. A low ratio will tend to entrap air pockets. High percentages of regrinds, powders, and other low bulk materials are usually processed by a high CR. A high CR can overpump the metering section (Tables 2.3 and 2.4).

A common misconception in engineering is that heat-sensitive plastics should use a low CR. This is true only if it is decreased by deepening the metering section, and not having a more shallow feed section. The problem of overheating is more related to channel depths and shear rates than to CR. As an example, a high CR in polyolefins can cause melt blocks in the transition section, leading to rapid wear of the screw and/or barrel. For thermosets the CR is usually 1, so that accidental overheating does not occur and cause the plastic to solidify in the barrel. Their barrels are usually heated using a liquid medium so that very accurate control of the melt occurs with no overriding of the maximum melt heat. With overheating thermosets melt solidifies. If

Table 2.2 *Extruder Output Rates for Various Plastics*

Diameter		$2\frac{1}{2}$ in. (64 mm)	$3\frac{1}{2}$ in. (89 mm)	$4\frac{1}{2}$ in. (114 mm)	5.12 in. (130 mm)	6 in. (152 mm)	8 in. (203 mm)
LLDPE	lb/h	275–340	550–650	900–1100	1150–1400	1600–1950	2840–3475
	kg/h	125–155	250–300	400–500	525–650	725–890	1290–1580
LDPE	lb/h	475–575	925–1125	1530–1870	1975–2400	2700–3300	4800–5900
	kg/h	210–260	420–510	695–850	900–1100	1255–1500	2200–2675
HDPE	lb/h	330–425	650–800	1080–1320	1390–1700	1920–2350	3400–4170
	kg/h	150–195	300–360	490–600	630–775	870–1065	1550–1895
PP	lb/h	360–440	700–865	1170–1430	1510–1840	2080–2540	3700–4500
	kg/h	160–200	320–390	530–650	685–835	945–1150	1680–2050
FPVC	lb/h	415–510	815–1000	1350–1650	1740–2125	2400–2930	4260–5200
	kg/h	190–230	370–450	610–750	790–965	1090–1330	1940–2370
RPVC	lb/h	220–270	435–530	720–880	925–1135	1280–1560	—
	kg/h	100–120	195–240	325–400	420–515	580–710	
ABS	lb/h	360–440	700–865	1170–1430	1510–11,840	2080–2540	3700–4500
	kg/h	160–200	320–390	530–650	685–835	945–1150	1680–2050
HIPS	lb/h	490–615	965–1200	1600–2000	2060–2580	2840–3550	5050–6320
	kg/h	225–280	440–550	725–900	935–1170	1290–1615	2300–2870
Acrylic	lb/h	415–510	815–1000	1350–1650	1740–2125	2400–2930	4260–5200
	kg/h	190–230	370–450	610–750	790–965	1090–1330	1940–2370
Polycarbonate	lb/h	285–350	560–680	925–1125	1200–1450	1650–2000	2900–3500
	kg/h	130–160	250–310	420–510	550–660	750–900	1325–1600

Courtesy of Davis-Standard

Note: Outputs given are typical for the extruder size and general polymer classification.

it solidifies, the CR of one also permits ease of removal by just "unscrewing" it from the screw. A CR ratio of 1 is also used for thermoplastics when the rheology so requires.

Table 2.3 *Examples of Compression Ratio for Thermoplastics*

Low-compression screw (1.2–1.8 compression ratio)	Acrylics Acrylic multipolymer ABS and SAN Polyvinyl chloride, rigid
Medium-compression screw (2.0–2.8 compression ratio)	Acetal (Delrin 100) Cellulosics (acetate, propionate) Nylon (low melt index) Phenylene oxide-based resin (Noryl) Polycarbonates Polyethylene (medium to low melt index) Polypropylene (medium to low melt index) Polystyrene (crystal and impact) Polyvinyl chloride (flexible)
High-compression screw (3–4.5 compression ratio)	Acetal (Delrin 500 and 900; Celcon) Fluoroplastics (Teflon 110) Nylon (high melt index) Polyethylene (high density) Polyethylene (high melt index) Polypropylene (medium to high melt index)

Depending on melt index and heat (shear) sensitivity of material, compression ratios may differ from those indicated.

2.3.3.4 Screw Length-to-Diameter Ratio

Usually just called L/D (see also earlier section in this chapter on barrel length-to-diameter), it is the length to diameter ratio of a plasticating screw. The diameter of the ratio is reduced to 1 so that the ratio is defined evenly, for example, a 24 : 1 screw that has a screw length 24 times its diameter. Based on the melt characteristics, there are general reasons for having short or long L/Ds.

Advantages of a short screw are:

- Less residence time in the barrel so that heat sensitive plastics are exposed to heat for a shorter time, thus lessening the chance of degradation.
- Plasticator occupies less space.
- Requires less torque, making strength of screw and amount of horsepower less important.
- Less investment cost initially and for replacement parts.

Table 2.4 Guide to Selecting Optimum Screw Design

Material	Transition length[a]	Compression ratio[b]	Meter depth[c]
ABS	Long	Low	Deep
Acetal	Short	Medium	Shallow
Acrylic	Long	Low	Deep
Cellulosics	Medium	Low	Deep
FEP	Short	Medium	Medium
HDPE	Medium	Medium	Shallow
LDPE	Medium	Medium	Medium
Nylon	Short	High	Shallow
PAI	Long	Low	Deep
PBT	Medium	Low	Shallow
PC	Medium	Low	Deep
PEI	Medium	Medium	Medium
PES	Medium	Medium	Medium
PET	Medium	Medium	Medium
PMP	Medium	Medium	Medium
PP	Medium	Medium	Shallow
PPO	Medium	Low	Medium
PS	Medium	Low	Medium
PVC (F)	Medium	Low	Medium
PVC (R)	Long	Low	Deep
SAN	Medium	Low	Medium

[a] Short—4 turns or less; medium—5 to 7 turns; long—8 or more turns
[b] Low—less than 2.5 : 1; medium—2.5 to 3.5 : 1; high—3.5 : 1 or more
[c] degree of depth; example—medium for 2 diameter 0.100 to 0.125

Advantages for the long screws are:

- Allows for greater output and melt recovery rates.
- Screw can be designed for greater mixing and more uniform output.
- Screw can be designed to operate at higher pressures.
- Screw can be designed for greater melting with less shear and more conductive heat from the barrel.

2.3.3.5 Screw Type

The configuration of the screw segments significantly affects the capabilities of the plasticization process. To meet different plastic melt processing requirements, different screw segments are used. Figure 2.9 shows different types of screw design configurations to provide better mixing, pumping, and so forth.

2.3.3.5.1 Single Flight

A screw having a single helical flight.

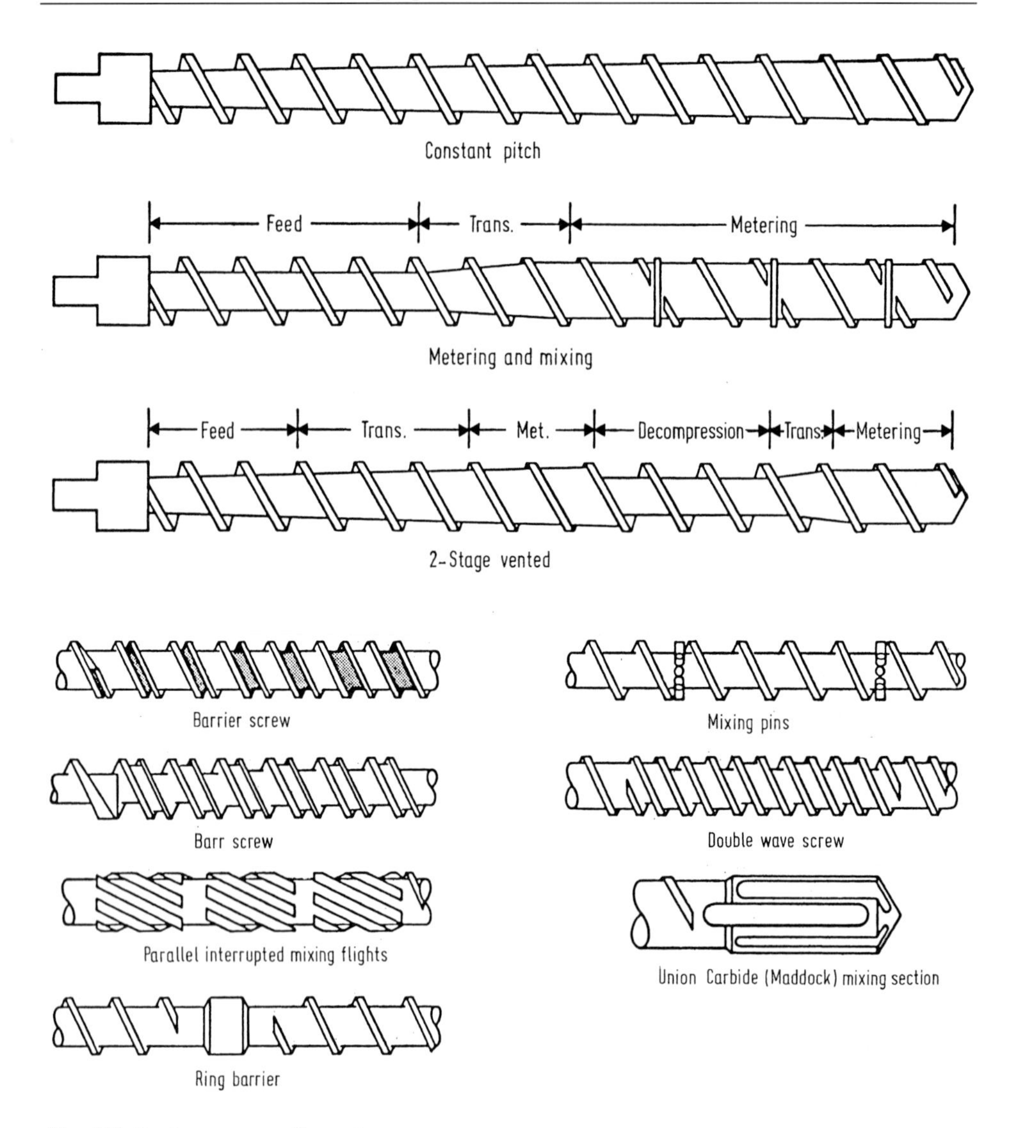

Fig. 2.9 *Basic screw configurations*

2.3.3.5.2 Square Pitch

Many screws have a pitch equal to the diameter of the screw (maximum diameter of the flight). This screw is called a square pitch screw with a helix angle of 17.8°. The flight pitch is the distance in an axial direction from the center of the screw flight at its periphery to the center of the next flight. In a single flighted screw, pitch and lead will be the same, but they will differ in a multiple flighted screw. The location of measurement is specified.

2.3.3.5.3 General Purpose

The general-purpose single screw configuration is designed to suit as wide a range of plastics as possible. They will not be the ideal answer for a specific plastic. As an example, a screw designed for a semicrystalline material must provide initially at least greater heat input than for an amorphous thermoplastic. Thus, when a specific material is going to be used for a long run, it becomes economically very beneficial to use a dedicated screw.

2.3.3.5.4 Barrier

These screws are designed to accelerate and provide more efficient melting. They have been developed along different and sometimes radically opposed concepts. Usually, the melting rate is controlled by providing a barrier between the solid bed and the melt pool to ensure that the solid bed does not break up prematurely and become encapsulated in the melt.

Figure 2.10 provides a schematic on the basic operation of a barrier screw where:

- The feed section establishes the solids conveying in the same way as a conventional screw.
- At the beginning of the transition (compression) zone, a second flight is started. This flight is called the barrier or intermediate flight, and it is undercut below the primary flight's outer diameter. This barrier flight separates the solids channel from the melt channel.
- As melt progresses down the transition, melting continues as the solids are pressed and sheared against the barrel, forming a melt film. The barrier flight moves under the melt film and the melt is collected in the melt channel. In this manner, the solid pellets and melted polymer are separated and different functions are performed on each.
- The melt channel is deep, giving low shear and reducing the possibility of overheating the already melted polymer. The solids channel becomes narrower and/or shallower, forcing the unmelted pellets against the barrel for efficient frictional melting. Breakup of the solids bed does not occur to stop this frictional melting.
- The solids bed continues to get smaller and finally disappears into the back side of the primary flight.

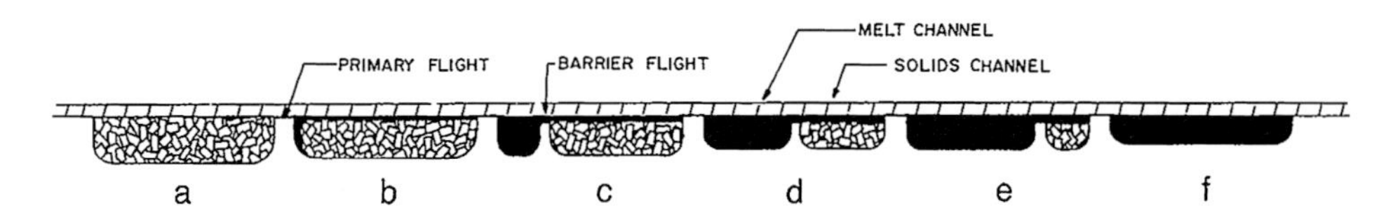

Fig. 2.10 *Melt model for barrier screw*

- All of the polymer has melted and gone over the barrier flight. Melt refinement can continue in the metering section. In some cases mixing sections are also included downstream of the barrier section. In general, the melted plastic is already fairly uniform on exit from the barrier section.

2.3.3.5.5 Double Wave

The double wave mixing screw has two equal width channels separated by an undercut barrier flight. The roots of each channel go up and down like a wave. The channel depth on one is shallow while the channel across the barrier is deep. This continual channel reversing action forces melt back and forth across the barrier, subjecting the melt to high and low shear. Usually these mixing sections are located in the metering zone.

2.3.3.5.6 Pin Mixing

About 1960, several companies started to place radial mixing pins in the screw root. These pins tend to interrupt the laminar melt flow and do a better job of mixing than the regular screw. Because these pins improve mixing, it is also possible to design the screw a little deeper to obtain higher output with the same degree of mixing. Many patterns and shapes (streamlined, etc.) of pins have been used. In general, they are usually placed in rows around the screw. They are located in the metering section after most of the melting has taken place.

A typical arrangement would have three rows, with one row at the beginning of the metering section, another row one flight back from the end, and the third row halfway between the other two rows. Pins are usually staggered from row to row. The pins should be hardened and have an interference fit to prevent dislodgment. Pins, unlike other mixing devices, are easy to install as an afterthought. This is usually done after the screw has been running and found to need more mixing capability.

2.3.3.5.7 Dulmage

A Dulmage section can be incorporated as an integral part of the screw. This design was one of the first mixing screws and was developed by Fred Dulmage of Dow Chemical Co. It has a series of semicircular grooves cut on a long helix in the same direction as the screw flights. There are usually three or more sections interrupted by short cylinder sections. This design interrupts the material flow where the melt follows the screw channel. It divides and recombines the melt many times. In this way, it works something like a static mixer. It is still used on screws that are processing certain materials such as foamed plastics.

2.3.3.5.8 Maddock

The Union Carbide mixer, also referred to as the Maddock mixer, was offered to the public without patent royalty charges. It consists of a series of opposed, semicircular grooves along the screw axis. Alternate grooves are open to the upstream entry, with other grooves open to the downstream discharge. The ribs or flutes that divide the alternating entry and discharge grooves also alternate. These flutes are called mixing flutes and wiping or cleaning flutes. The plastic is forced over the mixing flute that has an undercut from the screw outer diameter. The cleaning flute is narrower and has a full diameter.

This mixer does an effective job of mixing and screening unmelted plastic. The plastic is pumped into the inlet groove, and as the screw rotates, the undercut mixing flute passes under it. The melted plastic ends up in the outlet or discharge groove. As it goes over the undercut mixing flute, it is subjected to high shear but for a very short time interval. The plastic is then pumped out of the discharge groove as new plastic enters over the full-diameter cleaning flute.

2.3.3.5.9 Marbleizing

There are marbleizing or mottling screws where no mixing or very little is desired. One designed screw will have only a low compression ratio with a good portion of the screw consisting of the feed section followed by a short taper and usually in the metering section with few flights. The screw permits colorants basically to stay in their own channels until exiting. Usually a worn out screw can provide marbleization. They are used to obtain decorative effects such a swirling or grainy effects in the plastic coloration. This conventional system does not reproduce the same pattern; it may be close but it will not duplicate. In order to duplicate, coextrusion or coinjection processing is used.

2.3.3.5.10 Multiple Flighted

Multiple flighted screws have more than one helical flight such as double flighted, double lead, double thread, or two starts, and triple flighted.

2.3.3.5.11 Planetary

A planetary is a multiple screw device in which a number of satellite screws, generally six, are arranged around one longer and larger diameter screw. The portion of the central screw extending beyond the satellite screws serves as the final pumping action as in a single screw extruder. This screw system provides special compound mixing actions such as the discharge of volatiles toward its hopper end when processing powders such as dry blended PVC.

2.3.3.5.12 Pulsar

In the Pulsar mixer (from Spirex Corp.), the metering section is divided into constantly changing sections that are either deeper or shallower than the average metering depth. This requires all the plastic to alternate many times from shallower depth and somewhat higher shear to deeper channels with lower shear. During this changing action, it experiences a gentle tumbling and massaging action. This design interrupts the undesirable laminar flow and causes excellent mixing, distribution, and melt uniformity without high shear.

2.3.3.5.13 Reverse Flight

The reverse flight is a type of extruder screw with left hand flights on one end and right hand flights on the other, so that material can be fed at both ends of the barrel and extrude from the center.

2.3.3.5.14 Saxton

The Saxton has a plurality of minor flights and channels on a helical angle that is normally greater than that of the primary flight. The minor flights and channels are interrupted by major channels that are cut with an opposite hand lead. One drawback of these nonflighted interruptions is that the melt no longer has the positive forward conveyance other than pressure flow, although the interruptions on a helix are much better than those on tangential grooves as some barrel wiping action still takes place.

2.3.3.5.15 Conical or Involute

The two basic types of screws are conical and the usual involute (or spiral), each with different mechanisms based on the theories used. The conical transition has a root that is cone shaped and is not parallel to the axis of the screw. The involute transition has a root that is always parallel to the screw axis, and the channel depth varies uniformly. Actually, the use of the word involute is not correct in geometric terms, but most people working with screws understand this. In the involute screw one side is deeper than the opposite side, causing an imbalance that at high pressures causes rapid wear. Surges also can occur since solid plastic blocks are formed. The disadvantage of the conical is that it is more difficult to machine and more expensive.

2.3.4 Screw Torque

Screws always run inside a stronger and more rigid barrel. For this reason, they are not subjected to high bending forces. The critical strength requirement is resistance to torque. This is particularly true of the smaller screws with diameters of 2.5 in. (6 cm) or less. Unfortunately, the weakest area of a screw is the portion subjected to the highest torque. This is the feed section which has the smallest root diameter. A rule of thumb is that a screw's ability to resist twisting failure is proportional to the cube of the root diameter in the feed section. Finite element analysis (FEA) software has been used to obtain a more accurate determination of the stress levels.

Rotating the screw in a stationary barrel does the work of melting. The rotational force called torque is the product of the tangential force and the distance from the center of the rotating member. For example, if a 1-lb (0.454-kg or -N) weight was placed at the end of a 1-ft (0.035-m) bar attached to the center of the screw, the torque would be 1 ft × 1 lb or 1 ft-lb (1.36 Nm). Torque is related to horsepower (hp) which equals [torque (ft-lb) × rpm]/5252 or kW = [torque (Nm) × rpm]/7124. The torque output of an electric motor of a given hp depends on its speed. A 30 hp (22 kW) motor has the following torque at various speeds: 87.5 ft lb (119 Nm) at 1,800 rpm, 133 at 1,200, and 175 at 900 [2, 5, 6].

The screw torque speed of a given hp motor drive system is built into a motor. Changes in torque and speed can also be accomplished by changing the output speed of the motor by using a gear or belt train. The change in torque varies inversely with the speed. As an example, if an AC motor is used it will develop a starting torque of almost twice the running torque. The screw has to be protected against overload to prevent screw breakage. This is not a problem with hydraulic drives. The drive must supply enough torque to plasticize at the lowest possible screw speed, but not enough to mechanically shear the metal screw.

Different torque requirements are used to meet the requirements of the different plastics. As an example, a much higher torque is required to plasticize PC (polycarbonate) than general-purpose polystyrene (GPPS). The strength used limits the input hp. Using little torque to turn the screw can indicate that the heater bands are providing too much of the energy required to melt the plastic, usually as a result of poor or no temperature control. Plastication efficiency suffers in these conditions, and mixing problems and/or long, inconsistent recovery times are symptomatic of this condition.

2.3.5 Screw Cooling

Various extrusion processors, particularly those using PVC, sometimes use screw cooling. Some improvement in plastic extrusion is possible by circulating cooling water or oil through the cored center section(s) of the screw, at least through the feed section. The amount of cooling required in this ‘pipe’ is dependent on screw design and operating parameters. Cooling is more critical for larger diameter screws, because the larger volume of melt flow requires more cooling (heat control). Superior extrusion may be achieved by optimizing cooling, but reduced output rates and/or surging may result unless proper processing temperatures are maintained.

When extruding thermosets (TSs), the screw cannot be permitted to overheat in the barrel; if overheated, the thermoset plastic will solidify. TSs, being very heat-sensitive, usually use a water-cooled barrel that may have cooling zones to provide positive temperature control.

2.3.6 Static Mixer

The static mixer is also called a motionless mixer. Many plasticizing operations involve either the processing of regrinds with virgin plastics and/or the addition of color, regrind, etc. Uniform mixing of the feed mix is important to achieve acceptable product properties; as output rates increase, it becomes more difficult. Head pressure and stock temperature must be held constant to maintain extrudate control. Thus more machines, particularly extruders, are being equipped with a static mixer to overcome some of these problems. In an extruder if a gear pump is used, the static mixer is located between the screw and gear pump.

Static mixers are designed to achieve a homogeneous mix by flowing one or more plastic streams through geometric patterns formed by mechanical elements in a tubular tube or barrel. They contain a series of passive elements placed in a flow channel (Fig. 2.11) that cause the plastic compound to subdivide and recombine so as to increase the homogeneity of the melt.

There are no moving parts and only a small increase in the screw energy is needed to overcome the resistance of the mechanical baffles. The installation of a static mixer increases the effective L/D of the extruder and usually results in some increase of the melt temperature and head pressure. The overall result is a more homogeneous melt and a more stable extrusion process with less output surge. When using static mixers sufficient heatup time should be provided with startups of the extruders. To prevent excessive pressures and damage to the extruder during startups, the rpm of the screw should be kept low until plastic flows uniformly from the die.

2.3.7 Screw/Barrel Bridging

When an empty hopper is not the cause of machine output failure, plastic might have stopped flowing through the feed throat because of screw bridging. An overheated feed throat, or startup followed by a long delay, could build up sticky plastics and stop flow in the hopper throat. Plastics can also stick to the screw at the feed throat or just forward from it. When this happens, plastic just turns around with the screw, effectively sealing off the screw channel from moving plastic forward. As a result, the feed throat is said to be "bridged" and stops feeding the screw. The common solution is to use an appropriate rod such as brass rod to break up the sticky plastic or to push it down through the hopper.

2.3.8 Screw/Barrel Clearance

The screw/barrel clearance is the difference in the diameters of the screw and barrel bore (diametrical clearance) or more commonly one-half the diametrical clearance that is referred to as the radial clearance.

2.3.9 Screw-to-Diameter Size

Screws do not have the same outside continuous diameter. On receiving a machine or just a screw, it is a good idea to check its specified dimensions (diameters vs.

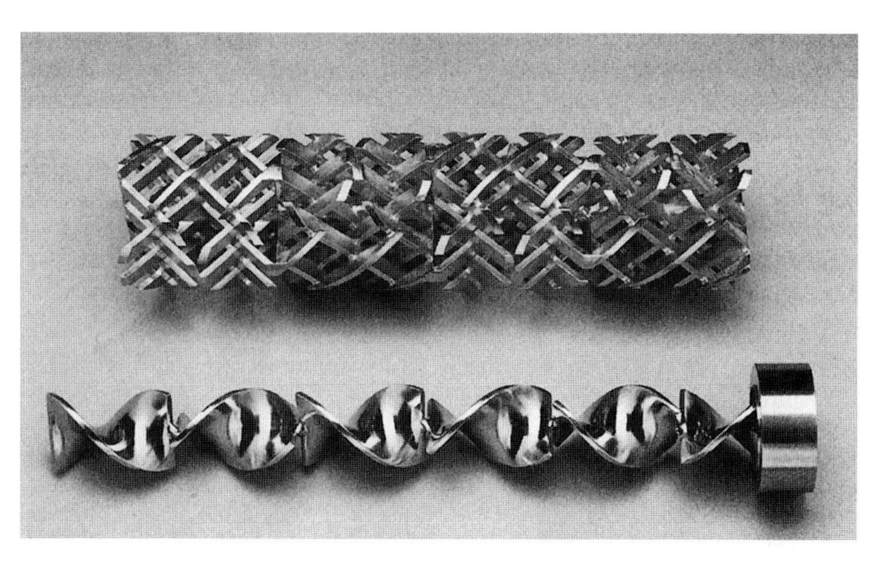

Fig. 2.11 Two different styles of static mixers

locations, channel depths, concentricity and straightness, hardness, spline/attach-attachment, etc.) and make a proper visual inspection. This information should be recorded so that comparisons can be made following a later inspection. Some special equipment should be used other than the usual methods (micrometer, etc.) to ensure that the inspections reproduced accurately. Such equipment is readily available and actually simplifies inspections; it also takes less time, particularly for roller and hardness testing (Chapter 12). Details on conducting an inspection and important processing behaviors are available from screw and equipment suppliers such as Techware Designs, a subsidiary of Spirex Corp. with a computer software package called the "Extruder's Technician". To ensure proper performance, different parts of the barrel can be checked to meet tolerance requirements (usually set up by the manufacturer) and determine if any wear has occurred such as inside diameter, straightness and concentricity, and surface condition.

2.3.10 Screw Output

The rate of output (throughput) or the speed at which plastic is moved through the plasticator continue to be pushed higher as a result of design advances in equipment and materials. Output rates generally range from a few kilograms to over 5 tons per hour on single screw machines. Twin screw extruder diameters have output rates ranging from a few kilograms to at least 30 tons per hour. A rough estimate for output rate (OR) in lb/h can be calculated by using the barrel's ID in inches and using the following equation:

$$OR = 16\ ID^2\ (lb/h \times 0.4536 = kg/h).$$

However, the output of a screw is somewhat predictable, provided that the melt is under control and reasonably repeatable. With a square pitch screw (conventional screw where the distance from flight to flight is equal to the diameter), a simplified formula for output is

$$R = 2.3D^2h \cdot N$$

where $R =$ rate or output in lb/h (kg/h), $D =$ screw diameter in inches (mm), $h =$ depth in the metering section in inches (mm) (for two-stage screw use the depth of the first metering section), and $N =$ screw rpm.

This formula does not take into account back flow and leakage flow over the flights. These flows are not usually a significant factor unless the plastic has a very low viscosity during processing, or the screw is worn out. The formula assumes pumping against low pressure, giving no consideration to melt quality and leakage flow of several worn OD screws.

There are basically two methods of evaluating output loss caused by screw wear; one is rather accurate and the other is a rough approximation. The accurate method is to compare the current worn screw output with a production benchmark reference output established when the screw was new. The approximate method involves determining screw wear by measuring the screw, calculating the resultant added screw to barrel clearance, and estimating the output loss from the added clearance. There are three major problems with the approximate method. First, about 24 h of machine downtime occurs. Second, it may understate the extent of output loss by as much as $2\frac{1}{2}$ times. Third, it does not account for the commonly encountered problem where the screw rpm is simply increased, yet the output loss due to increased melt temperature is not well established.

With a change in barrel temperature, one is able to predict the change in output. It is necessary to study the interaction of the die and extruder to predict the overall change in output from a temperature increase.

2.3.11 Screw Material and Melt Flow

Solid conveying is always enhanced by anything that increases the friction coefficient between the plastic and the internal surface of the barrel or decreases the friction between the plastic and the surface of the screw. In other words, it feeds well if it adheres to the barrel and slips on the screw. Reduction of friction on the screw surface can be achieved by improved surface conditions or by chrome plating, which will improve feeding and melt flow. If a screw has a pitted or rough surface in the feed section a better polish will usually help. The brightest mirror-like finish is not always the best for low coefficient of friction. Sometimes a fine matte finish obtained by a fine grit blast provides better release and improved sliding.

Chrome or chrome-based platings can help to maintain the screw finish so that feeding conditions do not change rapidly. For materials that are very difficult to feed it may be necessary to provide a barrel that has axial grooves in the internal diameter from the beginning of the feed pocket (throat) to a position three to four flights forward. See Tables 2.5 and 2.6 for materials of construction and protection of screws.

Poor sliding on the screw surface can be caused by melted material sticking to the root of the screw channel or the sides of the flights as a result of heat traveling back from the hotter front portion of the screw. Most often, it occurs when the machine is allowed to stand for a period of time. This problem can sometimes be cleared up by inserting larger pieces of plastic such as tabs, or sliced up parts directly into the feed throat. The hopper is removed and caution exercised to keep hands out of the screw. The larger pieces will usually clean the melted material from the feed section sufficiently so that the pellets can do the remainder of the job. In extreme cases, this can also happen while the screw is turning and high frictional heat generated in the front is conducted back to the feed section, melting the plastic on the screw. In that case, the continuous supply of cold unmelted pellets cannot continually clean the melted material from the screw surfaces of the feed section. This can be eliminated by water cooling the screw in the feed section. Screw cooling requires a rotary union, a cooling tube, and a drive system that includes a through hole large enough to accommodate this hardware.

The only significant control the operator has over feeding is barrel temperature settings in the rear of the barrel. These barrel temperatures can play an important role in the feeding characteristics of a screw, barrel, and material combination. The goal is to set the temperatures to maximize the frictional force of the solid plastic against the inner wall of the barrel. This will inhibit sliding and promote feeding. If the temperature is too cold, the frictional force will be too low and slipping will occur. If the temperature is too hot, the solid will melt and it will slide easily along the very fluid plastic resulting in poor feeding. To aid in mixing and melt uniformity barrier screws can be used.

This phenomenon is shown in Fig. 2.12, which simulates a hypothetical plastic. The plastic feeding occurs at the point of maximum friction, point C on the graph. At this point the melt film is sufficient and has high enough viscosity to cause sticking on the barrel surface. At point A the inner surface of the barrel is not hot enough to form sufficient melt to cause sticking. At point B the barrel is too hot, causing the melt to have a lower viscosity and an easier circumferential movement and poorer feeding. All of this also explains why feeding is sometimes improved by raising barrel settings and sometimes by lowering them. This hypothetical review is good for only one situation involving a certain screw speed, specific plastic and lot, operating pressure, and other parameters.

2.3.12 Screw Replacement

When you want to change a screw to see how your process is running, it can be expensive. Consequently many people tend to run a poorly performing screw long

Table 2.5 *Construction Materials for Screws*

Material	Finished flight hardness	Base material hardness	Base tensile strength	Resistance to		Comments
				Abrasion	Corrosion	
Carbon steels						
4140	50–55 Rc	28–32 Rc	150,000	Poor	Poor	Principal screw material used in the United States
4340		38–42 Rc	180,000	Poor	Poor	
1020		79 Rb	69,000	Poor	Poor	Used in place of 4140 for high torsion strength.
1035		90 Rb	85,000	Poor	Poor	Used only with hard faced flights: where low torsion strength required.
4130		90 Rb	85,000	Poor	Poor	
Nitrided steels						
Nitralloy						
135M	65 Rc	30 Rc	145,000	Good	Fair	Substitute for 4140 flame hardened. Offers hardness on root as well as flight lands for abrasive applications.
4140	50 Rc	30 Rc	150,000	Fair	Poor–fair	
Tool steels						
D-2	64 Rc	64 Rc	240,000	Good	Poor	Provides high torsion strength and good abrasion resist, on full surface area. Limited to smaller sizes.
H-13	51 Rc	51 Rc	260,000	Good	Poor	
Stainless steels						
304		81 Rb	170,000	Poor	Fair–good	Used almost exclusively in corrosion resist, applications: in FDA certified processing.
316		78 rb	118,000	Poor	Good	
17.4 PH		38 Rc	170,000	Fair	Good	Hard surfaced in all plastics applications.
Duranickel						
301		32–42 Rc	180,000	Poor	Excel.	Good resist. to hot HF and other acids. Most screws are hard surfaced.
Hastelloy						
C-276		94 Rb	134,000	Poor	Excel.	Excellent corrosion resist. to almost all chemicals and environments. Most screws are hard surfaced.

Table 2.6 *Materials for Abrasion or Corrosion Protection of Screws*

Material	Finished flight hardness	Base material hardness	Base tensile strength	Resistance to Abrasion	Resistance to Corrosion	Comments
Cobalt base						
Stellite 6	49 Rc[a]	37 Rc	105,000	Good–excel.	Good	Widely used hard facing for abrasion resist. Can be applied to most screw materials.
Stellite 12	50 Rc[a]	41 Rc	76,000	Good–excel.	Good	
Stellite 1	48 Rc	48 Rc	47,000	Good–excel.	Good	
Nickel base colmonoy						
56	50–55 Rc	50–55 Rc	45,000	Good–excel.	Good–excel.	Provides excellent abrasion resist. and resists galling. Application to carbon steel base results in some cracking.
5	45–50 Rc	45–50 Rc		Good	Good–excel.	
6	56–61 Rc	56–61 Rc	30,000	Excel.	Good–excel.	
Bimetallic coatings						
UCAR						Provides excellent abrasion resist; can be applied to all screw materials.
WT-1	70 Rc	N/A	N/A	Excel.	Poor	
Nye-carb	60–65 Rc	N/A	N/A	Excel.	Good	
Ceramic coatings						
Chrome oxide	80 Rc	N/A	N/A	Excel.	Poor	Most abrasion resist. materials currently used. Coatings are fragile.
Aluminum oxide	80 Rc	N/A	N/A	Excel.	Poor	
Chrome plating						
Hard chrome	70–72 Rc	N/A	N/A	Good–excel.	Good	Used mostly for corrosion resist. If applied in suitable thickness, offers abrasion resist.
Nickel plating						
Electroless nickel	45–50 Rc	N/A	N/A	Poor–fair	Excel.	Can be applied more evenly than chrome.

[a] Work hardened

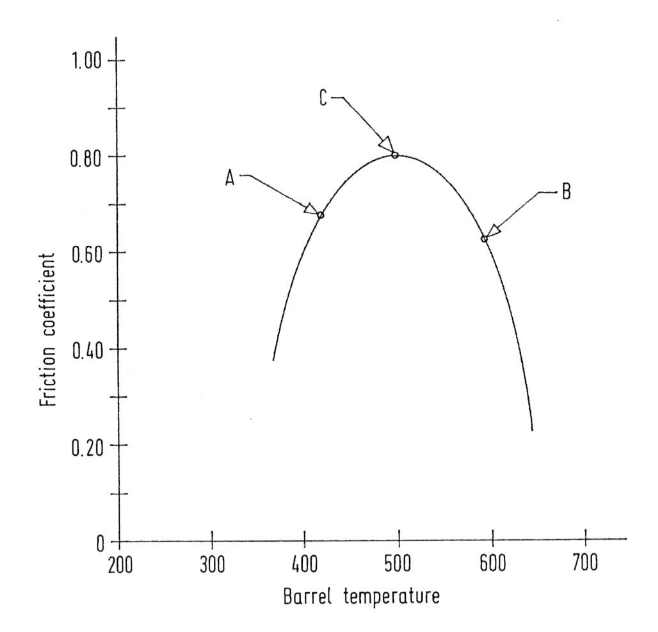

Fig. 2.12 Hypothetical plot of solid plastic friction coefficient vs. barrel temperature

after it should have been changed. Converting the cost of a screw into an equivalent volume of plastic or into a profit per day will determine payback. Assume a screw cost $30,000 to $40,000 each with an output at 3,000 lb/h (1,400 kg/h) of a $0.40/lb plastic. If you waste 1,000,000 lb (454,000 kg) of plastic, you justified the cost of the new screw. A new screw would represent 33 h of processing. It pays to replace the screw.

2.4 Plasticator Venting

In a plasticator, melt must be freed of gaseous components that include moisture and air from the atmosphere and from plastics, plasticizers, and/or other additives as well as entrapped air and other gases released by certain plastics. Gas components such as moisture retention in and on plastics have always been a potential problem for all processors. All kinds of problems develop on products (splay, poor mechanical properties, dimensions, etc.). This situation is particular important when processing hygroscopic plastics (Chapter 7).

One major approach to preventing plastic degradation is to use plasticators that have vents in their barrels to release these contaminants. Venting can be used with or without drying (Chapter 7). It is very difficult to remove all the gases prior to fabrication from particularly contaminated powdered plastics, unless the melt is exposed to vacuum venting (typical of most vented screws, a vacuum is connected to the vent's exhaust port in the barrel).

The standard machines operate on the principle of melt degassing. The degassing is assisted by a rise in the vapor pressure of volatile constituents, which results from the high melt heat. Only the free surface layer is degassed; the rest of the plastic can release its volatile content only through diffusion. Diffusion in the nonvented screw is always time dependent, and long residence times are not possible for melt moving through a plasticator. Thus, a vented barrel with a two- or three-stage melting screw is used (Figs. 2.13 and 2.14).

As shown in Fig. 2.13 the basic operation of a vented barrel is as follows: (1) Wet plastic enters from a conventional hopper; (2) The pellets (or other plastic form) are conveyed forward by the screw feed section, and are heated by the barrel and by some

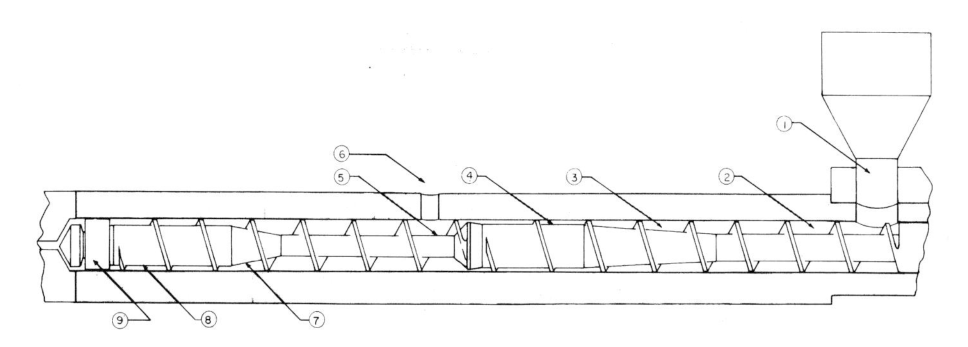

Fig. 2.13 *Example of a two-stage vented screw*

frictional heating. Some surface moisture is removed here; (3) The compression or transition section does most of the melting; (4) The first metering section accomplishes final melting and even flows to the vent section; (5) Plastic is pumped from the first metering section to a deep vent or devolatilizing section. This vent section is capable of moving quantities well in excess of the material delivered to it by the first metering section. For this reason, the flights in the vent section run partially filled and at zero pressure; (6) It is here that volatile materials such as water vapor and other nondesirable materials escape from the melted plastic. The vapor pressure of water at 500 ft is 666 psi. These steam pockets escape the melt, and travel spirally around the partially filled channel until they escape out the vent hole in the barrel [6]. Water vapor and other volatiles escape from the vent; (7) The plastic is again compressed and pressure is built in the second transition section; (8) The second metering section evens the flow and maintains pressure so that the screw will be retracted by the pressure in front of the nonreturn valve; (9) A low resistance, sliding ring, nonreturn valve works in the same manner as it does with a nonvented screw.

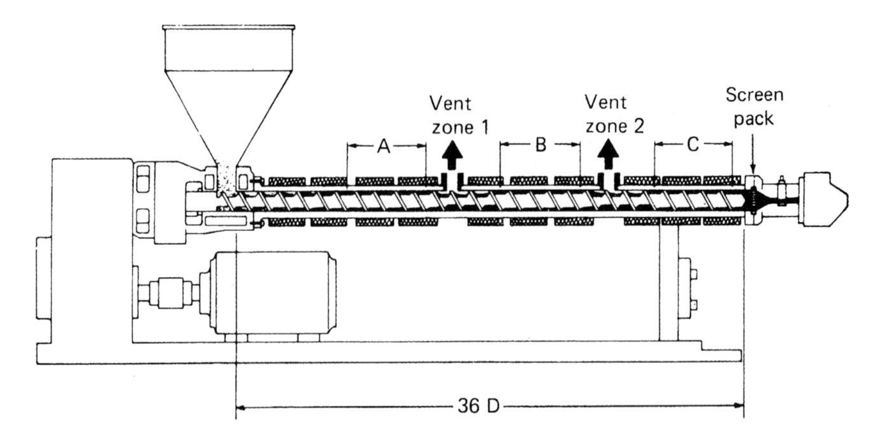

Fig. 2.14 *Example of a three-stage vented screw*

Barrels with one vent use a two-stage screw that basically looks like two single screws attached in series. Where the two meet, there is a very shallow channel section so that when the melt reaches that section, no melt pressure exists (Fig. 2.15). In turn gaseous (diffusion) materials are released through a port opening. In barrels having two vents, a three-stage screw is used that provides another stage to eliminate contaminants.

The first stages of the transition and metering zones are often shorter than the sections of a single-stage conventional screw. The melt discharges at zero pressure into the second stage, under vacuum instead of pressure. The first-stage extrudate must not be so hot as to become overheated in the second stage. And the first-stage must not deliver more output per screw rotation at discharge pressure than the second-stage can pump through the barrel under the maximum normal operating pressure. This usually means that the second-stage metering section must be at least 50% deeper than the first-stage section. The decompression zone exists between the first and second compression zones, allowing venting of volatiles without the escape of melt even with vacuum applied.

In practice the best metering-section depth ratio (pump ratio) is about 1.8 : 1. The ratio to be used depends on factors such as screw design, downstream equipment, feed stock performance, and operating conditions. There is likely to be melt flow through the vent (avoid this situation) if the compression ratio is high or the metering depth ratio is slightly too low. If the metering depth ratio is moderately high, there is a gradual degradation of the output. With the screw channel in the vent area not filling properly, the self-cleaning action is diminished, and the risk of plate-out increases. In any case, sticking or smearing of the melt must be avoided or degradation will accelerate.

A controlled material feeding device (also called a starve feeder) may be necessary to maximize the operation of a vented system. Regardless of its necessity, it is certainly

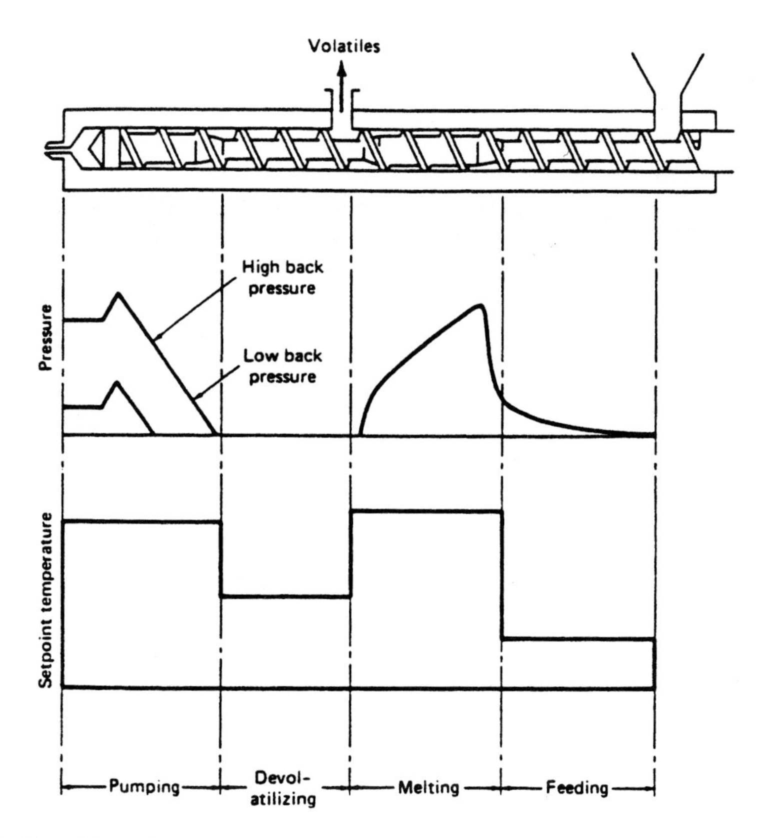

Fig. 2.15 Vented barrel pressure and temperature profiles

a useful device for many reasons. The controlled feeding device determines the amount of plastic that is being fed into the screw, thereby controlling the output of the first stage of a two-stage screw. This action should eliminate all cases of possible vent bleeding and plugging of the vent hole.

In addition, by partially filing the screw flight channels, one allows the surface moisture that is being driven off the plastic a place to evaporate to the atmosphere. Other advantages include its ability to govern the amount of shear and energy that is delivered to the plastic via the screw geometry, different shear history can be provided, and so forth.

The exhaust from vented plasticating barrels can show a dramatic cloud of swirling white gas; almost all of it is condensed steam proving that the vent is doing its job. However, a small portion of the vent exhaust can be other materials such as byproducts released by certain plastics and/or additives and could be of concern to plant personnel safety and/or plant equipment. Purifiers can be attached (with or without vacuum hoods located over the vent opening) to remove and collect the steam and other products. The purifiers include electronic precipitators.

2.5 Purging

It has always been a necessary "evil" that at the end of a production run the plasticator has to be cleared of all its plastics in the barrel/screw to eliminate barrel/screw corrosion or contamination; this is also required when a change of material occurs. This process consumes substantial nonproductive amounts

Table 2.7 Guidelines for Plastic Changes

Material in machine	Material changing to	Mix with rapid purge and soak	Temperature bridging material	Follow with
ABS	PP	ABS	—	PP
ABS	SAN	SAN	—	SAN
ABS	Polysulfone	ABS	PE	Polysulfone
ABS	PC	ABS	PE	PC
ABS	PBT	ABS	PE	PBT
Acetal	PC	Acetal	PE	PC
Acetal	Any material	PE	—	New material
Acrylic	PP	Acrylic	—	PP
Acrylic	Nylon	Acrylic	—	Nylon
TPE	Any material	PE	—	New material
Nylon	PC	PC	—	PC
Nylon	PVC	Nylon	PE	PVC
PBT	ABS	PBT	PE	ABS
PC	Acrylic	PC	—	Acrylic
PC	ABS	PC	PE	ABS
PC	PVC	PC	PE	PVC
PE	Ryton	PE	PE	Ryton
PE	PP	PP	—	PP
PE	PE	PE	—	PE
PE	PS	PS	—	PS
PETG	Polysulfone	PETG	—	Polysulfone
Polysulfone	ABS	Polysulfone	PE	ABS
Polysulfone	ABS	Cracked acrylic	—	ABS
PP	ABS	ABS	—	ABS
PP	Acrylic	Acrylic	—	Acrylic
PP	PE	PE	—	PE
PP	PP	PP	—	PP
PS	PP	PP	—	PP
PVC	Any material	LLDPE or HDPE	—	New material
PVC	PVC	LLDPE or HDPE	—	PVC
PPS	PE	PPS	PE	PE
SAN	Acrylic	Acrylic	—	Acrylic
SAN	PP	SAN	—	PP

Table 2.8 Guidelines for Purging Agents

Material to be purged	Recommended purging agent
Polyolefins	HDPF
Polystyrene	Cast acrylic
PVC	Polystyrene, general-purpose, ABS, cast acrylic
ABS	Cast acrylic, polystyrene
Nylon	Polystyrene, low-melt-index HDPE, cast acrylic
PBT polyester	Next material to be run
PET polyester	Polystyrene, low-melt-index HDPE, cast acrylic
Polycarbonate	Cast acrylic or polycarbonate regrind; follow with polycarbonate regrind; do not purge with ABS or nylon
Acetal	Polystyrene; avoid any contact with PVC
Engineering resins	Polystyrene, low-melt-index HDPE, cast acrylic
Fluoropolymers	Cast acrylic, followed by polyethylene
Polyphenylene sulfide	Cast acrylic, followed by polyethylene
Polysulfone	Reground polycarbonate, extrusion-grade PP
Polysulfone/ABS	Reground polycarbonate, extrusion-grade PP
PPO	General-purpose polystyrene, cast acrylic
Thermoset polyester	Material of similar composition without catalyst
Filled and reinforced materials	Cast acrylic
Flame-retardant compounds	Immediate purging with natural, non-flame-retardant resin, mixed with 1% sodium stearate

of plastics, labor, and machine time. It is sometimes necessary to run hundreds of pounds of plastic to clean out the last traces of a dark color before changing to a lighter one; if a choice exists, process the light color first. Sometimes there is no choice but to pull the screw for a thorough cleaning.

There are few generally accepted rules on the purging agents to use and how to purge. The following tips should be considered: (1) try to follow less viscous with more viscous plastics; (2) try to follow a lighter color with a darker color plastic; (3) maintain equipment by using preventative maintenance; (4) keep the materials handling equipment clean; and (5) use an intermediate plastic to bridge the temperature gap such as that encountered in going from acetal to nylon (Table 2.7).

Ground/cracked cast acrylic and PE-based (typically bottle grade HDPE) materials generally are the main purging agents. Others are used for certain plastics and machines (Table 2.8). Cast acrylic, which does not melt completely, is suitable for virtually any plastic. PE-based compounds containing abrasive and release agents have been used to purge the "softer" plastics such as other olefins, styrenes, and certain PVCs. These type of purging agents function by mechanically pushing and scouring residue out of the extruders.

3 Extrusion Blow Molding

3.1 The Basic Process

The basic extrusion blow molding (EBM) machine consists of an extruder, crosshead die with single or multiple parison (possible accumulator), clamping arrangement, and mold. Variations include multiple extruders for coextrusion of two or more materials, parison programmer to shape the parison to match complex blown product shapes, and multiple station clamp systems to improve output through the use of multiple molds.

In EBM, an extruder melts, mixes and forms the plastic tube used in blowing the hollow product. Figures 3.1 and 3.2 show the basic steps in the EBM process. The plastic is homogenously melted and passed through the heated barrels by the shearing action of the plasticator screw. At the end of the plasticator is the extruder head and die unit, which forms/extrudes the hot melted resin into a tube called a parison. At a specified length, the hollow cavity mold closes over the resin tube. Compressed air is blown into the tube from an opening at the top or bottom of the parison. The air expands the walls of the resin tube against the sides of the mold cavity. The mold is cooled and the resin solidifies to the shape of the mold cavity. Finally, the blown container is ejected from the mold and trimmed appropriately.

The parison transfer system (Fig. 3.3) may use several heads to transfer several parisons. This system can be used to produce containers with finished necks that do not require post-forming operations other than detabbing. The schematic includes hand transfer systems; however, machines usually use automatic electromechanical controlled transfer systems. Figure 3.3 illustrates the intermittent EBM process. The extruder operation is interrupted after the parison has reached the desired length and before the mold halves have closed around it. Operation is resumed at the end of the ejection step.

Turning continuously, the extruder screw feeds the melt through the diehead as an endless parison or into an accumulator. The size of the part and the amount of material necessary to produce the part (shot size) dictate whether or not an accumulator is required. The non-accumulator machine offers an uninterrupted flow of plastic melt.

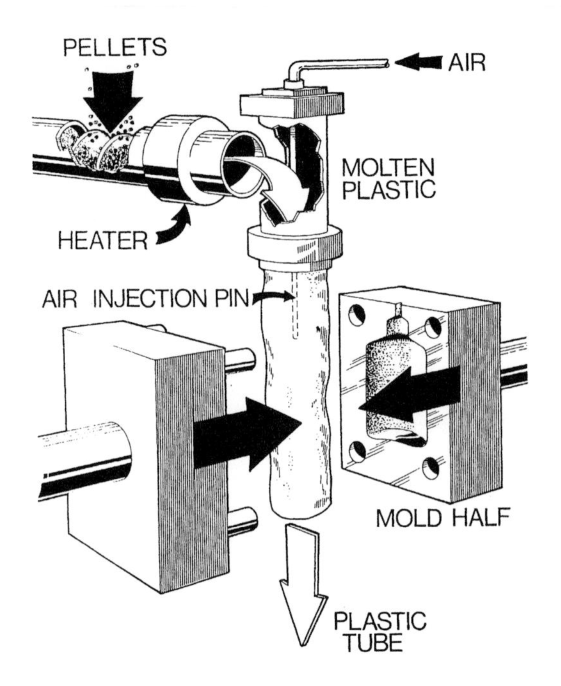

Figure 3.1 *Melt flow schematic for EBM*

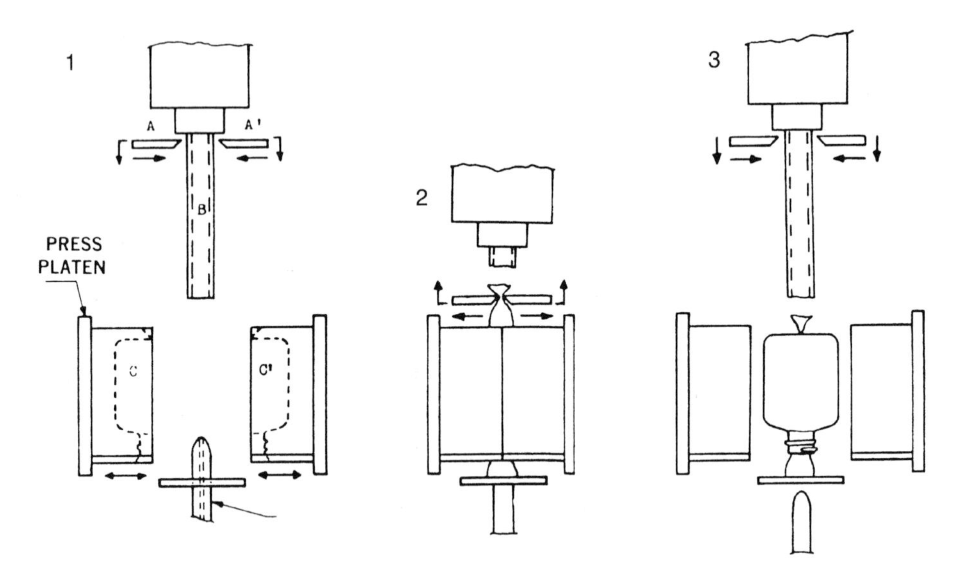

Figure 3.2 *Basic EBM process stepwise: (1) parison being extruded, (2) compressed air inflates parison, (3) blown container being ejected, and view of the process*

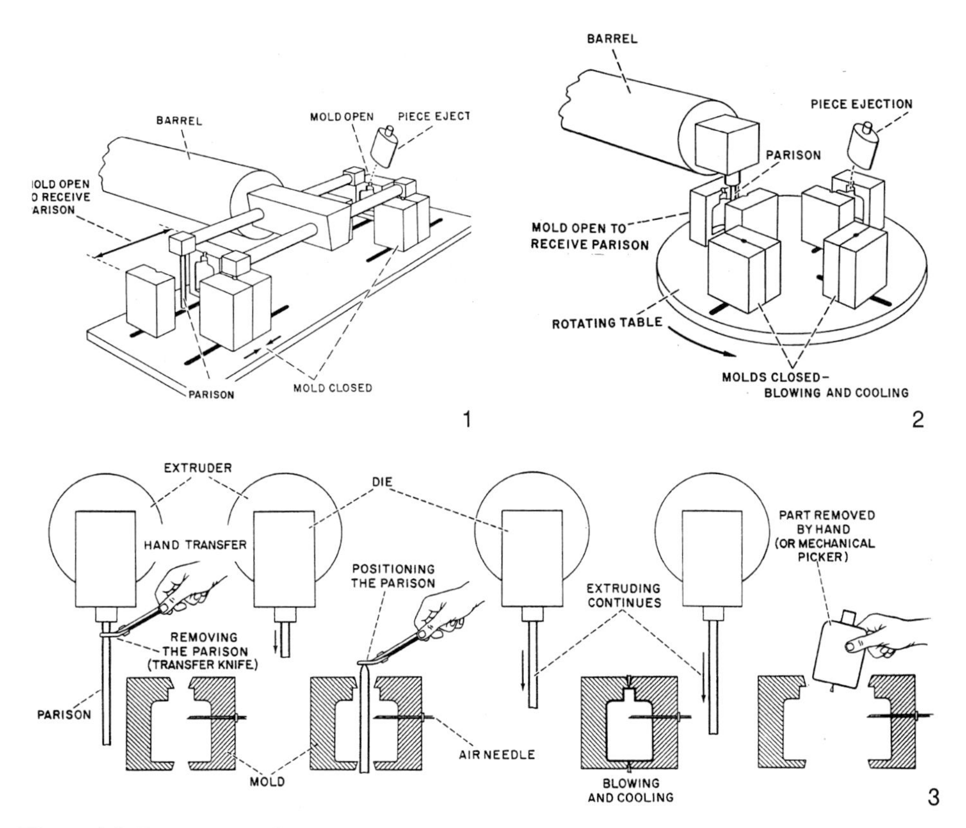

Figure 3.3 *Parison transfer system with continuous extrusion of the parison*

With the accumulator, flow of parison through the die is cyclic. The connecting channels between the continuously operating extruder and the accumulator and within the accumulator itself are designed rheologically to prevent restrictions that might impede the flow or cause the melt to hang up. Flow paths should have low resistance to melt flow to avoid placing unnecessary load on the extruder. To ensure that the least heat history is developed during processing, the design of the accumulator should provide that the first material to enter the accumulator is the first to leave when the ram empties the chamber and the chamber should be close to totally emptied on each stroke.

Large industrial parts require an accumulator head, as the parison is too heavy and it will sag if continuous extrusion is used. The plastic accumulates inside the head and a moving piston forces the plastic through the tubing die whenever the press is ready for BM.

When the parison or tube exits the die and develops a preset length, a split cavity mold closes around the parison and pinches one end. Compressed air inflates the parison against the hollow blow mold surfaces that cool the inflated parison to the

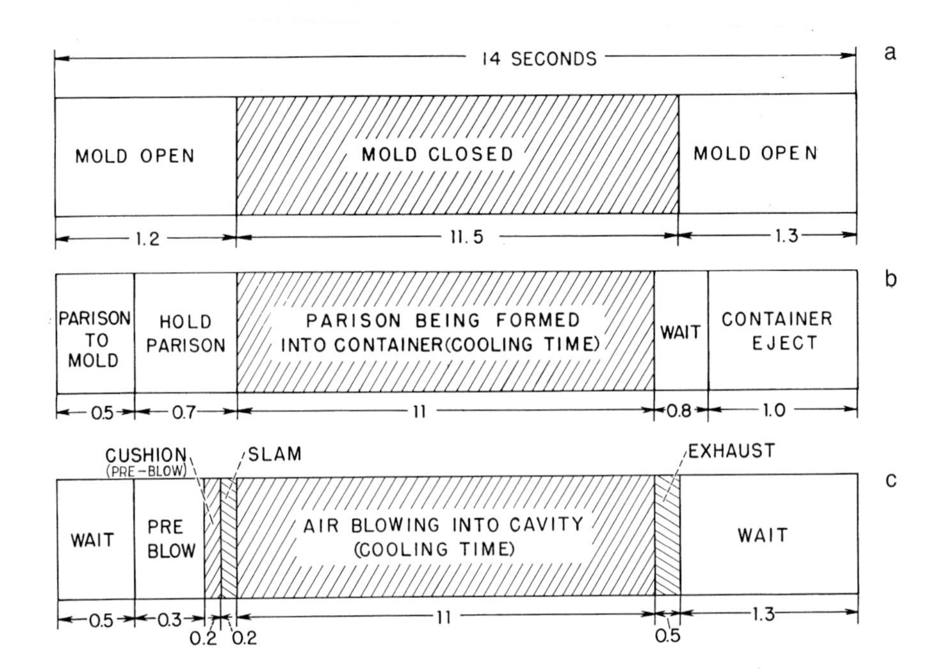

Figure 3.4 *EBM cycle*

blow mold configuration. On contact with the cool mold wall, the plastic cools and sets the part shape. The mold opens, ejects the blown part, and closes around the next parison to repeat the cycle. Figure 3.4 shows the stages of the molding cycle.

Various techniques are used to introduce air into the parison: through the extrusion die mandrel, through a blow pin over which the end of the parison has dropped, through blow heads applied to the mold, or through blowing needles that pierce the parison. Parison programming, blow ratio, and the part configuration usually control the wall distribution and thickness of the blown part. Devices are available from EBM machine manufacturers and process control manufacturers to control the parison wall thickness and thereby the container wall thickness (Chapter 9).

BM machines with several parisons are very common (Fig. 3.5). The parisons may be fed with the melt from one or a number of extruders, an experience that has led to easy understanding of the concept of multilayer or coextruded parisons.

BM machines for continuous multilayer coextrusion are available as both single and twin station machines. A twin station double-headed installation allows for nests of mold cavities to be installed. This arrangement is particularly applicable for the production of three-layer coextruded tubes. For the five- to seven-layer systems, manufacturers tend to prefer rotary machines with 18 stations or more.

Coextrusion BM techniques can readily be used to produce containers from 0.2 to 5.0 liter (6.8 to 16.9 oz) capacity. Available are much larger containers, for example,

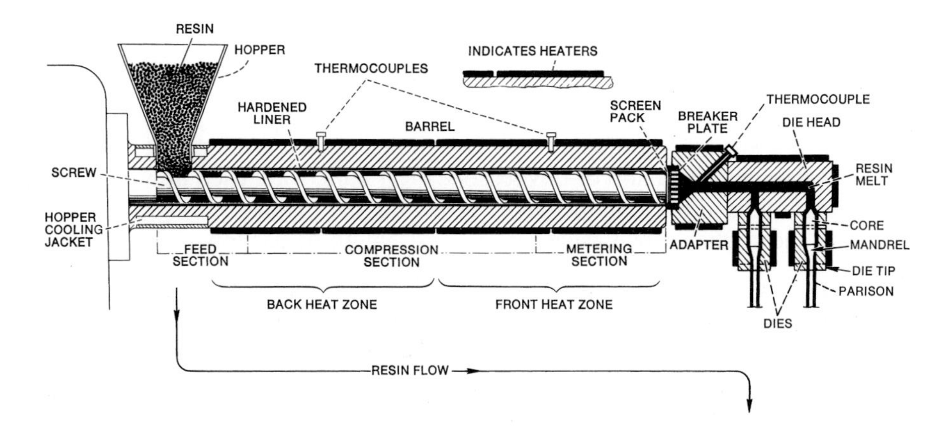

Figure 3.5 *Schematic of a two parison die*

automobile fuel tanks, using this basic process. Products having a sandwich wall (multilayer) construction are produced by coextrusion BM methods. Insert molding and two shot molding techniques are also used to produce multiwall injection molded preforms that can later be BM in conventional equipment [6].

Other approaches used or in development include the stacking of thermoformed and injection molded parts before reheating and blowing, and the use of cut lengths of multiwall extruded tubing [2, 8]. Developments abound that are designed to facilitate or improve the productivity of EBMs. In product cooling that may take up to 80% of the molding cycle, for example, air chillers reduce the temperature of the compressed blowing air to around −70 °C (−94 °F). The effect is to improve productivity by between 15% and 30%. The control and monitoring functions are carefully regulated.

3.2 Processing Requirements

It is required that the EBM parison be separated from the die, usually by means of a knife. In this operation only low knife speeds are required (peripheral speed < 1 m/s, 3.3 ft/s). By correctly positioning of the knife and suitable choice of the material from which it is made, it is possible to ensure that the die edges are not damaged. Cross sectional distortion of the cut parison can be countered by the use of support air. The ensuing parison emerges with an open, nondistorted free end. Because this method requires a flush fit for the core with the die orifice, it is used only in special cases. In cutting with a cold knife it is essential for the knife to rotate at sufficient speed (peripheral speed $< 5\,m/s$; 16.4 ft/s). Cutting should preferably involve a "drawing" action. Tensioning of the parison (e.g., by a rise-and-fall movement of the extruder or by movement of the mold or transfer arm) helps to obtain a clean cut.

In cutting with a shearing arrangement, although the parison is not pulled to one side, it is nevertheless pressed flat. For this reason, shear cutters are used in extrusion BM only when the parison is cut immediately below the die (while blowing with support air) and the shears then move aside for the next length of parison.

Heated blade cutters are made of nonscaling metal and usually have a sword-shaped cross section. They are heated to a dark or cherry red by an electric current. The blade carrier must be insulated from the device that moves it, and be grounded from possible leak voltage. For heating, the main voltage is stepped down to a 24- or 6-V supply. It may be necessary to resharpen or replace heated blade cutters after about a week in service. When heated blade cutters are used to cut parisons containing inorganic pigments, plate-out or crust formation frequently develops on the blade. To avoid these problems, it is necessary to adjust the blade temperature upwards, and it is therefore advisable to provide an infinitely variable or graduated temperature control.

In the processing of plastics such as unplasticized polyvinyl chloride (PVC), the heated blade cutting causes undesirable thermal degradation in the cutting region. It is therefore necessary to optimize processing conditions by varying blade speed and temperature. In acetal processing a low blade temperature should be set to avoid igniting the melt.

The mold clamping methods are hydraulic and/or electrical. Sufficient daylight in the mold platen area is required to accommodate parison systems and removal of the blown product. Clamping systems vary based on part configuration and exist in three types. The "L-shape" style has the parting line at an angle of 90° to the centerline of the extruder. The "T-shape" has the parting line in-line with the extruder centerline. The mold opening is perpendicular to the machine centerline. The third method is the "gantry" type, in which the extruder/die unit is arranged independently of the clamping unit. This arrangement permits the clamp to be positioned in either the "L" or "T" shape without being tied directly into the extruder [86].

Monitoring and control of single to multiple parison heads for continuous and accumulator extruder systems include 100- to 200-point versions. There are those with up to at least six different profiles that can be run simultaneously across the parison heads. Features include real-time display of actual vs. set profile, a parison editor for adjusting the setup, simple calibration of tolling stroke, and recipe storage and retrieval. Weight, profile range, and parison drop time are among the parameters checked. The controller's production monitor can shut down the BMM when the required number of products has been produced (see Chapter 9).

Based on the extruder capability, type of plastic processed, and process control conditions, BM product performances can vary. Examples are shown in Table 3.1.

Table 3.1 *Influence of PE Properties vs. Processing Conditions on BM Products*

	Increase in these factors ⟶ affects these properties ↓	Resin properties		Extrusion conditions		
		Density	Melt index	Melt temperature	Annular die clearance	Extrusion rate
Bottle Properties	Stiffness	Large increase	NSE[a]	NSE	—	Slight decrease
	Environmental stress crack resistance	Decrease	Large decrease	—	—	—
	Gloss	Large increase	Increase at 0.915 density No effect at 0.932 density	Increase	NSE	—
	Overall appearance[b]	Improvement	NSE	Slight improvement	NSE	NSE
	Wall thickness uniformity	NSE	NSE	Slight decrease	Decrease	NSE
	Weld-line thickness	Large decrease	NSE	NSE	Large increase	NSE
	Parting-line difference	NSE	No effect at low blow pressure Large increase at high blow pressure	Slight increase	NSE	NSE
Mold cycle time		Large decrease	NSE	Increase	Increase	—
Range of tests limited to ⟶		0.915–0.932 g/cm[b]	1.3–4.7 g/10 min	300°–340 °F (150°–170 °C)	54.5 to 79.5 mil (1.4–2.0 mm)	0.2 to 0.4 in./s (5–10 mm/s)

NSE, no signifant effect
[a] Tests on 4 oz (115 g) Boston round PE bottles
[b] Quantitative rating of gloss, surface smoothness, and flow lines in bottles

3.2.1 Extrusion Head

The plastic melt is kept at its lowest temperature to maintain high melt strength so that the vertically extruded tube will support itself. Table 3.2 provides examples of the melt temperature and pressure of some plastics for EBM [69]. The melt is then fed into an extrusion head (Fig. 3.6). The melt from the extruder enters the head from the side port, where it contacts a mandrel. The melt stream divides, flows around the mandrel, welds together, and makes a 90° turn so the flow is vertically downward. The molten plastic then flows through an annulus formed by a mandrel (pin) and a die (bushing). The mandrel and die are sized to produce BM products having the desired wall thickness.

There are two types of mandrel and die bushings: the convergent type (Fig. 3.7) and the the divergent type (Fig. 3.8). The selection of the mandrel and die bushing

Table 3.2 *Melt Temperature and Pressure for EBM*

Material	Melt temperature (°C)	Melt pressure (bar)
PE- LD	140/150	100/150
PE- HD	160/190	100/200
PP	230/235	150/200
PVC-U	180/210	100/200
PVC-P	160/165	75/150

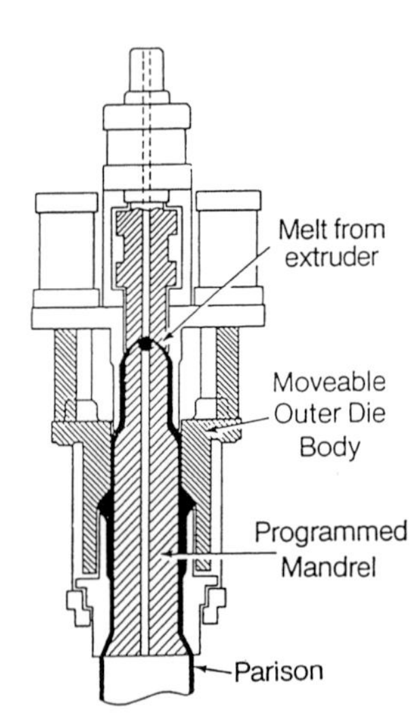

Figure 3.6 *Example of a polyolefin extrusion head with programmed mandrel*

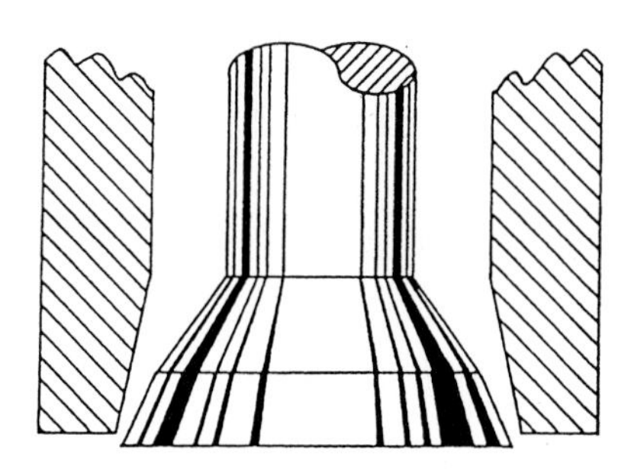

Figure 3.8 Divergent style head tooling

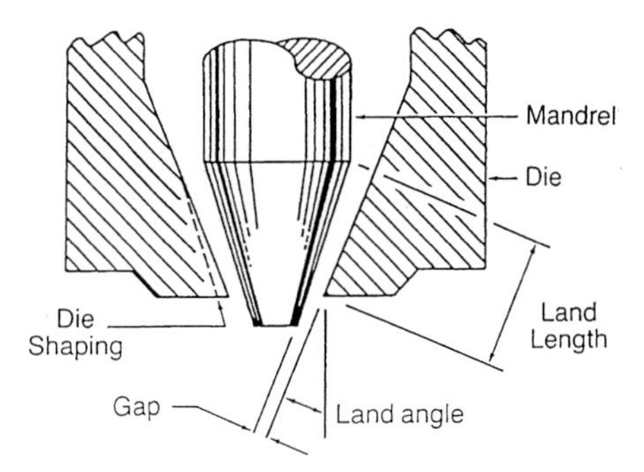

Figure 3.7 Convergent style head tooling

(commonly known in BM as head tooling) is based on the size of the parison necessary. Convergent tooling is the easiest to control and is used whenever possible. Divergent tooling is used when the parison needs to be larger for bottles such as handleware or industrial type containers, as the divergent tooling causes the parison to flare out as it is extruded. Both convergent and divergent mandrels are mounted directly to the programming mandrel and are programmed up and down to open or close the gap between the pin and bushing (Fig. 3.9). Thus the parison wall thickness can be adjusted to allow for more material in certain spots of the parison for bottles that are oval shaped and require more material to be blown in one area than another.

To eliminate the weld line and the inherent stagnation where the melt stream welds together, programming mandrels can be designed similar to in-line extrusion pipe dies [5]. Thus, the melt from the extruder is conveyed through an adapter that makes a 90° turn so the flow is vertically downward. The melt enters the head where it passes around two or more spider legs that support the mandrel. Dual offset spider legs stagger the weld lines so one layer reinforces the other (Fig. 3.10). This example eliminates the stagnation at the weld lines and thus PVC and other heat-sensitive plastics can be processed with this system. Furthermore, colder parisons can be used (less sagging), as welding of the melt is easier to achieve compared to the side-fed system.

3.2.2 Blow Molding Press

Monsanto was the first company in the United States to produce EBM plastic bottles. They had to design their own equipment, as no machinery was commercially

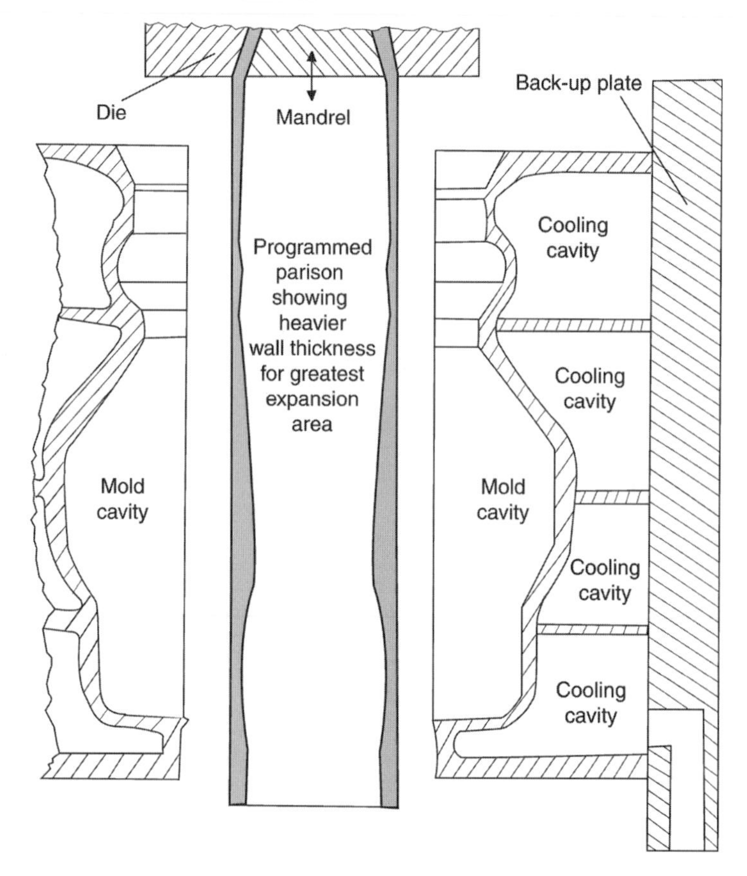

Figure 3.9 *Examples of programmed parisons for oval shaped bottles*

available at the time. Monsanto developed horizontal rotary indexing wheels using a continuous extruded parison. The molds were mounted and indexed to the extrusion head where the molds picked off the parison and transferred it to a second station for the introduction of the pin to blow the bottle. This process allowed for multiple molds to be mounted on as large a wheel as was necessary for production and the use of one extruder and one parison only (Fig. 3.11).

Other large companies in the glass and metal can business followed with vertical rotary wheels for BM plastic containers [48]. They used the principle of one continuous extruder and a multitude of molds that picked off the parison and moved around the axes of the wheel to blow the container, cool it, and discharge it (Fig. 3.12).

The shuttle type machines were developed to meet commercial requirements (Fig. 3.13). The extruder and parison head are located in the center and molds are mounted to each side. The molds shuttle to the parison, pick off the parison, and move back to their original position. The shuttle method led to the development of the

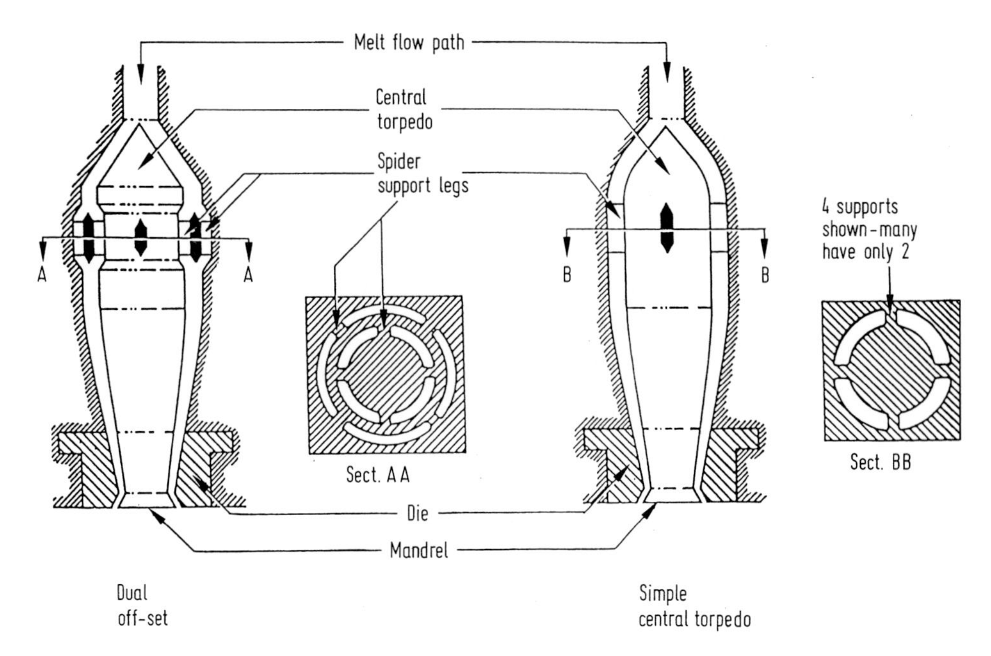

Figure 3.10 *Example of a PVC centerflow extrusion head*

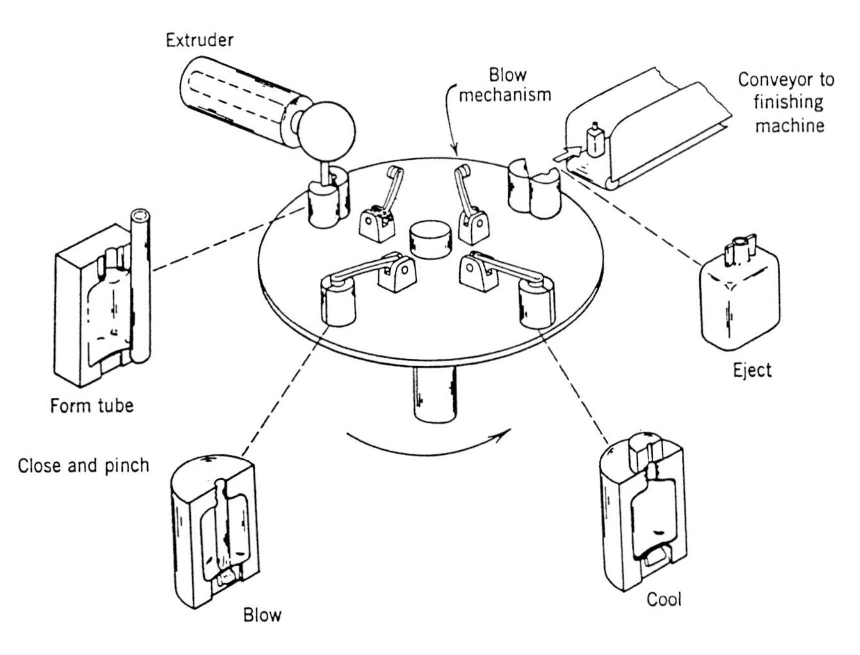

Figure 3.11 *Example of a horizontal rotary wheel*

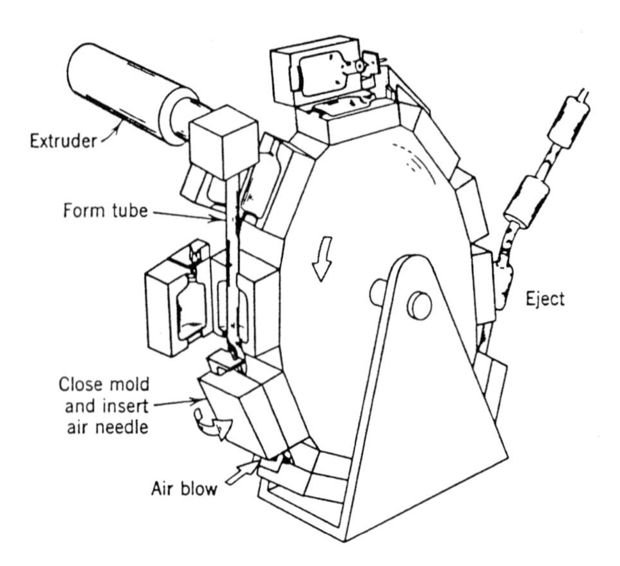

Figure 3.12 Schematic of a vertical rotary wheel

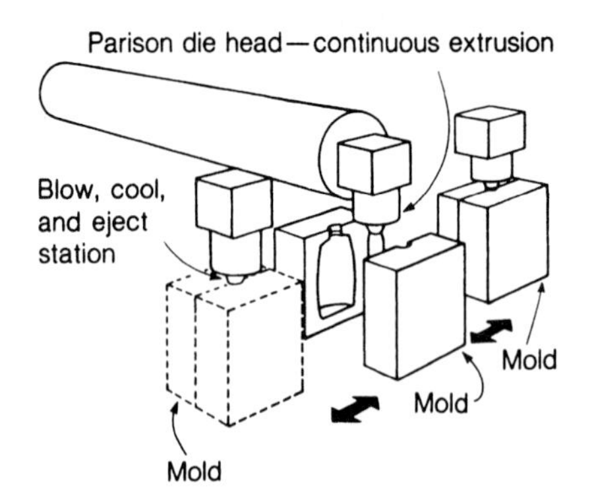

Figure 3.13 Example of a shuttle continuous EBM

calibrated neck finish, otherwise known as the compression molded neck finish (Fig. 3.14).

The calibrated or compression neck is formed by inserting a calibrated blow pin with a tip that is sized to form the inner diameter (ID) of the bottle and compresses the plastic into the thread area that is cut into the mold. This action provides a solid thread, rather than a blown hollow thread as is produced on a rotary wheel type machine. The top of the bottle is formed by a combination of a striker plate on the mold, sitting on a 60° angle, and a cutting sleeve mounted to the blow pin that shears the plastic against the cutting sleeve and forms the top of the bottle to its correct dimension.

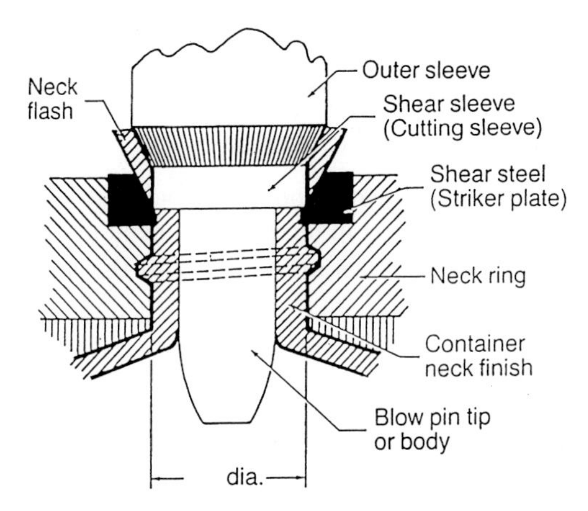

Figure 3.14 *Example of a calibrated or compression molded neck finish*

It has been reported that worldwide there are more than 25,000 of these shuttle type machines in production. They can be used for any type of plastic that is blow moldable because they utilize continuous extrusion. Multiple cavities, with multiple parison heads, are used for high-speed production [48].

3.3 Melt Flow

Several aspects of melt flow in the die are important. Those directly related to die flow include pressure drop, forces acting on the mandrel, and the distribution of melt velocity around the annular flow gap. Die flow also influences the swell behavior of the melt after it leaves the die. Approximate methods have been proposed to calculate the important aspects of flow in a BM die. It has been found that the calculation of the major forces and velocity distributions requires only a knowledge of the viscous properties of the melt, that is, the viscosity as a function of shear rate and temperature. There have been suggestions that the neglect of the temperature change in the die leads to a large error in the calculated pressure drop in the case of rigid PVC. Models proposed for polyolefin flow generally assume the flow to be isothermal (Chapter 7).

Such models have been used, for example, to calculate the total force exerted by the melt on the mandrel. For a converging die, the normal pressure acts in a direction opposite to the shear force. This makes it possible to design a die so that the net force on the mandrel is quite small, thus simplifying the die design so that a stepper motor can be used rather than a servohydraulic actuator to drive a movable mandrel for purposes of parison programming.

Use has been made of a power law viscosity equation to calculate the distribution of velocity around the die gap. The resulting model was then used to design dies to provide a uniform velocity distribution, and dies fabricated based on these designs were found to have very good flow distribution.

In summary, the forces acting in the die and the velocity distribution are governed by the viscosity function, and variations in temperature need to be taken into account, at least for polyolefins with melt indexes in the range of 0.5 to 1.5 (Chapter 7).

3.3.1 Parison Formation

Die design also has important effects on the behavior of the parison that is formed at the die lips. First, if there is a spider to center the mandrel, weld lines will be formed as the melt flows around the legs. The number of molecules bridging the weld lines to return the melt to a homogeneous state increases rather slowly with time, and for this reason, there will be some lateral weakness in the parison and in the molded product in the neighborhood of a weld line.

A second way in which die flow influences the parison properties is through the process of shear modification or shear refining. The effects of this process are most pronounced in highly branched and high molecular weight linear polyolefins. The effect can be particularly pronounced in the case of blends. The molecular mechanisms underlying the process are not fully understood, but shearing at high rates alters the structure of the melt, decreasing the strength of the interactions between molecules, a process sometimes referred to as disentanglements. The effect is a reversible one, and if a shear modified plastic is annealed for a period of time at a temperature above its melting point, it will recover its preshear structure. However, the time required for this recovery is often much longer than the time a parison is formed and the time it is inflated. Thus, the melt that forms the parison and is inflated into the mold may be a material highly altered by shear in the die. Shear modification reduces elasticity and can also result in a lower viscosity, especially in the case of branched polymers.

A third way in which die flow influences parison behavior is molecular orientation, which manifests itself at the die exit as extrudate swell. The non-Newtonian behavior of a plastic (Chapter 7) makes its flow through a die somewhat complicated. Flow at the inlet of a die, where the streamlines are converging rapidly, involves a high rate of stretching in the flow direction. This produces a high degree of molecular orientation and if the melt is permitted to exit through an orifice plate immediately, there would be a very high degree of swell. However, if a long straight section (for example, capillary or straight annular die) follows the entrance, molecular relaxation processes will lead to a disappearance of the orientation generated at the entrance; as the die is made longer, the degree of swell is reduced (Fig. 3.15). At the same time, however, the shearing in the die produces some axial orientation, and for a very long die, there is still significant swell resulting from this shear flow (Table 3.3). If the die has an expansion or contraction section, this will introduce stretching, which will create an

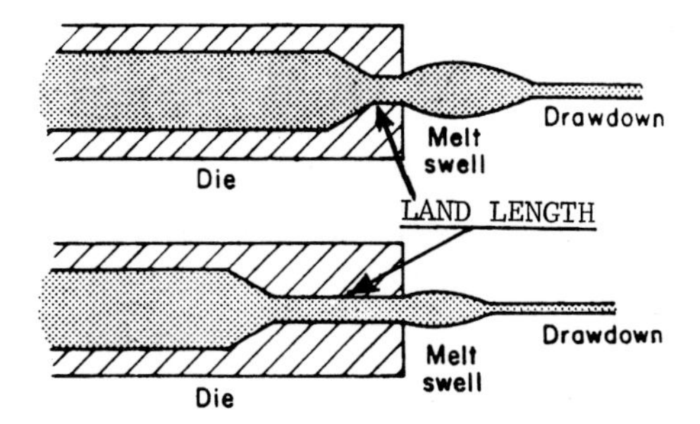

Figure 3.15 *Effect of land length on swell*

orientation whose direction depends on the details of the die design. For these reasons, parison swell is very sensitive to die design.

Another important effect on the melt is the orifice shape (Fig. 3.16). The effect of the orifice is related to the melt condition and the die design (land length, etc.). A slow cooling rate can have a significant influence, especially with thick parts. Cooling is more rapid at the corners; in fact, a hot center section could cause a part to 'blow' outward and/or include visible or invisible vacuum bubbles. After exiting the die, it usually stretches down to a size equal to or smaller than the die opening. The dimensions are reduced proportionally so that, in an ideal plastic, the drawn-down section is the same as the original section but smaller proportionally in each dimension. The effects of melt elasticity mean that the material does not draw down in a simple proportional manner; thus the drawdown process is a source of errors in the profile. The errors are significantly reduced in a circular extrudate, such as in a BM parison. These errors must be corrected by modifying the die and takeoff equipment [5].

Finally, the melt leaving the die may exhibit sharkskin or melt fractures, which are irregularities in the surface of the extrudate that can affect the surface finish of an extrusion BM container. This effect occurs above a critical shear stress in the die and

Table 3.3 *General Effect of Shear Rate on Die Swell of Various Thermoplastics*

Shear rate, s^{-1} / Plastic	Die swell ratio at 392 °F (200 °C)			
	10	100	400	700
PMMA-HI	1.17	1.27	1.35	—
LDPE	1.45	1.58	1.71	1.90
HDPE	1.49	1.92	2.15	—
PP—copolymer	1.52	1.84	2.1	—
PP—homopolymer	1.61	1.9	2.05	—
HIPS	1.22	1.4	—	—
HIPVC	1.35	1.5	1.52	1.53

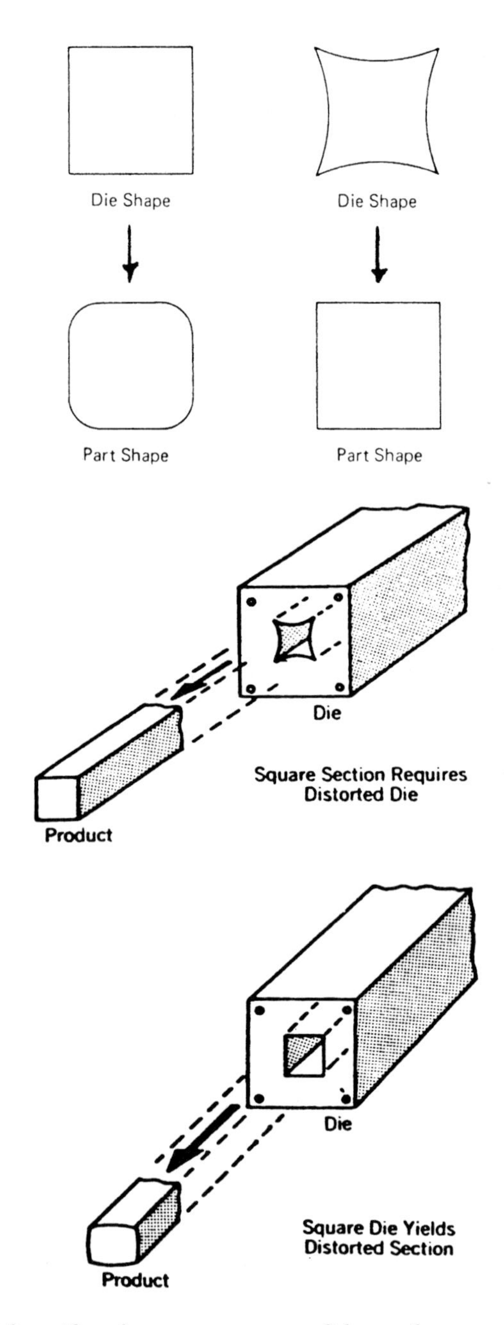

Figure 3.16 *Effect of die orifice shape on a square solid extrudate*

is often the factor that limits the speed of an extrusion process. Extrudate distortion is most severe in the case of narrow molecular weight distribution (MWD), high viscosity plastics. Increasing the temperature or reducing the extrusion rate can sometimes eliminate it, but either of these actions will increase the cycle time. The use of an internal heater in the mandrel is a method to reduce the tendency toward melt fracture on the inside of the parison. The detailed origins of this phenomenon are not fully understood, but the shape and material of construction of the die and the formulation of the plastic are known to be contributing factors.

Clearly, the cross sectional shape and surface appearance of an extrudate are governed by many factors in addition to the dimensions of the die lips, and there is currently no reliable method for predicting these characteristics for a given die, resin, and operating conditions [5]. A further complication is that as soon as the melt leaves the die to form part of a parison, it becomes subject to the force of gravity, which leads to sag or draw down. This tends to make the parison smaller at the top than at the bottom. The shape of the parison at the moment of inflation is thus the result of the simultaneous processes of sag and swell.

Because of the complexity of this situation, together with the complexity of the rheological properties of the melt, it is not possible to design a blow molding machine to produce prescribed parison shape and dimensions at the moment of inflation. However, these dimensions are of crucial importance, as they govern the thickness distribution in the finished product. For this reason machines being used to produce large or irregularly shaped objects are usually equipped with parison programming devices to alter the geometry of the die (annulus) during parison extrusion to produce a parison having a prescribed shape and dimension. However, parison programming cannot compensate for plastic properties that are basically unsuited for the process, and for this reason it will be useful to examine in more detail the processes of parison swell and sag.

3.3.2 Parison Swell

The wall thickness of the BM product can be related to the swelling ratio of the parison. Referring to Fig. 3.17, the swelling thickness of the parison is given by

$$B_t = h_p / h_d$$

and the swelling of the parison diameter

$$B_p = D_p / D_d.$$

Using the relationship

$$B_t = B_p^2$$

it follows that

$$h_p = (h_d)(B_p^2).$$

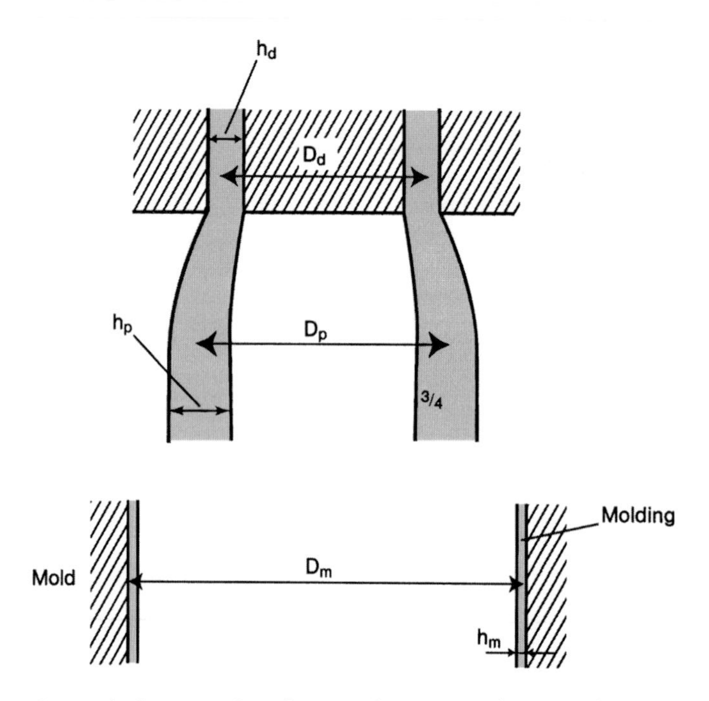

Figure 3.17 Relating thicknesses of swell ratio of parison and BM product

Thus the swell ratio B_p depends on recoverable strain that can be measured [69].

Examples of average parison swell for some plastics are given in Table 3.4.

Both the quality and cost of a BM container are strongly dependent on the parison swell ratios. If the diameter swell is too small, incomplete handles, tabs, or other unsymmetrical features may result. On the other hand, if the diameter swell is too large, polymer may be trapped in the mold relief or pleating may occur. Pleating, in turn, can produce webbing in a handle. Weight swell governs the weight, and thus the material cost of the molded product. What is desired is the minimum weight that provides the necessary strength and rigidity (Fig. 3.18).

Table 3.4 *Examples of Plastic Melt Parison Swell*

Polymer	Swell %
PE-HD (Phillips)	15/40
PE-HD (Ziegler)	25/65
PE- LD	30/65
PVC- U	30/35
PS	10/20
PC	5/10

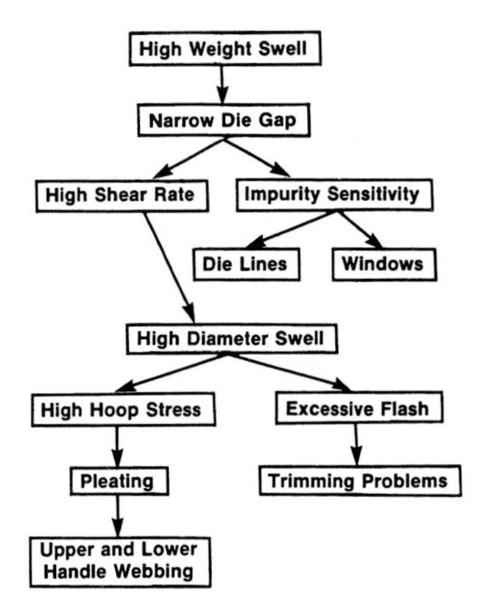

Figure 3.18 *Problems encountered in "countering" high weight swell*

Because swell is a manifestation of the viscoelasticity of the melt, it is time dependent. For example, for high-density polyethylene (HDPE) at 170 °C (338 °F), 70% to 80% of the swell occurs during the first few seconds after the melt leaves the die, but the remainder can occur over a period of 2 to 3 minutes. For polypropylene at 190 °C (374 °F) only 50% of the swell occurs in the first few seconds, while more than 10 minutes are required to reach an ultimate value. Another source of time dependency in the case of intermittent extrusion is the so-called cuff effect. The first melt to be extruded after a no-flow period has had an opportunity to relax in the die and experiences relatively little shear just prior to being extruded. Thus, there is less molecular orientation, and thus less swell, than for melt that has just flowed through the entire length of die. Owing to the combined effects of the cuff and the natural viscoelasticity of the melt, the swell is least at the bottom of the parison (the cuff), increasing to a maximum and then decreasing. Of course sag will accentuate the decrease in swell toward the top of the parison.

Swell increases as the temperature decreases, although it takes place somewhat more slowly. It has been suggested that this effect might be used to control swell by the manipulation of the power for the die heater. Because the time during which the parison is exposed to the air before the mold closes is rather short, and the air surrounding the parison is warm and relatively still, the decrease in temperature that occurs during parison formation is generally rather small, usually less than 5 °C (9 °F). For this reason, parison formation is generally assumed to be an isothermal process.

Swell increases as the flow rate increases owing to the enhanced orientation resulting from molecular orientation in the die. Swell varies greatly from one plastic to another and is strongly affected by the extent of branching and the molecular weight distribution (MWD). For linear plastics, a broader MWD generally results in a higher swell. However, plastics with very similar measured MWDs can have significantly different swell behavior, and this probably reflects the fact that swell is highly sensitive to amounts of high molecular weight material that are too small to be detected using standard analytical methods. Highly branched plastics tend to swell more, but it is not possible to generalize when both branching and MWD are changing.

Because swell is an elastic recoil process that results from molecular orientation in the die, the shape of the die channel has a strong effect on both diameter and thickness swells. The simplest case is a long straight annular die. Here we have only shear flow with no stretching, and we would expect to see some orientation in the axial direction. This suggests that there would be no preferential direction of swell and that the diameter swell should be equal to the thickness swell.

For the more complex BM dies the swell ratios are strongly influenced by two die geometry features, the angle of divergence or convergence of the outer die wall, and the variation of gap spacing along the flow path. A diverging die stretches the melt in the hoop direction, and this should reduce diameter swell by counteracting the axial orientation generated by the shear flow [75].

3.3.3 Parison Sag

Sag (drawdown) can cause a large variation in thickness and diameter along the parison, and in an extreme case can cause the parison to break off. For a Newtonian fluid, sag could be kept under control simply by using a material with a sufficiently high viscosity. The fact that plastics are non-Newtonian and generally have a melt index of less than 1 causes sag; sag becomes more severe as the temperature is increased. Because plastic melts are viscoelastic, resistance to sag cannot be quantitatively correlated with viscosity (see Chapters 7 and 8).

3.3.3.1 Viscoelasticity

A number of proposals have been made as to which viscoelastic property of a melt governs sag. The linear creep compliance, as calculated from a tensile relaxation modulus, can predict sag. Another proposal suggests that the extensional stress growth function can be correlated with sag behavior. J. M. Dealy (McGill University) points out that the sag process is neither a constant stress nor a constant strain rate process, so that neither of these material functions is directly relevant [47]. Noting that the strain rate is small, it is proposed that the use of a relatively simple general model of a viscoelastic liquid be used.

Clearly, reliable methods for predicting sag behavior on the basis of well defined rheological properties are still lacking. In their absence, empirical techniques for evaluating sagging tendency are employed. An example is to make video records of parison length vs. time. A simpler procedure that might be useful is to determine the extensibility of film plastics. They simply clamp a weight to a specified length of extrudate from a melt indexer and monitor the resulting behavior.

When we consider the combined effects of swell and sag, the situation becomes quite complex from a rheological point of view. Figure 3.19 is a sketch of parison length

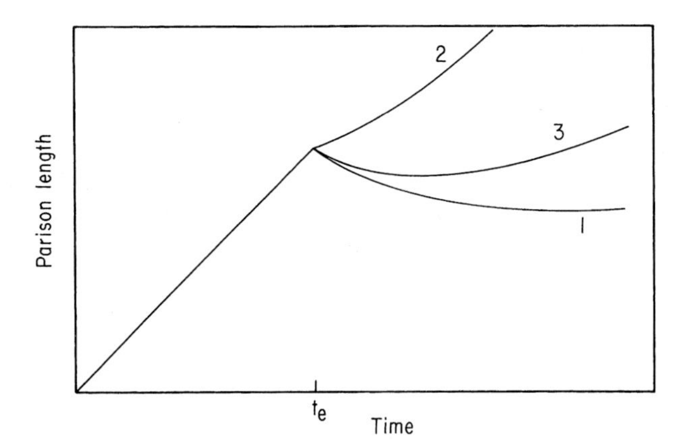

Figure 3.19 *Parison length vs. time curves for three different situations*

vs. time curves for various cases. The first part of the curve reflects the extrusion period during which the parison is formed. Once extrusion stops, the length is governed entirely by swell and sag. Curve 1 shows the case where there is swell but not sag, while curve 2 is for the case of sag with no swell. In the case of an actual parison, curve 3, there is initial recoil followed by a slower increase in length, reflecting the complex time dependency of the swell.

A number of models of the parison formation process have been proposed. All of these models assume that swell and sag are in some way additive. Unfortunately, none of them can be used to predict with confidence the behavior of the parison for a given die, resin, and operating conditions.

3.3.4 Pleating

Pleating (also called draping or curtaining) is a buckling of the parison that occurs when the melt at the top of the parison is unable to withstand the hoop stresses due to the weight of the parison suspended from it. Pleating is undesirable, because it can cause webbing in the blown product.

Obviously, large diameter swell and a small die gap are factors that will increase the probability of pleating. It is not clear exactly which rheological property governs the ability of a melt to resist pleating, but the viscosity is a rough guide to the melt stiffness that is desired. Thus, high viscosity plastics are less likely to pleat, and increasing the temperature increases the likelihood of pleating.

3.3.5 Parison Inflation

The behavior of the closed parison during the inflation process is a manifestation of its extensional flow rheological properties. It has been observed that the parison does not inflate uniformly and tends to bulge out in the center. Blowouts can occur if the ratio of the mold diameter to the parison diameter (the blowup ratio) is too high.

The deformation is not the same as that which occurs in the uniaxial extension experiment, but the results of such an experiment are still thought to be relevant to inflation performance [18]. In particular, it is thought that extension thickening behavior implies that a plastic will be easy to inflate and unlikely to exhibit blowout, even when the blowup ratio and inflation pressure are high. Extension thinning, on the other hand, is thought to imply unstable inflation and an increased likelihood of blowouts. This hypothesis is consistent with the observation that low-density (branched) polyethylene (LDPE) is easier to inflate than high-density polyethylene (HDPE).

3.3.6 Melt Fracture

Melt fracture is a loosely defined term that has been applied to various forms of extrudate roughness or distortion encountered at high extrusion rates for all plastic melts. In some cases there may be a small-scale roughness, rippling, or sharkskin; in others a very regular helical screw-thread extrudate, and generally at very high rates the extrudate becomes completely irregular.

The term melt fracture originated from early observations of marker threads or of flow birefringence at the entrance of a die. A line that was continuous at low extrusion rates could be seen to snap suddenly and to recoil when the flow rate was increased, and visible evidence of roughness was noted simultaneously. These observations suggested a mechanism of literal breakage of melt that had been stretched or oriented beyond its natural limit in flow. Other mechanisms that have been proposed include slippage at the die entry or along the die wall and an instability of the circulating vortex flow often seen at the comers of a flat entry inlet to a die. The small-scale roughness has been attributed to stick-slip flow at the die exit.

It should be noted that in some cases the flow curve (viscosity vs. shear rate) is not affected at all by the onset of melt fracture. Some plastics, however, do exhibit a pronounced discontinuity of the flow curve. The most important of these is linear PE. The onset of visible roughness occurs near, but not necessarily at, the flow discontinuity. The discontinuity has been ascribed to slippage at the wall, leading to increased flow and lower apparent viscosity.

The effect of slippage on the extrudate depends on the nature of the extrusion apparatus. A very stiff apparatus, such as a piston driven by a motor, will give a constant extrusion rate but severe pressure fluctuations. A soft apparatus, for example, a gas-driven rheometer, will produce large variations in output. The most serious effect occurs when the extrusion is done at a rate very close to the critical one for the onset of slippage. The output can then oscillate between the two branches of the flow curve as shown in Fig. 3.20, causing large variations in flow rate and gross distortions of the extrudate. The severity of distortion is also markedly dependent on die geometry.

These effects may be rationalized by considering that in the absence of slippage the melt is highly oriented by the flow. When the melt slips suddenly, it creates room for

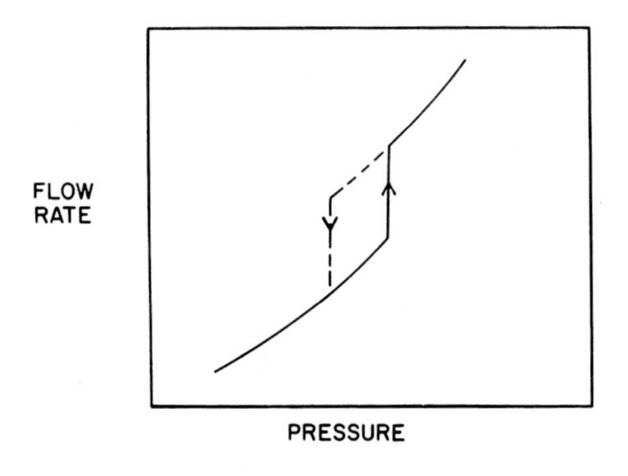

Figure 3.20 *Oscillating melt flow rate near slip discontinuity of flow curve*

the relatively unoriented stagnant melt that has been circulating in the entry vortexes to enter the die and to be extruded without the usual stretching at the die entry. The alternating of highly oriented extrudate, with large elastic recovery, with that of the unoriented melt accounts for the extrudate distortion.

Also, there is a pronounced effect of inlet taper angle on the extent of the distortion. A correctly tapered inlet suppresses the formation of a stagnant vortex. Therefore, even if slippage occurs there is no stagnant material to cause nonuniformity and the external appearance of the extrudate is relatively smooth. It has also been reported that the use of a slightly tapered conical die, rather than a cylindrical one, delays the onset and minimizes the severity of melt fracture.

It is generally agreed that shear stress is a better criterion for the onsets of melt fracture than shear rate. For example, when the extrusion temperature of LDPE is varied from 130 °C to 230 °C (266 to 446 °F), the shear stress at the onset of melt fracture varies by only about 30%, whereas the critical shear rate varies by a factor of 100. The critical stress generally ranges from about 10^6 dynes/cm^2 (14.5 psi) for nonpolar polymers [such as polystyrene (PS), polypropylene (PP), and polyethylene (PE)] to about 10^7 dynes/cm^2 (145 psi) for polar polymers such as nylon, polyacetal, and polyethylene teraphthalate (PET).

The effects of molecular weight and MWD are not as clear cut. The most comprehensive study indicates that the critical stress increases linearly with the reciprocal of the weight average molecular weight, that is, the critical stress is lowest for high molecular weight polymers.

The effect of MWD on the critical stress is less well defined, with many conflicting reports in the literature. Because of the difficulty of measuring the extremes of the MWD, it is probably not practical to use such a correlation for quantitative predictions in any case. As with die swell, it is easier and more reliable to measure the critical shear stress for the polymer of interest.

3.3.7 Extensional Flow

Historically, shear flow is important because it is the simplest way to deform a fluid, particularly a low viscosity fluid, in a controlled manner that can be analyzed and used to measure the viscosity. And of course it is a flow that occurs in nearly all plastic melt processing operations. However, shear flow is not the only deformation that can be applied to a fluid. In general the flow that occurs in melt processing machines may be quite complex. In addition to simple shear flow, extensional flow usually occurs. The most elementary type of extensional flow is viscosity, and end correction, needed in any case to characterize the processing parameters, thus specifies the die swell for any extrusion condition. It should be noted, of course, that a die swell curve depends on the molecular weight and distribution of the plastic, as do rheological properties in general.

Next is a review on the dependence of die swell on molecular parameters if the situation is not completely clear. With one exception the literature indicates that die swell increases with increasing breadth of MWD. The breadth of the MWD is usually specified in terms of a ratio of two types of average molecular weight. However, these averages are strongly affected by the shape of the MWD, especially at the extremes of high and low molecular weight. It has been found that a blend containing a very low molecular weight fraction shows unexpectedly high swell. Conversely, it has been found that the addition of as little as 0.1% of ultrahigh molecular weight PE increased the die swell of HDPE.

Perhaps it is best to say that die swell varies with the MWD in a manner not predictable by a simple parameter, but depends in some complicated way on the entire shape of the MWD. Another possible complication is that HDPE may contain small amounts of long chain branching (LCB), and the amount can vary with the details of the polymerization process. It is known that LCB has a strong influence on melt rheology, especially elasticity.

Fillers, particulate or fibrous, have the effect of reducing or suppressing die swell. Small amounts of crystallinity have a similar effect, and this may account for the surprisingly small die swell of rigid PVC.

3.4 The Milk Bottle

In the late 1950s the milk industry set its eye on the plastic milk jug to replace the breakable glass containers and the bulky cartons that were currently on the market. The gallon size was readily acceptable because the plastic container incorporated a handle. Various types of BM machines were built to make the plastic milk gallon, but because of the neck finish requirements, the standard trimming operation outside the machine was not acceptable. The milk industry used a very low cost plastic or cardboard snap cap and an aluminum foil cap that was only crimped on the bottle. The use of expensive screw caps was not accepted by the dairy industry because of cost.

Uniloy (now part of Milacron) at the time had entered the BM machine business in the United States as a joint venture with Dow Chemical Co. Plants were built by Uniloy to produce bottles for Dow using Dow's HDPE. Dow in turn sold the bottles for bleaches and detergents. With Uniloy's entry into the machinery business for BM the milk bottle, already accepted as a viable package, was taken into heavy development by Uniloy. A pull-up neck finish that was done in the mold, with a special blow pin, was developed entirely by Uniloy and produced an acceptable neck for the existing closures on the market. Uniloy has since sold many hundreds of BM systems both to custom blow molders and in-house filling operations using this patented neck finish for the gallon and half-gallon milk containers. The machines that Uniloy developed used a reciprocating screw with stationary platens that opened and closed only. The reciprocating screw was acceptable for PE and plastics were developed that used a 0.7-melt index that created a high swell when shot through the heads. This high swell made the production of handleware much easier. These machines are currently on the market and are now upgraded with microprocessors and parison programmers. The bottles were removed from the mold with bottom tail grippers and placed on a cooling conveyor and transferred to a lay down trimmer for deflashing of the handle area, neck area, and the tails.

Simultaneously with the development of the Uniloy milk bottle process, BM machines were being developed in Europe for the production of PVC bottles. The market requirements for mineral water and items such as edible oil and vinegar required a clear plastic bottle. The 1-liter size became dominant for the market and three German-based companies, Bekum, Kautex, and Fischer, all developed BM machines for production of PVC bottles. PVC, being a heat-sensitive material, required the use of continuous extrusion and preferably would not be run on a reciprocated screw type machine or an accumulator head type machine. PVC also is a very rigid material and does not trim well outside the machine after it is cooled, which prompted the development of the calibrated neck finish by Bekum Plastics Machinery, Inc. (patent filed in 1959). The calibrated neck is where the thread finish on the bottle is compression molded with a sized blow pin and a cutting sleeve is attached to the blow pin to form the top surface of the bottle.

All three machine builders in Germany produced machines using continuous extrusion and moved molds to the extruded parison from the head and back to a fixed position that allowed for a calibrated blow pin to enter and blow the finished bottles. A single-sided version of these machines would produce a limited number of bottles, and in 1963 the double-sided high-speed version of this method was developed and patented by Bekum. Today the same types of machines are still produced by the German-based companies and others have entered the market. Some 25,000 to 30,000 continuous extrusion, moving platen machines have been placed into the market in the last 35 years.

EBM has enjoyed an approximately 10% growth for the last 30 years, and this trend should continue with the conversion of many new products from glass and metal to plastic bottles. The state of the art in EBM now consists of finishing bottles inside the machine behind the safety gate. This includes handle ware and wide-mouth containers. These operations were done outside the machine in a post-trimming operation. The newer machinery consists of completely finished bottles inside the machine, with the bottles discharged from the machine in an oriented upright position for packing and/or filling and packing.

In the past, labeling of plastic bottles had always been a bottleneck in the production plants. The appearance of post-labeled bottles has never been satisfactory. In time the procedure of in-mold labeling developed. The label is applied to the plastic during the BM process, requires no flame treating and less labor and in most cases allows for a reduction of weight in the bottle.

3.5 Industrial Blow Molding

Large containers ranging from 5-gallon (18.9-liter) to 55-gallon (208.2-liter) have replaced those made of metal. The US Department of Transportation continues to approve more plastic containers for hazardous material. The light weight of the plastic as well as the number of trips that it can make provide an attractive 55-gallon drum for the industry. No reconditioning is necessary, such as in steel drums, and plastic containers up to the 55-gallon size can be made on site as opposed to metal containers that are made and shipped to the consumer. This type of EBM is done on accumulator type machines, as the large weight of the parison precludes the use of continuous or intermittent type machines. The accumulator machines extrude the parison at a high speed so that the parison does not sag (Fig. 3.21).

Blow molding of industrial products or large containers above 5-gallons differs from bottle blowing, as the materials cannot be continuously extruded because of the weights required to produce these parts or containers. Accumulator heads are used and the material is ejected through the head at high rates so that the parison drag weight is not a deterrent to the operation. The industrial part market has been growing throughout the world; a major market is in the automotive area. Industrial containers from the 5-gallon through the 55-gallon drum have penetrated the market. The production of automotive gas tanks started in Europe decades ago and has begun in the United States. As an example, in the United States the Environmental Protection Agency (EPA) requires that HDPE gas tanks are either coated or extruded with a form of barrier. Fluorination has been a process available for inside coating. Different approaches are used to meet EPA requirements such as coextruded structures.

The production of industrial products has continually led to the development of new EBM techniques because high molecular weight PEs are generally used to keep the weight of the container at a minimum with its highest tensile strength. High molecular weight PEs are much more difficult to process than standard high-density PE, and as a

Figure 3.21 Accumulator head for EBM fabricating large drums, etc

result groove barrel extruders and zero compression screws (CR = 1) have been used. This combination of a groove barrel extruder and zero compression screw also provides for high output at a low melt temperature on small extruders. As much as 40% more output can be achieved than with a standard compression screw.

Parison heads have also made a change, as industrial drums go through a stringent drop test for transporting chemicals, and so forth. New heads have been developed that completely eliminate a weld line through the plastic that has been common since the beginning of EBM. Parisons have been made without the weld line going through the parison and as a result the containers can withstand the drop test at a lighter weight (see Chapter 5). With microprocess control systems and special tooling designs, the wall thickness distribution of industrial products has been significantly improved. Full automated systems for producing large containers and industrial parts have been developed using post-cooling. Finished products or containers come out of the machine with the highest quality and highest efficiency. In the past, industrial part BM was always considered a sloppy operation but with the automotive requirements and US Department of Transportation specifications for containers, this business did become quite efficient and sophisticated.

3.6 Multilayer Manufacturing

Coextrusion blow molding methods are used for sandwich wall (multilayer) products. Figure 3.22 shows a multilayer bottle produced with coextrusion blow molding. Coextrusion BM techniques can readily be used to produce containers from 0.2 to 5.0 liter (6.8 to 16.9 oz) capacity. Larger containers, for example automobile fuel tanks, are also available using this basic process. Insert molding and two-shot molding techniques are also used to produce multi-wall injection molded preforms that can later be blow molded in conventional equipment [6].

There are two basic approaches to the production of multilayer structures: coextrusion and coinjection. Other methods include coating (Chapter 7).

With the coextruded (or coinjection) technology well established, the variety of achievable properties is readily extendable through the correct combination of different materials. The potential for BM packaging products to be much more than simple bottles has been known for many decades. The know-how and economics are such that many previously futuristic ideas continue to be much closer to reality. The potential in using coextruded and coinjection bottles, coupled with the machin-

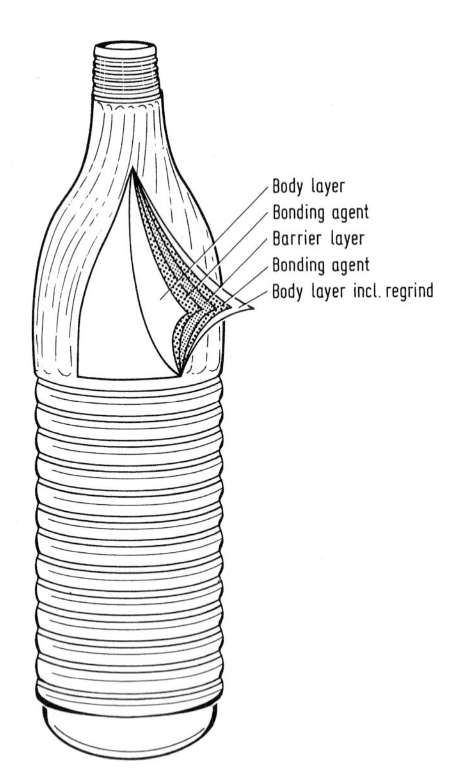

Figure 3.22 *Example of a five-layer multilayer EBM product*

ery platform established, means that glass and metal will continue to come under increasing pressure. Often the higher initial cost of the plastics is counterbalanced by the total packaging economics that can be obtained.

In this section, coextrusion multilayer manufacture is reviewed. Other techniques, such as coating, can be used to meet certain requirements such as performance and/or cost. All these types of manufacture offer certain basic advantages that include meeting specific product performance requirements that may not be available with one plastic, or if one plastic could be used the multilayer system permits reducing the cost. Another important benefit is that in many applications regrind/recycled plastic can be used as an interlayer or on a surface depending on product performance requirements, and many others.

3.6.1 Coextrusion Blow Molding

Coextrusion is basically an extrusion BM process using several extruders instead of one (Fig. 3.23). The production of a plastic bottle with coextrusion does not differ much from using a single layer. Proprietary wheel type machines are in production for coextrusion as well as platen type machines. Polypropylene (PP) is mostly used as the base material since most food products being packaged in coextrusion bottles need to be hot-filled and PP can withstand the hot fill temperatures. PP has contact clarity but does not produce a crystal clear container. Coextruded bottles are produced for high-heat processes such as retorting. Polycarbonate combined with EVOH, for example, can produce a container that is crystal clear, has an oxygen barrier, and can be retorted.

The viewing stripe bottle is another form of coextrusion product that has been around for many years. This process of applying a clear viewing stripe in a colored container is a form of coextrusion using a small mini-extruder in combination with the base extruder for the body material. With the motor oil business converted almost exclusively to plastic bottles, the viewing stripe in the container has its application.

Acrylonitrile is an example of a plastic material in BM that can be used on the inside of a plastic bottle to provide a barrier for chemicals and pesticides. This market has been traditionally in glass and cans; it is being penetrated with plastic treated with, for example, the fluorination process for a barrier (through either a DuPont Selar program or others). The DuPont Selar program is not a coextruded type process but an additive to polyethylene (PE) to provide a barrier against certain chemicals. Selar provides a barrier that effectively prevents permeation of many hydrocarbons and organic solvents. The BM plastic is prepared by dry mixing the Selar in a 5% to 20% mixture with the desired polyolefin and it is run on standard PE type machines. This mixture provides up to a 140-fold improvement in hydrocarbon permeability compared to HDPE.

The advantage of coextrusion is that the equipment and technology is both readily available and is relatively fast and economic. A disadvantage can arise from the subsequent converting process in which significant amounts of scrap may be

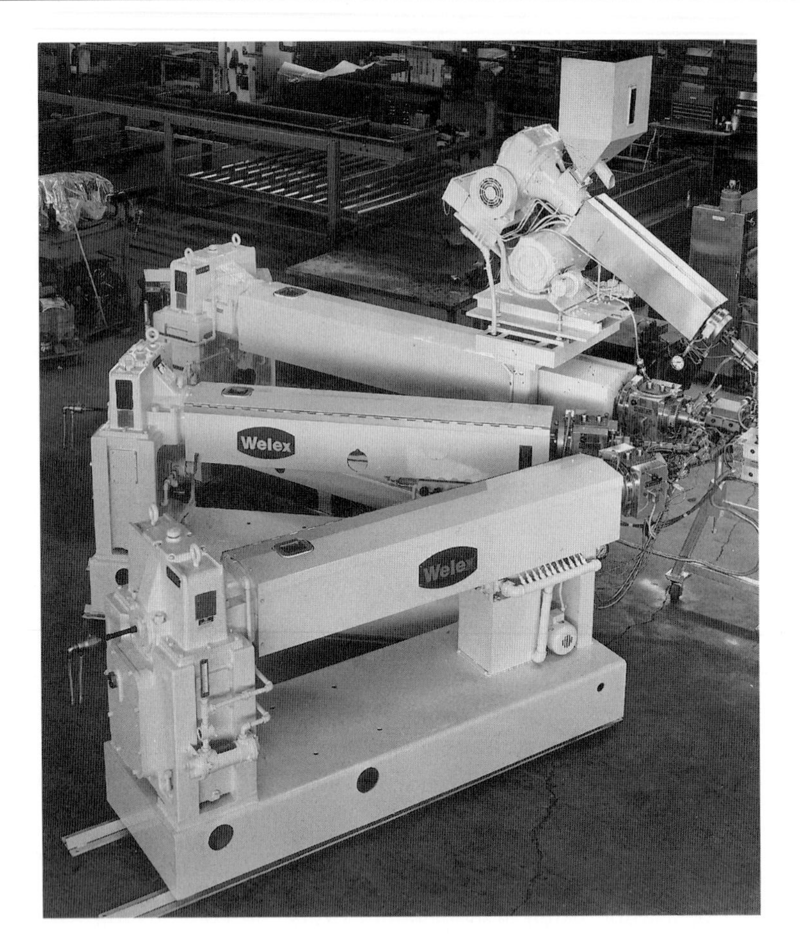

Figure 3.23 *Example of four Welex extruders delivering a four-layer structure*

generated, and wall and flange dimensions can be difficult to maintain. If not properly designed and processed there is the potential of exposed cut edge, which may be critical in situations where a water-sensitive barrier material, such as EVOH, is exposed to long periods of high relative humidity. Various approaches are used to resolve the scrap problem and cut edges. However, coextrusion offers a greater variety of conversion options than does coinjection, at lower processing cost.

3.6.2 Coextrusion Applications

Coextrusion of several materials has been around for some time and entered the blow molding industry over twenty years ago. With the success over the past half-century and particularly since 1985 in the conversion from glass and metal to plastic, the industry continues its search for plastics and processes to convert more products into plastic packaging. Multi-layer BM products seem to penetrate the higher-volume glass and metal container markets for those applications that can be switched to a plastic container. The billions and billions of containers used each year for food and chemical packaging require barrier properties that are not common with most plastic resins. The food industry needs either an oxygen barrier or a moisture barrier in their containers so that their products can be placed on the shelves without refrigeration. In addition, the chemical industry needs a chemical barrier for packaging chemicals in plastic containers. This can only be accomplished through the multi-layer combination of plastics used to produce a coextruded or coinjected bottle (Table 3.5).

Coextruded bottles were introduced in the late 1960s and early 1970s in Japan in the cosmetic and toiletry market. The cosmetic and toiletry marketing people want a pleasing appearance in their containers at a low cost. Conventional PE and PP do not provide a good surface for printing and do not provide the appearance of a high-priced product. Coextrusion was used to place a thin layer of nylon over a heavy body layer of PE or PP. This thin layer of nylon produced a container that had a glossy finish, required no flame treating and was virtually scuff proof. Besides containers, coextruded three-layer tubes have become quite popular for the cosmetic industry. This type of container was readily accepted by the cosmetic industry and was basically the beginning of coextrusion blow molding. An adhesive was used to bind the two materials because they were not chemically compatible. This development led to the introduction of containers using a PE, PP, or nylon as barriers for certain food or chemical products.

3.6.2.1 Chemical Market

Coextrusion or coinjection for the chemical market usually only requires three layers as the barrier material can either be nylon or acrylonitrile placed on the inside as the chemical barrier for the container and inexpensive PE or PP on the outside with a tie layer in between. This type of three-layer container produces a plastic bottle that replaces the cans that are used in the chemical and pesticide market as well as glass containers. The three-layer typical coextrusion system is based on either shuttle-type or wheel-type machines with the inclusion of additional extruders and a head system through which the plastic is brought in at different points and layered to form a parison. The remainder of the process, blowing the bottle, etc., is the same as running a single layer container.

3.6.2.2 The Food Market

The food market normally requires an oxygen barrier because oxygen causes deterioration in most food products. The normal requirement is 1 year of shelf life without refrigeration. Many oxygen-sensitive food products are either hot filled or retorted. In these cases the container must have the requisite thermal as well as barrier properties. Further, if the container is to be subjected to elevated temperatures during filling or processing it must be designed to either resist or anticipate paneling. Both of these factors considerably increase the complexity of developing satisfactory rigid barrier containers.

Table 3.5 Guide to Plastic Bonding in Coextrusion

	Olefins			Styrenes						PVCs		Misc.					Adhesives		
Material	LDPE	HDPE	PP	low-impact PS	medium-impact PS	high-impact PS	crystal PS	ABS[b]	ABS	rigid PVC	flexible PVC	Polycarbonate	Polyurethane	Acrylic	Nitriles	Nylon 6	EVA	Ionomers	SBS[c]
Olefins																			
LDPE	G[a]	G	G	P	P	P	P	P	P	P	P	P	P	P	P	P	G	G	G
HDPE	G	G	G	P	P	P	P	P	P	P	P	P	P	P	P	P	G	G	G
PP	G	G	G	P	P	P	P	P	P	P	P	P	P	P	P	P	G	P	G
Styrenes																			
low-impact PS	P	P	P	G	G	G	G	G	P	P	P	P	F	P	P	P	G	?	G
medium-impact PS	P	P	P	G	G	G	G	G	P	P	P	?	F	P	P	P	G	?	G
high-impact PS	P	P	P	G	G	G	G	G	P	P	P	?	F	P	P	P	G	?	G
crystal PS	P	P	P	G	G	G	G	G	P	P	P	?	F	P	P	P	G	?	G
ABS[b]	P	P	P	G	G	G	G	G	G	G	G	?	F	G	P	P	G	?	G
ABS	P	P	P	P	P	P	P	G	G	G	G	?	F	G	P	P	G	?	G
PVCs																			
rigid PVC	P	P	P	P	P	P	P	G	G	G	G	U	G	G	P	U	G	?	?
flexible PVC	P	P	P	P	P	P	P	G	G	G	G	U	G	G	P	U	G	?	?
Misc.																			
Polycarbonate	P	P	P	P	?	?	?	?	?	U	U	G	?	G	P	?	?	?	?
Polyurethane	P	P	P	F	F	F	F	F	F	G	G	?	G	?	P	?	?	?	F
Acrylic	P	P	P	P	P	P	P	G	G	G	G	G	?	G	P	?	?	?	?
Nitriles	P	P	P	P	P	P	P	P	P	P	P	P	P	P	G	P	F	?	F
Nylon 6	P	P	P	P	P	P	P	P	P	U	U	?	?	?	P	G	P	G	P
Adhesive																			
EVA	G	G	G	G	G	G	G	G	G	G	G	?	?	?	F	P	G	?	?
Ionomers	G	G	P	?	?	?	?	?	?	?	?	?	?	?	?	G	?	G	?
SBS[c]	G	G	G	G	G	G	G	G	G	?	?	?	F	?	F	P	?	?	G

[a] G = good, F = fair, P = poor, ? = unknown, U = undesirable.
[b] Less than 20% acrylonitrile content.
[c] Styrene butadiene styrene.

There are a few materials that are excellent oxygen barrier plastics but it is very expensive to convert them to a single layer container. Coextrusion of lower-cost materials with a barrier layer is used to produce a container that competes with glass or metal. The most common barrier resin used today is ethylene-vinyl alcohol (EVOH), which was FDA approved. A layer of EVOH of only 0.001-in. (0.00254-cm) can provide a barrier against oxygen for most foods for at least one year.

In the beginning, more than 50 lines were installed throughout Europe and Japan for three-layer production of coextrusion bottles. EVOH was used as the barrier material for food and the base material varied from PE to PP. An adhesive was used as a tie layer. The EVOH was placed on the outside and produced a container with a limited shelf life, as EVOH is hygroscopic and absorbs moisture (Chapter 6). As the moisture content increases, the barrier properties are lost. Placed on the inside of the container, EVOH will absorb moisture from the product and also loses its barrier properties.

In the early 1980s, a five-layer system came on the market where the EVOH was placed in the middle of the bottle with two tie layers and two body layers that prevented the EVOH from exposure to the inside of the bottle and the outside atmosphere. It produced a bottle with good oxygen barrier properties and shelf lives of at least 1 year. A six-layer system was then developed so that regrind could be placed between the EVOH and the outside body layer. Today there are several commercial products in five- or six-layer coextrusion, such as ketchup, jelly, tomato juice, and barbecue sauce containers.

PP is used because most food products require hot filling at 185 °F. PE and PVC cannot withstand this temperature. PP is also a relatively inexpensive plastic, and when combined with the more expensive EVOH does provide a reasonably priced container. Furthermore, PP has good contact clarity for products such as ketchup bottles. The industry has geared up for high-speed production of coextruded bottles since the large food companies and the consumer have accepted the plastic bottle in place of glass. The fact that these containers are nonbreakable and can be handled by children has led to their overwhelming success. A side benefit is the squeezability of the plastics, which in ketchup, jelly, salad dressing containers gives it another major advantage.

To summarize the situation in the United States, the first nationally marketed coextrusion BM bottle was the Gamma bottle, introduced in late 1983 by American Can for Heinz catsup. Gamma utilized PP both as a structural layer and to provide hot fill capacity. The barrier layer was EVOH. The innovative technology incorporated in the Gamma bottle was the development of an appropriate adhesive to combine PP and EVOH. American Can, in addition, developed a modified extrusion head for use on their existing BM wheel. The technology developed by American Can was similar to that previously developed by Toyo Seikan in Japan; therefore, American Can Co. and Toyo entered into a patent exchange giving each the royalty fee right to exploit the others patents. No know-how was exchanged.

The PP/EVOH bottle has three primary drawbacks: the inadequate cold impact strength of PP; the less than water clear appearance of the bottle; and its cost, which is 20–30% higher than the bottle it replaced. In addition, when used to replace glass in a narrow neck configuration for a viscous liquid like catsup, it substantially reduced filling line speeds. Regardless of these negatives, the consumer reaction was positive, based on the added dispensing convenience provided by the bottle.

Considerable development work has been done to improve the clarity of the bottle. Attention has been focused primarily on PET and PC, which are the only other plastics that offer both clarity and the other performance characteristics required.

The critical performance problem with PET is its thermal instability at hot fill temperatures unless it is heat-set. Monsanto approached this problem by using a wide mouth heat-set stretch blow molded container utilizing technology licensed from Yoshino Kogyo in Japan. Presumably, they coated the container to achieve a serviceable bottle for hot-filled oxygen-sensitive products. Most of the other companies who are working extensively with PET, in particular Continental and Owens-Illinois, believe that a multilayer structure that either incorporates heat setting or some thermally stable polymer plus a barrier layer is the better solution.

3.7 The Extruder

To produce EBM products different factors are involved that include operating an extruder efficiently, process plastic properly (Chapter 7), set up efficient process control on the extrudated melt (Chapter 9), and so forth. This section provides information on the extruder that is to operate so that its extrudate (melt) will perform properly and most efficient during the EBM process. The extruder, which offers the advantages of a completely versatile processing technique, is unsurpassed in economic importance by any other process. This continuously operating process, with its relatively low cost of operation, is predominant in the manufacture of shapes such as blown containers, profiles, films, sheets, tapes, filaments, pipes, rods, in-line postforming, and others [5].

In the basic processing concept the plastic material passes from a hopper into a plasticating cylinder in which it is melted and moved forward by its rotating screw (Fig. 1.4). The screw compresses, melts, and homogenizes the material. When the melt reaches the end of the cylinder, it may be forced through a screen pack prior to entering a die that gives the desired shape with no break in continuity. The screen pack is a filter that restricts unmelted plastic and/or contaminants from entering the die.

Practically only thermoplastics go through extruders; no major markets have been developed to date for extruded thermosets.

A major difference between extrusion and injection molding (IM) is that the extruder processes plastics at a lower pressure and operates continuously. Its pressure usually ranges from 1.4 to 10.4 MPa (200 to 1,500 psi) and could reach 34.5 or 69 MPa (5,000 or possibly 10,000 psi). In IM, pressures range from 14 to 210 MPa (2,000 to 30,000 psi). However, the most important difference is that the IM melt is not continuous; it experiences repeatable abrupt changes when the melt is forced into a mold cavity. With these significant differences, it is actually easier to theorize about the extrusion melt behavior as many more controls are required in IM.

Good quality plastic extrusions require homogeneity in terms of the melt-heat profile and mix, accurate and sustained flow rates, a good die design, and accurately controlled downstream equipment for cooling and handling the product. Four principal factors determine a good die design: internal flow length, streamlining, construction materials, and uniformity of heat control. Heat profiles are preset via tight controls that incorporate cooling systems in addition to electric heater bands.

On leaving the extruder for BM, the melt (extrudate) is drawn down. This is an important aspect of downstream control if tight dimensional requirements exist and/or conservation of plastics is desired. The processor's target is to determine the tolerance required for the drop rate and to see that the device meets the requirements. Even if tight dimensional requirements are not required, the probability is that better control of the drop speed will permit tighter tolerances so that a reduction in the material's output will occur.

In extrusion, as in all other processes, an extensive theoretical analysis has been applied to facilitate understanding and maximize the manufacturing operation. However, the real world must be understood and appreciated as well. The operator has to work within the limitations and variabilities of the materials and equipment. The interplay and interchange of process controls can help to eliminate problems and aid the operation in controlling the variables that exist. The greatest degree of instability is due to improper screw design, or using the wrong screw for the plastic being processed. Screws are designed to meet the melt requirements for the different types of plastics (Chapter 2).

For uniform and stable extrusion it is important to check periodically the drive system and other equipment, and compare it to its original performance. If variations are excessive such as surging [60], all kinds of problems will develop in the extruded product. An elaborate process-control system can help, but it is best to improve stability in all facets of the extrusion line. Examples of instabilities and problem areas include (1) nonuniform plastic flow in the hopper; (2) troublesome bridging, with excessive barrel heat that melts the solidified plastic in the hopper and feed section and reduces or stops the plastic flow; (3) variations in barrel heat, screw heat, screw speed, screw power drive, die heat, die head pressure, and the takeup device; (4) insufficient melting or mixing capacity; (5) insufficient pressure-generating capacity; (6) wear or damage of the screw or barrel; (7) melt fracture/sharkskin; (8) improper alignment of the extruder; downstream equipment.

3.7.1 Extruder Basics

Extruders can be classified as: (1) continuous with single screws (single and multistage) or multiscrews (twin screw, etc.); (2) continuous disk or drum that uses viscous drag melt actions (disk pack, drum, etc.) or elastic melt actions (screwless, etc.); and (3) discontinuous that use ram actions (very low viscosity thermoplastics (TPs), thermoset (TS) plastics, and rubbers/elastomers) and reciprocating actions (injection molding, etc.).

There are differences between the popularly used single-screw and multiscrew types (Fig. 3.24 and Table 3.6). Each has benefits, depending on the plastic being processed and the products to be fabricated. At times their benefits can overlap, so that either type could be used. In this case, the type to be used would depend on cost factors, such as cost to produce a quality product, cost of equipment, and cost of maintenance.

Similar extruders from different manufacturers, and even those from the same manufacturer processing the same plastic, will often require different operating settings to produce similar products. This is true even when screw designs are theoretically identical. The reasons for the differences include factors such as variability of plastic (most important), barrel nonuniform internal dimensions, control sensor locations, variable or limited available heater wattage, coolant flow rates, and so forth. To obtain a consistent performance for the same material from one extruder to another, one has to know the variables that exist in setting up the machine controls on both machines. Good quality extrusions require: (1) upstream equipment delivering properly controlled plastic to the hopper; (2) homogeneity by the extruder in terms of the melt heat profile and mix, with accurate and sustained flow rates; (3) a good die design; and (4) accurately controlled downstream equipment for cooling and handling the product.

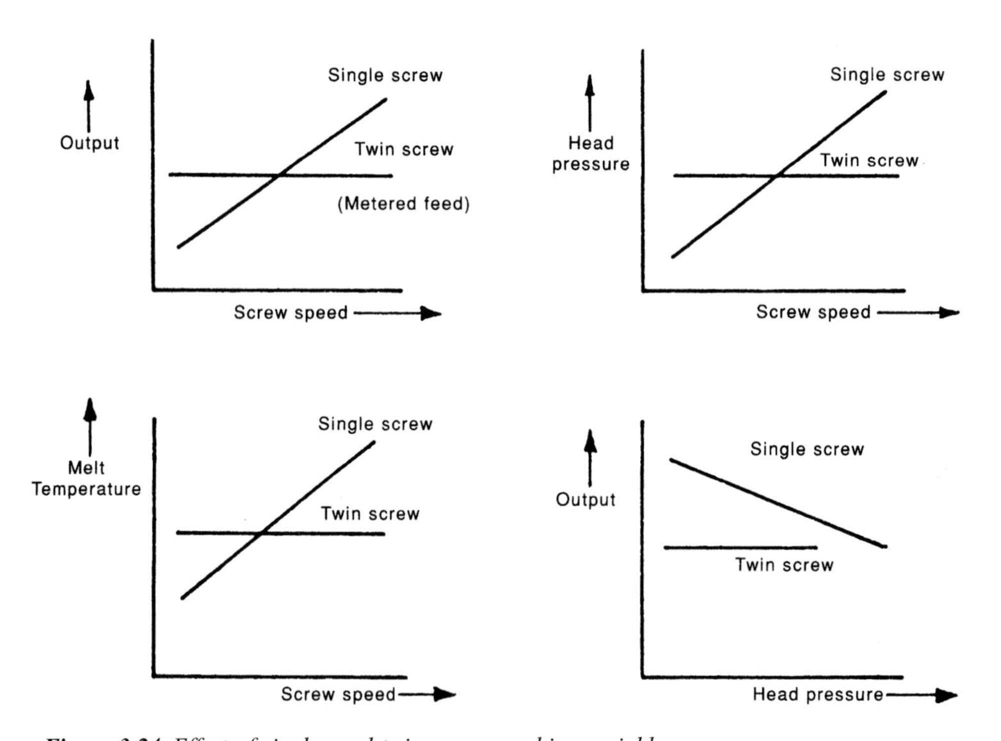

Figure 3.24 *Effect of single- and twin-screw machine variables*

Table 3.6 Comparison of Single and Twin Screw Extruders

	Single screw	Twin screw
Flow type	Drag	Near positive
Residence time and distribution	Medium/wide	Low/narrow (useful for reaction)
Effect of back pressure on output	Reduces output	Slight/moderate effect on output
Shear in channel	High (useful for stable polymers)	Low (useful for PVC)
Overall mixing	Poor/medium	Good (useful for compounding)
Power absorption and heat generation	High (may be adiabatic)	Low (mainly conductive heating)
Maximum screw speed	High (output limited by melting, stability, etc.)	Medium (limits output)
Thrust capacity	High	Low (limits pressure)
Mechanical construction	Robust, simple	Complicated
First cost	Moderate	High

3.7.1.1 Single Screw Extruder

Most of the BM lines use a single screw extruder. Features of this machine are shown in Fig. 2.1. Examples of temperature guides for extruders are given in Table 3.7 and Fig. 3.25.

When an extruder requires an improvement in melt, different methods can be used [5] such as the inclusion of a gear pump (Fig. 3.26).

The essential parameter in the extruder's pumping process is the interaction between the rotating flights of the screw and the stationary barrel wall. For the plastic material to be conveyed, its friction must be low at the screw surface but high at the barrel wall. If this basic criterion is not met, the plastic will probably rotate with the screw and not move in the axial/output direction.

In the output zone, both screw and barrel surfaces are usually covered with the melt, and external forces between the melt and the screw channel walls have no effect except when processing extremely high viscosity materials such as rigid PVC and ultrahigh molecular weight PE. The flow of the melt in the output section is affected by the coefficient of internal friction (viscosity), particularly when the die offers a high resistance to the flow of the melt.

The usual and more popular single screw types use conventional designs with basically uniform diameters of the screw and barrel. Examples include extruders that have decreasing screw channel volume, continuous variable speed, pressure control, and a venting (devolatilization) system. Special designs use conical or

Table 3.7 EBM General Temperature Processing Guide

Suggested starting points only

Material	Zone 1 temp °F	Zone 2 temp °F	Zone 3 temp °F	Zone 4 temp °F	Head (A) temp °F	Melt temp °F	Mold temp °F
ABS	380–440	390–440	400–440	400–440	400–440	400–440	110–170
Acrylic	370–460	380–460	390–460	400–460	400–460	400–460	130–200
LDPE	320–340	320–350	330–360	340–370	330–380	340–380	40–100
HDPE	350–390	360–400	370–410	370–420	370–420	370–420	40–100
Nylon 6	440–480	450–480	460–480	470–490	440–510	450–510	140–200
Nylon 6/6	470–550	480–540	480–540	490–530	520–540	510–540	140–200
PC	440–480	450–480	460–480	470–490	460–510	460–510	140–220
PC/ABS	420–445	430–455	440–465	450–475	440–490	450–490	120–220
PP General purpose	340–430	350–440	360–450	370–450	390–460	390–450	40–120
PS General purpose	340–360	360–390	390–420	400–440	360–450	360–440	40–120
PS High impact	370–390	390–420	420–450	430–470	420–460	420–460	40–120
PVC Flexible	250–290	260–300	270–310	280–320	310–375	310–375	50–100
PVC Rigid	280–300	300–320	320–350	340–370	330–390	340–390	50–100
TPE (Hytrel)	300–400	310–410	330–430	350–450	370–460	370–450	60–140
TPU	360–450	360–460	360–460	360–470	370–475	340–465	60–140
TPU/PVC (Vythene)	300–320	310–330	320–340	330–360	340–365	350–365	60–120

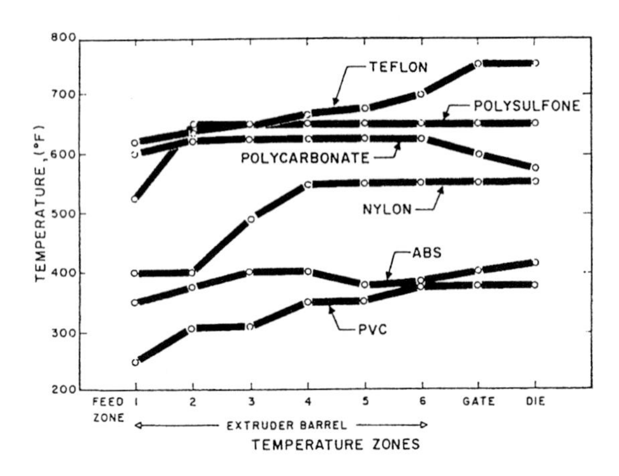

Figure 3.25 *Temperature profiles of different plastics going through an extruder having six temperature zones*

parabolic screws for special mixing and kneading effects. They can include eccentric cores, variable pitch superimposed flights of different pitch, kneading rotors, fitted core rings, and periodic axial movement. Barrels may have internal threads, telescopic screw shapes, and feeding devices.

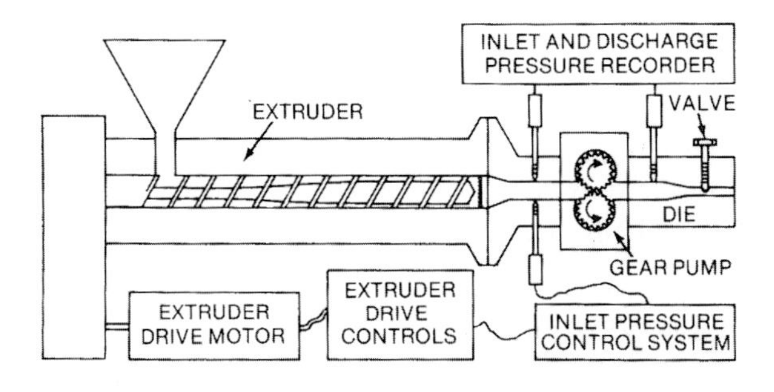

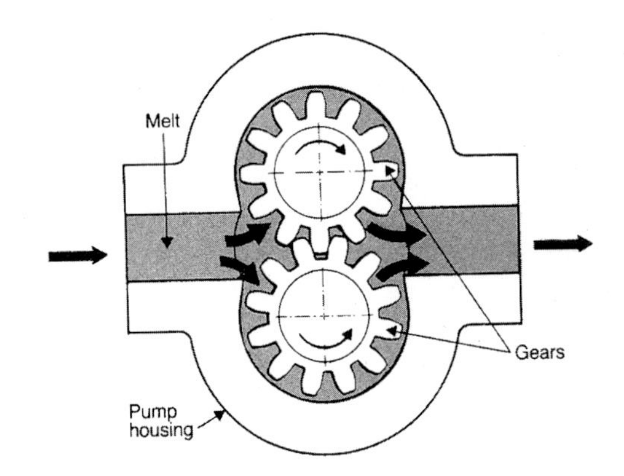

Figure 3.26 *Schematics of a gear pump location and operation*

3.7.1.2 Multiple Screw Extruder

With the development of extrusion techniques for newer thermoplastic materials, it was found in the past that some plastics with or without additives required higher pressures and needed higher temperatures. There was also the tendency for the material to rotate with the screw. The result was degraded plastics. The peculiar consistency of some plastics interfered with the feeding and pumping process. The problem magnified with bulky materials, also certain types of emulsion PVC and HDPE, as well as loosely chopped PE film or sticky pastes such as PVC plastisols.

During the early 1930s, twin and other multiscrew extruders were developed to correct the problems of the single screw extruder that existed at that time. The conveyance and flow processes of multiscrew extruders are very different from those in the single screw extruder. The main characteristic of multiscrew extruders include: (1) their high conveying capacity at low speed; (2) positive and controlled pumping rate over a wide range of temperatures and coefficients of frictions; (3) low frictional (if any) heat generation which permits low heat operation; (4) low contact time in the extruder; (5) relatively low motor power requirements for self-cleaning action with a high degree of mixing; and (6) very important, positive pumping ability which is independent of the friction of the plastic against the screw and barrel which is not reduced by back flow. Even though the back flow theoretically does not exist, the flow phenomena in multiscrew extruders are more complicated and therefore far more difficult to treat theoretically than single screw flow.

3.7.1.3 Twin Screw Extruder

Although there are fewer twin screw than single screw extruders, they are widely employed for manufacturing certain products, in particular specialty operations such as compounding applications. The popular common twin screw extruders (in the family of multiscrew extruders) include tapered screws with at least one feed port through a hopper, a discharge port to which a die is attached, and process controls such as temperature, pressure, screw rotation (rpm), melt output rate, and so forth.

For all types of extruders if the goal is to deliver a high quality melt at the end of the screw, the plasticating or melting process should be completed prior to reaching the end of the screws. Twin intermeshing counterrotating screws are principally used for compounding. Different types have been designed with basically three available commercially that includes corotating and counterrotating intermeshing twin screws; nonintermeshing twin screws are offered only with counterrotation. There are fully intermeshing and partially intermeshing systems and open- and closed-chamber types. In the past major differences existed between corotating or counterrotating; today they work equally well in about 70% of compounding applications, leaving about 30% in which one machine may perform dramatically better than the other. Figure 3.27 shows the different designs used with the twin screw extruders.

Similar to the single screw extruder, the twin screw extruder, including the multi-screw, has advantages and disadvantages. The type of design to be used will depend on performance requirements for a specific material to produce a specific product. With the multiscrews, very exact metered feeding is necessary for certain materials; otherwise output performance will vary. With overfeeding, there is a possibility of overloading the drive or bearings of the machine, particularly with counterrotating screw designs. For mixing and homogenizing plastics, the absence of pressure flow is usually a disadvantage. Disadvantages also include their increased initial cost due to their more complicated construction as well as their maintenance and potential difficulty in heating.

3.7.2 Extruder Operation

The extruder plasticates/melts the plastic solid and pumps the extrudate through an orifice in the die.

A successful extrusion operation requires close attention to many details, such as (1) the quality and flow of feed material at the proper temperature, (2) a temperature

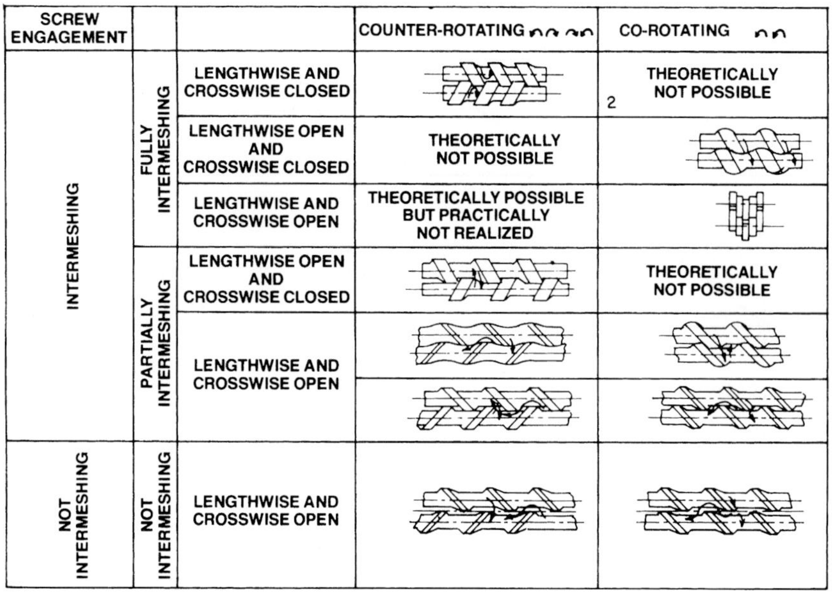

SCREW ENGAGEMENT			COUNTER-ROTATING	CO-ROTATING
INTERMESHING	FULLY INTERMESHING	LENGTHWISE AND CROSSWISE CLOSED		2 THEORETICALLY NOT POSSIBLE
		LENGTHWISE OPEN AND CROSSWISE CLOSED	THEORETICALLY NOT POSSIBLE	
		LENGTHWISE AND CROSSWISE OPEN	THEORETICALLY POSSIBLE BUT PRACTICALLY NOT REALIZED	
	PARTIALLY INTERMESHING	LENGTHWISE OPEN AND CROSSWISE CLOSED		THEORETICALLY NOT POSSIBLE
		LENGTHWISE AND CROSSWISE OPEN		
NOT INTERMESHING	NOT INTERMESHING	LENGTHWISE AND CROSSWISE OPEN		

Figure 3.27 *Examples of different twin-screw mechanisms*

profile adequate to melt but not degrade the polymer, and (3) a startup and shutdown that will not degrade the plastic.

Care should be used to prevent conditions that promote surface condensation of moisture on the plastic and moisture absorption such as by the pigments in the color concentrates. Surface condensation can be avoided by storing the sealed plastic container (for hygroscopic plastics) in an area at least as warm as the operating area for about 24 h before use. If color concentrate moisture absorption is suspected, heating for 8 to 16 h in a 250 to 300 °F (120 to 150 °C) oven should permit sufficient drying. Hopper dryers have proven very useful in drying plastics. They can also enhance extruder capacity in operations where the output is limited by the heating and melting capacity of the extruder. Figure 3.28 shows the relative difference in heat input required from the extruder to melt the plastic, with and without the hopper dryer used as a preheater.

The extruders form a homogeneous plastic melt and force it through a die orifice that is related to the shape of the product's cross section. The formed TP melt (extrudate) is cooled or the TS melt heated as it is being dropped downward away from the die exit or drawn through downstream equipment.

After the extruder has started up and is running, the target is to ensure that the melt temperature, pressure, and output rate (time) are consistent within a processing "window" of operation. These factors directly affect the final product. The time

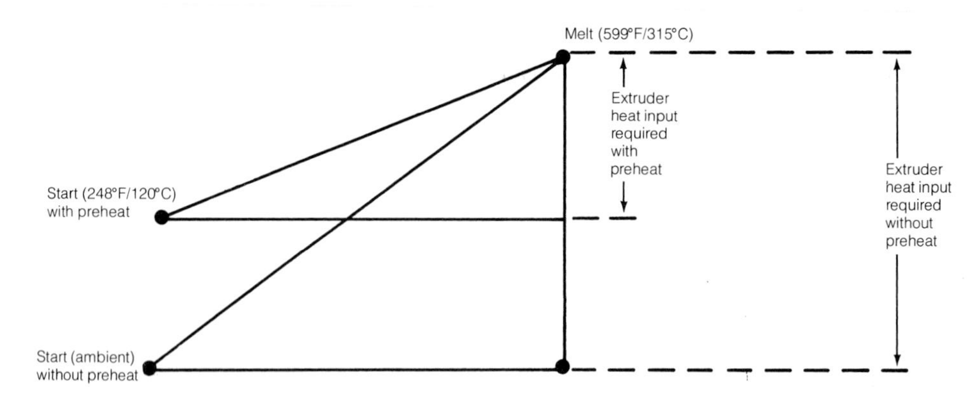

Figure 3.28 *Plastic heat input*

period for the melt exiting is influenced by such factors as the rotating speed of the extrusion screw. The usual plastic melts have rather wide operating windows. However, it is best to determine their ideal process control settings in order to obtain maximum performance/cost efficiency.

Within various types of plastics (PE, PVC, PP, or others), each type can have different profiles. Experience shows how to set the profile and/or obtain preliminary information from the material supplier. Degrading or oxidizing certain plastics is a potential hazard that occurs particularly when the extruder is subject to frequent shutdowns. In this respect, the shutdown period is even more critical than the startup period.

3.7.2.1 Extruder Checks

Prior to startup one must check certain machine conditions:

- Unless the same plastic is ready in the machine from a previous run, the entire machine should be cleaned and/or purged, including the hopper, barrel, breaker plate, die, and downstream equipment. If a plastic was left in the barrel for a while, with heat off, the processor must determine if the material is subject to shrink. It could have caused moisture entrapment from the surrounding area, producing contamination that would require cleanup (this situation could also be a source of corrosion in/on the barrel/screw).
- One must check heater bands and electrical connections, handling electrical connections very carefully.
- Check thermocouples, pressure transducers, and their connections very carefully.
- Be sure the flow path through the extruder is not blocked.
- Have a bucket or drum, half filled with water, to catch extrudate wherever purging or initial processing of plastics contained contaminated gaseous byproducts.

- Inspect all machine ventilation systems to ensure adequate air flow.
- Check operating manual of the machine for other startup checks and requirements that have to be met such as motor load (amperage) readings.

3.7.2.2 Startup

The following startup procedure provides a general review since each line has specific requirements. When starting up a new extrusion setup, start the screw rotation at about 5 rpm. Gradually look into the air gap between the feed throat and throat housing and make sure the screw is turning. Screws have been installed without their key in place, or the key has fallen out during installation. Also make sure that antiseize material is applied to the drive hub, to help installation and removal. Also if the key is left out and the drive quill is turning and the screw is not, the screw will not gall to the drive quill.

Details for operating the machine are based on what the plastic being processed requires, such as temperature settings, screw rpm, etc. available from the material supplier and/or experience. Startup procedures involve certain precautions:

- Starting with the front and rear zones (die end and feed section), one should set heat controllers slightly above the plastic melting point and turn on the heaters. Heatup should be gradual from the ends to the center of the barrel to prevent pressure buildup from possible melt degradation.
- Increase all heaters gradually, checking for deviations that might indicate burned-out or runaway heaters by slightly raising and lowering the controller set point to check if power goes on and off.
- After the controllers show that all heaters are slightly above the melt point, adjust to the desired operating temperature (based on experience and/or plastic manufacturer's recommendation), checking to ensure that any heat increase is gradual, particularly in the front/crosshead.
- The time required to reach temperature equilibrium may be 30 to 120 min, depending on the size of the extruder. Overshooting is usually observed with on/off controllers.
- Hot melts can behave many different ways, so no one should stand in front of the extruder during startup, and one should never look into the feed hopper because of the potential for blowback due to previous melt degrading, and so on.
- After set temperatures have been reached, one puts the plastic in the hopper and starts the screw at a low speed such as 2 to 5 rpm; some plastics, such as nylon, may require 10 to 20 rpm.
- The processor should observe the amperage required to turn the screw; stop the screw if the amperage is too high, and wait a few minutes before restart.
- When working with a melt requiring high pressure, the extruder barrel pressure should not exceed 7 MPa (1,000 psi) during the startup period.

- One should let the machine run for a few minutes, and purge until a good quality extrudate is obtained visually. Experience teaches what it should look like; a certain size and amount of bubbles or fumes may be optimum for a particular melt, based on one's experience after setting up all controls. If plastic was left in the extruder, a longer purging time may be required to remove any slightly degraded plastic.
- For uniform output, all of the plastic needs to be melted before it enters the screw's metering zone, which needs to be run full. When time permits, after running for a while, the processor should consider stopping the machine, let it start cooling, and remove the screw to evaluate how the plastic performed from the start of feeding to the end of metering. Thus one can see if the melt is progressive and can relate it to screw and product performances. Turn up the screw to the required rpm, checking to see that maximum pressure and amperage are not exceeded.
- Adjust the die with the controls it contains, if required, at the desired running speed. Once the extruder is running at maximum performance, set up controls for takeoff/downstream equipment, which may require more precision settings.
- Extrudate can start its tract from the die by "threading" through the downstream equipment to its haul-off.
- One may get into a balancing act of interrelating extruder and downstream equipment. Extruder screw speeds and haul-off rates may then be increased. Downstream equipment is adjusted to meet their maximum operating performance, such as having the vacuum tank water operate with its proper level and vacuum applied.
- The extruder can be "fine-tuned" to obtain the final required setting for meeting the desired output rate and product size (Chapter 8).

3.7.2.3 Shutdown

It is common to run the extruder to an empty condition when one is shutting down. This action ensures that there is no starting up with cold plastic, a condition that could overload the extruder if improper startup occurred. Some extruders, such as those processing PE film, are shut down with the screw full of plastic. This prevents air from entering and oxidizing the plastic. Because PVC decomposes with heat, to ensure that this material is completely removed at shutdown, purging material is used such as low melt PE that can remain in the barrel eliminating PVC decomposition. On startup, it is preferable to raise barrel heat slightly above its normal operating temperatures; the higher temperature ensures that unmelted plastic will not produce excessive torque in the screw.

Degrading or oxidizing certain plastics is a potential hazard that occurs particularly when the extruder is subject to frequent shutdowns. In this respect, the shutdown period is even more critical than the startup period.

Shutdown procedures vary slightly, depending on whether or not the machine is to be cleaned out or just stopped briefly. If the same type of plastic is to be run again, cleanout is generally not required. The goal is to avoid degradation by reducing exposure of the plastic to high heat. If cleanout is required, because a different type plastic is to be processed, it is necessary to disassemble the machine at a heat high enough to allow cleanout before material solidifies.

A procedure for shutdown without cleanout is as follows: (1) Empty the hopper as well as possible. (2) Reduce all heat settings to the melt temperature. (3) Reduce the screw speed to 2 to 5 rpm, purging the plastic if required into a water bucket or drum prior to reducing the melt heat. (4) When the screw appears to be empty, stop the screw and shut off the heaters and the main power switch (however, steps 2 and 3 must be completed before the melt drops significantly or a premature shutdown will occur with plastic remaining in the barrel). (5) If a screen pack with breaker plate is used, disconnect the crosshead (or die) from the extruder and remove the breaker plate and screen. If necessary, appropriate action is taken to clean them.

For cleanout of the extruder at shutdown, the first three steps are the same steps 1, 2, and 4 in the preceding paragraph. Then the procedure continues as follows. (6) When the extruder appears to be empty, stop the screw. (7) Shut off and disconnect the crosshead (or die) heaters. Reduce other heaters to about 170 to 330 °C (400 to 625 °F), depending on the plastic's temperature at the melt point. (8) Disassemble the crosshead and clean it while still hot. Remove the die, and gear pump if used, and remove as much plastic as possible by scraping with a copper spatula or brushing with a copper wire brush. Remove all heaters, thermocouples, pressure transducers, and so on. Consider using an exhaust duct system above the disassembly and cleaning area, even if the plastic is not a contaminating type; this procedure keeps the area clean and safe. (9) Push the screw out gradually while cleaning with a copper wire brush and copper wool. Care should be exercised if a torch is used to burn and remove plastic; tempered steel, such as Hastelloy, may be altered and the screw distorted or weakened as well as subjected to excessive wear, corrosion, or even failure (broken). (10) After screw removal, continue the cleaning, if necessary. (11) Turn off heaters and the main power switch.

Final cleaning of products, particularly disassembled parts, is best done manually, or much better, in ventilated burnout ovens, if available, operating at about 540 °C (1,000 °F) for about 90 min. For certain parts with certain plastics, the useful life could be shortened by corrosion; check with the part manufacturer. After burnout, remove any grit that is present with a soft, clean cloth. If water is used, air-blast to dry. With precision machined parts, water cleaning could be damaging because the potential of corrosion when certain metals are used.

3.7.2.4 Other Operating Details

Machine operation takes place in three stages. The first stage covers the running of a machine and its peripheral equipment. The next involves setting processing conditions to a prescribed number of parameters for a specific plastic, with a specific die in a specific processing line, to meet product performance requirements. The final stage is devoted to problem solving and fine tuning of the complete line using the FALLO approach that leads to meeting performance requirements at the lowest cost. A successful operation requires close attention to many details, such as quality and flow of feed materials, and a heat profile adequate to melt but not degrade the plastic. Processors must also become familiar with a troubleshooting guide.

- Screws for single screw extruders are usually righthand thread, and turn counter-clockwise when seen from the rear or power drive mechanism. Their rotation appears as if to unscrew itself backward out of the material. The machine's thrust bearing prevents this backward action. Twin screws may turn in either or both direction, but the principle is the same.
- Most of the heat to melt the plastics comes from the motor drive system, not the heaters. The exceptions are very small extruders, slow moving twin screws, and processing at high temperature.
- Motor speeds are generally reduced from 10 : 1 to 30 : 1 to obtain reasonable screw speeds, which are lowest for very small machines because they need residence time and very large machines that have to avoid excessive shear rates as the screw diameter increases.
- Process controls (PCs) provide ease of extruding products (Chapter 8).
- The major cooling action in the extruder is due to the plastic material entering it. The plastic absorbs the heat from the motor and heaters so that it can melt. When the feed is stopped, the system equilibrates. The result with this stop is usually that the front end transfers heat to the rear feed end of the screw. In turn, there is danger of material sintering, bridging, and/or sticking to the screw root.
- Pellets must stick to the barrel and slip on the screw root in its feed zone, and must stick to each other as much as possible. This action permits maximum conveying through that zone. More of this action is not always better, though, as some screws can "bite off" more than their front ends can "chew".
- Material is by far the biggest component of the manufacturing cost. Therefore, almost anything that saves material can be justified; material management is an important criterion to follow even at the expense of equipment development.
- Energy is a very small proportion of manufacturing cost, as an extruder is basically an energy-efficient machine when compared to other processes, such as injection molding. Moreover, excess energy would overheat the material and not make it extrudable. However, there is always room to reduce energy losses in the extruder and the complete line.
- Pressure at the screw tip is important, as it relates to thrust bearing wear, mixing efficiency, personal safety, screen contamination, melt temperature, and so on. This pressure is the cumulative demand of the head, from the screens to the die lips, and is not something generated independently by the extruder.
- Production rate is the drag flow displacement of the last screw flights of the extruder, less the effect of resistance (pressure demand of the head), plus the effect of overbite at the feed end. Leakage over the screw flights may also have an effect, sometimes in a positive direction if there is a pressure peak along the barrel.

- Effective shear rates are around 100 to 500 rev/s in single screw extruders, as well as in most die lips, compared to values of from 0.1 to 10 in a typical ASTM melt index (MI) test (Chapter 6). Proper comparison of materials demands at least two viscosities (not a ratio) with measurement or extrapolation into the practical 100 to 500 range.
- In the numerical simulation of an extrusion process, it is essential to use the physical data that is as accurate as possible before one can evaluate the usefulness of the program.

4 Injection Blow Molding

4.1 The Basic Process

Injection blow molding requires two sets of molds, an injection and a blow mold and has basically three stages (Figs. 4.1 and 4.2). In the first stage, plastic melt from an injection molding machine is injected into a split steel mold with a male and a female cavity to produce a preform, which in turn is temperature-conditioned (second stage) to be blow molded in a female cavity mold later (third stage).

The most common method of forming the preform is by using a conventional reciprocating screw injection molding machine (IMM) [6]. In the first stage, the IMM injects hot melt through the nozzle into a mold with one or more cavities and core pins to produce the preform(s). The plastic is plasticized by the conventional IM procedure. An exact amount (shot size) of plastic enters each cavity. The preform is shaped much like a test tube and can include a screw finish at the top. This screw finish is the final finish of the bottle and is molded to very close tolerances. The process results in scrap-free products. After injection of the melt into the mold cavity(ies), the two-part mold opens while the plastic remains at a temperature level sufficiently low to prevent sagging but still high enough to be blown in the second phase of a single-stage BM system.

The preform is transferred on a core rod to the blow molding stage (second stage). Here, a two-part cooled mold provides the desired mold cavity for BM. When the mold closes in this second stage, air is introduced via the core pin to expand the temperature-conditioned preform to the desired blown shape. Compared to EBM, the preform mold cavity is designed so that upon blowing, it develops a more precise and uniform overall thicknesses.

Controlled chill water usually 40 to 50 °F (4 to 10 °C) circulates through pre-designed mold channels around the mold cavities and solidifies the blown product. The mold opens and the core pins carry the blown product to the third stage.

In the third stage, the products are ejected. Ejection can be done by using stripper plates, air blowing, combination of stripper plate and air, robots, and so forth.

Some injection blow molders have a fourth stage (Fig. 4.3). This stage is added in either one of three places: between the preform injection stage and blow mold stage to provide extra temperature conditioning time; between the blow mold stage and

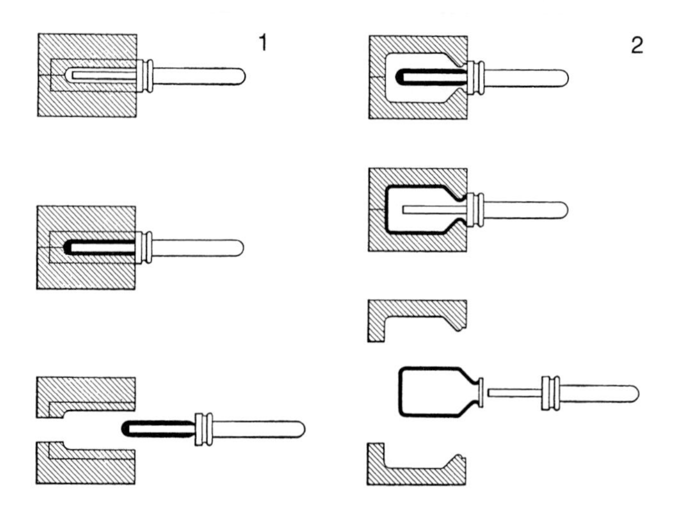

Fig. 4.1 *Basic IBM process: (1) injection preform, (2) BM and ejection*

ejection stage to provide time for some secondary operation, such as hot stamping or labeling; or between the ejection stage and the preform injection mold stage to provide time to detect any improperly ejected containers still attached to the core rod.

The cycle time of the complete operation is determined by the preform production by injection molding, generally by the time required in the injection mold if large swept volumes are involved, but also by the required metering time of the plasticizing

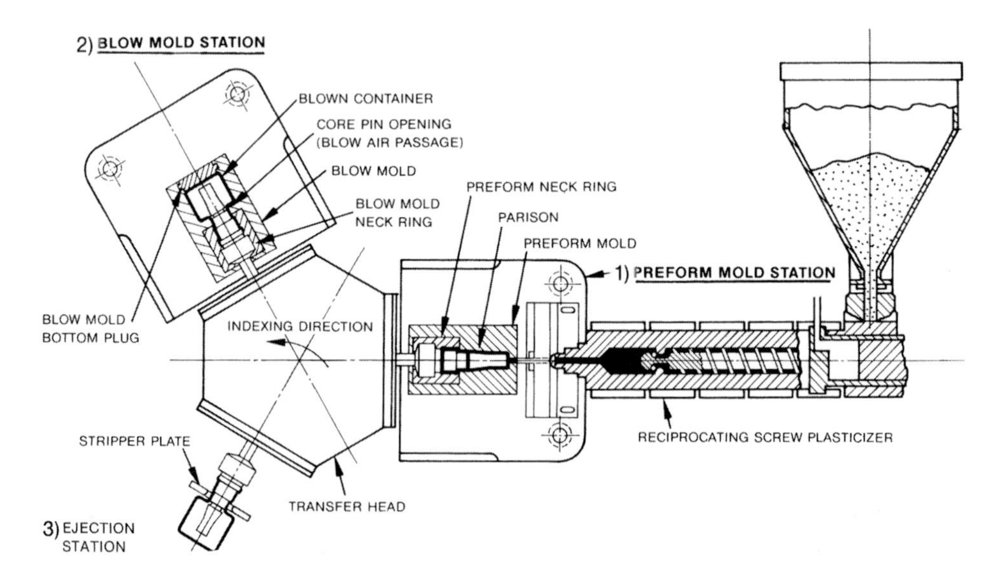

Fig. 4.2 *Layout of three-station IBM that includes the IMM*

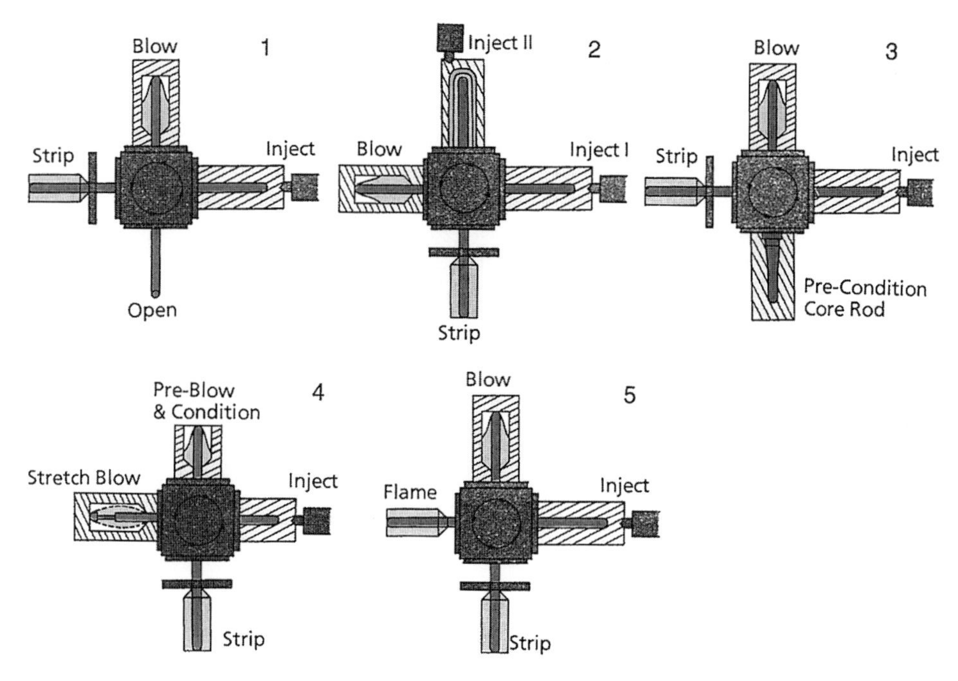

Fig. 4.3 *Four-station IBM: (1) basic mode, (2) coinjection mode, (3) pretreatment mode, (4) biaxial orientation mode, and (5) post-treatment mode*

injection molding machine. An example of typical IBM cycle times is given in Fig. 4.4.

The process parameters in preform production which determine the quality of the article are the injection speed, injection pressure, hold-on pressure, temperature control of the preform, cooling time of the blown article, and balance of these various factors (Chapter 8).

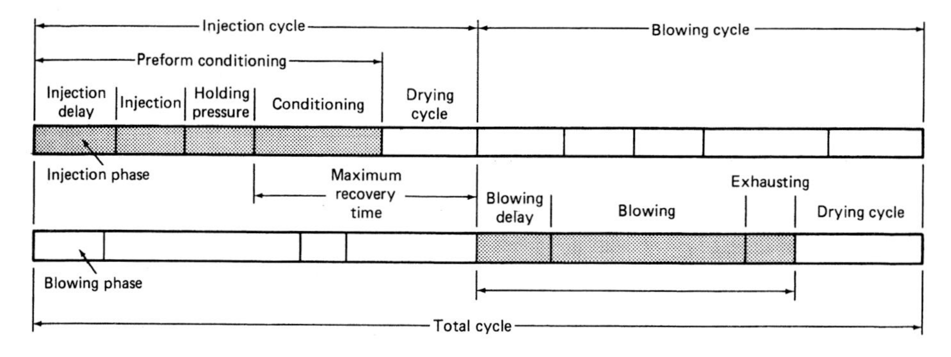

Fig. 4.4 *Basic IBM cycle*

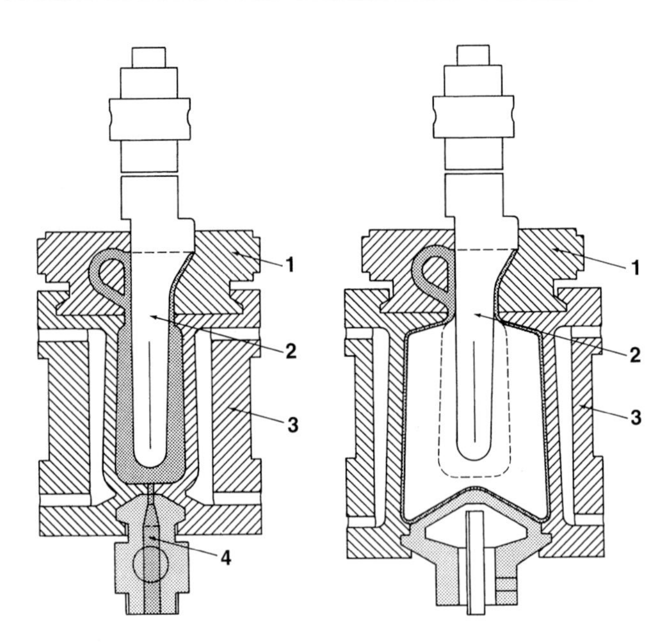

Fig. 4.5 *Injection blow molded bottle with solid handle*

The IBM process can be used for containers that have very close-tolerance threaded necks, wide mouth openings, solid handles (Fig. 4.5), and highly styled shapes. IBM features good materials distribution and the production of finished containers without the need for trimming and reaming.

Several other types of machines are available with different methods of transporting the core rods from one station to another. These include the shuttle two-parison rotary, axial movement, and rotary with three or more stations used in conventional injection molding clamping units [6].

A variation of IBM is displacement BM. Used for small non-handled containers, the process has the advantage of reducing the stress the melt is subjected to. The process differs in that a pre-measured amount of melted plastic is deposited into a cupel, the shape of preform. A core rod is inserted into the cupel, displacing the resin, packing it into the neck finish area. The preform is held on the rod as the cupel is moved away. A blow mold cavity then closes on the preform where air expands it against the cavity.

4.2 Coinjection: Fabricating Multilayer

With the coinjection (or coextruded) technology well established, the variety of achievable properties is readily extendable through the correct combination of different materials. The potential for BM packaging products to be much more than simple bottles has been known for many decades. The potential in coextruded and coinjection bottles, coupled with the machinery platform being established, means that glass and metal will continue to come under increasing pressure. Often the higher initial cost of the plastics is counterbalanced by the total packaging economics that can be obtained. Table 4.1 shows compatibility of various plastic materials for coinjection.

Table 4.1 *Compatibility of Plastics for Coinjection*

Materials	ABS	Acrylic ester acrylonitrile	Cellulose acetate	Ethyl vinyl acetate	Nylon 6	Nylon 6/6	Polycarbonate	HDPE	LDPE	Polymethylmethacrylate	Polyoxymethylene	PP	PPO	General-purpose PS	High-impact PS	Polytetramethylene terephthalate	Rigid PVC	Soft PVC	Styrene acrylonitrile
ABS	+	+	+				+	−	−	+		−		−	−	+	+	0	+
Acrylic ester acrylonitrile	+	+		+								−			0				+
Cellulose acetate	+		+	−															
Ethyl vinyl acetate		+	−	+				+	+			+		+			+	0	
Nylon 6					+	+		−	−			−			−				
Nylon 6/6					+	+	−	−	−			−			−				
Polycarbonate	+					−	+							−	0				+
HDPE	−			+	−	−		+	+	−	−	0							
LDPE	−			+	−	−		+	+	−	−	+		−				−	
Polymethylmethacrylate	+							−	−	+		−		−	0		+	+	+
Polyoxymethylene								−	−		+	−							
PP	−	−		+	−	−		0	+	−	−	+	−	−	−		−	−	
PPO												−	+	+	+				−
General-purpose PS	−			+			−	−	−	−		−	+	+	+			−	−
High-impact PS	−	0			−	−	0			0		−	+	+	+			−	−
Polytetramethylene terephthalate	+															+			+
Rigid PVC	+			+						+		−					+	+	
Soft PVC	0			0			+	−		+		−		−	−		+	+	+
Styrene acrylonitrile	+	+								+			−	−	−	+		+	+

Source: Battenfeld.

Note: + = good adhesion, − = poor adhesion, 0 = no adhesion, blank indicates no recommendation (combination not yet tested). The addition of fillers or reinforcements leads to a deterioration of adhesion between raw materials for skin and core.

4.2.1 Coinjection Blow Molding

Coinjection, like coextrusion, creates multilayer walls that offer advantageous properties to meet specific product performance requirements. Coinjection offers several distinct advantages over coextrusion as an initial step in the manufacture of rigid plastic barrier BM containers [6]. Finish and/or flange dimensions can be tightly controlled permitting double seaming on existing closing lines, scrap is eliminated, and wall thickness can be controlled (Figs. 4.6 and 4.7).

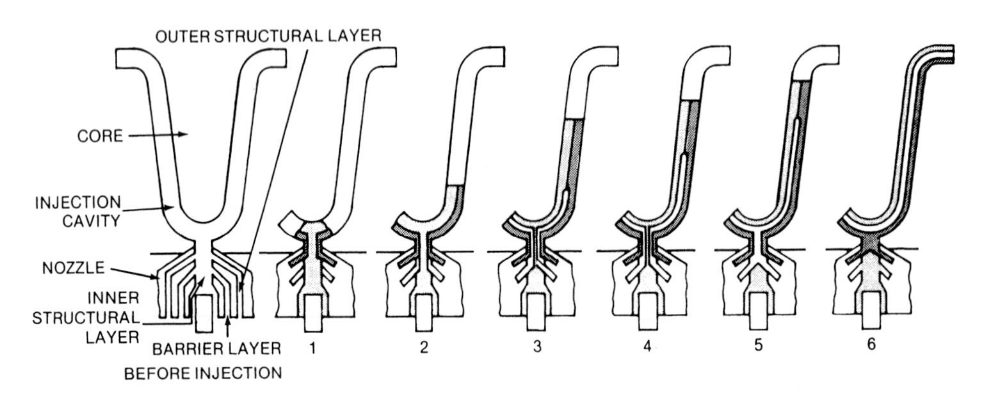

Fig. 4.6 Diagram of a 5-layer coinjection system by American Can

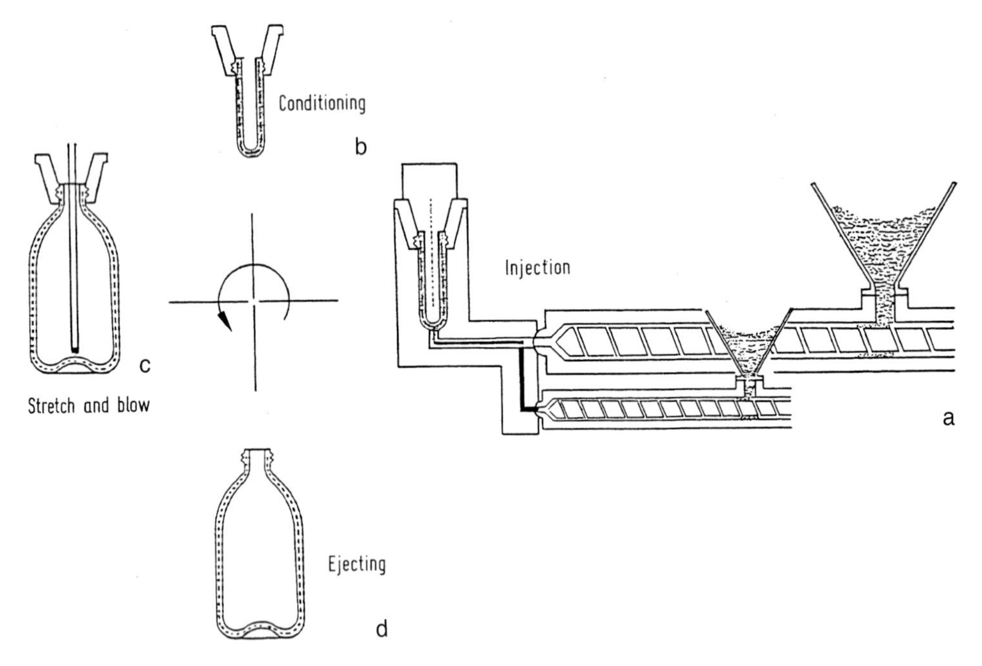

Fig. 4.7 Example of a coinjection blow molded container

4.3 Injection Molding

Injection molding (IM) is ranked as a major part of the plastics industry and big business worldwide, consuming approximately 32 wt% of all plastics. Of all the plastics consumed about 10% are processed through BM, of which almost 75% is EBM and almost 25% is IBM.

In the industry, IM is not identified as an extruder operation; however, it is basically a noncontinuous extruder and in a few operations it is a continuous operating machine. It has a screw plasticator, also called a screw extruder, that prepares the melt that is delivered to a mold (Chapter 2) [6].

There are many different types or designs of IMMs that permit molding many different shaped products based on factors such as quantities, sizes, shapes, product performances, and/or economics. Both small and large size IMMs have their advantages. If several small machines are used rather than one large one, a machine breakdown or shutdown for routine maintenance will have less effect on production rates. However, the larger machine is usually much more profitable. Because there are fewer cavities in molds for the small machines, they may permit closer control of the molding variables in the individual cavities.

The IM process can be identified by its two most basic popular methods of operation—the single-stage and the two-stage molding systems; there are also three-stage molding units, and so forth [6]. The single-stage system is also known as the reciprocating screw IMM. The two-stage system is known by other terms such as the piggyback IMM that can partially be related more to a continuous extruder.

These machines provide the following functions: (1) **plasticating**, heating and melting of the plastic in the plasticator; (2) **injection**, injecting from the plasticator under pressure a controlled volume shot of melt into a closed mold with solidification of the plastics beginning on the mold's cavity wall; (3) **after-filling**, maintaining the injected material under pressure for a specific time period to prevent back flow of melt and to compensate for the decrease in volume of melt during solidification; (4) **cooling**, cooling the thermoplastic (TP) molded part in the mold until it is sufficiently rigid to be ejected or **heating**, heating the thermoset (TS) molded part in the mold until it is sufficiently rigid to be ejected; and (5) **molded product release**, opening the mold, ejection of the part, and closing the mold so it is ready to start the next cycle with a shot of melt.

The IMM features the three basic components which are the injection unit, mold, and clamping system. The injection unit, also called the plasticator (Chapter 2), prepares the proper plastic melt and via the injection unit transfers the melt into the next component, the mold. The IMM's clamping system closes and opens the mold. This operation or cycle tends to be more complex than other processes such as extrusion, as it involves moving and then stopping the melt into the mold, rather than having a continuous flow of melt without interruptions. The IM process is extremely useful as it permits the manufacture of simple to particularly very complex and intricate three-dimensional (3D) extremely small to large BM shapes. When required these shapes/products can be molded to extremely tight tolerances, very thin, and relatively small in weight. The process needs to be thoroughly understood so as to maximize its performance and mold products, meeting performance requirements at the least cost and with ease.

4.3.1 Machine Characteristic

IMMs are characterized by their shot capacity. A shot represents the maximum usable volume of melt that is injected into the mold. It is usually about 30% to 70% of the actual available volume in the plasticator [6]. The difference relates to the plastic material's melt behavior, and provides a backup safety factor to meet different mold packing conditions. Shot size capacity may be given in terms of the maximum weight that can be injected into a mold cavity(ies), usually quoted in ounces or grams of general purpose polystyrene (GPPS). Because plastics have different densities, the better way to express shot size is in terms of the volume of melt that can be injected into a mold at a specific pressure. The rate of injecting the shot depends on the speed of the IMM and also the process control capability of cycling the melt to move fast-slow-fast, slow-fast, etc. into the mold cavity(ies).

Injection pressure in the barrel can range from at least 2,000 to 45,000 psi (14 to 310 MPa). The characteristic of the plastic being processed defines what pressure is required in the mold to obtain good products. Based on what cavity pressure is required, the barrel pressure has to be high enough to meet pressure flow restrictions going from the plasticator into the mold cavity(ies).

Clamp force on the mold halves required in the IMM also depends on the plastic being processed. A specified pressure is required to retain the pressure in the mold cavity(ies). It is basically the cross sectional area of any melt located on the parting line of the mold that includes the cavities and mold runner(s) located on the mold parting line. With a TP hot melt runner that is located within the mold half, its cross sectional area is not included in the parting line area. By multiplying the pressure required on the melt and the melt cross sectional area, the clamping force required is determined. To provide a safety factor consider a 10% to 20% increase [6]. Figure 4.8 shows the basic function of the injecting molding process.

4.3.2 Molding Plastic

Most of the literature on IM (or BM) processing specifically identifies or refers primarily to TPs with very little, if any at all, consideration of TS plastics. At least 90 wt% of all IM plastics are TPs. IM products include combinations of TP and TS as well as rigid and flexible TPs, TP and TS elastomers, and so forth. During IM the TPs reach maximum heat during plastication before entering the mold, whereas TS plastics reach maximum temperatures in the heated molds (Chapter 7).

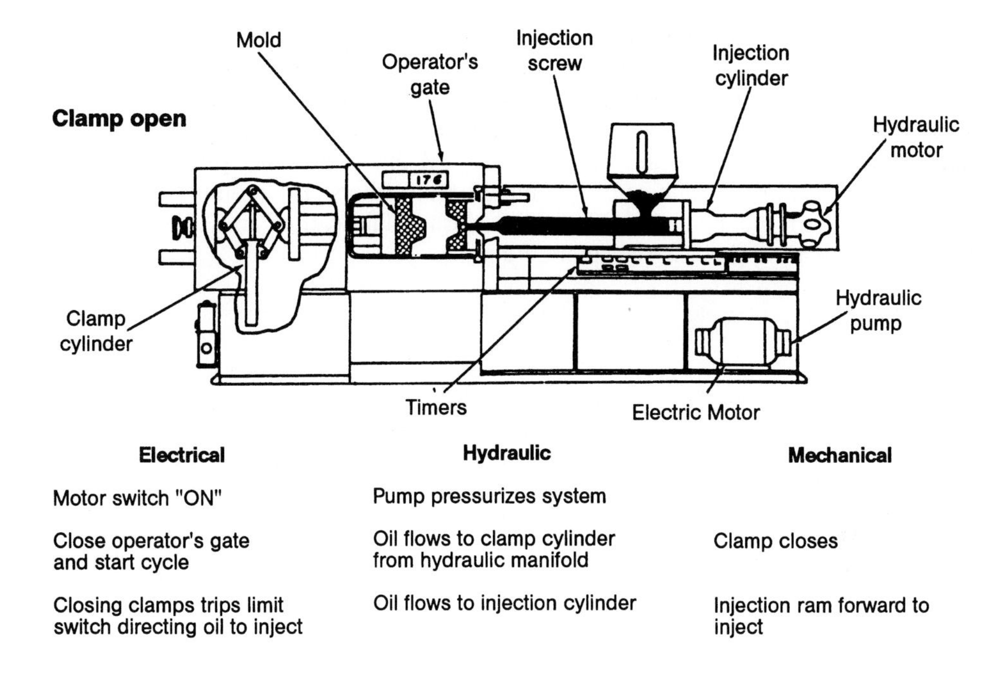

Clamp closed

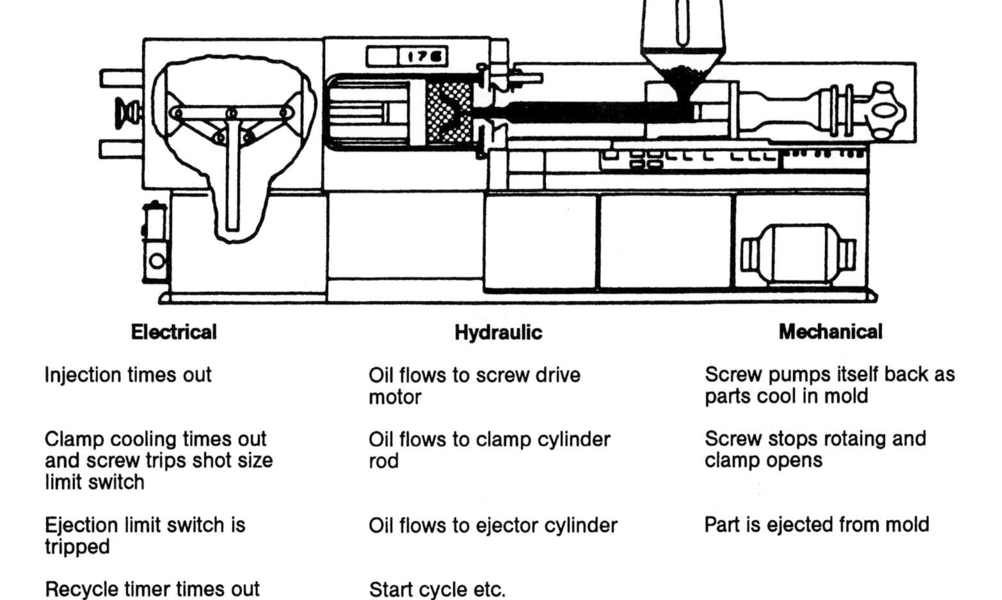

Fig. 4.8 Molding machine functions

4.3.3 Product Design Concept

The IM process, which includes IBM, is greatly preferred by designers because the manufacture of products in complex 3D shapes can be more accurately controlled with regard to dimensions and tolerances with IM than with other processes. As its method of operation is much more complex than others, IM requires a thorough understanding of the behavior of plastics during fabrication and control of the complete fabricating process (Chapters 7 and 9).

4.3.4 Molding Basics

IM is a repetitive process in which melted (plasticized) plastic is injected or forced into a mold cavity(ies) where it is held under pressure until removed in a solid state, basically duplicating the cavity of the mold. The mold may consist of a single cavity or a number of similar or dissimilar cavities, each connected to flow channels or "runners" that direct the flow of the melt to the individual cavities. Three basic operations exist: (1) raising the plastic temperature in the injection or plasticizing unit so that it will flow under pressure, (2) allowing the plastic melt to solidify in the mold, and (3) opening the mold to eject the molded product.

In these three steps the mechanical and thermal inputs of the injection equipment must be coordinated with the fundamental properties and behavior of the plastic being processed; different plastics tend to have different melting characteristics, with some being extremely different. They are also the prime determinants of the productivity of the process as manufacturing speed or cycle time (Figs. 4.9) will depend on how fast the material can be heated, injected, solidified, and ejected. Depending on shot size and/or wall thicknesses, cycle times range from parts of a second to many minutes. Other important operations in the injection process range from feeding the IMM, usually gravimetrically through a hopper, to the plasticator barrel thermal profile system to ensure that high quality products are produced (Chapter 11).

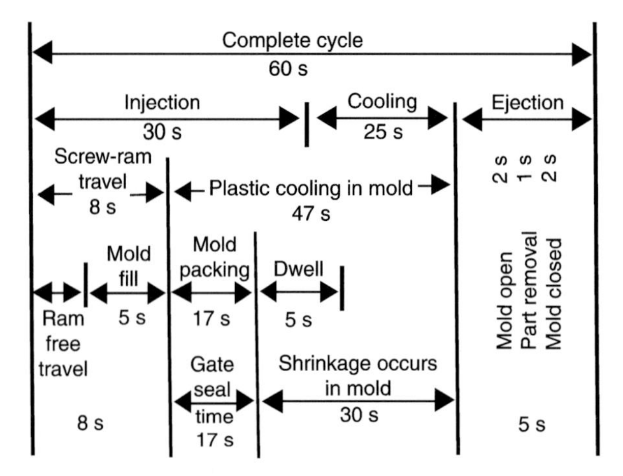

Fig. 4.9 The basic cycle

4.3.5 Plasticating

Plasticating is the process that melts or plasticates the plastics (Chapter 2). Different methods are used; the most common are the single-stage (reciprocating screw) or two-stage. As shown in Fig. 4.10, with a reciprocating screw the basic operations goes from (a) where the melt travels to the front of the screw while it is retracting. When the required melt shot size is obtained, the melt is forced forward as shown in (b). Following injecting the shot, the reciprocating screw melts the next shot as the screw retracts (c).

In Fig. 4.11 (a) and (b) are the ram (also called plunger) systems used in the original IMMs (since the 1870s and now used to process plastics with very little melt flow such as ultrahigh molecular weight PE). They use a piston with or without a torpedo to plasticate the material. View (c) is the single-stage reciprocating screw plasticator, and (d) is the two-stage screw plasticator.

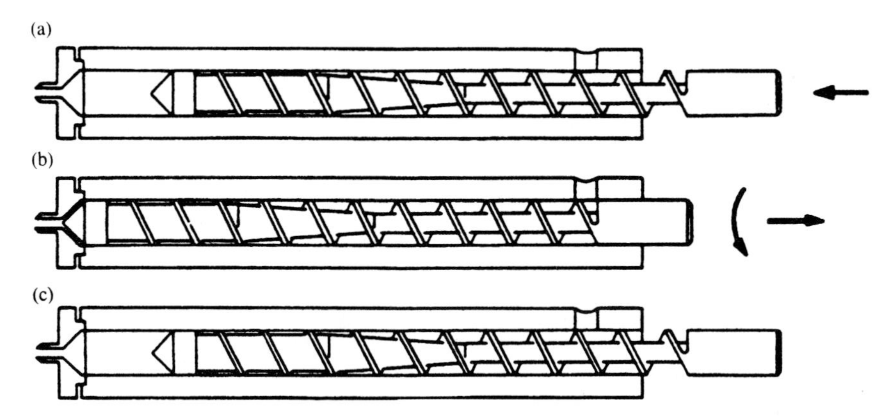

Fig. 4.10 *Reciprocating screw action*

Different IMM operating designs are used, namely all hydraulic, all electrical, and hybrids (combination of hydraulic and electrical). Each design provides different advantages such as reducing product weight (reducing plastic consumption), reducing molding cycle, reducing electric energy requirements, eliminating or minimizing molded-in stresses, molding extremely small to very large products, and/or improving performances [6].

Another design is injection-compression molding, also called injection stamping or more often coining. It uses a compression type mold having a male plug that fits into a female cavity (Chapter 1). After a short shot (relatively no pressure in cavity) enters the pre-opened-closed cavity mold, its stress-free melt is compressed to mold the finished product [6].

4.3.5.1 Screw Design

The primary purpose for using a screw, located in the plasticator barrel, is to take advantage of its mixing action and its ability to generate frictional heat to plasticate

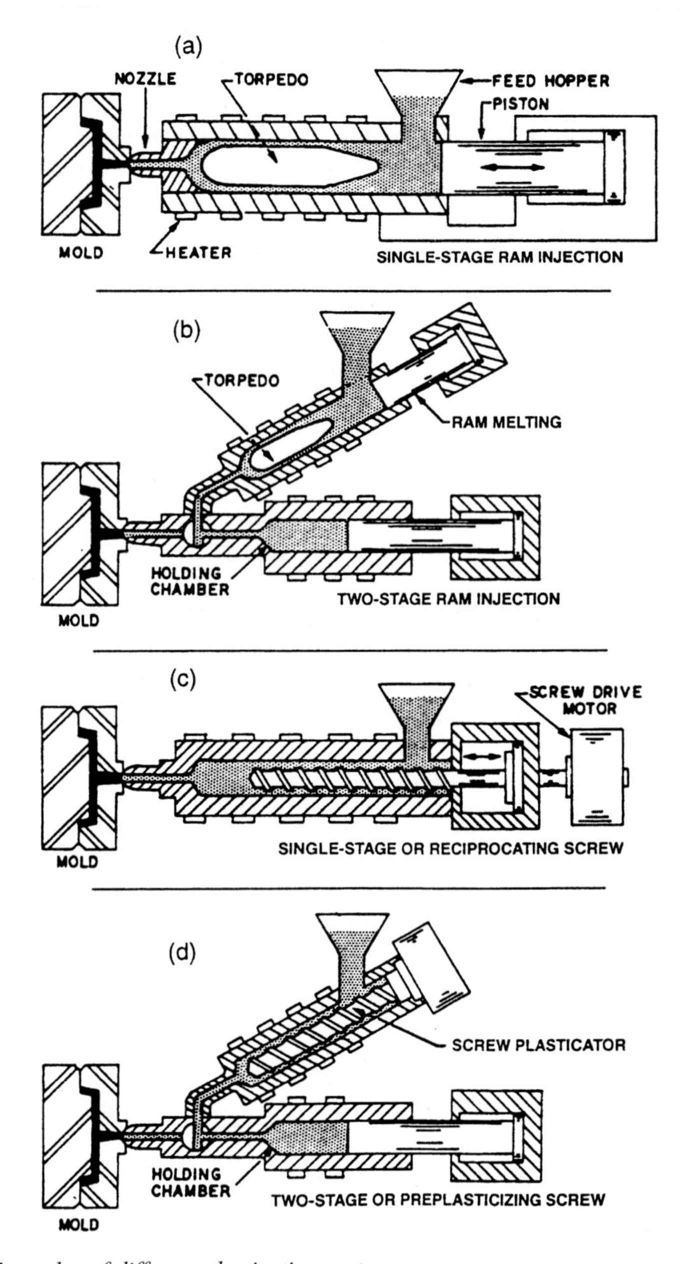

Fig. 4.11 *Examples of different plasticating systems*

the plastic. The motion of the screw keeps the IMM's process controls operating at their set points. The usual difference/variation in melt temperature, melt uniformity, and melt output is kept to a minimum prior to the plastic entering the mold. Heat is applied from heater bands around the barrel and the mixing action that occurs when plastic moves via the screw. Both conduction and mechanical friction heating of the plastic occur during screw rotation. The different controls used during IM, such as back pressure and screw RPM, influence melt characteristics [6]. Generally most IMMs use a single constant-pitch, metering-type screw for handling most of the plastics.

4.3.6 The Mold

A mold for the preform(s) must be considered a very important part of IBM. It is a controllable complex and expensive device providing different functions or capabilities to permit molding the desired product. If not properly designed/operated/handled/maintained, it will be a costly and inefficient operating device. Under pressure, hot melt moves rapidly through the mold. During the injection into the mold, air in the cavity is vented from the cavity(ies) to eliminate melt burning and/or forming voids in the products. Basically temperature-controlled water (with ethylene glycol if the water has to operate below its freezing point) circulates in the mold to remove heat from TPs (Chapter 6).

The mold consists of a properly designed sprue, runner, cavity gate, and cavity. The sprue is the channel located in the stationary platen that transports the melt from the plasticator nozzle to the runner. In turn, melt flows through the runner, gate, and into the cavity. With a single cavity mold usually no runner is used so melt flows from the sprue to the gate. In use are different runner systems to meet different processing requirements. The more popular are hot runners for TP (Chapters 6 and 7).

4.3.7 Processing

Different machine requirements or material conditions are met to ensure use of the most efficient IM process. It is important to understand and properly operate the basic IMM as well as auxiliary equipment. Figure 4.12 shows the relationship between manufacturing process and properties of the plastic products. In practically all operations it is necessary to pay attention to such factors as using screws that are not damaged or worn and properly dried plastics. Special dryers and/or vented barrels are required in drying the TP hygroscopic materials such as polycarbonate (PC), polymethyl methacrylate (PMMA), PUR, polyethylene terephthalate (PET), and so forth.

Use of TP regrind may have little effect on product performances (appearance, color, strength, etc.). However, reduction in performance could occur with certain TPs such as material with even one heat history through the IMM (Chapter 7). Granulated TSs cannot be remelted but can be used as additives or fillers in plastics; they are not remeltable.

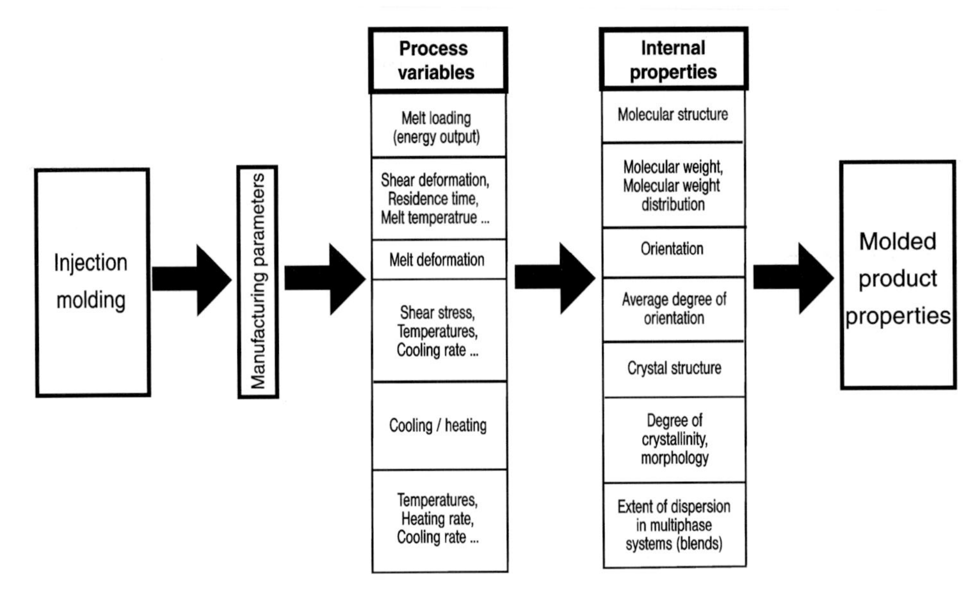

Fig. 4.12 Relationship between manufacturing process and properties of products

The process of heating and cooling many TPs can be repeated indefinitely by granulating scrap, defective products, and so on. During these cycles the plastic develops a "time-to-heat" history or residence time. This action can significantly reduce processing advantages and properties. Loss may require compensation of product design, process setup, and/or material modification by incorporating certain additives, fillers, and/or reinforcements (Chapter 7).

For all types of plastics, IM troubleshooting guides, which can be incorporated in process control systems (Chapter 12), are set up to take fast, corrective action when products do not meet their performance requirements.

4.3.8 Process Control

Proper injection of plastic melt into the mold is influenced by several process control conditions. Any one or a combination of conditions can affect different performance parameters from the raw material being fed into the IMM, flow of melt, packing of mold cavity(ies), cycle time, to product performances. As an example parameters that influence product tolerances involve: (1) product design, (2) plastics used, (3) mold design, (4) IMM capability, and (5) molding cycle time (Chapter 8).

4.3.9 Machine Type

The IMM has been one of the most significant and rational forming methods for processing plastic materials. The most recent examples are the all electric operating and improvements in hybrid IMM. A major focus continues to be on finding more rational means of processing the endless new plastics that are developed and also produce more cost-efficient products.

Where for years so-called product innovation was the only source of new developments, such as reducing the number of molded product components by making them able to perform a variety of functions or by making full use of a material's attributes, process innovations have also been moving into the forefront. This development includes all the means that help tighten, reorganize, and optimize the manufacturing process. All activity has been targeted for the most efficient application of production materials, a principle that must apply throughout the entire process from plastic materials to the finished product.

Even though modern IMM with all its ingenious microprocessor control technology is in principle suited to perform flexible tasks, it nevertheless takes a whole series of peripheral auxiliary equipment to guarantee the necessary degree of flexibility. Examples of this action include: (1) raw material supply systems; (2) mold transport facilities; (3) mold preheating banks; (4) mold changing devices that include rapid clamping and coupling equipment; (5) plasticating cylinder changing devices; (6) molded product handling equipment, particularly robots with interchangeable arms allowing adaptation to various types of production; and (7) transport systems for finished products and handling equipment to pass molded products on to subsequent production stages.

There are different types and capacities of IMMs to meet different product and cost–production requirements. The types are principally horizontal single clamping units with reciprocating and two-stage plasticators processing plastics. They range in injection capacity or shot sizes from less than an ounce (micro-IMM) to at least 400 oz (the usual is from 4 to 100 oz) and clamp tonnage up to 10,000 tons (usual from 50 to 600 tons). Other factors to include when specifying an IMM include clamp stroke, clamping speed, maximum daylight, clearances between tie rods, plasticating capacity, injection pressure, injection speed, and so on [6].

The type and size IMM to be used is dependent on the dimensions and size of the molded product with the processing requirements of the plastic material such as shot size and material pressure requirements. Examples of product dimensions that directly influence the size machine required include all part dimensions, number of parts to be molded in a single cycle, mold runner system to produce required number of parts, mold width, mold length, mold stack height, mold opening distance, and ejector rod spacing. This information will then determine the IMM preliminary requirements to produce the products.

4.3.10 Reciprocating (Single-Stage) Screw Machine

These machines are called reciprocating or single-stage injection molding machines. There are conventional types of IMM where plastic is melted using a combination of conductive heat from heater bands surrounding the barrel and frictional heating of the plastic created by a rotating screw inside the barrel (Fig. 4.13). The screw reciprocates

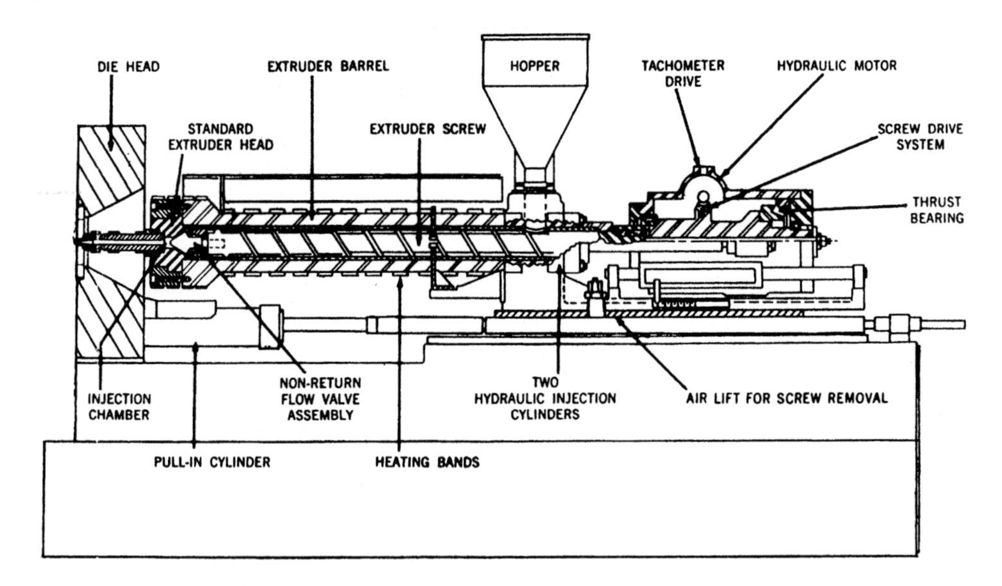

Fig. 4.13 *In-line reciprocating screw unit with hydraulic drive schematic*

(moves) back to allow melted plastic to accumulate ahead of it, then moves forward injecting the melt into the mold, all in a single stage. The accumulation of melt at the screw tip forces the screw toward the rear of the machine until enough melt is collected for a shot. The required low back pressure on the screw is used during this plasticating action and when the shot size is produced, the screw usually stops rotating.

With the mold halves closed, the nonrotating screw acts as a plunger and "rams" the melt into the cavity(ies) using controlled injection pressure and rate of travel. After injection of the melt is complete and after gate freeze, the screw rotates to prepare the next shot. The advantages of the reciprocating screw IMM compared to the two-stage IMM include the following: (1) reduced residence time, (2) self-cleaning screw action, and (3) responsive injection control. These advantages are key to processing heat-sensitive plastics.

Figure 4.14 provides a simplified reciprocating screw sequence of operations. (a) The shot (melted plastic) is in the front of the retracted screw that is being used as a ram to force the shot into the mold cavity(ies). (b) After the shot has completely filled the cavity(ies) and the plastic melt in the mold gate(s) is sufficiently solidified (frozen) so melt will not travel back into the plasticator, the screw starts rotating and retracts (reciprocates) to prepare the next shot. (c) The IMM processing cycle may have a soak or idle time period prior to the shot being forced into the mold cavity(ies). One complete cycle of the IMM operation is shown in Fig. 4.14

In the single-stage IMM melt is fed into a shot chamber (front of screw). This motion generates controllable low back pressure [usually 50 to 300 psi (0.34 to 2.07 MPa)] that causes the screw to retract at a pressure-controlled rate. When a preset device (such as a screw position transducer) is activated, the shot size is met and the screw is set to stop rotating. When an IMM does not have sufficient shot capacity, the screw is set to continue rotating, permitting additional melt to enter the cavity(ies) prior to having the shot enter. With the melt characteristics of certain plastics, however, melt flow problems could develop.

At a preset time, the screw acts as a ram to push the melt into the mold. Depending on the plastic's melt flow characteristics, injection pressure at the nozzle ranges from 2,000 to 45,000 psi (14 to 310 MPa). The required pressure is basically determined by the plastic being processed and melt pressure required in the cavity(ies) taking into account pressure dropoffs as the melt travels through the mold. Molds are designed to meet different requirements. With the shot being injected into the mold an adequate

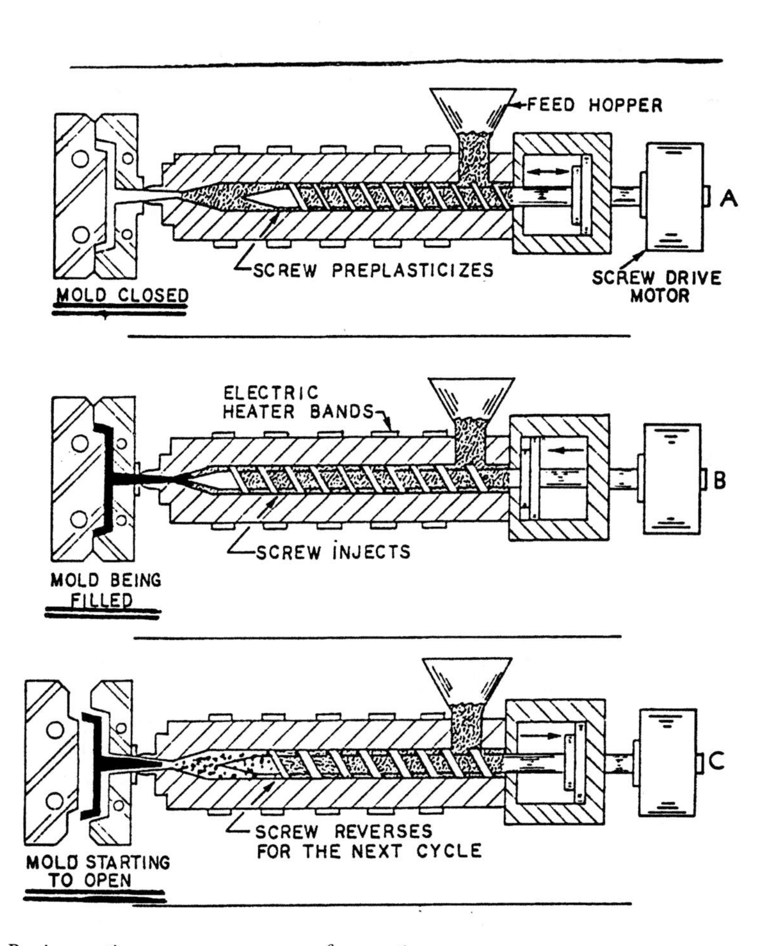

Fig. 4.14 *Reciprocating screw sequence of operations*

clamping pressure must be used to eliminate the mold from opening (flashing) during and after the filling of the cavity(ies).

4.3.11 Two-Stage Machine

Another very popular IM method is the arrangement of the two-stage screws. This machine is also called a preplasticizing IMM or piggy-back IMM. The two-stage IMM uses a fixed plasticating screw (first stage) to feed the required melted plastic through a valve mechanism into a chamber or accumulator (second stage). This screw does not require a reciprocating action (as in a single-stage IMM), as it only conveys melts by means of some type of diverter mechanism/valve into a "holding" or injection accumulator cylinder (Figs. 4.15 and 4.16). It basically operates as a continuous extruder.

When a sufficient quantity of melt has been transferred, the diverter valve again shifts to create a flow path at a prescribed time cycle from the accumulator cylinder into the mold. The second stage (ram injection stage) provides the pressure and rate of injection of the melt (shot size) into the mold cavity(ies). After injection is completed, the diverter valve shifts to direct the melt flow path from its first-stage into the second-stage holding cylinder and this operating cycle is continually repeated. During this process the first-stage extruder is continuously rotating without causing problems to the melt when it may be slightly restricted (building up pressure) and not flowing into the second stage.

Thus the diverter or shuttle valve has three positions. One position is in the closed mode during which time the "extruder" is only preparing the melt. The next position directs the melt from the extruder into an accumulator (second stage). Its third position directs a shot of melt from the accumulator into the mold cavity.

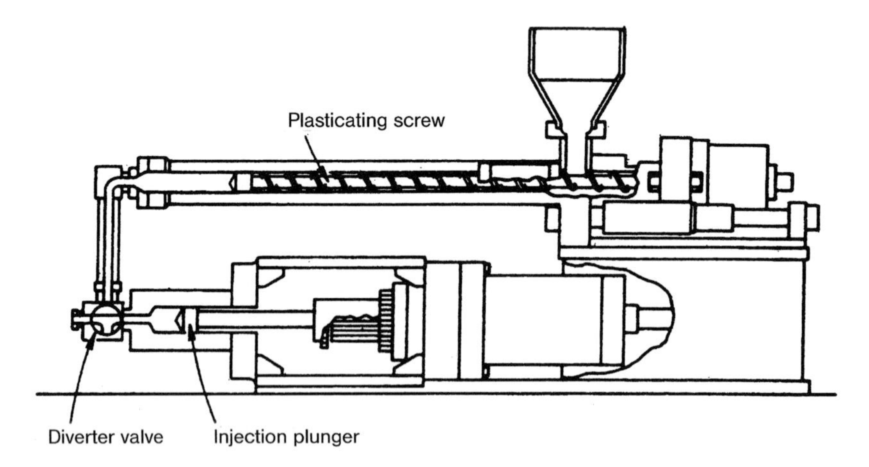

Fig. 4.15 *Schematic of a two-stage screw injection molding machine with parallel layup*

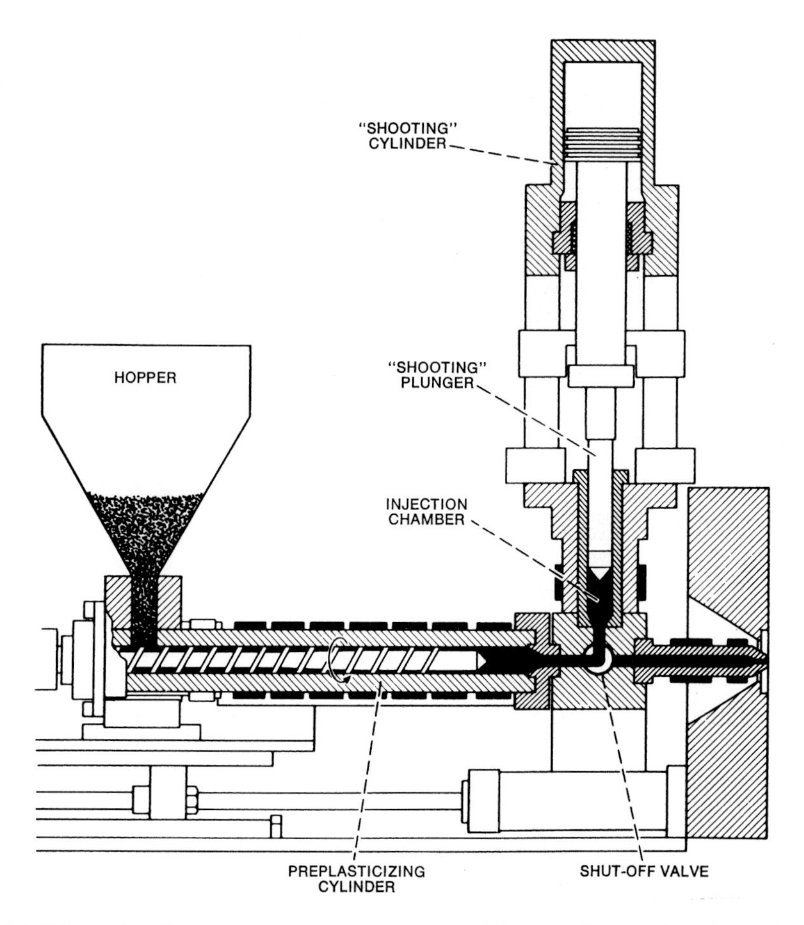

Fig. 4.16 *Schematic of a two-stage screw injection molding machine with right angle design*

When compared to the reciprocating screw IMM, advantages of a two-stage machine include: (1) consistent melt quality; (2) ram action in the accumulator provides high injection pressure very fast; (3) very accurate shot size control; (4) product clarity; and (5) easily molding very thin-walled parts. Disadvantages include higher equipment cost and potential increased maintenance.

4.3.12 Injection Hydraulic Accumulator

The injection hydraulic accumulator is a system for increasing the speed of the melt injected into a mold using a conventional IMM (Fig. 4.17). The hydraulic accumulator is a cylindrical pressure vessel that is precharged (filled), for example, with an inexpensive inactive gas (usually nitrogen) to a predetermined pressure level. Hydraulic fluid is pumped into the accumulator opposite the contained gas with an

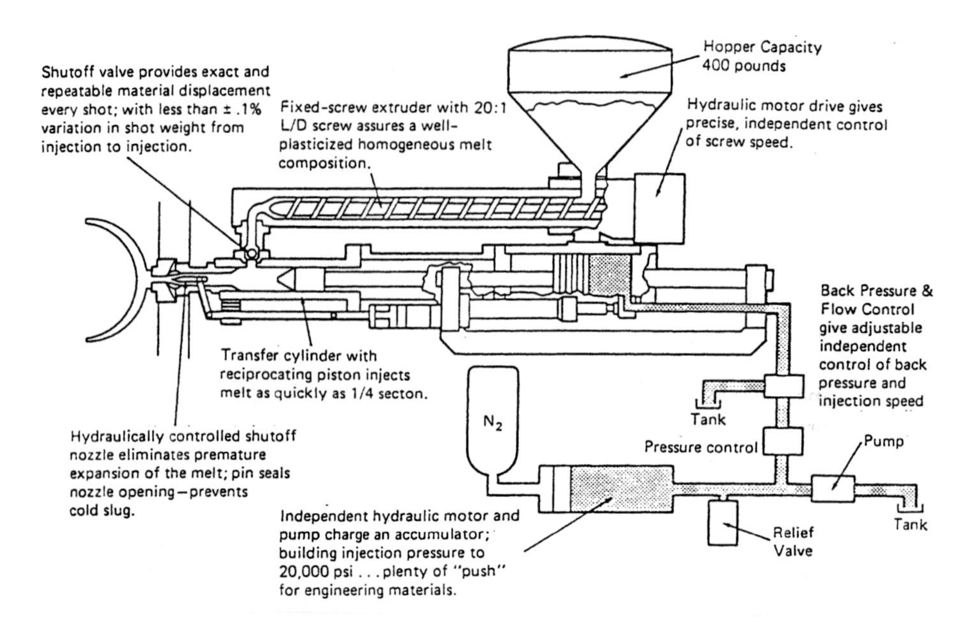

Fig. 4.17 *Example of a two-stage unit with a fast second-stage injection pressure system*

internal floating piston serving as a gas–oil separator. When the IMM signals to inject, the fluid in the accumulator is directed by controlled valving into the injection cylinder.

During this action, which occurs within seconds, the "extruder" (first stage) continues to operate, producing melt. When it is not directing melt into the accumulator, its melt remains in the barrel that could be building up slight pressure for a short time period. The extruder is designed so that the screw can have some motion in moving back for melt to accumulate in the front of its barrel without any major buildup of pressure. Designs are used so that controls can be set whereby no damage occurs to the melt.

4.3.13 Reciprocating vs. Two-Stage Machine

Both type of machines are operated hydraulically, electrically, or in hybrid combinations. The reciprocating screw design, which has many advantages in a hydraulic power environment, to date has limited the potential of an all-electric machine by requiring large and cost-impractical electromechanical drives as shot weights exceed 80 oz.

An example of a recent development with an electrical operating machine for the larger shot sizes is shown in Fig. 4.18. This Milacron machine has patented features for a two-stage design that allow first-in/first-out melt handling, quick and easy color change, and precision minishot control down to 2%–3% of barrel capacity. It also provides melt quality, compounding, and venting advantages unique to those of freestanding extruders [3]. Simultaneously it satisfies the need for high throughput and pressure on an all-electric IMM. It eliminates all the guesswork about sizing an injection unit. Introduced in 150, 110, and 80 oz (4,250, 3,100, and 2,300 g) capacities, they bring economical large-shot up to 30,000 psi (210 MPa or 2000 bar) pressure capability to higher tonnage all-electric IMMs. Figure 4.18 is Milacron's Powerline 935 ton that includes features of a very quiet machine with a noise level of just 68 dB, dry cycle time of 2.9 s, shot size up to 150 oz, injection pressure up to 30,000 psi (207 MPa), and tie bar spacing 1,250 mm wide × 1,000 mm tall.

Simple physics gives the two-stage electric unit gains. It takes less power to generate injection pressure with a small diameter screw, while lengthening the stroke to obtain the required melt volume. There are certain limits on the allowable parameter changes with a reciprocating unit because the plasticizing and injection functions are interdependent while the two-stage completely separates these functions.

Hydraulic IMMs share a drive for the screw motor and injection unit, using the extrusion screw as an injection plunger to lower machine cost. This capability of the reciprocating screw unit significantly expanded its use over two-stage units since the 1960s. However, as injection volumes increased, the diameter of its screw became larger quickly because there is an inherent limit on how far you can move the screw to obtain additional volume. These larger diameter screws are no problem to push with a

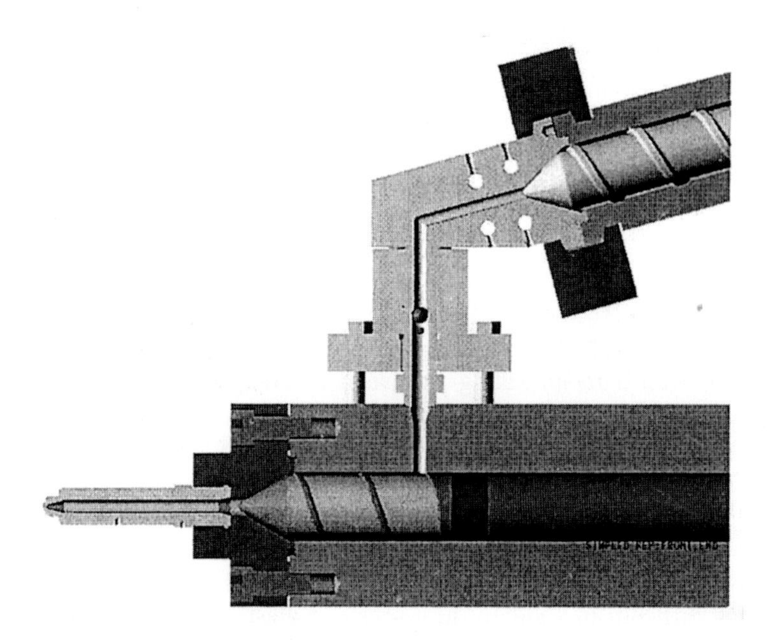

__Fig. 4.18__ Milacron's closeup of their two-stage machine with a special plasticating screw tip that allows first-in/first-out melt permitting longer stroke with smaller diameter second-stage barrel

hydraulic system, but to date it became cost prohibitive with the original electro-mechanical drive designs.

Other tradeoffs with the reciprocating design include the fact that increasing screw diameter to add volume results in limits at the low end of the shot range. Because the stroke gets so short, it is difficult to have precise melt control. The reciprocating injection unit is usually oversized for the actual molding requirement because the effective diameter for plasticating loss occurs with the screw stroke. As an example, a standard has evolved over time for sizing molding operations to 30% to 70% of a reciprocating injection unit's capacity. This sizing keeps the IMM in the best operating range for the larger shot, which is typically 300 oz (8,500 g), to do what a 150 oz (4,250 g) two-stage unit processes.

The two-stage unit design is basically a complete departure from the past reciprocating unit design. It frees the design of the injection function from dependence on the plasticizing function because it uses an independent "shooting" chamber. This permits use of a smaller diameter injection barrel and longer injection stroke for a given volume. The result is to make it easier to generate high injection rates, pressures, and volumes with smaller, precise, and proven electromechanical drives. In the newer two-stage injection unit the shot can reach its full volume, unlike the reciprocating units which are usually sized twice as large. The two-stage unit evolved as a practical, effective way to dramatically extend the performance range of their electric IMMs, while meeting cost targets.

Milacron used a derivation of its single-screw extruder to melt plastic and meter it into the injection (second-stage) barrel through a port in front. With extrusion as a separate function, plasticizing rates are sized to exact requirements. Injection control is much more precise when compared to the nonreturn valve in line with the injection screw plunger that has to seat before control of the shot occurs. Its longer stroke of a smaller diameter injection piston is what enables, as an example, the 150 oz (4,250 g) two-stage shoot down to 4 oz (110 g). This action is far smaller than would be possible with a 300 oz (8,500 g) reciprocating screw. The generally accepted practice in the industry is to avoid shooting less than 10 vol% of shot capacity for a reciprocating unit or the screw stroke becomes so short that it is difficult to control.

The separate "extruder" allows molders to take actions that would be more difficult or impossible with a reciprocating unit such as compounding glass fiber in-line, changing screw *L/D*, putting additives in the melt phase, and venting. The melt from the extruder is also more consistent and higher in quality because each pellet passes down the entire length of the screw, unlike a reciprocating screw where some of the feed end of the screw may be behind the hopper.

While the two-stage has its advantages, it created challenges to the machine designers. Most important is its first-in/first-out handling of the melt that made color change difficult. In addition, heat-sensitive plastics could stick to the plunger tip. The electrically driven ballscrew behind the injection piston (Milacron patent pending tip design) allows a new way to handle melt as it enters the shot chamber, overcoming these challenges. A screw-type tip is used on the injection piston. A

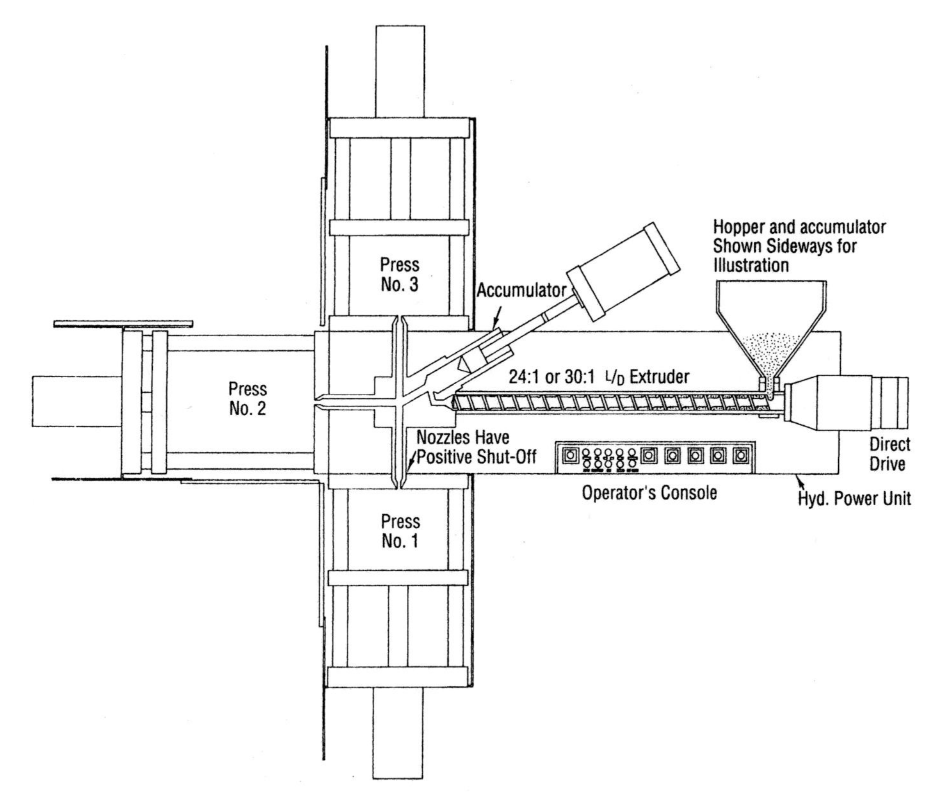

Fig. 4.19 *Two-stage IMM with three clamping presses*

one-way clutch rotates the tip while building a shot and retracting the piston, pushing melt forward over the tip (Fig. 4.18). Its first-stage "extruder" does not move melt directly into the front of the shot chamber at the piston tip. It travels through the screw thread to maximize the mixing and forward flow. Depending on the shot size, this tip gives first-in/first-out, middle-in/last-out, or last-in/middle-out. Even when the piston tip is back past the melt entry port, the rotation of the tip continues to wipe the plunger tip against the melt pool. With each shot, the tip starts by pushing new melt forward over the front again, maintaining its cleaning action. Other types of IM machines include machines with plasticators in different positions in addition to those reviewed. These other variations include multiple clamping units (Fig. 4.19), clamping for different mold motions (shuttle, rotary, ferriswheel, etc.), ram/plunger plasticator with one or more rams instead of screw systems.

5 Stretch Blow Molding

5.1 The Basic Process

Molten plastic that is not stretched is similar to a pile of spaghetti; the molecule strands are placed randomly with large and small gaps between them. When a bottle is blown, the stretching action automatically orients the molecules in a horizontal or uni-axial direction. Performance properties of the material are improved in the direction of orientation. Thus, to capitalize on this behavior, the container industry started to look to devise a mechanical system to stretch the parison vertically while it is extrusion or injection blown. The result was a process called Stretch Blow Molding (SBM), also termed biaxial orientation BM.

In the SBM process, the heated parison from the extrusion blow molding (stretched EBM) or the heated preform (stretched IBM) from the injection blow molding operation is inflated. The parison or preform can be stretched lengthwise by an external gripper, or by an internal stretch rod, and then stretched radially by blown air to form the finished container against the mold walls (Fig. 5.1).

The result is that the molecules are aligned along two planes, an arrangement that adds strength to the finished container. Strength is not the only advantage of SBM. Better clarity, increased impact strength, improved gas and water vapor barriers, and reduced creep are also major advantages. The process also allows wall thicknesses to be more accurately controlled and also allows weights to be reduced, resulting in lowered material costs.

SBM processes can be classified as one-stage or two-stage processes. In the one-stage (single step) process, the parison is produced at the required temperature profile. A rod or pulling device is introduced and stretches the preform while it is blowing. The SIBM uses two sets of molds in one machine.

In a two-stage machine, one machine extrudes a parison or injection-molds a preform. This parison or preform is then placed into a second machine that is commonly

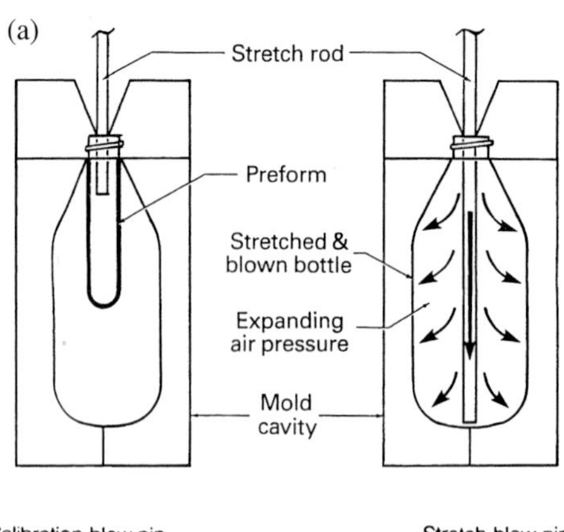

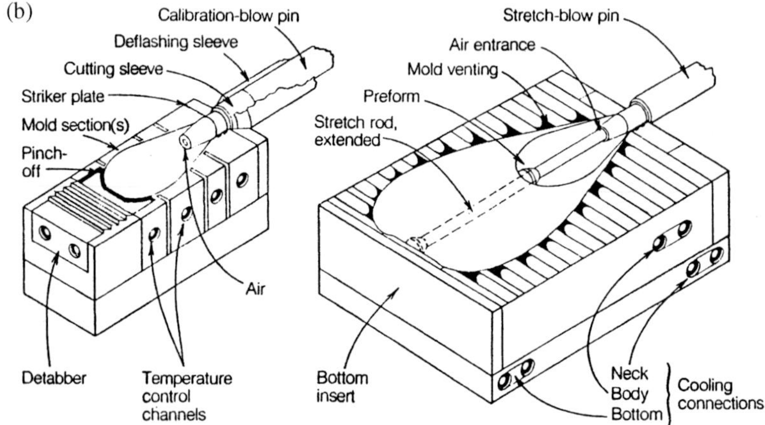

Fig. 5.1 *(a) Example of a temperature conditioned injection preform being rapidly stretched using a rod and air pressure. (b) Example of a temperature conditioned preform inserted in a SIBM cavity followed with stretching: (1) left view only air pressure being applied to obtain stretched bidirectional properties, (2) right view shows using a rod to stretch the preform axially with air pressure to form in the radial direction*

known as a reheat and blow machine. In this machine the parison or preform is brought up to an orienting temperature; the actual profile temperature depends on the type of plastic being processed. It can then be physically stretched vertically and circumferentially while it is being blown.

Proper temperature control of the parison or preform is required to develop the proper orientation. The preform is not heated to a temperature above the plastic melting point but only to a temperature sufficient to produce a rubbery consistency (Table 5.1). Examples of approximate temperatures at which materials yield maximum properties are PET at 107 °C (225 °F), PP at 160 °C (320 °F), PS at 125 °C (257 °F), AN at 120 °C (248 °F), acetal at 160 °C (320 °F), and PVC at 100 °C (212 °F). Details on the behavior of plastic materials during temperature-controlled molecular orientation are provided in Chapter 7.

Table 5.1 Example of Data on Stretched Blow Molded Plastics

Polymer	Melt temperature, °C	Stretch orientation temperature, °C	Maximum stretch ratio
PET	250	88/116	16:1
PVC	200	99/116	7:1
PP	170	121/136	6:1
PAN	210	104/127	9:1

Proper selection of the time, which governs the inflation temperature, is of critical importance. The essential feature of the process is the generation of strong molecular orientation during the inflation step. In order to accomplish this, it is necessary to ensure that the stretching deformation is taken up by the plastic as elastic strain rather than as viscous work. This usually occurs in a fairly narrow range of temperatures between 88° and 116 °C (190° to 104 °F) for PET, for example.

Changing the formulation of the plastic can alter the rubbery state temperature range. In the case of PP, a comonomer can be used, and in the case of PET, diethylene glycol can be added. Absorbed water can also act as a plasticizer, reducing the optimum temperature for orientation. However, such changes will also affect other characteristics of the plastic and can sometimes result in a deterioration of mechanical properties. Table 5.2 provides examples of the volume shrinkage after SBM.

Table 5.2 Examples of How Volume Shrinkage of SBM Is Influenced by Type of Plastic

Type of bottle	Percent
Extrusion blow molded PVC	—
Impact-modified PVC (high orientation)	4.2
Impact-modified PVC (medium orientation)	2.4
Impact-modified PVC (low orientation)	1.6
Non-impact-modified PVC (high orientation)	1.9
Non-impact-modified PVC (medium orientation)	1.2
Non-impact-modified PVC (low orientation)	0.9
PET	1.2

The use of SBM is possible for thermoplastic materials. The amorphous materials with a wide range of thermoplasticity are easier to stretch than the partially crystalline types. Certain plastics such as polyethylene (PE), polypropylene (PP), polyethylene terephthalate (PET), and polyvinyl chloride (PVC) have been receiving the most attention with respect to SBM, because they are easily stretched. PET has experienced a tremendous growth in SIBM of carbonated beverage bottles, toiletry, cosmetic, detergent, and other container markets. Several years ago, PET completely replaced the 0.5 gallon (2-liter) bulky glass beverage bottles.

PET is nearly Newtonian (Chapter 7) above its melting point and is therefore considered unsuitable for extrusion blow molding. However, it has been suggested that the use of a multifunctional comonomer, which produces long chain branching and broadens the molecular weight distribution, can render PET suitable for extrusion blow molding.

Several different processing techniques were placed in production in the early and mid 1970s. PET became FDA-approved immediately as a very clean plastic for food products, whereas PVC and acrylonitrile were reported to have some carcinogenic components. As a result, acrylonitrile was banned for the use in the beverage industry in the mid 1970s. Later this ruling was overturned, and acrylonitrile was approved for the beverage and other markets. PVC entered the liquor market in the early and mid 1970s, but because of the vinyl monomer scare at that time it was banned for liquor in USA (Europe used PVC liquor bottles); however, PVC is used in other food container applications.

PP, because of its low cost, was investigated quite intensively for biaxial orientation EBM procedures. It has a processing window that allows for orientation. When compared to the other plastics that are used in SBM, it has a relatively long conditioning time until it can be blown into a container. With the crystalline type (such as PP), if crystallization is too rapid, the bottle is virtually destroyed during the stretching process.

5.1.1 Blowing Stages

Stretch blow processing can be separated into two categories: in-line and two-stage. In-line processing (or one-stage) is done on a single machine, while two-stage processing requires either an extrusion or injection line to produce parisons or preforms, with a reheat in a blowing machine to make the finished container.

In the in-line process an extruded or injection molded parison or preform passes through conditioning stations that bring it to the proper orientation temperature. A rather tight temperature profile is held in the axial direction of the preform that also controls temperature thicknesswise (Fig. 5.2). Advantages of in-line systems are that heat history is minimized (crucial for temperature-sensitive materials) and the parison or preform can be programmed for optimum material distribution.

With the two-stage process (Fig. 5.3), extruded or injection molded parisons or preforms that have been cooled are conveyed through an oven that reheats them to the proper orientation blow temperature. The advantages of this process are higher output rates of the blown containers and the capability to stockpile parisons or preforms until containers are required (saving storage space and cost to store).

In-line injection stretch blow systems, while offering more flexibility from a material standpoint, do not give the degree of parison programming available in extrusion systems. An extruded parison can be heat stabilized, and then the parison is held externally and pulled to give axial orientation while the bottle is blown radially.

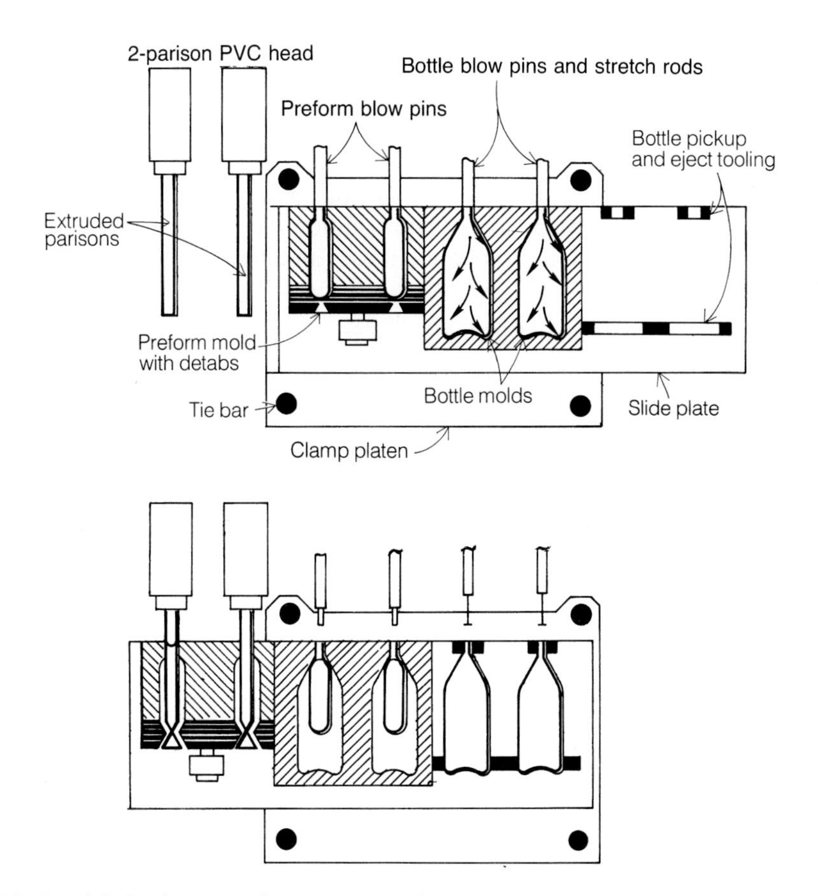

Fig. 5.2 *Simplified schematic of a one-step (in-line) stretched IBM machine*

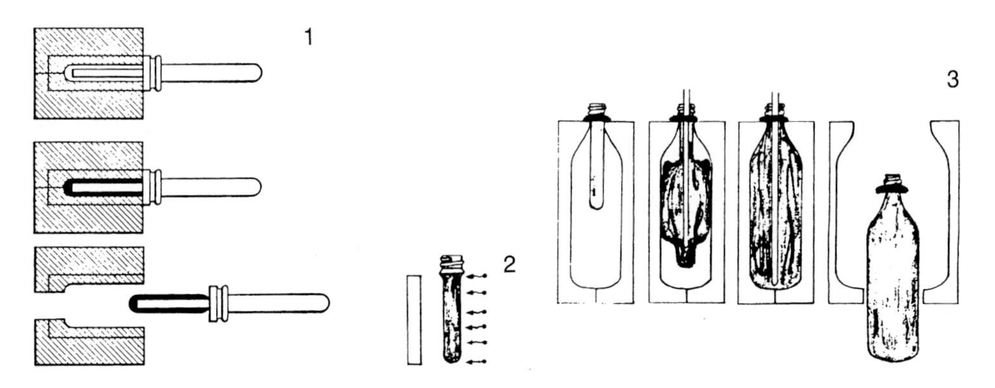

Fig. 5.3 *Basic stretched IBM two-step process: (1) from injecting the preform to transferring action, (2) reheat or apply uniform heat, (3) stretch BM*

Another technique stretches and preblows the parison before the blowing operation is completed. In both these processes, scrap material is produced at the bottleneck and at its base. The process orients the entire bottle, including the threaded area. The threads are then post-mold finished.

Extrusion processes also use programmed parisons and two sets of molds. The first mold is used to blow to temperatures that condition the preform which cool to the orientation temperature, and the second mold uses an internal rod to stretch and blow the bottle.

When the two-stage extrusion SBM process is used, the preforms used are open ended. They are reheated, then stretched by pulling one end while the other end is clamped in the blow mold. Minimum scrap is produced, and the compression molded threads provide a good neck finish. This system utilizes a conventional extrusion line and a reheat-blow machine whose oven contains quartz lamps to provide the proper temperature profile in order to obtain the proper container wall distribution.

The designed two-stage machines have higher output rates than single-stage, while keeping scrap at a minimum. Preforms are injection molded on multicavity molds including the thread finish, then cooled to ambient temperatures before going to the reheat BM machine. Preforms are heated in a reflective radiant heat oven with designs that shield the preform thread areas.

After the oven heat exposure, preforms pass through an equilibration zone to allow the heat to disperse evenly through the preform wall. At the blow station, a center rod is inserted into the preform, stretching it axially, and to properly center the preform in the bottle mold. High-pressure air is then introduced to blow the preform to shape against the mold walls and give the bottle its radial orientation.

In 2000, Uniloy Milacron provided a new two-step, electric PET blow molding system. The machine produced half-gallon milk bottles at a rate of approx. 4000/h, and single-serve 16-oz bottles at a rate of approx. 7800/h. A 48-cavity preform mold produced 24-g, 16-oz bottles with a cycle time of approx. 15-s. The PET two-stage injection unit used a 120-mm screw extruder capable of processing more than 1000 lb/h, feeding a shooting pot with 110-oz capacity. To maximize throughput, the injection unit featured a separate “packing pot” to address the longer pack time required for PET. The packing pot provides pack pressure, freeing up the shooting pot for immediate refill and reducing cycle time.

Figure 5.4 shows the Orbet Process (Phillips Chemical Co.), which is a two-step process designed to gain the advantages of material orientation in the blown container. Continuous extrusion of tubing for the parison is extruded and cut to length separately. These cold parison tubes are loaded on spindles that travel through a carefully heat controlled oven. After passing through a programmed heating procedure, the hot plastic tube-parisons are picked up by a clamp that transports them to the blow spindle where the end is clamped in the neck ring. At this point, the picker rises to stretch the parison; following the stretch that provides axial orientation, the mold halves clamp together and the “blow” expands the form into the finished

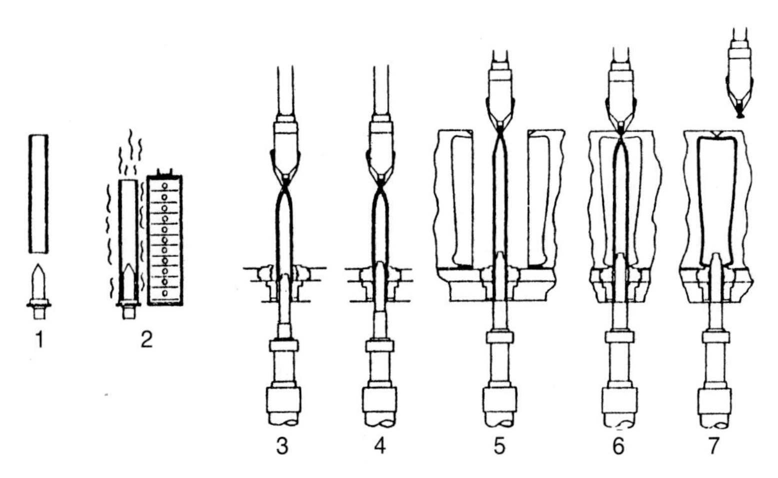

Fig. 5.4 Detailed schematic of the Orbet stretched blowing process by Phillips Chemical Co.

product with orientation perpendicular to the axis. The top and bottom of these containers are not biaxially oriented, so these parts do not have the same physical properties as the sidewalls. This process is particularly suitable for orienting polypropylene (OPP) and the bottle produced has particularly desirable properties.

Based on the original work conducted in SBM, 2-liter containers of PVC, PET, and AN with average wall thicknesses of 0.012 to 0.015 in. (0.3 to 0.4 mm) have been used. Normal BM would use an average wall thickness of 0.018 to 0.022 in. (0.46 to 0.56 mm) with no less than 0.009 to 0.010 in. (0.23 to 0.25 mm) in the thinnest point. The latter is used in either stretch blown or standard blow molding.

The relationship of wall thickness to degree of orientation is important. Control of the bottle's wall thickness can be as important, if not more, than the amount of biaxial orientation. Soft drink bottles are pressurized to about 60 psi (0.4 MPa) so 0.009- to 0.015-in. (0.23- to 0.38-mm) side walls cannot collapse due to internal pressure. However, a product such as mouthwash does not have internal pressure, and thin 0.009- to 0.015-in. walls would experience paneling or wall collapse due to permeation losses.

5.2 Advantages of Stretch Blow Molding

SBM for SEBM and SIBM is the process of plastic orientation that consists of a controlled system for stretching TP molecules in unioriented (unidirectional) or bioriented (biaxial direction) to improve their strength, stiffness, optics, electrical, and/or other properties with the usual result that improved product performance-to-cost occurs. Depending on the properties of a specific plastic, the stretch ratio may vary from $2\frac{1}{2}:1$ to at least as high as $10:1$.

Practically all TPs can undergo orientation, although it is particularly advantageous in certain types such as PET, PP, PVC, PE, PS, PVDC, PVA, and PC. Mechanical properties depend directly on the relationship between the axis of orientation of the plastic molecules and the axis of mechanical stress on the molecules. Modulus, strength, and so forth increase in the direction of stretch and decrease in the perpendicular direction. This is the mechanism of pseudoplastic and thixotropic rheology typical of a non-Newtonian plastic flow behavior [9]. After processing, some loss in properties may occur when the plastic is subjected to heat during further processing, such as thermoforming, heat sealing, and solvent sealing.

Biaxial orientation of crystalline plastics generally improves clarity of plastics. This occurs because stretching breaks up large crystalline structures into ones smaller than the wavelength of visible light. With uniaxial orientation, the result is an anisotropic refractive index and thus birefringence, especially in crystalline plastics.

Orientation decreases electrical dissipation factors in the direction of the orientation, and increases occur perpendicular to orientation. Since modulus changes in the opposite way, this indicates that polar vibrations along a stretched plastic molecule are decreased, while transverse vibrations between the stretched molecules are increased.

5.3 Orientation Behavior

The technique of stretching (orientation) is used during the processing of many different products. As reviewed in this chapter, an important and major product is thermoplastic (TP) stretched blow molded bottles, containers, and drums.

The molecular orientation of a TP can be accidental or deliberate. An accident can occur during the processing of TPs where excessive frozen-in stresses develop; however, with the usual proper process control, there is no accidental orientation. The frozen-in stresses with certain TPs can be extremely damaging if products are subjected to environmental stress cracking or crazing in the presence of heat, chemicals, and so forth. Initially during the melting stage the molecules are relaxed; molecules in the amorphous regions are in random coils; those in crystalline regions are relatively straight and folded (Chapter 7).

During processing the molecules tend to be more oriented than relaxed, particularly when the melt is subjected to excessive shearing action. After temperature–time–pressure is applied and the melt goes through restrictions (mold, die, etc.), the molecules tend to be stretched and aligned in a parallel form. The result can be undesirable changes in the directional properties and dimensions immediately when processed and/or thereafter when in use if stress relaxation occurs.

The amount of change depends on the type of TP, the amount of restriction, and, most important, its rate of cooling. The faster the rate, the more retention there is of the frozen orientation. After processing, products could be subject to stress relaxation, with changes in performance and dimensions. With certain plastics and processes there is an insignificant change. If changes are significant and undesirable, one must take action to change the processing conditions during and/or after processing. As an example, during processing increase the cooling rate. Annealing of the product is a post-processing condition.

By deliberate stretching, the molecular chains of a plastic are drawn in the direction of the stretching, and inherent strengths of the chains are better realized than they are in their naturally relaxed configurations. Stretching can take place with heat during or after processing by BM, extruding film, thermoforming, and so forth. Products can be drawn in one direction (uniaxially) or in two opposite directions (biaxially), in which case many properties significantly increase uniaxially or biaxially.

Molecular orientation results in increased stiffness, strength, and toughness, as well as resistance to liquid and gas permeation, crazing, microcrack, and others in the direction or plane of the orientation. The orientation of fibers in reinforced plastics causes similar positive effects [9, 10]. Orientation essentially provides a means of tailoring and improving the properties of plastics.

As an example consider a fiber or thread of nylon-6/6, which is an unoriented glassy polymer. Its modulus of elasticity is about 2,000 MPa (300,000 psi). Above the glass transition temperature, (T_g) its elastic modulus drops even lower, because small stresses will readily straighten the kinked molecular chains. However, once it is extended and has its molecules oriented in the direction of the stress, larger stresses are required to produce added strain. The elastic modulus increases. The next step is to cool the nylon below its T_g without removing the stress, retaining its molecular orientation. The nylon becomes rigid with a much higher elastic modulus in the tension direction [15,000 to 20,000 MPa (2 to 3×10^6 psi)]. This is nearly 10 times the elastic modulus of the unoriented nylon-66 plastic. The stress for any elastic extension must work against the rigid backbone of the nylon molecule and not simply unkink molecules. This procedure has been commonly used in the commercial production of manmade fibers since the 1930s via the development by DuPont.

Some general rules for orienting by stretching are: (1) the lowest temperature will give the greatest orientation (tensile strength, etc.); (2) the highest rate of stretching will give the greatest orientation at a given temperature and percent stretch; (3) the highest percent stretch will give the greatest orientation at a given temperature and rate of stretching; and (4) the greatest quench rate will preserve the most orientation under any stretching condition. With orientation, the thickness is reduced and the surface enlarged. If film is longitudinally stretched, its thickness and width are reduced in the same ratio. If lateral/transverse contraction is prevented, stretching reduces the thickness only.

5.4 Orientation and Crystallinity

The temperature is normally 60% to 75% of the range between the plastic's glass transition temperature (T_g) and melting point (T_m) (Table 5.3). Equipment can be used in-line or off-line to gain output yield and properties. Generally the orientation temperature is 60% to 75% between the plastics glass transition temperature and melting point (Chapter 7).

Table 5.3 *Examples of SBM Processing Characteristics*

	PVC	PET	AN
Melting, °F	400–500	475–510	475–525
Glass-transition, °F	170–180	150–180	220–230
Orientation, °F	175–225	180–210	260–290
Specific gravity	1.4	1.4	1.1

All plastics have some amount of elasticity associated with them. As long as the plastic stretches within its elastic limit, it will eventually return to its natural shape. If the plastic is overstressed, it reaches what is known as plastic deformation, meaning the plastic will not return to its original shape. It is important to keep the drag force from feed mechanism friction as low as possible and to keep acceleration of the feed material low enough so the product does not permanently deform or tear.

Sinusoidal acceleration profiles reduce the load on the plastic by providing gentle forces near zero speed and top speed while making up for lost ground in the middle, with the result of diminishing stretching. Stretching also contributes to reduced accuracy. This situation can be overcome by either compensating for the stretch in the controller when the amount of stretch is predictable or by installing a position verification device between the feed mechanism and the post-feed process.

The orientation processing hardware is fairly expensive and increases the cost per unit weight of the product. However, the yield increases considerably and its quality improves greatly, resulting in cost savings based on the product weight. Many products are made much stronger, flexible, and tougher, resulting in significant overall cost advantages. The maximum ultimate economic value depends on the relative cost of stretching vs. the increase in yield and properties.

5.5 Coinjection SBM

Nissei ASB Company has used coinjection SBM to improve specific barrier characteristics inherent in the barrier plastics. (Nissei had successfully tested containers using both ethylene vinyl alcohol (EVOH) and nylon as a barrier layer.) The Nissei container had three layers only (the Omni Can had five layers). Nissei was able to dispense with the adhesive layer so long as the barrier is fully encapsulated and the separate layers are in intimate contact throughout the container. This obviously held considerable attraction in the United States for products such as soft drink containers that must be recycled.

Unlike the vast majority of barrier containers, American Can Co.'s (ACC's) Omni can was an example of a manufacturing process developed by the firm. ACC took the coinjection process, normally used in large, single-cavity structural foam parts (such as business machine housings), and adapted it to multicavity production of five-layer containers at high production rates. The advantage of the process was threefold: a precise and sturdy flange to apply a double-seamed end to, deep draw capabilities, and a highly productive process. It has been apparent that the machine is capable of near-thermoforming rates and has the added advantage of no inherent scrap.

ACC modified a large injection molding machine by adding three extruders and a proprietary multicavity mold system. The material from each extruder travels through its own runner system to individual multichannel nozzles for each cavity. There, a programmed injection sequence ensures that the inner barrier and tie layers are totally encapsulated by the outer structural layers, leaving no barrier-less regions at the sprue.

As Fig. 4.1 shows, a preform is injection molded and shuttled over to a BM station. A split ring locks the flange of the container to the core pin, keeping flange thickness variance down to ± 0.003 in. In the BM station, an air ring at the base of the core pin blows the container to full size.

In contrast to thermoforming, it is possible to vary the thickness of one layer while maintaining the thickness of others. This is key to the hinged bottom of the Omni can, a critical component for successful retort performance. The PP layers are thinned to allow expansion, while the EVOH layer is unchanged, preserving the container's barrier properties. This feature allows the container to balloon out quite a bit during retorting. Generally, thermoformed containers have difficulty surviving the continuous retort systems that use steam. Rather than try to get food processors to convert to water-cook retorts with overriding pressure, ACC designed a container that will withstand the steam pressure.

The first major company commercial with this technology was ACC with their coinjection BM Omni Can. ACC utilizes a die head that incorporates intermittent injection of the barrier material. This permits total encapsulation of the barrier layer at the gate and flange cut edge. In addition, ACC can program the flow to maintain consistent barrier thickness even in places where the total wall thickness is decreased.

The Omni Can incorporates two additional novel features. The first, as already noted, is the ability to use a double seamed lid that ACC attributes to the close control on flange dimensions achieved by injection molding. The second is a desiccant system that is utilized in the adhesive layer to maintain the EVOH barrier layer below the 85% relative humidity (RH) level during retort at which it would otherwise lose much of its oxygen barrier properties. The Omni Can was originally test marketed in January 1985 by Campbell and Delmonte. The present major production, however, is by Hormel with a single service meat product.

5.6 Molding a PET Bottle

The molding method basically increases the overall density (or crystallinity) of a PET bottle. It had been common knowledge that the density of a biaxially oriented, stretch blown PET bottle could be increased by secondary process treatment. One can increase density in a continuous molding by using high temperature in the blow cavity without causing deformation.

Neck deformation is a more serious problem than deformation of the bottle body. The neck deformation due to heat not only renders the bottle useless but also causes a capping problem at the same time. For example, some bottles are filled to the brim with very hot contents and capped. In such a case, the high temperature of the content softens the neck which results in disruption of a chain of functions involving capping machine pressure block, thread, and pilfer proof roller. There are several methods to make the neck heat resistant: (1) by increasing the crystallinity of the entire neck portion to make it heat resistant; (2) by insert-molding the neck portion with polycarbonate or other suitable heat resistant resins, or mold double layer heat resistant neck finish so that the PET itself becomes an integral part of the heat resistant neck in either case; (3) by molding the entire neck by the biaxial orientation stretch blow method and crystallizing it at the same time to make it heat resistant.

The first method has a shortcoming because the white crystallized neck portion becomes yellowish as time elapses, which spoils the clean bottle appearance. The whiteness of the neck portion also does a disservice to the image of the transparent PET bottle. Because the neck portion is in an amorphous state and then is crystallized, the threads shrink a great deal. This results in a capping problem. Since this is an additional process, it adversely affects economy and productivity. On the other hand, this method increases the surface hardness, which is a desirable attribute.

The second method was tried as part of the whole developmental project that enabled development of the molding method to be reviewed. It was observed that PC and PET plastics, with completely different processing properties, have an affinity for each other. PC products deform at temperatures from 130 to 140 °C (266 to 284 °F) and undergo boiling well. The melting point of PC is almost equal to that of PET. PC products are hygienic and completely transparent.

As for the neck molded by the third method, it could lack the most important precise finish. Some experiments have disclosed that even theoretically possible heat performance could not be achieved by this method because of the low hoop blow ratio.

Draw ratios that are used to achieve the best properties in a PET bottle are usually 3.8 in the hoop and 2.8 in the axial direction. This will yield a bottle with a hoop tensile of approximately 29,000 psi (200 MPa) and an axial tensile of 14,000 to 16,000 psi (96.6 to 110.3 MPa). Moisture barrier, CO_2 barrier, and drop impact are improved as a result of the process.

Useful were the accumulated experiences that included molding 20-liter PC drinking water bottles with the Nissei ASB-650 SD machine, a special double layer molding machine, as well as molding PC nursing bottles with the ASB-50 machine. The developmental project progressed smoothly thanks to the further advantage in having double layer molding machines ranging in sizes from ASB-150 D to 650 D.

A method for molding with a standard ASB machine includes the following steps: (1) Injection mold a large quantity of PC insert chips. (2) Put PC insert chips in the automatic insert machine. (3) At the fourth station of ASB machine, the lip mold opens to eject the bottle (Fig. 5.5). (4) The robot on the automatic insert machine inserts the PC chip in the lip mold (Fig. 5.6). At this time, the PC chip is positioned so that its male thread perfectly fits the female thread of the lip cavity. (5) When the completion of PC chip insertion is confirmed, the lip mold closes (Fig. 5.7). (6) The

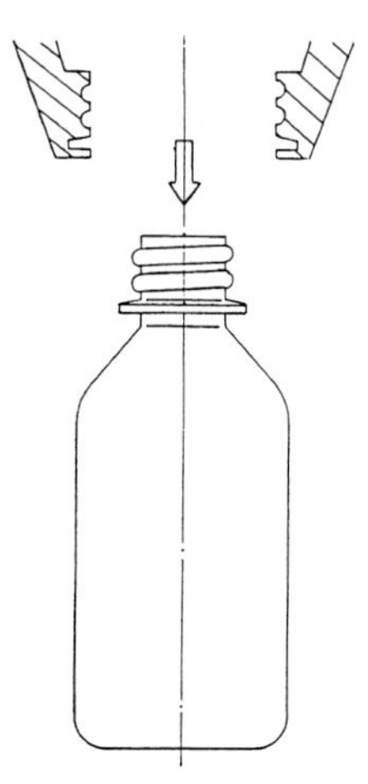

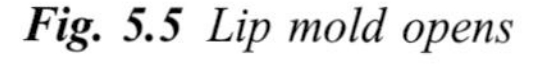

Fig. 5.5 Lip mold opens

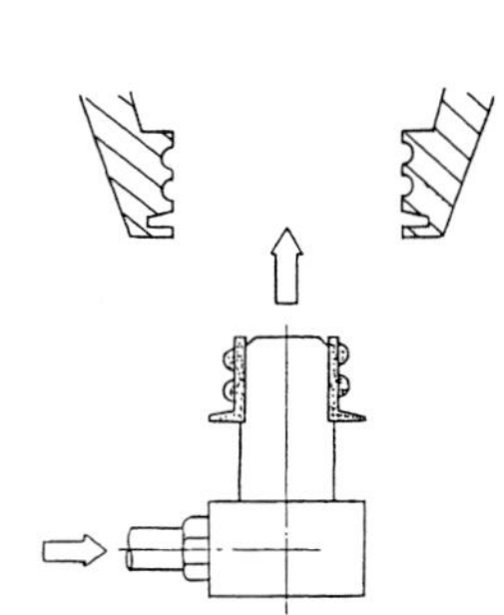

Fig. 5.6 Automatic insert

robot returns to the position where it picks up another PC insert chip and holds it in the proper position for the next insertion operation. (7) The lip mold, which is holding the PC insert chip, rotates and moves to the injection station where the mold clamping takes place. (8) Melted PET resin is injected into the space between the PC insert chip, the lip mold, and the core pin to integrally mold the neck with PET and PC (Fig. 5.8). (9) The bottle is blown with a temperature control process (Fig. 5.9).

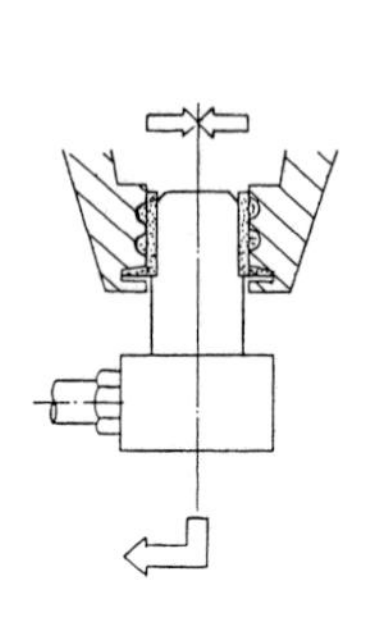

Fig. 5.7 *Lip mold closes*

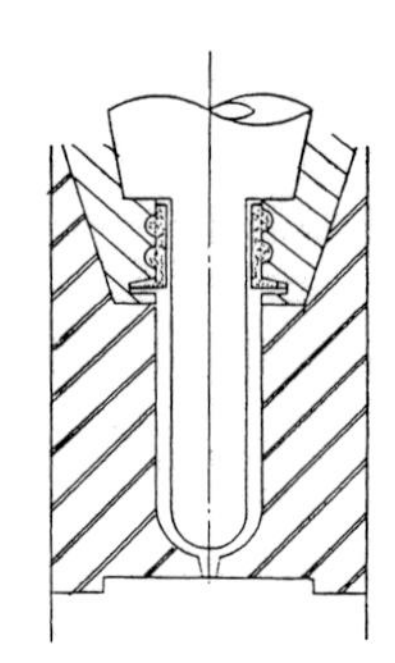

Fig. 5.8 *PET resin injected*

These steps form one continuous process, which is illustrated in Fig. 5.10. When this molding method is used, the advantage is that one can continue to use the existing machine and mold by adding only an automatic insert machine for inserting injection molded PC chips. PC chips must be kept clean because oil-stained PC chips, for example, reduce adhesion between PET and PC, causing debonding or peeling.

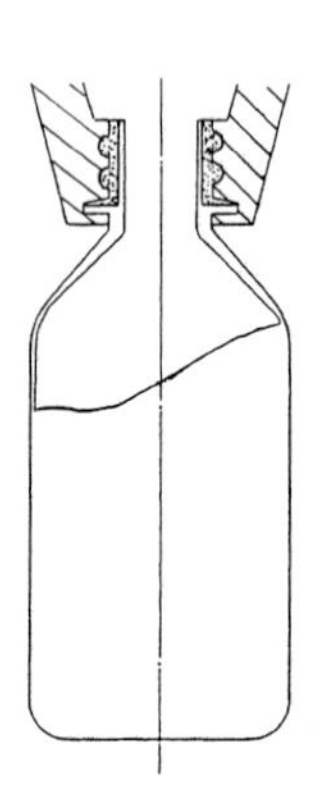

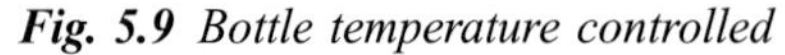

Fig. 5.9 *Bottle temperature controlled*

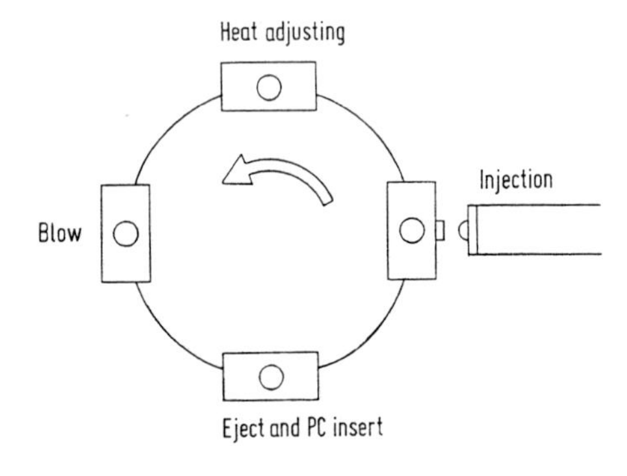

Fig. 5.10 *Diagram of the complete continuous process*

Another method does not require an insert machine. Here, the neck is molded in two layers with the PET layer inside and the PC layer outside. Figure 5.11 shows the first station. PC threads, each weighing 2 g, are molded in eight cavities with an injection unit of 2 oz shot capacity. Figure 5.12 shows the PC injection core pin and the hot runner in the opened injection mold. Only the lip mold moves in a rotary motion to the second station where mold clamping takes place with the PET injection core pin and the injection cavity. Then, melted PET is injected in to mold the preform with a PET and PC double layer neck.

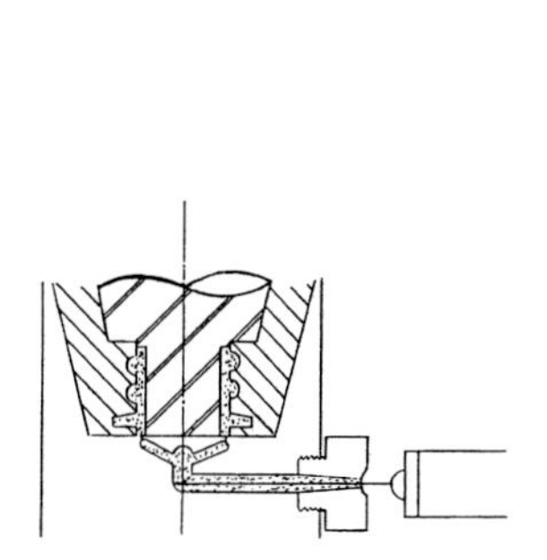

Fig. 5.11 *View of first station*

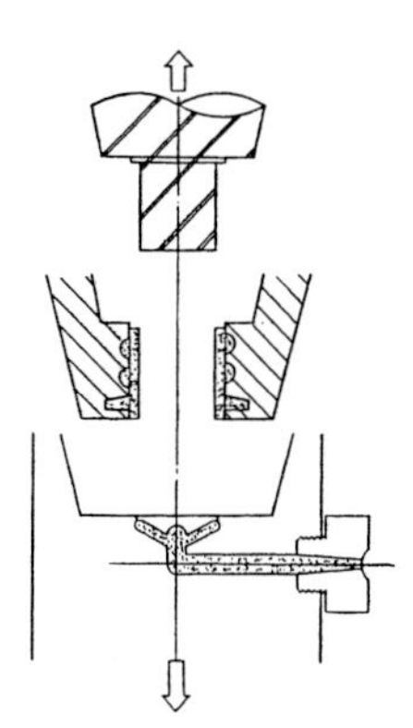

Fig. 5.12 *Diagram of first station*

After the heat adjusting process, blow and heat setting processes complete the molding of a heat-resistant PET bottle with double layer neck. The two methods described are for manufacturing heat-resistant PET bottles with the main emphasis on the neck finish.

5.6.1 Mold Temperature

Figure 5.13 shows the shrinkage rates of non-heat-resistant PET bottles molded in the blow mold at temperatures ranging from 60 to 105 °C (140 to 221 °F). The shrinkage of each bottle was measured before and after the bottle was filled to the fill point with hot water with temperatures ranging from 60 to 95 °C (140 to 203 °F). It was observed that the bottle blown at 60 °C shrank five times as much as the bottle blown at 105 °C. Bottles molded at blow mold temperatures in excess of 80 °C (176 °F) slightly deformed due to the shrinkage caused by the high heat.

This shows that the blow mold temperature is an important factor in fabricating heat-resistant PET bottles. When the heat setting process is applied to the molding of a 1-liter PET bottle, the shrinkage is much less than that of a 1-liter PET bottle that is not heat treated (Fig. 5.14).

When Figs. 5.13 and 5.14 are compared, it can be observed that the percent shrinkage of the heat set bottle before and after filling with hot water is 0.01% at 80 °C, 0.25% at 85 °C, and 0.78% at 90 °C. However, the ordinary bottle shrank as much as 4.5% at 80 °C. Although 1- and 2-liter bottles of slightly different shapes are used for illustration in Figs. 5.13 and 5.14, the importance of the blow mold temperature is

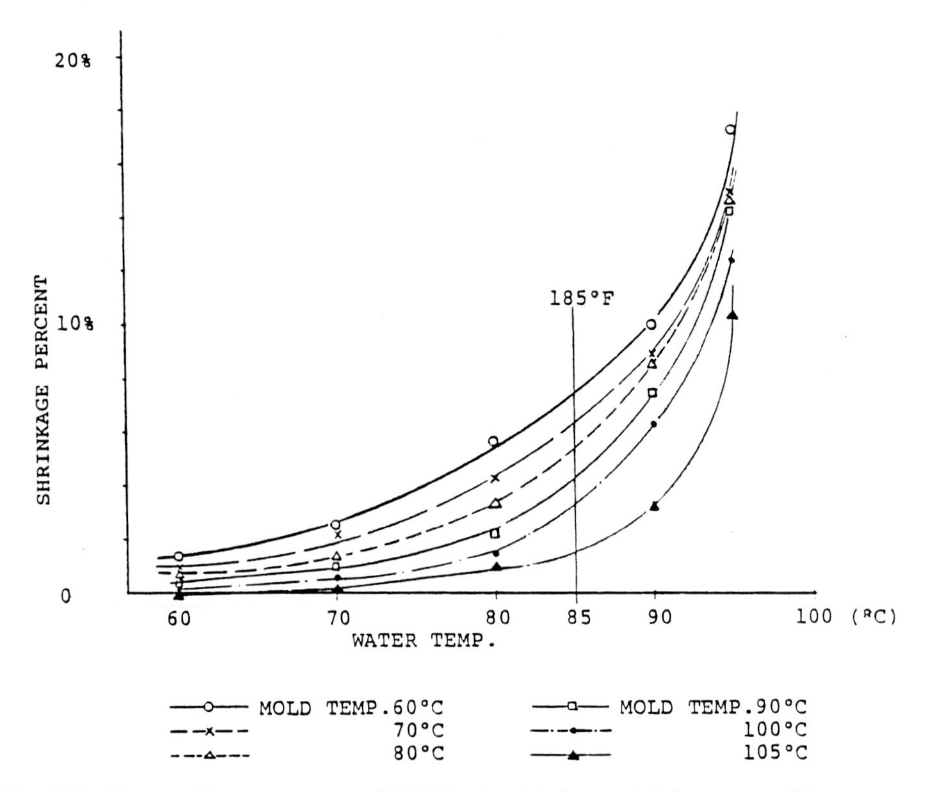

Fig. 5.13 *Blow mold temperature and PET bottle shrinkage; shrinkage vs. mold temperature and filling temperature*

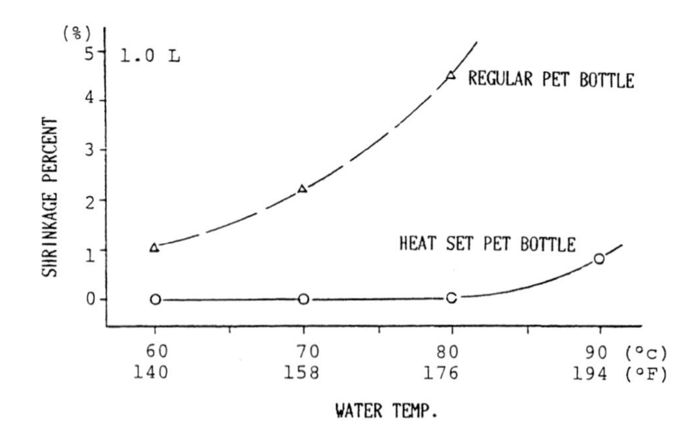

Fig. 5.14 *Volume change after filling with hot water*

underscored by the fact that Fig. 5.14 shows 2.8% shrinkage at 60 °C (140 °F) blow mold temperature.

5.6.2 Parison Temperature

One of the important considerations in the heat setting process is the parison temperature just before the parison is blown (Fig. 5.15). When molding ordinary PET bottles, the shrinkage would be 23% when the parison temperature is 90 °C (194 °F) just before blowing which was nearly three times as much as 8.5% shrinkage of the parison at 105 °C (221 °F) immediately preceding the blowing operation. In comparison, the heat set bottle shrinkage would be less than 1%.

Needless to say, a perfect bottle cannot be made if the preblow temperature of the parison is very high because it decreases the stretching stress. Therefore, the proper parison temperature and the stretching stress for heat setting process has to be determined.

5.6.3 Bottle Weight

Bottle weight has less effect on shrinkage than blow mold and parison temperatures. Figure 5.16 illustrates the difference in shrinkage when two bottles with 9 g weight

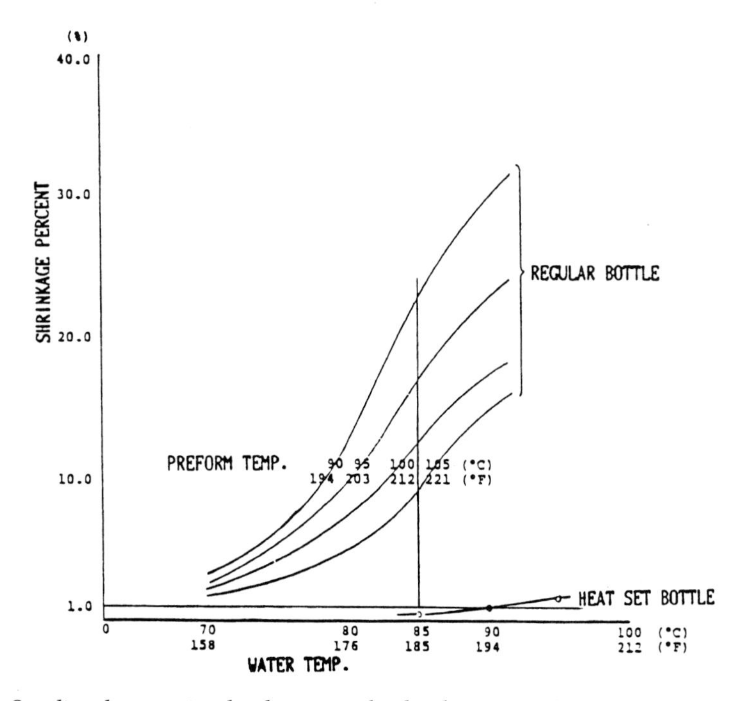

Fig. 5.15 One-liter heat setting bottle vs. regular bottle

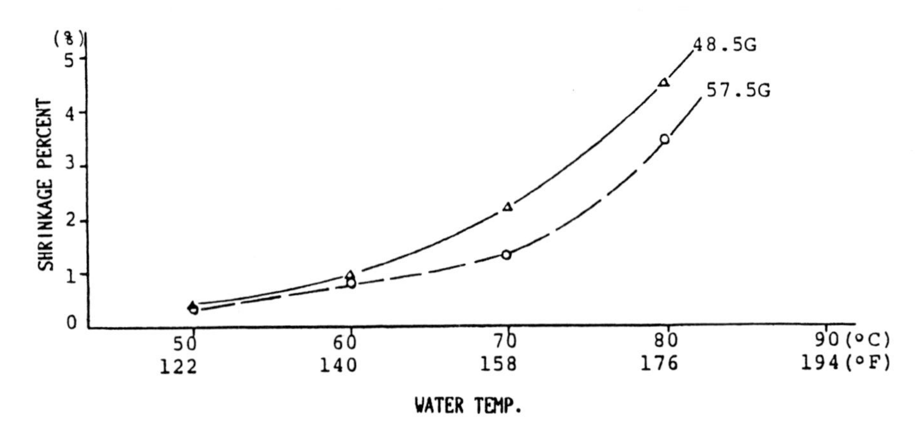

Fig. 5.16 Volume changes after filling hot water

difference are used to measure the shrinkage rates before and after the bottles are filled to the brim with hot water. In this experiment about 1% difference is observed. This experiment shows that it is better to slightly increase the bottle weight when molding heat-set PET bottles.

5.6.4 Neck Design

The heat performance of the neck with PC insert is compared with that of the standard neck (PET). Figure 5.17, a logarithmic graph, shows the measurements of the neck dimensions. The neck was filled with 85 °C water to support the ring and then the bottle was capped. The bottle was cooled down gradually to compare shrinkage of the bottle before and after filling.

This shrink graph and the numerical values listed clearly point out the differences in shrinkage between the PC insert neck and standard neck that is not heat treated. The numerical values for the PC insert neck are within the tolerances of the neck finish specifications. Furthermore, the sealing efficiency of the PC insert neck remains perfect without leaks and loosening of the cap after the hot filled and capped bottle is allowed to naturally cool off.

5.6.5 Shrinkage

The shrinkage of the heat-set PET bottle molded by the Nissei ASB method is reported to be far less than that of the ordinary PET bottle, see Figure 5.18. The graph indicates that the ordinary soft drink bottle is perfectly suitable for its intended use. The graph shows that a 25-day test in an oven at 38 °C (100 °F) proved that the shrinkage properties of the heat-set bottle are better than the ordinary soft drink bottle. An example of postmolding shrinkage is shown in Fig. 5.19.

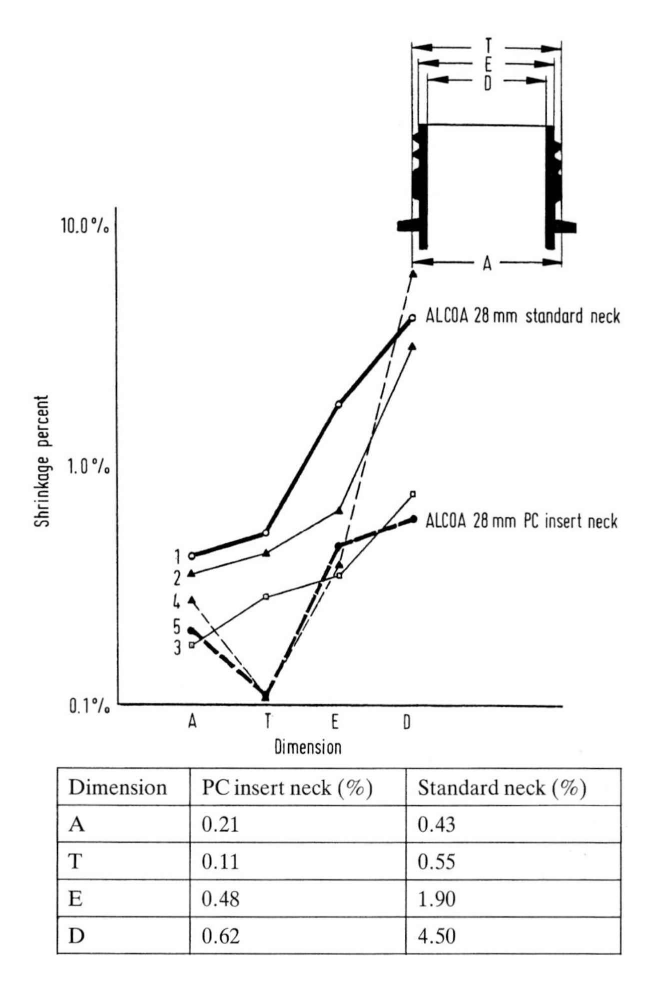

Dimension	PC insert neck (%)	Standard neck (%)
A	0.21	0.43
T	0.11	0.55
E	0.48	1.90
D	0.62	4.50

Fig. 5.17 *Shrinkage percent of neck*

5.6.6 Hot Filled PET Bottle

The term "temperature resistant PET bottles" covers all bottle qualities that are exposed to elevated temperatures for extended periods during filling or handling, including hot filled, refillable, and tunnel pasteurized bottles. Within these categories, temperature levels and exposure to temperature may vary depending on the bottler's requirements.

These two factors combined lead to the destruction of microorganisms that could otherwise contaminate the beverage. In addition to the type of microorganisms to be killed, the possible organoleptic modification of the product is decisive in determining

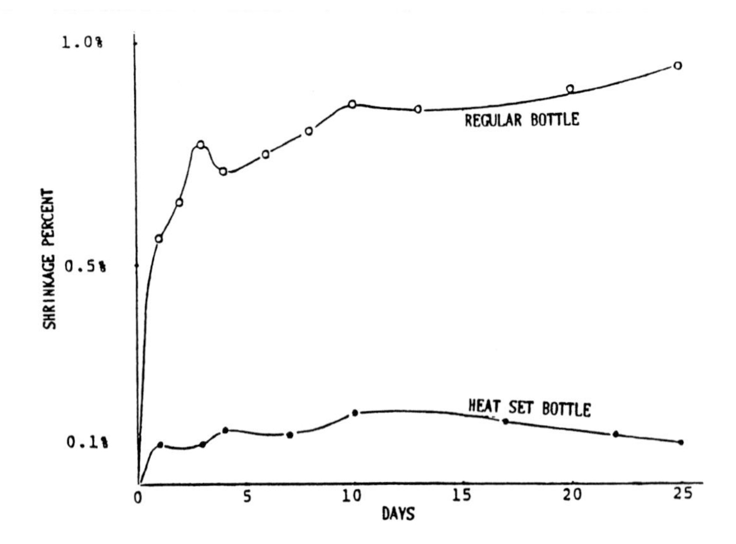

Fig. 5.18 Shrink properties in a 38 °C oven

the temperature level and the processing duration. Typical hot-fill products include fruit juices, teas, isotonic drinks, milk drinks, sauces, health drink vegetable juices, and water.

Refillable PET bottles are submerged and sprayed in the cleaning machine using heated caustic soda solution and hot water so as to render harmless any mold and fungus contamination and supply clean, sterile beverage bottles to the filler. Bottling is effected at temperatures between 4 °C and 20 °C. The following treatments are

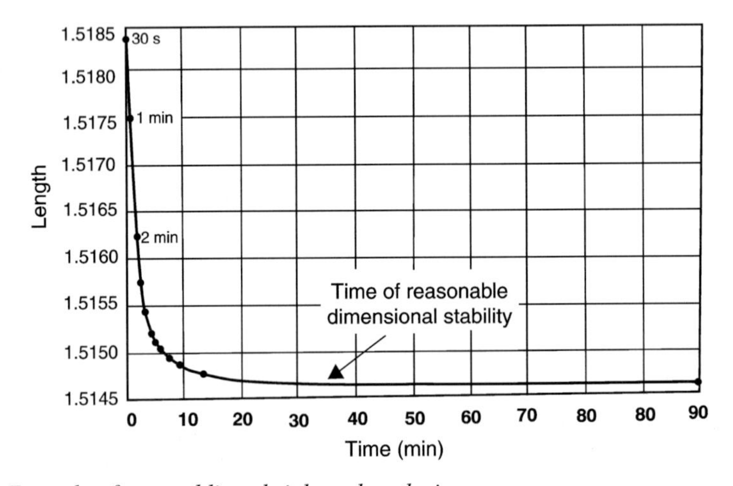

Fig. 5.19 Example of postmolding shrinkage lengthwise

typical of the cleaning process based on temperature/submerged treatment/spray treatment:

maximum 59 °C/ $\leq$ 25 min/ $\leq$ 45 s and maximum 75 °C/ $\leq$ 12 min/ $\leq$ 45 s.

In addition to withstanding the temperatures involved, it is important that shrinkage of refillable bottles be limited to a maximum of 1% up to 25 cleaning cycles, so as to

Table 5.4 *Density and Crystallinity*

	Heat set bottle		Regular bottle	
	Density (g/cm^3)	CRYS (%)	Density (g/cm^3)	CRYS (%)
Shoulder	1.3715	32.0	1.3580	21.7
Panel	1.3723	32.5	1.3590	22.6
Bottom	1.3705	31.0	1.3605	23.4

Table 5.5 *Degree of Orientation (X-Ray Analysis)*

Measuring position	Direction	Degree of orientation (%)
Upper panel	Axial	95.1
	Hoop	90.4
Lower panel	Axial	81.0
	Hoop	90.2

Table 5.6 *Tensile Test Data of PET Material*

		Yield point (kg/mm^2)		Break point (kg/mm^2)		Elongation at break (%)	
Position	Direction		Ave.		Ave.		Ave.
Upper panel	Axial	9.30		8.82		55.0	
			9.30		9.27		56.3
		9.47		9.72		57.5	
	Hoop	—	—	18.29		22.5	
					18.83		23.8
		—	—	19.37		25.0	
Lower panel	Axial	9.33		11.86		92.5	
			9.31		10.24		62.5
		9.28		8.62		32.5	
	Hoop	—		18.58		20.0	
			—		19.12		22.5
		—		19.66		25.0	

ensure that even after the final washing and filling cycle they are still of the quality required and meet the regulations of packaging codes. Tunnel pasteurization ensures that any germs in the product, bottle, or closure are killed. The process is applied in many bottling operations, especially those involving carbonated beverages where hot filling is not possible.

5.6.7 Other Characteristics

PET bottles that are heat treated by the Nissei ASB method are compared to regular PET bottles in Table 5.4 (the panel is the straight section between the shoulder and bottom bends). Table 5.5 provides the degree of orientation using X-ray analysis. Tensile strength of the PET plastic is provided in Table 5.6; tensile test operating speed was 50 mm/min.

6 Tooling (Dies and Molds)

6.1 Overview

In this chapter tooling is reviewed, with discussions on how melt flow is affected by the different tool designs used for blow molding different products. The tools used in processing plastics include dies, molds, manifolds, mandrels, jigs, fixtures, punch dies, and perforated forms for shaping and fabricating products. The terms are virtually synonymous in the sense that these devices have a female and/or negative cavity into or through which a molten or rigid plastic moves, usually under heat and pressure, or they are used in other operations such as cutting dies, stamping sheet dies, and so forth. Tool is the term that identifies all these devices, particularly molds and dies (Figs. 6.1 and 6.2).

Many tools have common assembly and operating parts so that the tool's opening or cavity can be designed to form the desired final shapes and sizes. They can be composed of many moving parts requiring high-quality metals and precision machining. To capitalize on advantages, many molds, particularly for injection molding, have been preengineered as standardized parts that can be used to include cavities, runner systems, cooling lines, and so forth.

All the BM processes used for the manufacture of the many different products produced worldwide require some type of a die and/or mold. Most tools have to be handled very carefully and must be properly maintained to ensure their proper operation. They are generally very expensive and can be highly sophisticated. Some molds and dies cost more than the processing machinery, with the typical cost approximately half that of the primary machine. As a general example 5% to 15% of the cost is for the material, design about 5% to 10%, mold building hours about 50% to 70%, and profit at about 5% to 15%.

Using low-cost material to meet high performance requirements will compromise the integrity of the mold. For more than 90% of molds and dies, construction material cost is 5% to 15% of the total cost. Thus it does not make sense to compromise mold integrity to save a few dollars; use the best material for the applications. The cost of labor, particularly manufacturing, is the bulk of the mold's cost.

High precision surfaces have a mirror finish. As the slightest scratches can produce flaws in the fabricated products, great care must be used during their installation,

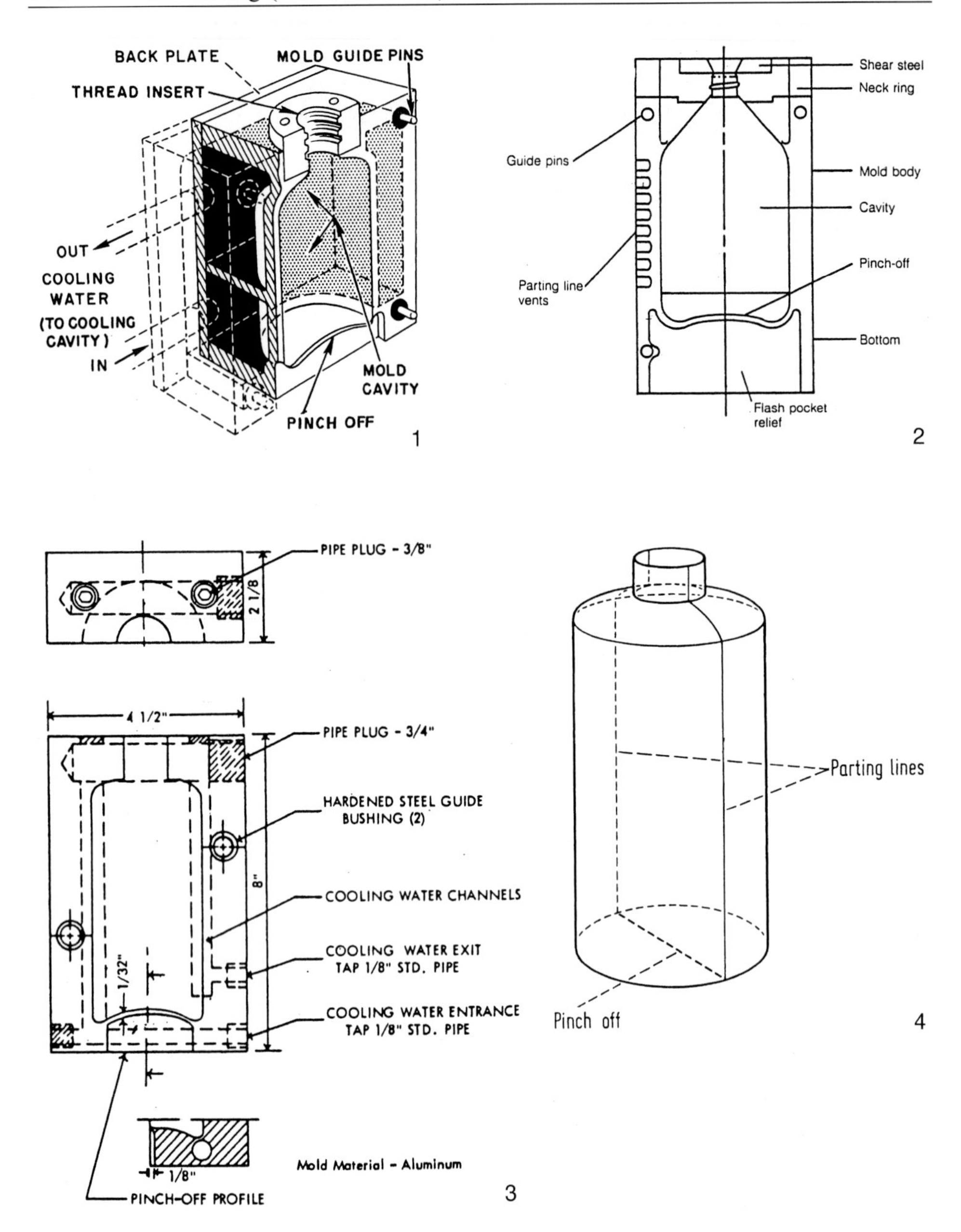

Fig. 6.1 Examples of BM molds; four views

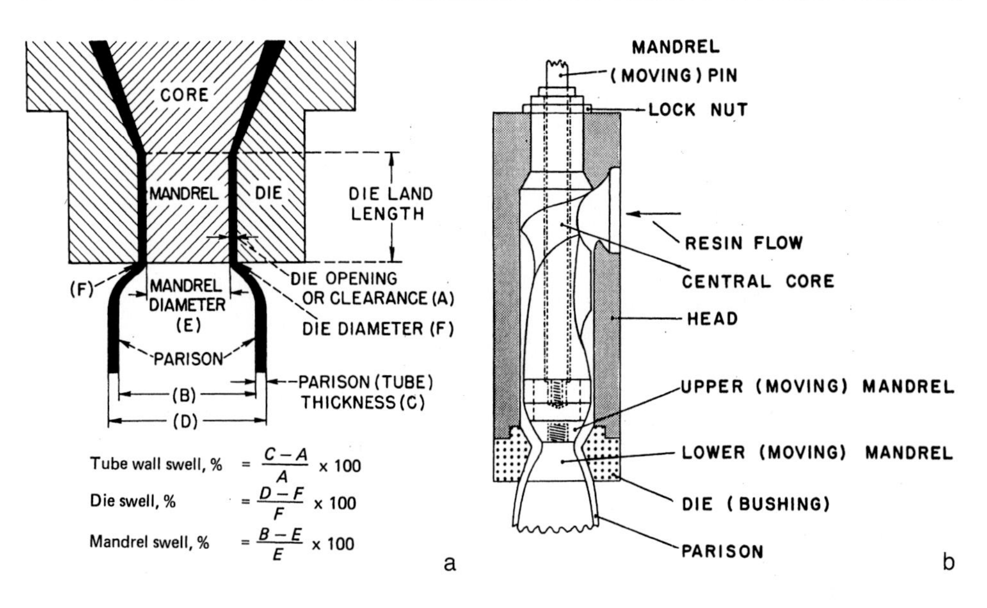

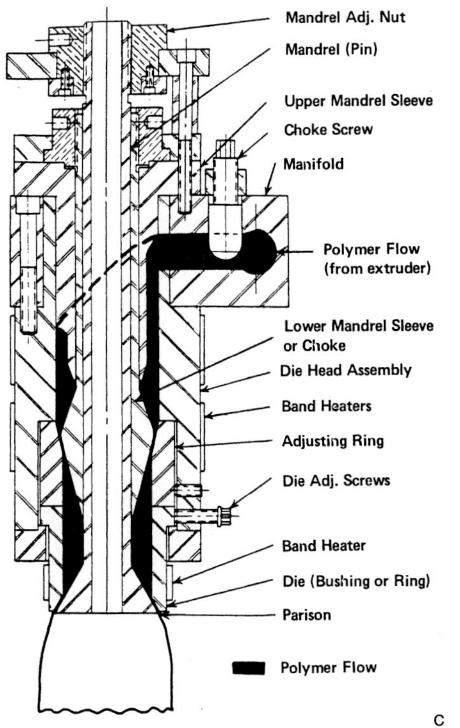

Fig. 6.2 *Examples of EBM continuous dies*

operation, removal, cleaning, and particularly storage. When designing components, the target is to use as few parts as possible; they should be easily handled for installation and easily dis-assembled, cleaned, and reassembled [87].

6.2 EBM Die Heads

All extrusion BM machines have an extrusion manifold head leading to a die from which the parison is extruded. The design of the EBM head is very important in determining the quality of the parison.

EBM dies have a specific orifice (opening) with a specific shape so that different products can be produced that range from cylindrical to complex shapes, uniform to nonuniform wall thicknesses, and so forth. Their function is to accept and control the available extrudated parison from an extruder and deliver it to the blow mold with minimum deviation in cross sectional dimensions, smooth class A surfaces, and at the fastest possible rate.

The function of the die, usually composed of steel, is to control the shape of the extrudated parison. To do this, the extruder must deliver melted plastic to the die at a constant rate, temperature, and pressure. Measurement of these variables is desired and needs to be carefully performed (Chapter 9).

Die efficiency analyses can be conducted and must include a careful examination of the compatibility of the die with the extruded products. If the geometry of the flow channel is optimized for a plastic under a particular set of conditions (heat, flow rate, etc.), a simple change in flow rate or in heat can make the geometry very inefficient. Except for possibly circular dies, it is essentially impossible to obtain a channel geometry that can be used for a relatively wide range of plastics and a wide range of operating conditions.

In designing a die: (1) minimize head and tooling interior volumes to limit stagnation areas and residence time; (2) streamline flow through the die, with low approach angles in tapered transition sections; and (3) polish and plate interior surfaces for minimum drag and optimum surface finish on the extrudate. Basically the die provides the means to “spread” the plastic being processed/plasticated under pressure to the desired width and thickness in a controllable, uniform manner. In turn, this extrudate is delivered from the die (targeted with uniform velocity and uniform density lengthwise and crosswise) to the blow mold. Pumping pressures on the melt entering the die heads differ to meet their melt flow patterns within the die cavities. The pressure can range from 500 to 4,000 psi (3.5 to 27.6 MPa), depending on plastic to be extruded and die design.

6.2.1 Plastic Melt Behavior

The non-Newtonian behavior of plastics melt makes its flow through a die somewhat complicated. The behavior of the melt as it flows from the extruder through the die(s) plays a vital role in creating the parison and in ensuring product quality.

The type of plastic being extruded affects flow behavior of the melt. For example, high-density polyethylene (HDPE) commonly used in BM is more sensitive to changes in direction of flow than is low-density polyethylene (LDPE). Other plastics also vary in their relationship between flow behavior and shear rate.

Differentials in the velocity of a plastic melt are caused by friction of the plastic on the extruder surfaces and to some extent by internal friction within the plastic mass. Frictional drag is directly proportional to the distance traveled. Pressure drop in the head of a BM machine caused by changes in cross section is particularly acute at the right-angle bend in going from a horizontal extruder into a vertical parison die. Pressure differentials between the incoming side of the head and the opposite side of the core pin can be as much 300 to 400 psi (2 to 2.8 MPa) in a symmetrical head design. This pressure differential causes changes in the shear rate and velocity within the plastic mass. The highest velocity and shear rate occur directly below the inlet channel from the extruder; maximum internal pressure also occurs at this point. Minimum shear rates and velocities occur at a point 180° away from the core pin. Pressures between these two points drop more or less uniformly in a concentric head design.

While the settings of the die control the inner and outer dimensions of the parison, balances in melt flow will cause parison wall variations even with a fully concentric die.

To bring this pressure back to a circumferential equilibrium by the time the melt enters the die without making the head excessively long, a restriction may be placed on the mandrel to increase and equalize back pressure. The restriction should have a continuous curvature to avoid any land length, thereby avoiding excessive back pressures. Short radii are also to be avoided to prevent stagnant areas and plastic degradation. The taper of the mandrel in relation to the head as the die is approached should be such as to constantly increase or at least maintain melt compression. The taper of the head approach to the vertical die lands should be under 35°.

A choke is usually added slightly below the inlet channel to create a high pressure and to slow the flow on the inlet side of the core pin. The design and positioning of a choke to produce uniform flow is dependent on factors such as overall machine design, the output rate of the extruder, and the back pressure. A properly designed choke will compensate for the various differences in pressures and shear rates. The choke must not create turbulence in the flow pattern. As there are no theoretical or empirical formulas that accurately predict the melt flow in the head, the actual flow patterns in a head design must be determined experimentally.

In addition to uniform flow conditions in the extrusion head, it is necessary to control the flow leading to the die orifice. A tapered approach angle to the die orifice is recommended for most applications. Each type of material will require a different approach angle. The ratio of gap opening of the die orifice to the length of die land is a controversial subject. For HDPE, for example, die lengths as high as 12 to 20 times the gap opening are recommended by some material suppliers, whereas some BM machines have operated successfully with HDPE using minimal die land lengths approaching knife edges in thickness. Die land specifications can, therefore, best be classified as dependent on molding shop practice.

6.2.2 Die and Extruder

An important approach to melt flow behavior is to recognize that the extruder and the die operate as a combined unit. Showing the dependence of the output on the melt pressure between the screw and die head usually represents the interaction between screw and die. The screw requires that the viscosity of the plastic does not change either in the metering zone or in the die. This means that temperature and/or pressure changes and other influences on viscosity have to be avoided as much as possible. The pressure drop through the die varies directly with the land length and inversely with the cube of the gap opening.

When a melt is extruded from the die, there is usually some degree of swelling (Chapter 3). Figure 6.3 shows the influence of die designs on free-hanging extruded parisons from viscoelastic melts.

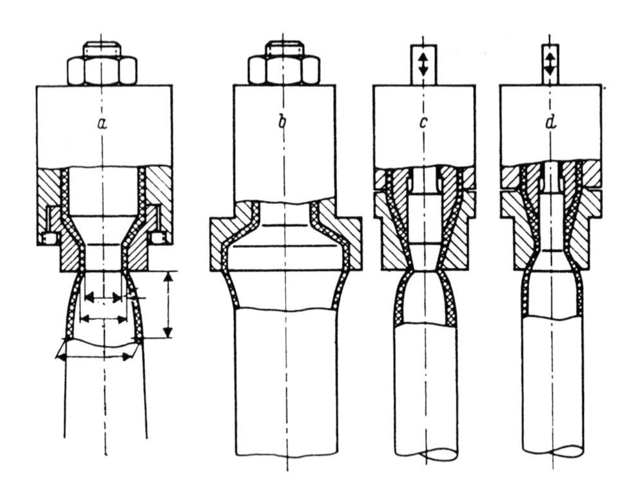

Fig. 6.3 *Examples of die designs affecting the swell of parisons. (a) cylindrically parallel die gap with conical (progressively narrowing) inflow, (b) cylindrically parallel die gap expansion die, (c) conical (progressively narrowing) die gap, and (d) conical (progressively widening) die gap.*

Ideally, after the melt exits the die, it is stretched or drawn to a size equal to or smaller than the die opening. However, the effect of melt elasticity is a source for error in the profile that can be corrected by modifying the die and the downstream equipment.

A substantial influence on the plastic is the flow orientation of the molecules, such as having different properties parallel and perpendicular to the flow direction. These differences have a significant effect on the performance of the product.

6.2.3 Manifold Head

The head receives the melt from the usual horizontal extruder and generally changes the flow to a vertical downward direction. The melt passes around a vertical mandrel, and on to the extrusion die in a steady state, all of which must be done in a short time. Failure to accomplish these objectives can result in weld lines, streaking, and lack of parison wall control by fixed die alignments. The head design therefore does play a vital role in determining parison quality and thus part quality.

In the extruder head, the melt flows around a forming pin or mandrel (sometimes called a flow diverter pin) to form the parison. In most BM machines, the melt changes direction so the parison is generally extruded downward. For technical processing reasons the melt or preform is extruded as an unsupported parison hanging vertically downwards; an exception to this is horizontal extrusion for rotating table and two-station plants with BM tools that collect the parison without causing separation.

It is desirable to have a restriction in the head at some point beyond the area where the plastic flows together around the mandrel forming pin. One purpose of the restriction is to overcome the "elastic memory" of the plastic and to allow it to knit or reform properly after having passed around the pin. The restriction also promotes more uniform flow distribution so that the wall thickness is uniform. The mandrel in some BM machines is slightly elliptical to compensate for pressure differences in the extruder head. It is recommended that before any modifications are made to the forming pin, the manufacturer of the particular machine be consulted. Very small changes in the design in this area of a machine will make large differences in the parison as it is extruded.

Plastic flowing from the extruder can be channeled to one or more heads to form parison(s) for blow molding a container, and so forth. This means channeling, called the extrusion manifold, functions much like the manifold of an automobile, which channels a gas/air mixture to each cylinder. The manifold may split the flow of the material into several paths that are directed to each head. These flow paths have either changing diameters (larger as the distance increases from the main feed) or the same diameters with choke screws (valves) incorporated. In either case, they ensure that each head receives identical amounts of melt. The choke screws generally are adjusted to open up more of the flow channel as one moves away from the main feed. This is necessary to maintain uniform parison lengths. Uneven parison lengths can cause unloading or trimming problems from tails being too short or too long.

6.2.3.1 Manifold and Die

Manifolds are either center fed (also called spider fed) or side fed. In a center fed manifold the melt flow is in line from the extruder and if a core exists in the die, melt distribution around the core is rather uniform. In the side fed manifold the melt flow is not as uniform; however, with proper design and enough length to travel through the die, a certain degree of uniformity will exist.

In the spider or axial flow head (Fig. 6.4) a central torpedo is positioned in the melt flow path, generally supported by two spider legs. This style is generally used for polyvinyl chloride (PVC) because the smooth and direct path offers little opportunity for plastic to adhere and degrade. Cross sectional areas of the melt flow path, leading from the extruder over the torpedo and the spider legs, are carefully balanced to ensure even and consistent pressure and flow.

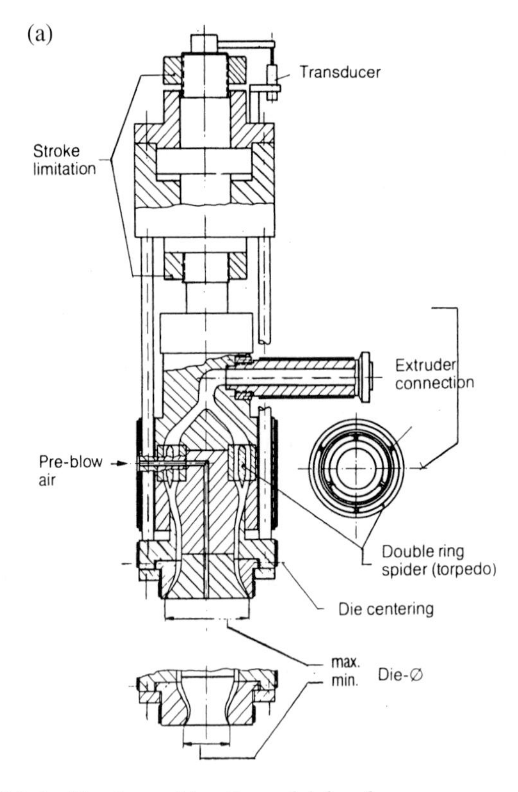

Fig. 6.4 (a) Center fed double ring spider (torpedo) head.

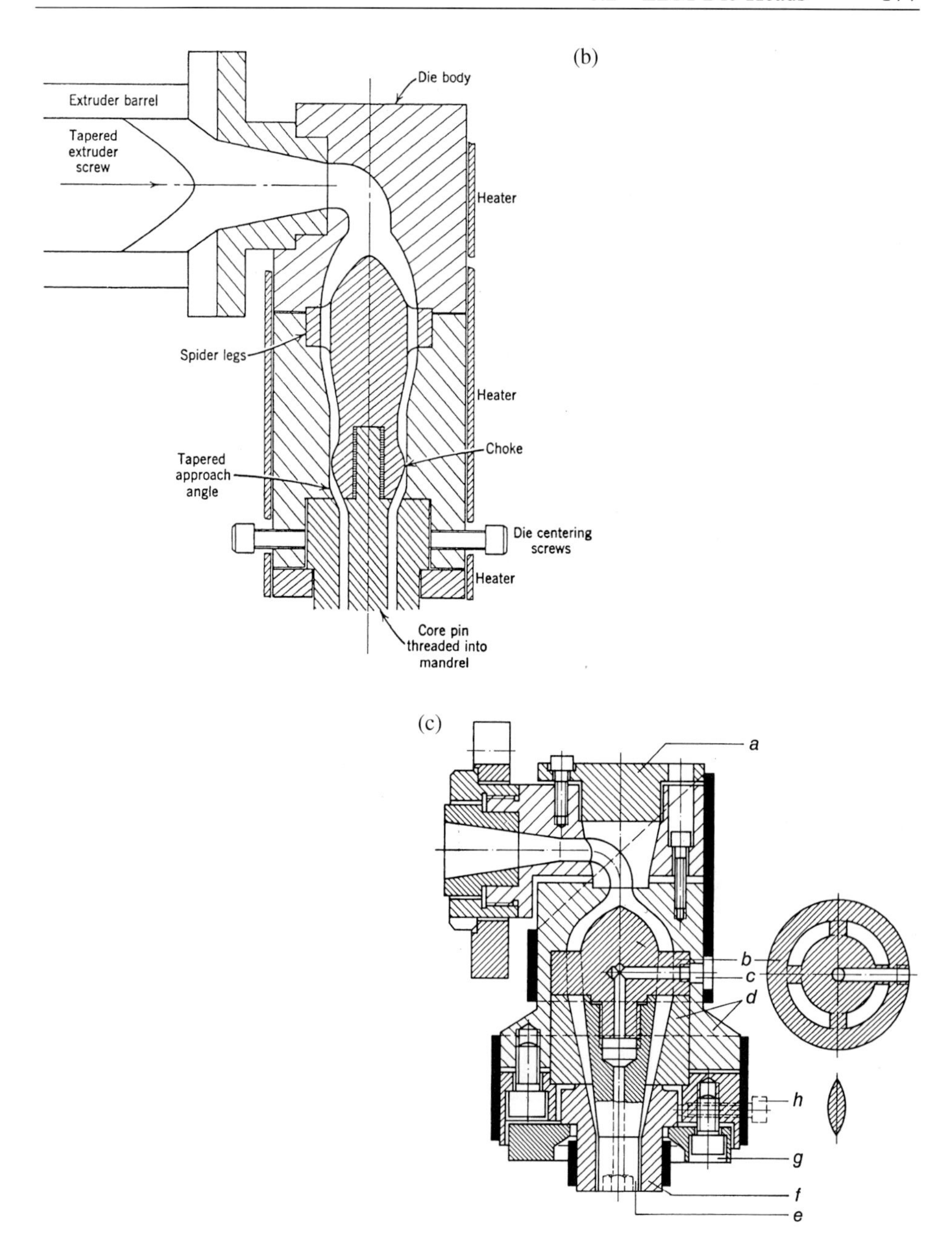

Fig. 6.4 *(b) Center fed spider or axial flow head. (c) Detailed analysis of a center fed spider or axial flow head*

Occasionally, the spider head style is used for polyethylene (PE), which offers the potential for rapid color changes. However, offset spider legs are required to mask the elastic memory of the PE to ensure a smooth and weld-free parison. This requirement increases the manufacturing cost of the head but provides targeting for zero-defect blown products.

Figure 6.5 shows the side fed manifold with different modified mandrels.

Figure 6.5b shows a modified mandrel with heart-shaped channel.

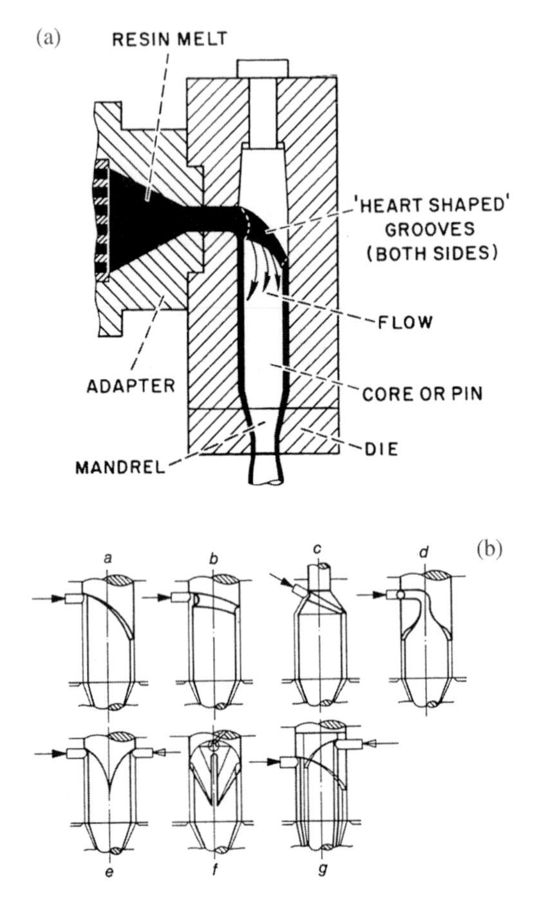

Fig. 6.5 (a) Simplified schematic of a side fed or radial flow head. (b) Modified mandrel with different side fed heart-shaped melt flow channels. a, mandrel with heart-shaped channel greatly elongated in vertical direction; b, distribution channel with wedge-shaped crosspiece above the restriction point opposite the flow channel; c, design with tapered cross section in the distribution zone; d, design with two diametrically opposed distribution channels; e, design with diametrically opposed heart-shaped channels greatly elongated in the vertical direction, and two flow channels for the production of a two-color parison; f, design with feed channel for a colored or transparent stripe; g, design for internal or external coating.

Figure 6.5c is a radial flow crosshead die with top mounted mandrel and circular groove for distributing the melt stream; the right side of the diagram shows the die gap adjustment in the intake zone.

The crosshead in Fig. 6.5d is bottom mounted through a mandrel. It has a flat heart-shaped distribution channel. The right side of the diagram shows an adjustable die gap.

The side fed type of design is preferred for PE. Melt enters the head from the side, divides around the mandrel, and is rewelded. As the melt moves downward it enters the pressuring area, creating high back pressure, which thus ensures rewelding. This head is simple and rugged, but unfortunately has relatively long color changes. Often, during a color change, a fine trail of the previous color, originating from reweld areas of the head, is noticeable for several hours. The problem has been reduced in modified designs by careful attention to streamlining and polishing of the flow path and, in some cases, by the insertion of a bleed screw to remove the trail of color.

Provisions should be made to accurately control temperatures in all parts of the head and die. To meet these requirements, it is important to measure melt temperature and

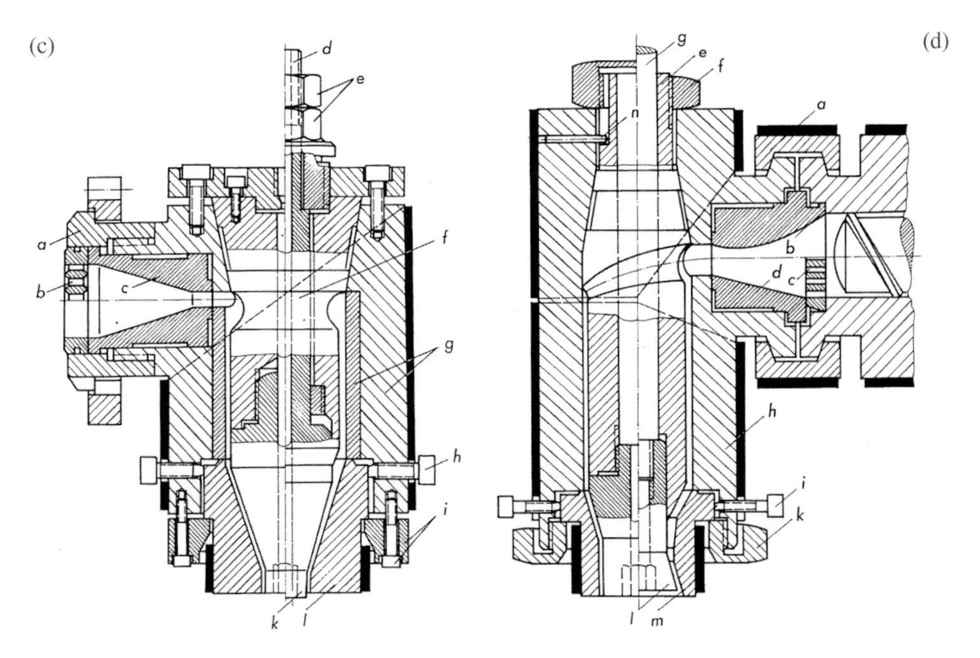

Fig. 6.5 *(c) Radial flow crosshead for parison dies with top mounted mandrel and circular groove for distributing the melt. (d) Crosshead (side fed) with bottom mounted through a mandrel and flat distribution in a heart shape. a, split clamping flange; b, restrictor; c, breaker plate for supporting a screen pack; d, adapter; e, bottom-mounted through mandrel; f, retaining screw; g, sliding rod for die adjustment; h, extrusion head housing; i, adjusting screw; k, retaining ring; l, die core; m, die orifice; n, adjusting pin*

pressure in the head via sensors such as stock thermocouples and pressure transducers. Plus or minus 1 °F is typically used in today's control systems. If there is a cold area in the die, the melt flow in that area will be slow and the result will be a thin gauge. A hot area results in more flow and the potential to burn (degradation) the exiting plastic.

A heat pipe is a means to transfer heat quickly and efficiently [76], either by removing or by adding heat in all kinds of fabricating equipment. Compared to metals, heat pipes have an extremely high heat transfer rate. They are capable of transmitting thermal energy at near sonic isothermal conditions and at near sonic velocity.

Dies are available for coextrusion to produce different products with two or more layers bonded together. They are designed to permit the proper melt flows of each plastic.

6.2.3.2 Accumulator Head

The size of the product and the amount of material necessary to produce the product (shot size) dictate whether or not an accumulator is required. The nonaccumulator machine offers an uninterrupted flow of plastic melt. When an accumulator is added to the system, the flow becomes cyclic. Thus, the connecting channels between the extruder and the accumulator and within the accumulator itself must be designed rheologically (Chapter 7) to prevent restrictions that might impede the flow or cause the melt to hang up. Examples of different accumulator melt flow heads are shown in Fig. 6.6.

Flow paths should have low resistance to melt flow to avoid placing unnecessary load on the extruder. But most important, the design of the accumulator should guarantee that the first material to enter the accumulator is the first to leave when the ram empties the chamber, and the chamber should be totally emptied on each stroke.

The accumulator head serves the functions of both an accumulator and an extruding die head, and is filled directly by a continuously operating extruder. The typical accumulator head offers a first-in/first-out melt flow path. This prevents excessive dwell of material in the head, with no dead spaces where melts can stagnate. Even the "mini melt leakage" that the technique requires is automatically removed from the die head zone and collected. This is of particular importance when BM high molecular weight PE or other high viscosity plastics used for large industrial products, drums, and automotive fuel tanks.

The system differs from the ram accumulator type (such as ram injection molding type in Fig. 6.7) in that the extruded melt flows directly into the head from the extruder. This results in lower stress transmitted to the plastic melt. This head is considered a subcategory of intermittent extrusion BM. It is the combination of an extrusion head with a first-in/first-out tubular ram-melt accumulator. It can also be called reciprocating type extrusion, providing an action similar to the reciprocating injection molding machine where the melt collects in front of the reciprocating screw, followed by the screw "ram" action with first-in/first-out melt to form the parison.

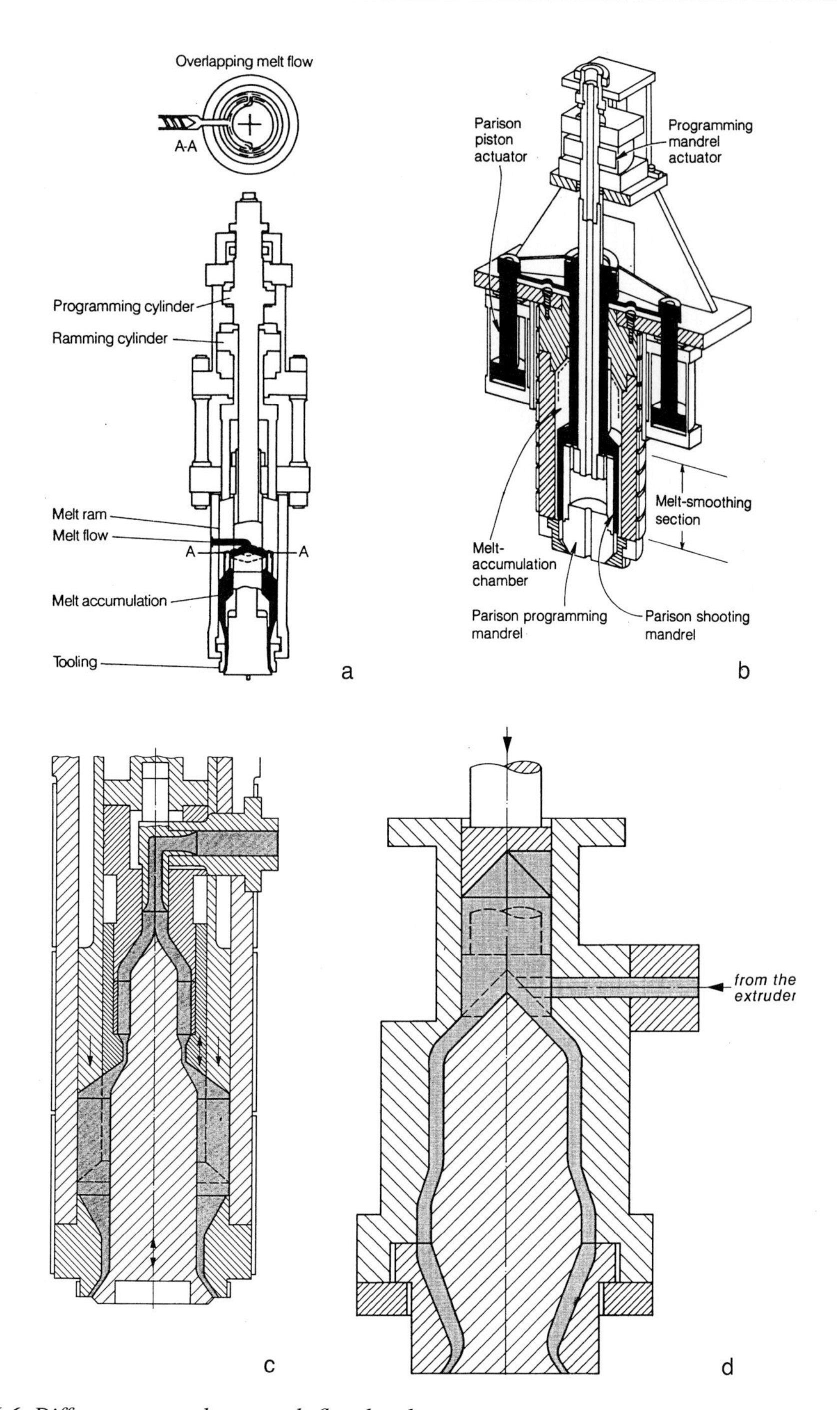

Fig. 6.6 Different accumulators melt flow heads

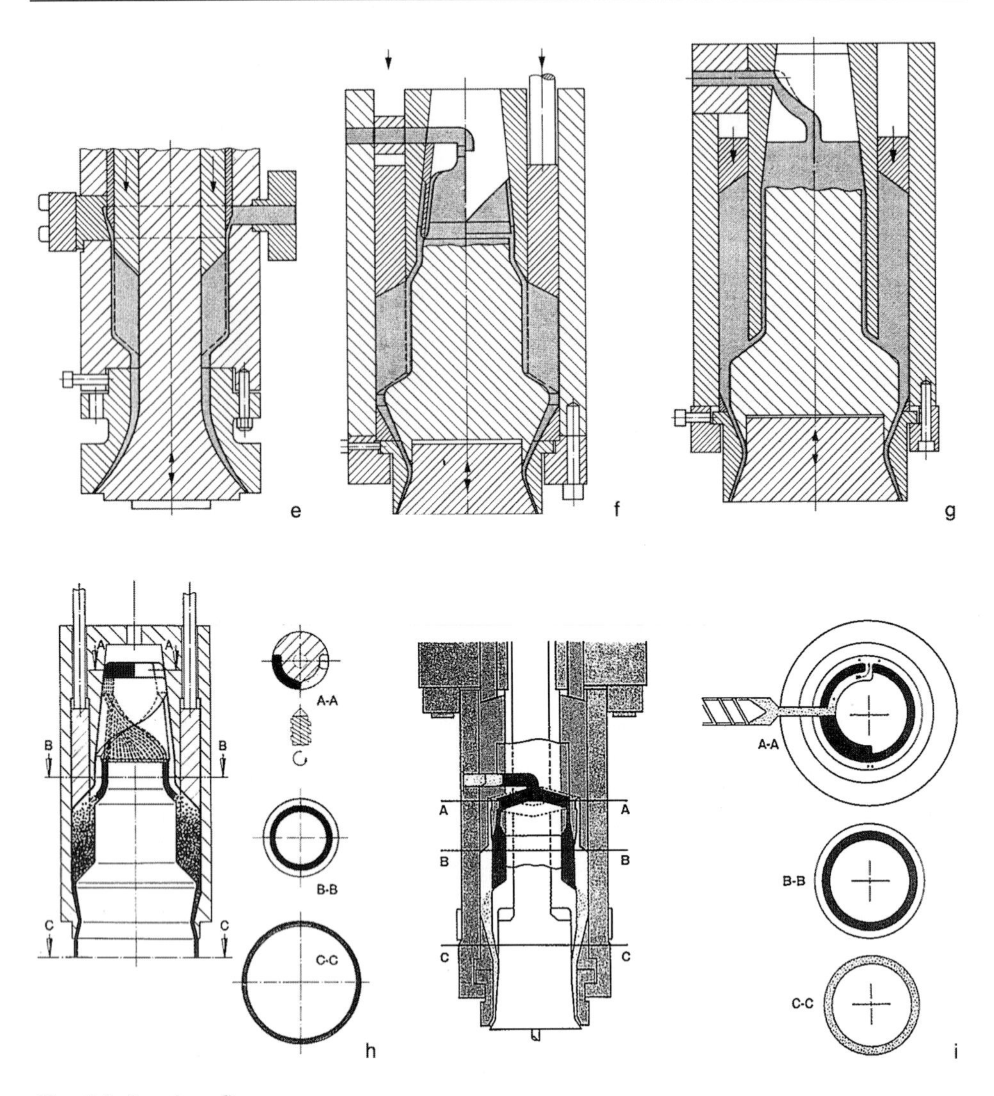

Fig. 6.6 (continued)

In a typical EBM accumulator head, as the material enters the head, the outer die body moves up to accumulate a predetermined shot size. The outer die body then moves down to extrude the melt as a parison. Because the accumulator head shoots the parison directly from the head annulus, excessive pressure drops are eliminated in the manifold and head entry systems. Melt stress and pressure are minimized, as the only resistance to flow is in the tooling; also the force pushing the melt out of the annulus is uniform around the die opening, providing a more uniform, stress-free parison.

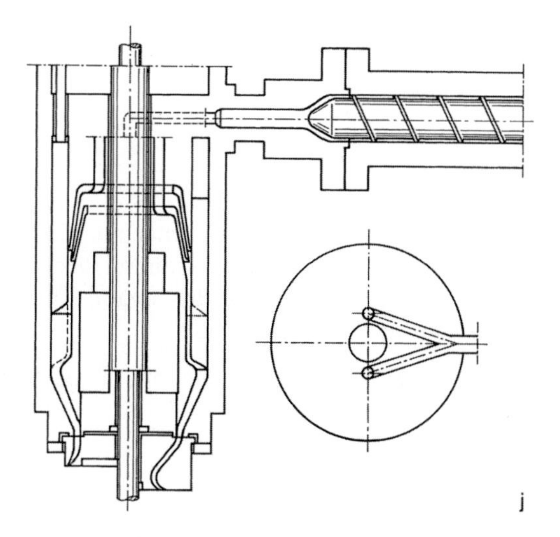

Fig. 6.6 (continued)

All material flow channels are designed to ensure that there is very little pressure loss. Accordingly, the deformation introduced into the material is very slight and the relative wall thickness differences are very low. The precise material distribution achieved in the head also means that color change is very rapid. Accumulator heads are also designed with minimum seam (weld line) weakness. They produce more uniform parison, streamline melt flow, and provide rapid drop rate. One design approach has the plastic melt entering the parison forming chamber from two different inlet points, each 180° away from the other (Figs. 6.6i and j). This design is generally referred to as an overlapping technique where melt from the extruder is diverted in the accumulator. This arrangement produces an inner and an outer ring of plastic melt so that the seam formed by the inner ring is directly opposite the seam in

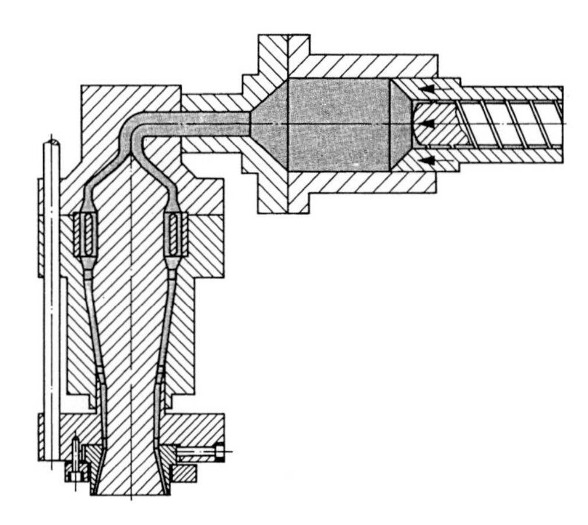

Fig. 6.7 *Example of an injection molding reciprocating barrel accumulator*

the outer ring. As a result there is no seam that goes all the way through the parison wall at any one point, and wall integrity is significantly enhanced.

6.2.4 Die Head Control

As the molten plastic leaves the manifold, it can enter the die head. This assembly forces the melt to split into two streams around a mandrel sleeve or flow divider and meet on the opposite side where it welds together again. This is the first step in forming a hollow tube or parison.

As the molten plastic moves down the head assembly, it meets the lower mandrel sleeve. The lower mandrel sleeve is adjusted vertically to control the amount of back pressure, but is generally left in the full up position.

In some cases an adjusting ring is incorporated in the die head to control the differences in the melt velocities. Plastic tends to flow slower down the weld line side of the head, causing curling of the parison on its exit. This ring can be adjusted eccentrically to restrict the flow on this side, thus resulting in a parison that drops straight without curling. In most cases the ring is centered around the mandrel sleeve or choke. As the melt leaves the lower mandrel sleeve (choke), it enters the die. This action is sometimes referred to as the bushing (or ring) and pin assembly. This is the final shaping orifice.

The inside surfaces of the flow channels within the head, including the die and the core, should be well polished to prevent any accumulation of the plastic flowing over these surfaces. If the surface of the channel is rough, stagnant layers, or "hangup," of the melt may form. This material will eventually degrade, causing dark streaks or black specs to appear in the parison. Roughness in flow channels may also cause the parison to rupture or blow out in affected areas during this expansion in the closed mold.

Streamlining the shape of the flow channels also helps to prevent areas of stagnation and aids in reuniting the plastic after it has flowed around both sides of the mandrel in the head. At that point where the material rejoins after flowing around the mandrel, a small pocket of stagnant material can form if the flow channel in the head has been improperly streamlined. This stagnant pocket of material will then continually bleed degraded material along the weld line of the parison.

Many variations of streamlining are used. A common method is to cut away the sidewall of the mandrel at the point of impact with the incoming plastic melt. In some designs the point of contact is reduced to almost a knife edge. The cutaway portion of mandrel is usually shaped into channels that flow downward as they circle the mandrel. It is at a point 180° from the entrance port and a little below the port the two streams of plastic melt from each side of the mandrel are brought together. The metal of the cutaway mandrel at the point of confluence is also machined almost to a knife edge to minimize stagnant areas. In passing around a mandrel the pressure of the plastic melt will decrease. To compensate for this pressure drop, the cross section of the melt channel on the reverse side is usually reduced at the point of confluence as compared to the cross section 180° around the mandrel. The lower cross section increases the pressure on the melt, compensating for the loss and aiding the reuniting of the two streams of plastic into a homogeneous melt.

Healing of the weld line in the flowing plastic depends on three factors: retention time in head, head temperature, and back pressure on the melt. Time is an important factor. Sufficient retention time must be provided to allow the hot plastic to intermix and cohere. Increases in head pressure and temperature will assist this process, but are less effective than an increase in time.

To increase head pressure, an annular restriction called a "choke" is often placed on the mandrel in the extrusion head downstream from the flow deflector. This system decreases the cross sectional area of the annular flow channel in the head and builds up melt back pressure in the weld area just past the flow deflector and upstream from the choke. The use of chokes, and increased head pressure without a flow deflector, is not effective in producing optimum weld conditions (elimination of weld line in the parison).

6.2.5 Die and Mandrel Design

The die and mandrel (core pin and ring) assembly in the head is the final determining factor in the size and shape of the parison.

The die and mandrel design is determined by many factors, such as container size and shape, container weight, neck finish, and plastic type. The two basic types of dies are the converging and diverging (Fig. 6.8). The diverging type is generally used for large containers, while the converging type is normally used for smaller ones in which the parison can be encapsulated by the neck, thus eliminating neck flash.

Two adjustments are provided to obtain the proper parison. The die bolts (Fig. 6.9) can be adjusted to move the die eccentrically around the mandrel and thereby produce a uniform wall around the parison, which should also result in a straight parison.

In addition, the mandrel can be moved up or down vertically, which varies the parison wall thickness and therefore the container weight. Adjustment can be done by turning a mandrel-adjusting nut or by means of a hydraulic and/or electrical controlled cylinder by a parison programmer. Parison programming is a means to produce a container with an even wall distribution vertically by moving the mandrel up and down during the extrusion of a parison. This movement produces a parison wall thicker and thinner vertically in areas needed for the shape of the container. The programmer can be adjusted to change the average wall thickness or weight of the container without changing the parison profile. As the parison(s) leave(s) the die, it is then clamped by the mold.

The die and mandrel are precision tools and should be cared for as such, as any nicks and scratches in them will end up as a flaw in the finished container. Care must be used not to drop them with abrasive tools. A copper scraper or wire brush will suffice for hand cleaning.

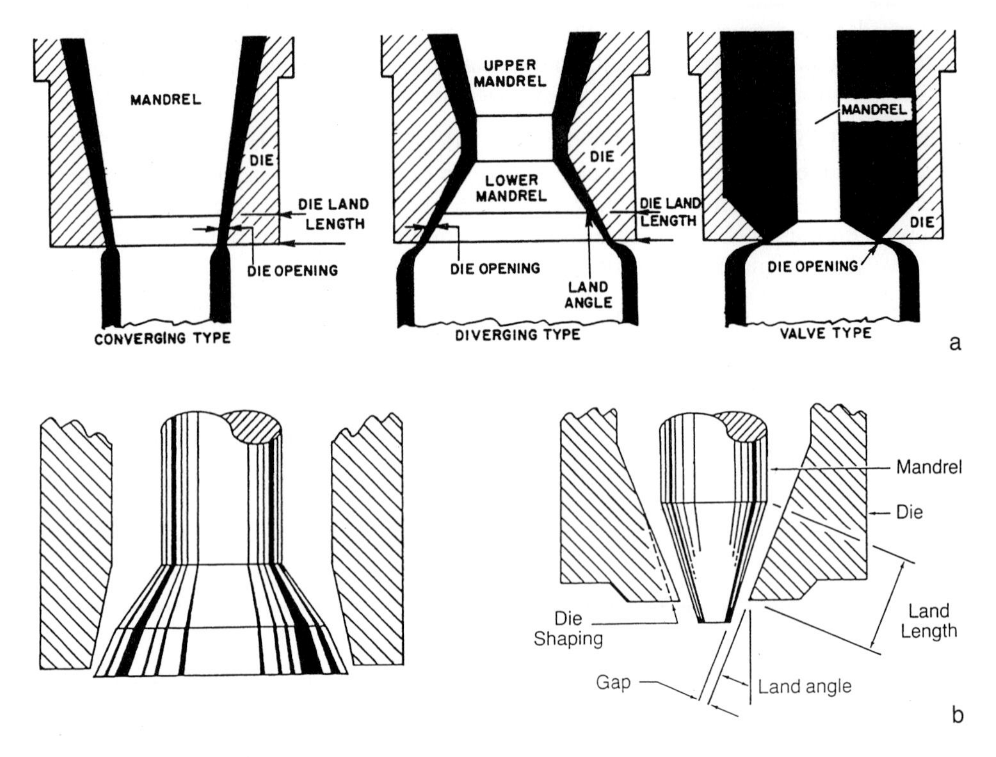

Fig. 6.8 *Basic type dies to form parisons*

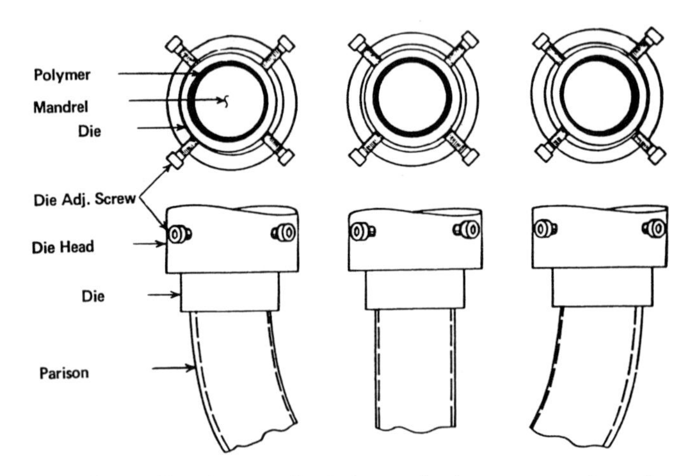

Fig. 6.9 *Use of mechanically adjusting die bolts to obtain proper parison shape; they can be electronically controlled*

In view of possible needs to change plastic, temperature, or pressure in a die, consider designing removable die pins rather than having them integral with the mandrel. A hex socket in the bottom of the pin will allow for rapid changes in pin diameters. The edges of the pin should have light radii to avoid material hangup resulting in die lines. A set of interchangeable dies along with pins to be used in varying combinations will minimize downtime for die changes and increase versatility of product weight and wall thickness control. Die centering is accomplished by centering the die bushing, preferably with a four-bolt adjustment.

The process parameters to be considered in BM will be conditioned by different factors such as the type of plastic used (e.g., making an acetal product would involve higher blow pressures than would be required for PE), the type of BM unit used, the product being made, and the length of production run.

6.2.5.1 Calculating Die and Mandrel Dimensions

Various mathematical formulas have been developed to permit the selection of die dimensions. Although these calculated dimensions are intended as approximations or starting points in die selection, they have been found to yield products, in the majority of cases, within ±5% of the design weight. In some cases, only slight changes in mandrel size or stock temperature and/or extrusion rate are necessary to obtain the desired weight.

The discussion below deals primarily with the EBM of HDPE bottles—a technique, material, and application in common use today. In selecting the die bushing and mandrel dimensions to be used for the production of a BM PE product, several features must be considered [18].

For bottles as well as other containers, the weight, minimum allowable wall thickness, and minimum diameter are important considerations. Also, the need, if any, to use a parison within the neck area and whether there may be adjacent pinch-offs has to be considered. The type and melt index of the resin used are factors because of swell and elasticity characteristics.

In using mathematical formulas, part of the die dimensions will also depend on processing stock temperature and extrusion rate anticipated for production. The basic formulas presented here are for use with long land dies, those having 20 to 30 : 1 ratio of mandrel land length to gap clearance between mandrel and bushing. Consideration must be given as to the anticipated blow ratio, the ratio of maximum product outside diameter to the parison diameter. Normally, ratios in the range of 2 to 3 : 1 are recommended. The practical upper limit is considered to be about 4 : 1.

The ratio for large bottles with small necks is generally extended as high as 7 : 1 so that the parison fits within the neck. In such a case, a heavier bottom and pinch-off results from the thicker parison. Also, less material is distributed in the bottle walls 90° from the parting line than in similar bottles with lower blow ratios.

When the neck size of a bottle or the smallest diameter of the item is the controlling feature (such as when the parison must be contained within the smallest diameter), the following approximations may be used to calculate the dimensions of the mandrel and die. For a free-falling parison assume:

$$D_d = 0.5N_d$$

$$P_d \cong \left(D_d^2 - 2B_d t + 2t^2\right)^{1/2}$$

where

D_d = diameter of die bushing, in.
N_d = minimum neck diameter, in.
P_d = mandrel diameter, in.
B_d = bottle diameter, in.
t = bottle thickness at B_d, in.

This relationship is useful with most PE BM resins, and is employed when bottle dimensions are known and a minimum wall thickness is specified. It is particularly useful for round cross sections. In this example the 0.5 figure presented for selecting the diameter of the die bushing may change slightly depending on processing conditions employed (stock temperature, extrusion rate, etc.), resin melt index, and die cross sectional areas available for flow. It may be slightly lower for a very thin die opening (small cross section) and higher for large openings.

If product weight is specified rather than wall thickness for a process employing *inside-the-neck* blowing, the following approximation may be employed:

$$P_d = \left[D_d^2 - 2(W/T^2)Ld\right]^{1/2}$$

where

W = weight of part, g
L = length of part, in.
d = density of the resin, g/cm^3
T = wall thickness, in.

This system is applicable to most shapes and is of particular advantage for irregularly shaped objects.

A controlled parison is one in which the dimensions are partially controlled through tension, that is, the rotary wheel, the falling neck ring, etc. Because of this, the following relationships are employed (assume the parison is not free falling):

$$D_d \cong 0.9N_d$$

$$P_d \cong \left(D_d^2 - 3.6B_d t + 3.6t^2\right)^{1/2}$$

$$P_d \cong \left[D_d^2 - 3.6(W/T^2)Ld\right]^{1/2}$$

When a polymer is forced through a die (blowing-inside-the-neck), the molecules tend to orient in the direction of the flow (nonisotropic). As the extrudate leaves the die, the molecules tend to relax to their original random order. Parison drawdown, the stress exerted by the parison's own weight, tends to prevent complete relaxation. This results in longitudinal shrinkage and some swelling in diameter and wall thickness. It has been found for most HDPE BM resins that:

$$D_d \cong 0.5N_d$$
$$A_d \cong 0.5A_b$$

where

D_d = die diameter
N_d = minimum neck diameter
A_d = cross sectional area of the die
A_b = cross sectional area of the part

$$A_d = \frac{\pi}{4}(D_d^2 - P_d^2)$$
$$A_b = \frac{\pi}{4}\left[B_d^2 - (B_d - 2t)^2\right]$$

where

P_d = mandrel diameter, in.
B_d = part diameter, in.
t = part thickness at B_d, in.

Therefore,

$$A_d = 0.5A_b = 0.5\frac{\pi}{4}\left(B_d^2 - B_d^2 + 4B_dt - 4t^2\right)$$
$$\frac{\pi}{4}(D_d^2 - P_d^2) = 0.5\frac{\pi}{4}\left(-4t^2 + 4B_dt^2\right)$$

Dividing through by $\frac{\pi}{4}$ and rearranging terms:

$$P_d^2 = D_d^2 - 2B_dt + 2t^2$$

or

$$P_d = \left(D_d^2 - 2B_dt + 2t^2\right)^{1/2}$$

Also

$$A_b = \frac{W}{Ld}$$

where

W = part weight, g
L = part length, in.
d = resin density, g/cm^3

Therefore, since

$$A_\mathrm{d} = 0.5A_\mathrm{b}$$

$$\frac{\pi}{4}(D_\mathrm{d}^2 - P_\mathrm{d}^2) = 0.5\frac{W}{Ld}$$

$$D_\mathrm{d}^2 - P_\mathrm{d}^2 = \frac{4}{\pi}0.5\frac{W}{Ld}$$

$$P_\mathrm{d}^2 = D_\mathrm{d}^2 - \frac{2}{\pi}2\frac{W}{LD}$$

$$P_\mathrm{d} = \sqrt{(D_\mathrm{d}^2 - 2W/\pi Ld)^{1/2}}$$

The same derivation is employed for controlled parisons except that

$$D_\mathrm{d} = 0.9N_\mathrm{d}$$
$$A_\mathrm{d} = 0.9A_\mathrm{b}$$

As stated previously, the size selected for die bushing and mandrel depends on wall thickness of the finished blow molded part, the blow ratio, and certain resin qualities included in the above formulas for various PE BM resins. These qualities are parison swell (increase in wall thickness as the parison exits the die) and parison flare (ballooning of the parison as it exits the die). Both depend on processing conditions. It has been shown that calculations can be made for the general die dimensions.

The other dimensions of the die (approach angles and lengths) vary widely with machinery capabilities and manufacturer's experience. Calculations for these dimensions therefore are not given here. Generally, the land length of the die should be eight times the gap distance between the pin and the die. With tabular form, dimensions are:

Gap Size (in.)	*Land Length (in.)*
Above 0.100	1 to 2
0.030 to 0.100	$\frac{3}{4}$ to 1
Below 0.030	$\frac{1}{4}$ to $\frac{3}{8}$

Notice that the land length is at least $\frac{1}{4}$ in. (6 mm) regardless of gap size. This land length is necessary to get the desired parison flare.

The die should be streamlined to avoid abrupt changes in flow which could cause polymer melt fracture. When no further changes are expected in die dimensions, the die mandrel and bushing should be highly polished and chrome plated. This keeps the surface cleaner and eliminates possible areas of resin hangup. Finally, the edges of the pin (mandrel) and die should have slight radii to minimize hangup within or at the exit of the die area. The face of the mandrel should extend 0.010 to 0.020 in. (0.254 to 0.508 mm) below the face of the die to avoid a doughnut occurring at parison exit.

6.2.6 Die Shaping

It is sometimes necessary to deviate from round dies and mandrels for BM, even when finished products are cylindrical, to provide uniform wall thickness. In these instances, it is more desirable to leave the mandrel round and modify the design of the die. The reason is that in assembling the head and installing the die and mandrel in many BM machines, the screw threads that hold the mandrel in the machine are such that there can be no assurance that the mandrel will always be in the same position in relation to the die. Also, in most machines, weight adjustments for BM products are made by adjusting the mandrel either up or down in the head. In certain machines, the mandrel can turn during this adjustment. If the mandrel is made elliptical and rotation does occur, the finished product would be distorted. However, if the die is nonsymmetrical, it can always be installed in the same position in the machine.

An example of this would be a container with a diameter of 5 in. (127 mm) in which the pinch-off is to be kept inside the chimes of the container. In this case, the die diameter might be somewhere around 0.900 in. (23 mm) and the mandrel about 0.750 in. (19 mm), using converging tooling with an included angle in the die of 25° ($12\frac{1}{2}°$ per side). To have a uniform wall thickness around the part near the bottom it would be necessary to make the die (ring) elliptical by approximately 0.007 in. (0.18 mm) per side at right angles to the mold closing line.

Differences in extrusion pressure between BM machines may make it necessary to ovalize the die by more or less than the amount indicated. The only sure way to know that the die has the proper design and is ovalized enough is to “cut and try.” It is always better to remove metal cautiously, as it is easier to remove metal from the die than it is to add metal to it. It is suggested that before the die is cut in any way, a line be scribed across the die at the molding parting line and the front of the die be marked (one easy method is with a punch). In this manner, the die can always be installed in the machine in the same relative position.

While programming varies the entire wall thickness of the parison, die shaping introduces variations in the cross sectional area of a parison. A well designed die head extrudes a parison that is round and that has a uniform wall thickness. Whenever a round uniform walled parison is blown to form a square-shaped item, the wall thickness of the blown product will be less in the edges and corners than in the flat side surfaces. This occurs because the parison must stretch farther to reach the edges and corners.

To overcome this problem the parison is tailored or made thicker in that section that stretches the farthest so that the wall thickness in the edges and corners of the molded product is increased. Thick areas in the parison are made by removing metal from the corresponding section of the die. The metal can be removed from either the mandrel or the die ring (bushing). A square-shaped item with and without die shaping is shown in Fig. 6.10. Figure 6.11 shows the effects of ovalized die tooling on wall thickness uniformity. The die mandrel or bushing can be easily shaped by machining on the lathe or on a milling machine.

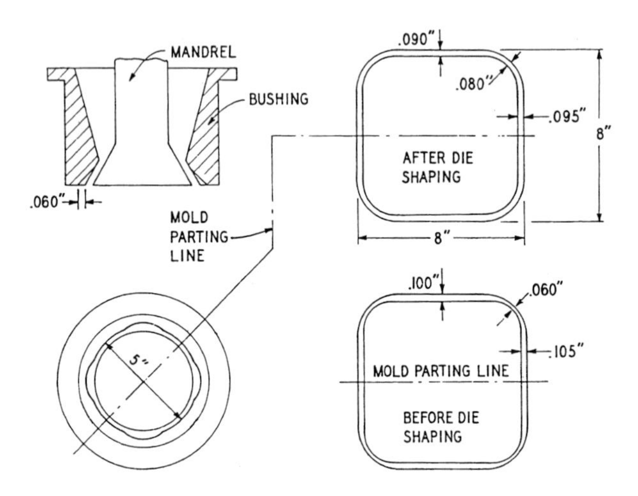

Fig. 6.10 *Wall distribution with and without a shaped die*

Fig. 6.11 *Ovalized tooling promotes wall thickness uniformity: (a) standard round tooling and (b) ovalized die tooling*

An example of a practical method to tailor a die is shown in Fig. 6.12. Here, three lines are scribed across the face on the die—one on the mold closing line, and one 30° on either side of the mold closing line. Metal is then removed on the 30° lines, using the "cut and try" method until the corners are sufficiently strong. The area between the 30° lines is then blended to a smooth radius. Owing to the pressure differential within the head of some BM machines, it has been found that in a situation such as this, turning the die 180° in the machine can make a difference in wall distribution in the corners of the bottles.

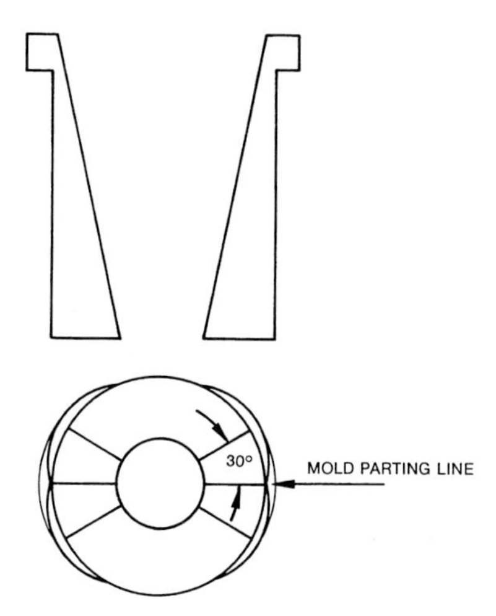

Fig. 6.12 Example of another tailored die

There are some applications, particularly in using diverging angles in the die and mandrel to make larger products, when it is more convenient to shape the mandrel than to shape the die. There are still other applications in which it is necessary to shape both the die and mandrel.

6.2.7 Die Orifice

Another important characteristic is the effect of the orifice shape (Fig. 6.13) on the melt. It is related to the melt condition and the die design (land length, etc.), with a slow cooling rate having a significant influence, especially in thick products. Cooling is more rapid at the corners; in fact, a hot center section could cause a product to blow outward and/or include visible or invisible vacuum bubbles.

Other factors are considered, such as the angle or taper of entry and the parallel length of the die land. For most thermoplastics (TPs), the entry angle must be as small as possible to ensure good product quality, particularly at high output rates. The abrupt changes in the direction of melt flow tend to cause rough or wavy surfaces and principally internal flaws; the condition is called melt fracture. Figure 6.14 shows examples of mandrel/die designs.

The approach used for shaping cavities in the dies is important, requiring three-dimensional (3D) evaluation that includes streamlining. Where possible, all dies should be designed to promote streamlined melt flow and avoid the obvious pitfalls associated with the areas that could cause stagnation: right angle bends, sharp corners, and sections in which flow velocities are diminished and are not conducive to streamlined flow. In certain cases such as profiles, complex shapes do not lend

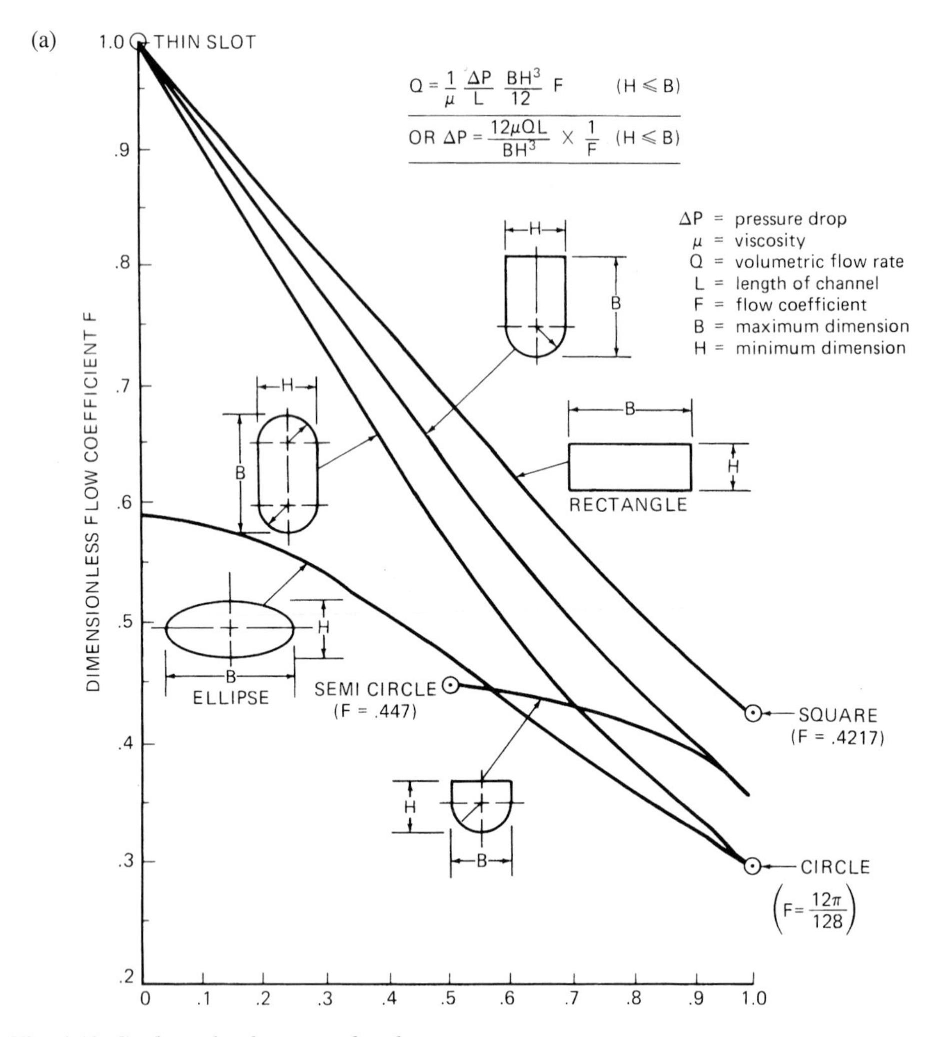

Fig. 6.13 Guide to developing orifice die opening

themselves to absolute streamlining. In these instances the stability of the plastic melt must be watched much more closely than one with a clean flowing die.

There are different approaches to developing the streamlined shapes. They range from totally trial-and-error to finite element analysis (FEA) (Chapter 8). The trial method usually involves gradually cutting or removing the die orifice metal. Between cuts an examination is made of the extrudate and the metal cavity surface to check on melt hangups, melt burning, streaks, and other stagnating problems. With FEA one can easily determine streamline flow patterns for some simple, well balanced wall thickness shapes using appropriate rheological plastic data and for others it provides a guide.

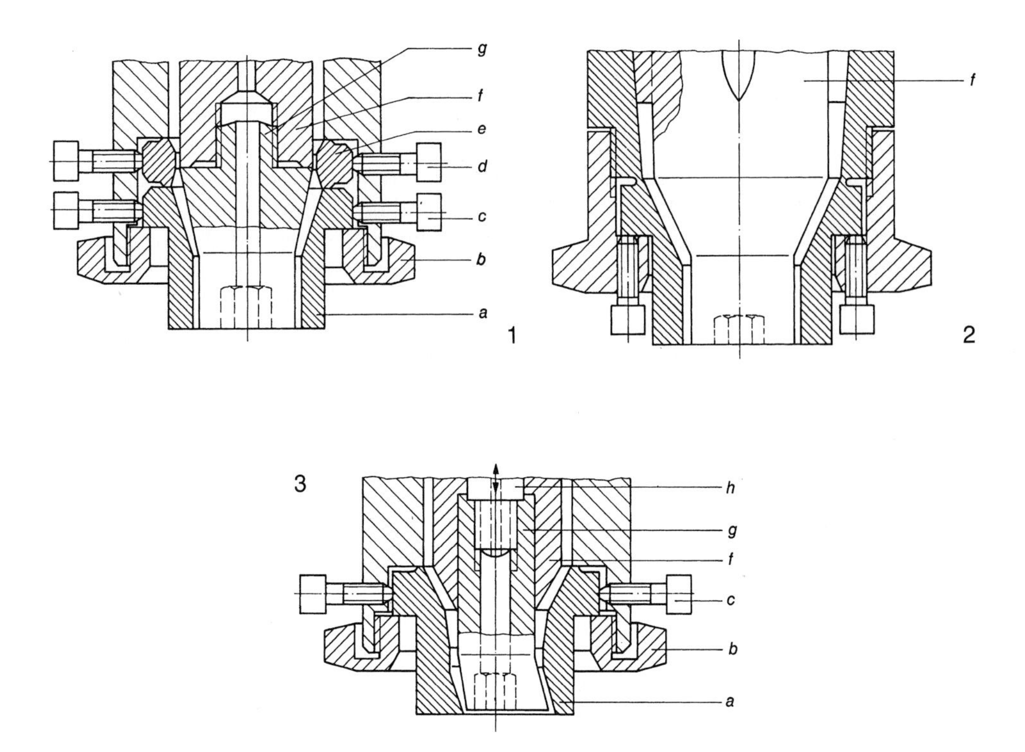

Fig. 6.14 *Typical mandrel/die designs: (1) die with conical inflow zone and adjustable choke ring; (2) die with torpedo, spider supports, and elastically deformable extrusion die head housing for die alignment; and (3) conical die for programmed adjustment of die gap. a, orifice; b, retaining ring; c and d, adjusting screws; e, adjustable choke ring; f, mandrel (or, second view, torpedo); g, core or interchangeable core, and h, sliding rod for programming die gap adjustment*

Streamlining can provide a variety of advantages: (1) Dies can operate at higher outputs; (2) pressure drops are lower and more consistent over a range of melt temperatures and pressures; (3) generally the melt uniformity across the extrudate is more uniform and shape control is enhanced; and (4) streamlining is sometimes crucial for high production output rates where plastics have limited stability and cause hangups/degradation going through nonstreamlined dies.

The land is the parallel section of the pin and bushing just before the exit of the die head in the direction of the melt flow. Its length is usually expressed as the ratio between the length of the opening in the flow direction and the die opening, for example, 10 : 1. It is vital to shaping the extrudate parison and providing thickness dimensional control. A very important dimension is the length of the relatively parallel die land, which in general should be made as long as possible. However, the total resistance of the die should not be increased to the point where excessive power consumption and melt overheating occur.

6.2.8 Coextrusion Die

The die is as important as the material for the coextrusion BM process. It has to receive the required amount of melt stream from each extruder and form it into a coherent parison with perfectly concentric material layers. As would be expected, the uniformity and the wall thickness distribution depend on precise rheological sizing of the die channels and very exact fabrication methods. Die design, fabrication and process control parameters are well advanced with both continuous and accumulator

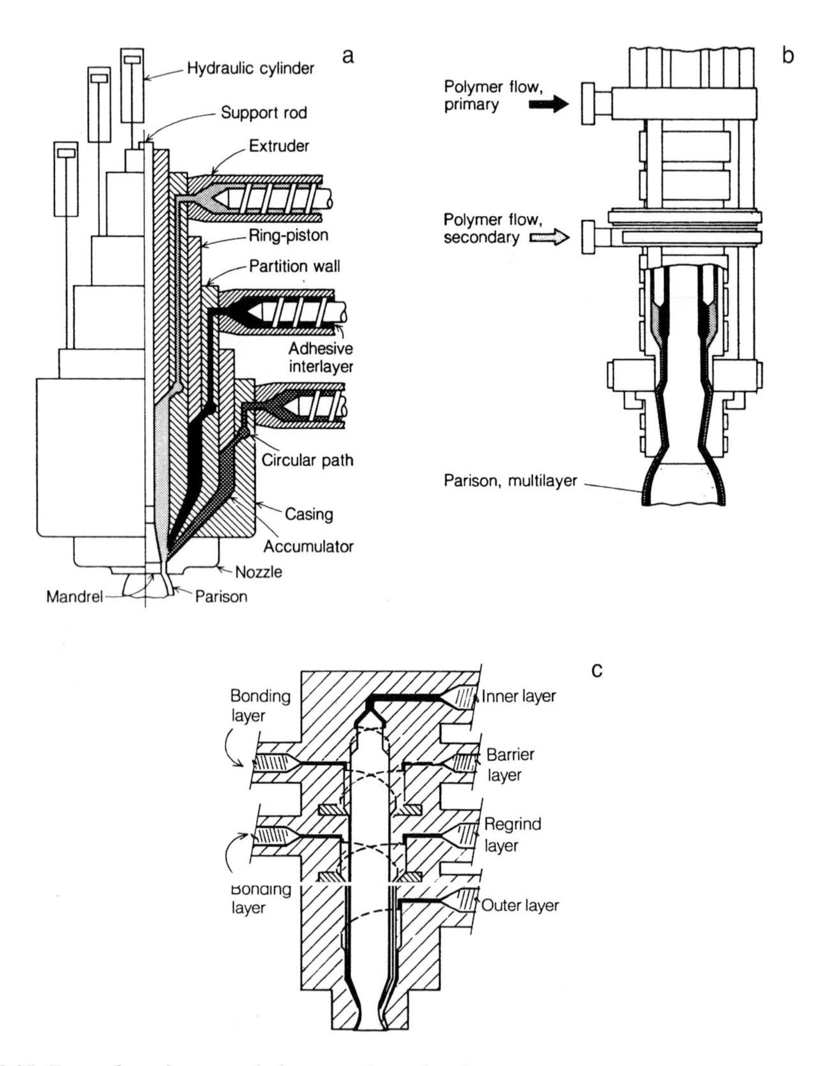

Fig. 6.15 Examples of coextruded accumulator head

extruders (as well as in the injection blow molding (IBM) process). Coextruded blown containers are in production throughout the world.

Accumulator head designs are capable of forming up to at least seven layers. An example of the coextruded accumulator die head is shown in Fig. 6.15. Multiple extruders feed resin into separate chambers within the head. Ring-piston rams extrude the materials to form a multiparison. Layer thickness is controlled by programming the pressure adjustment of the pistons.

The processing of thermally sensitive materials such as ethylene vinyl alcohol (EVOH), with a melt temperature of 180 to 200 °C (356 to 392 °F) between two layers of polycarbonate (melt temperature 240 to 260 °C (464 to 500 °F)), or PE having lower melt temperatures, is straightforward using presently available die heads and process controllers (Chapter 9).

The molds are generally light in weight and require only light clamping pressures. This allows great scope to the machine designers. Individual molds may close like books; be mounted together; or be fitted on reciprocating racks, rotating tables, or vertically arranged wheel-like structures [6]. Each approach has advantages and the selection should be considered only in relation to the product type and production volumes involved.

All must allow free access for cleaning and for component extraction. In-mold or external trimming may also be required to take place as the molding is removed. The mechanisms required for these operations are reasonably simple.

6.3 Extrusion Blow Molding

In EBM, plastic melt flows through the die to create the parison. The mold clamps around the parison and air is injected in the hollow center to fill the mold cavity. Several tooling elements and considerations need to be combined to create the final finished product. These considerations include not only the head but also the blowing, necking, and trimming actions.

6.3.1 Blowing the Parison

The air used for blowing serves to expand the parison tube against the mold walls, forcing the material to assume the shape of the mold cavity, thus creating on the surface details such as raised lettering, surface designs, and so forth. The air performs three functions: it expands the parison against the mold, exerts pressure on the expanded parison to produce surface details, and aids in cooling the parison. The BM pin can be located in different positions (Fig. 6.16).

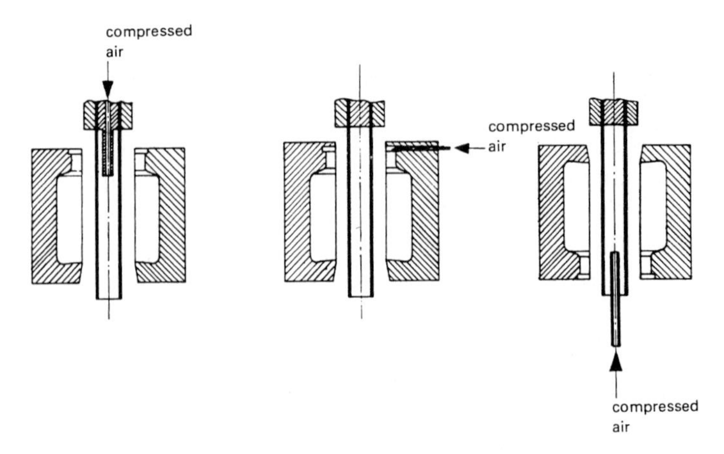

Fig. 6.16 *Example of three basic locations for the BM pin*

6.3.1.1 Air Flow

Blowing bottles and other containers requires air pressures that depend on the plastic being processed (Table 6.1). Air is introduced into the parison by various techniques: through the extrusion die mandrel, through a blow pin over which the end of the parison has dropped, through blow heads applied to the mold, or through blowing needles that pierce the parison. Parison programming, blow ratio, and product configuration usually control the wall distribution and thickness of the blown product.

As an example, extrusion heads can have an air path that allows for air to be introduced in the parison as it is being extruded to eliminate collapse of the parison. This allows for certain functions of blowing products in which the parison can be closed after it is cut and inflated for certain processes.

During the expansion phase of the blowing process it is desirable to use the maximum allowable air volume so that the expansion of the parison against the mold walls is

Table 6.1 *Example of Air Blowing Pressures for Different Plastic Melts*

Polymer	Pressure (bar)
Acetal	6.9/10.34
PMMA	3.4/5.2
PC	4.8/10.34
PE-LD	1.38/4.14
PE-HD	4.13/6.9
PP	5.2/6.9
PS	2.76/6.9
PVC-U	5.2/6.9
ABS	3.4/10.34

accomplished in the minimum amount of time. The maximum volumetric flow rate into the cavity at a low linear velocity can be achieved by making the air inlet orifice as large as possible. In the case of blowing inside the neck this is sometimes difficult. Small air orifices may create a venturi effect, producing a partial vacuum in the tube and causing it to collapse. If the linear velocity of the incoming blow air is too high, the force of this air can actually draw the parison away from the extrusion head end of the mold. This results in an unblown parison. Air velocity must be carefully regulated by control valves placed as close as possible to the outlet of the blow tube.

Normally gauge pressure of the air used to inflate parisons is between 30 to 300 psi (0.21 to 2.1 MPa). Often, too high a blow pressure will blow out the parison. Too little, on the other hand, will yield end products lacking adequate surface detail. As high a blowing air pressure as possible is desirable to give both minimum blow time (resulting in higher production rates) and finished products that faithfully reproduce the mold surface. The optimum blowing pressure is generally found by experimentation on the machinery with the product being produced.

Air blowing against the hot plastic can result in freeze off and stresses in the bottle at that point. Air is a material just as is the parison and, as such, it is limited in its ability to blow through an orifice. If the air entrance channel is too small, the required blow time will be excessively long, or the pressure exerted on the parison will not be adequate to reproduce the surface details of the mold. General rules to be used in determining the optimum air entrance orifice size when blowing are given in Table 6.2.

Table 6.2 *Air Orifice Sizes*

Orifice diameter		Part size (vol.)	
in.	mm		
$\frac{1}{16}$	1.6	Up to one quart	(0.95 liter)
$\frac{1}{4}$	6.4	1 quart–1 gallon	(0.95–3.8 liter)
$\frac{1}{2}$	12.7	1 gallon–54 gallons	(3.8–205 liter)

The pressure of the blowing air will vary the surface detail in the molded product. Some PE products with heavy walls can be blown and pressed against the mold walls by air pressures as low as 30 to 40 psi (0.18 to 0.25 MPa). Low pressure can be used because items with heavy walls cool slowly, giving the plastic more time at a lower viscosity to flow into the indentations of the mold surface. Thin-walled products cool rapidly: therefore, the plastic reaching the mold surface will have a high melt viscosity. Higher pressures (in the range of 50 to 100 psi (0.3 to 0.7 MPa)) will be required. Large items such as 1-gallon bottles require increased air pressure (100 to 150 psi (0.7 to 1 MPa)). The plastic has to expand further and takes longer to get to the mold surface. During this time the melt temperature will drop somewhat, producing a more viscous plastic mass in which it requires more air pressure to reproduce the details of the mold.

A high volumetric air flow at a low linear velocity is desired. A high volumetric flow gives the parison a minimum time to cool before coming in contact with the mold and provides a more uniform rate of expansion. A low linear velocity is desirable to prevent a venturi effect from collapsing a portion of the parison while the remainder is expanding. Volumetric flow is controlled by line pressure and orifice diameter. The flow control valves that are located as close as possible to the orifice control linear velocity.

Blow time for an item may be computed from Table 6.3.

Table 6.3 *Discharge ft^3/s at 14.7 psi and 70 °F with Blow Time Formula*

	Orifice diameter in. (mm)			
Gauge (psi)	$\frac{1}{16}$ (1.6)	$\frac{1}{8}$ (3.2)	$\frac{1}{4}$ (6.4)	$\frac{1}{2}$ (12.7)
5	0.993	3.97	15.9	73.5
15	1.68	6.72	26.9	107
30	2.53	10.1	40.4	162
40	3.10	12.4	49.6	198
50	3.66	14.7	58.8	235
80	5.36	21.4	85.6	342
100	6.49	26.8	107.4	429

This is the free air per minute, but since there will be a pressure build up as the parison is inflated, the blow rate of air has to be adjusted.

$$\text{Blow time} = \frac{\text{Mold vol. ft}^3}{\text{ft}^3/\text{s}} \times \frac{\text{Final mold psi} - 14.7\ \text{psi}}{14.7\ \text{psi}}$$

The ft^3/s is converted from Table 6.2, with line pressure and orifice diameter known, to cubic feet per second. For purposes of the calculation, the final mold pressure is assumed to be the line pressure.

Actually the blow air is heated by the mold, raising its pressure. Calculations ignoring this temperature effect will be satisfactory when blow times are under 1 s (for small to medium items), but if blow times are longer, the air will have time to pick up heat, resulting in a more rapid pressure buildup and shorter than calculated blow times.

The air that expands the parison against the mold cavity enters through blow pin tooling, which may be a simple tube, needle, ram-down or calibrated prefinish, or pull-up prefinish. When the tube style is employed, the mold is closed around the parison, which is closed and sealed around the tube. The tube usually enters the cavity through the center of the extrusion mandrel. When the needle style is employed, a fine needle is inserted through the parison from a remote position in the mold after the mold has been closed. The hole is usually inconspicuous or hidden.

Both the tube and the needle are used mainly for industrial products. If used for bottles, a post-machining or facing of the neck finish is required. Plastic chips falling into the bottle led to the development of bottle prefinishing systems. In these systems the mold neck ring and blow pin form the neck finish to specification before the bottle is removed from the mold cavity.

Figure 6.17 provides an example of BM with escaping blowing air. The blowing mandrel is fitted to the mandrel shaft in such a way that it can be easily replaced. With multicavity molds, it is usual to employ a crosshead on which the blowing mandrels are arranged in series. Mandrel chains can also be used for product ejection. The channel for the blowing medium should be as large as possible. Angled radial orifices at the tapered end of the mandrel achieve a fast distribution of the cooling media and turbulence.

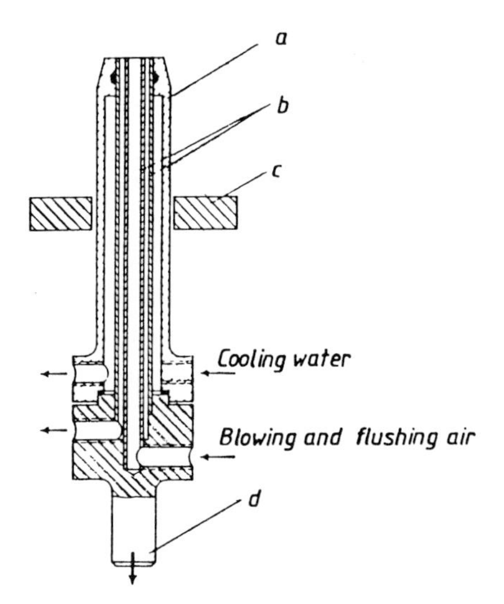

Fig. 6.17 Blow molding with escaping blowing air: (a) replaceable blowing mandrel, (b) channel for the blowing air, (c) stripping yoke (bubbler for the water has not been illustrated), (d) shaft

To aid in blowing the parison, the spreading of the parison to a predetermined width and its correct positioning in the mold can be accomplished using spreading mandrels, hooks, claws, and suction devices. The upper end in this instance is held and the lower, open end is spread from inside. The spreading mandrels can furthermore be used to perform the calibration of the neck opening. The device can be swivelled sideways, if necessary, with an angled mold parting.

In addition, preblowing and support air can be used to produce blown items. The parison is first welded together by pincers and cut off semicircularly if necessary. Preblowing has to be initiated with a consideration of the required width and also provides a uniform wall thickness distribution. As discussed, possible surface defects can result due to premature contact with mold areas. When extruding an open parison, support air is also used to prevent the collapse of the parison and adhesion of the internal walls to themselves. Calibration with an inserted stepped mandrel is also initiated with support air.

6.3.1.2 Needle Blowing

BM by hollow needle insertion at the mold parting line is considered when a very small opening is required that may even need to be closed or if the hollow object must be blown up on the side (Fig. 6.16). The needle is led right up to the mold cavity by the retainer and moved by a cylinder. This device should be fastened to one mold half, and the arrangement should be provided close to a firmly held section of the parison.

Needles can be introduced practically anywhere on a parison. For example, if the needle is required in an area adjacent to water lines or cooling areas, or at the top or bottom of the product, movement of the needle during the cycle is required. The type of molding machine available has a strong influence on the blow opening used. Some machines, for instance, use needle blowing exclusively, excess material being machined off to provide larger openings when required.

The blowing chamber above the neck of the hollow article is known as the lost head, which always becomes a necessity when the width of the neck no longer allows for calibration. In these instances one is concerned with wide neck containers. A height should be chosen that ensures uniform wall thickness of the neck. A ratio of diameter to height of 2 : 1 should be used. When blowing with escaping air and for fast venting it is recommended that a tear-off line in the blowing chamber be provided.

When troubleshooting, recognize that moisture in the blowing air can cause pock-marks on the inside product surface. This defective appearance is particularly objectionable in thin walled items such as milk bottles. A system of separators and traps to dry the air taken into the air compressor can prevent this problem.

6.3.2 Pinch-Off

The pinch-off is the section of the mold where the extruded parison is squeezed and welded together. It must: (1) have structural soundness to withstand the pressure of the plastic material and to withstand the repeated closing cycles of the mold; (2) push a small amount of plastic material into the interior of the product to slightly thicken the weld area; and (3) at the same time, cut through the parison to provide a clean break point for later flash removal. To accomplish this, most molds use a double angle pinch-off or, in special cases, a compression pinch-off to fulfill these criteria (Fig. 6.18).

The double angle pinch-off is rugged (Fig. 6.18c, right view). The 0.010 in. (0.254 mm) land will stand up to the pressures and still provide a clean break point. The 30° angle provides a slight back pressure and causes the melt to thicken the weld area. The length of this 30° area is controlled by the second angle at 45°.

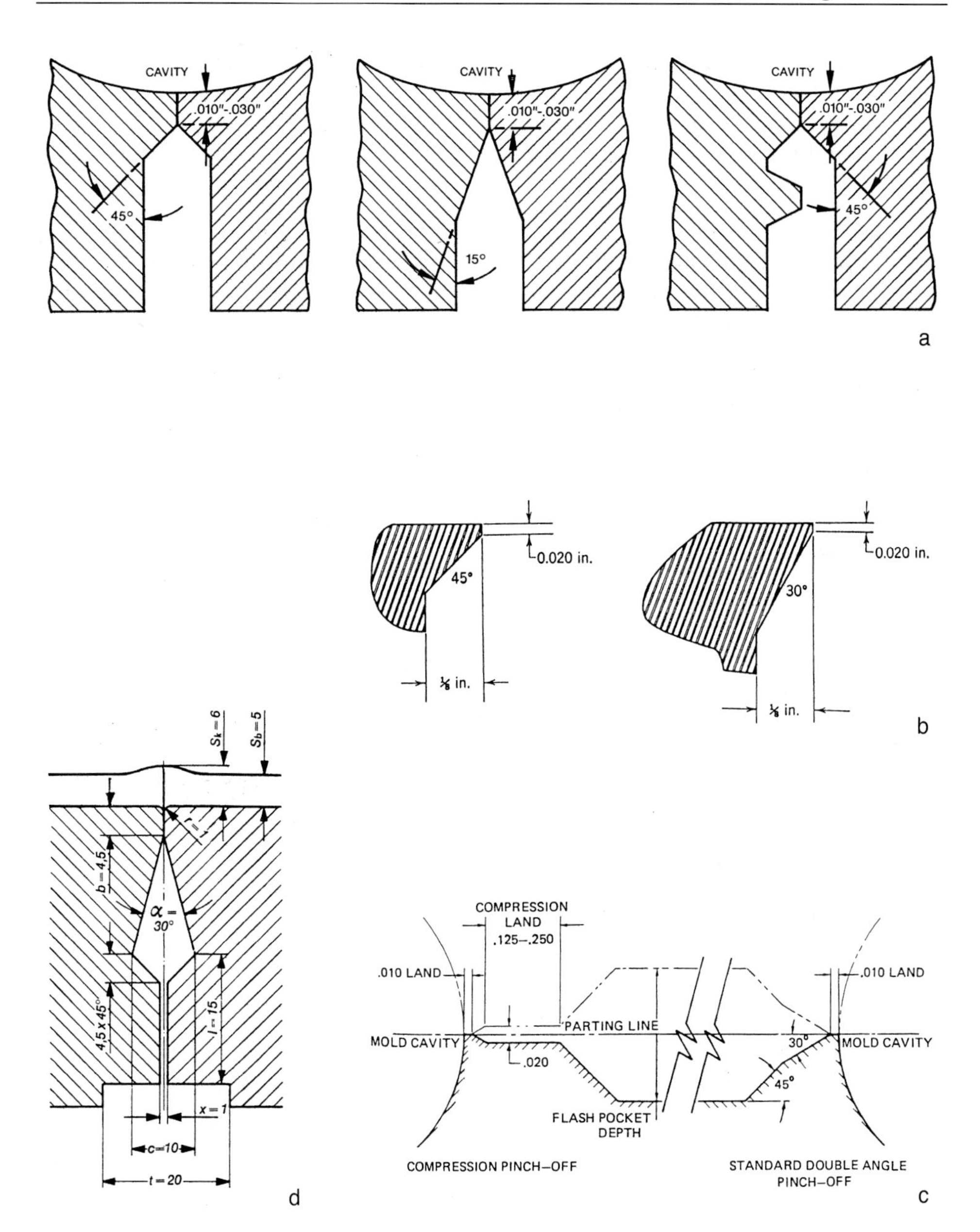

Fig. 6.18 *Examples of pinch-off designs (view d has metric units)*

A compression pinch-off is used in special cases in which the parison is stretched a great deal by the mold just before the squeeze and weld, which will most often occur in the sharp radius of the handle opening in handled containers. The weld is usually very thin and weak. To overcome this problem, a compression pinch-off with a land of $\frac{1}{8}$ to $\frac{1}{4}$ in. (3 to 6 mm) is used to push material into the interior of the product.

A design is shown in Fig. 6.19 that provides the formation of a holding flange on the open parison to facilitate parison transfer. The left view is in the starting position, the center view shows pinching-off and flange formation, and the right view has the parison being transferred.

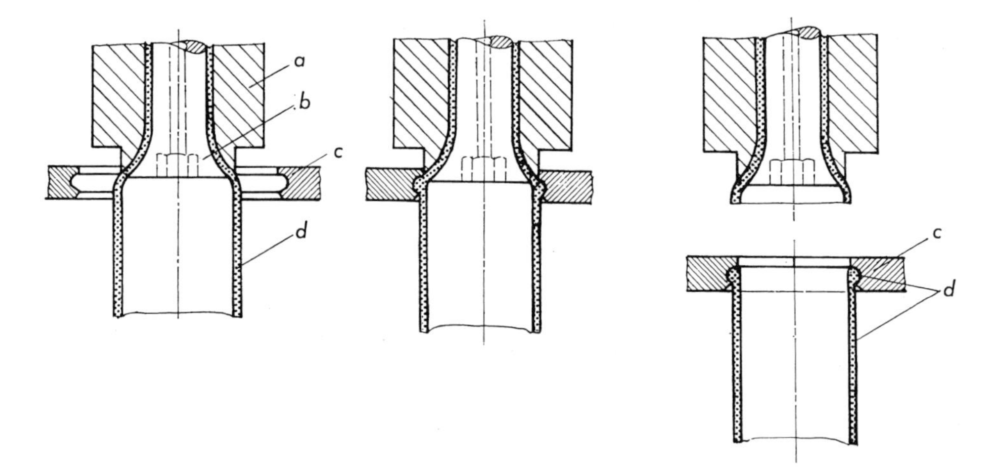

Fig. 6.19 *Parison pinch-off design a, die orifice; b, die core; c, transfer jaw or claw with annular groove; d, parison*

The flash pocket depth is related to the pinch-off area and is very important for proper molding and automatic trimming of the part. Three factors determine flash pocket depth: the approximate density of the plastic material to be used, the weight of the part and the approximate weight of the parison, and an estimate of parison diameter.

A gross miscalculation of pocket depth can cause severe problems. For example, if the pocket depth is too shallow, the flash will be squeezed with too much pressure, putting undue strain on the molds, mold pinch-off areas, and machine press section. The molds will be held open, leaving a relatively thick pinch-off that will be difficult to trim properly. If pocket depth is too deep, the flash will not contact the mold surface for proper cooling. Between the molding and the autotrimming of the container, heat from the uncooled flash will migrate into the cool pinch-off, softening it. In the trimming cycle, the softened container pinch-off will stretch instead of breaking free.

Knife edged cutters sever the parison. The width of the pinch-off edge depends on the plastic material to be processed and on the wall thickness of the BM product as well on the size of the relief angle, the closing speed, and the time blowing starts. For small products in the volume range of up to 10 ml, the width may be 0.1 to 0.3 mm (0.004 to 0.012 in.).

When processing LDPE the narrowest edge is sufficient. The rule of thumb, width of edge in mm = cube root of the volume in liters, can be applied using the equation $t = (v)^{1/3}$.

Edge tapering starting from the center can also prove advantageous. The internal edge should be machined smooth and possibly broken slightly to prevent surface damage on the parison or the molded product during closing and opening.

The mold can contain a pinch relief behind the pinch-off that is of uniform or increasing depth toward the outside. The depth of this relief should amount to nine tenths of the parison wall thickness per mold half so that it is truly compressed and the flash is cooled sufficiently. Besides this customary design, molds have been known to contain one-sided pinch relief.

A relief angle should be chosen that prevents the pinch-off edges from entering the parison during mold closing but allows sufficient melt to be pushed into the seam during the welding process so that the surplus can enter the pinch relief. Relief angles of 15° to 45° are the result, which generally speaking should increase with increasing edge width. It must be noted that too great a relief angle can hinder the adequate cooling of the edge zones, which act as barrier between flash and the wall of the blow molded product. This in turn could result in dull spots next to the weld line. With too small a relief angle, a cooled-off seam might be pushed into the mold cavity and folded under sideways during blowing, resulting in a deep groove along the weld line.

Outside of the pinch-off blades, the mold should be relieved on each side to compress and cool the tail to avoid sticking between ejected bottles so that the mold completely closes. For easy removal of the cooled tail, the pinch lands may be relieved 2° on each blade away from the cavity. The pinch blade should be at least $\frac{1}{4}$ in. longer than the width of the flattened parison, as any portion of the parison pinched outside the blade area will have poor strength. Good pinch design forces a bead into the closed mold while poor pinch design creates a thin grooved pinch. Processing conditions, irrespective of pinch design, can also influence pinch integrity.

When employing dams, an escape for the overflow of surplus material must remain open; otherwise the tool might be damaged. It becomes increasingly more difficult to achieve the required weld line quality with increasing wall thickness. Apart from dams and processing measures, the possibility exists with many hollow articles to externally reinforce the weld in the shape of a bead. To ensure that the compression provides a good weld, the width of the bead in the transition zone should not be substantially larger than the wall thickness. Delayed blowing is recommended to prevent a tearing to the weld line. To avoid notch sensitivity, all edges should be rounded off appropriately. Triangular and trapezoidal beads, as well as rounded beads, are acceptable.

Knife inserts can be employed in the pinch relief to facilitate the regrinding of solid or bulky flash. The occasional sticking of the still warm flash, which may fold over when ejecting the molded product, can be avoided in individual cases by designing the mold parting line in the flash area in the shape of an arch or with multiple angles.

A good pinch-off insert, as shown in Figs. 6.20 and 6.21a, has a land before it flares outward in a relief angle. This pinch-off insert should not have knife edge as these could tend to yield a grooved weld for certain plastics as shown in Figs. 6.21b and 6.22. The weld line should be smooth on the outside and form an elevated bead on the inside. Other uses for inserts include replaceable neck finishes to allow making

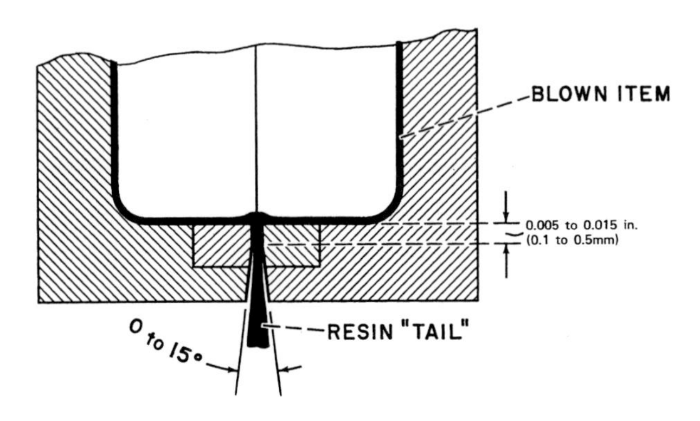

Fig. 6.20 *Pinch-off insert*

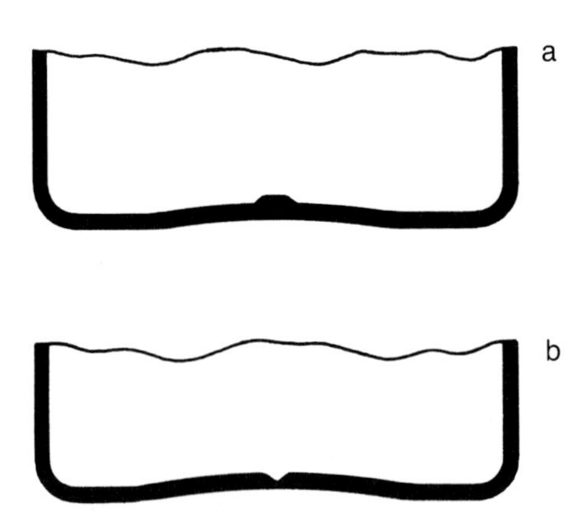

Fig. 6.21 *Good weld line (a) and weaker line (b)*

bottles with different neck sizes from a single mold, and also to incorporate in products textured areas or lettered information or trademarks that can be changed, making some molds more versatile.

A badly designed pinch-off can lead to a poor weld and excessive thinning at the weld. In extreme cases containers may be incompletely blown owing to the formation of a hole in the pinch-off area. The pinch-off must also be designed to permit easy removal of flash. Figure 6.22 shows a bad pinch-off design for linear PE. The same or similar designs have been found to be satisfactory with PVC, polystyrene (PS), and other common thermoplastic materials. The exact width of the land may require alteration to meet a specific molding condition or product design. For example, tough engineering type materials such as the acetals and PCs require much smaller pinch-off widths. Pinch-off for these materials should be in the range of 0.002 to 0.005 in. (0.05 to 0.12 mm). The relief area should be shallow enough to cool the plastic, but if it is too shallow it may prevent the mold from closing completely. A thickness of about 80% to 90% of the total thickness of plastic in the parison tube has been found

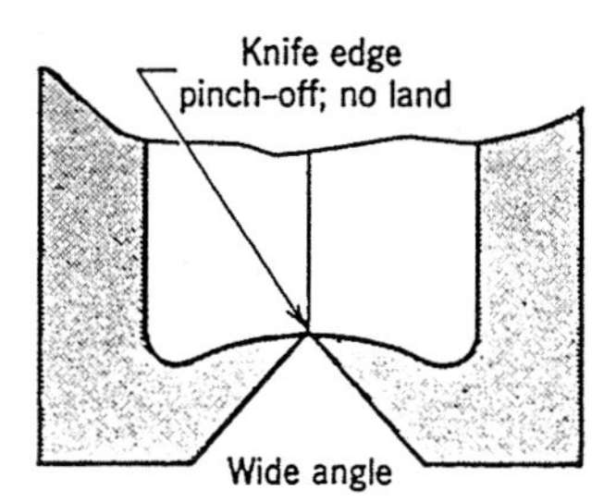

Fig. 6.22 *Expect an unacceptable pinch-off*

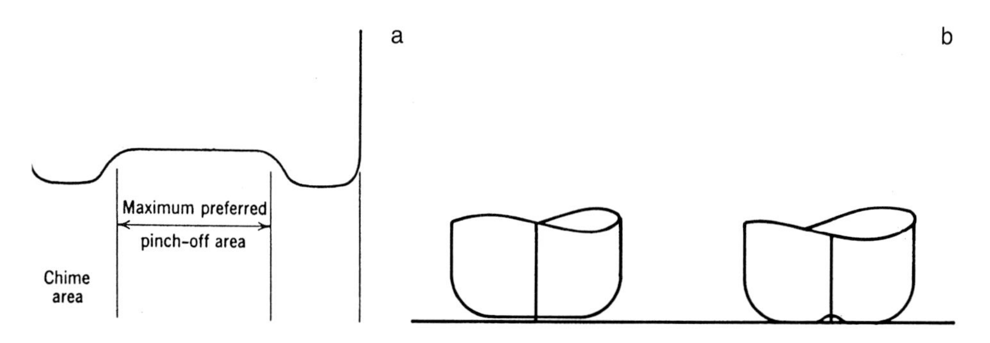

Fig. 6.23 *(a) Chime area where pinch-off should be located. (b) Bottom of bottle standing on hard surface. Left: Pinch-off in chime bears load and contributes to stress cracking. Right: Pinch-off no longer bears load or contributes to stress cracking.*

to be most satisfactory. If possible, pinch-off for bottles should be designed so as not to extend into the chime (edge or rim) of base (Fig. 6.23) to avoid weakening the chime against drop impact and environmental stress-cracking conditions.

6.3.3 Molded Neck Prefinishing

For smaller parts and containers automated processing can be used to deflash and finish the neck. Prefinishing of container necks in the mold is the most efficient technology in container production. Neck, shoulder, and tail flash can be removed in a secondary deflashing or finishing operation. An additional facing operation on the container lip also can be performed at this finishing station. The containers can then be conveyed directly to decorating and filling or pack-off stations.

Machines that use two separate stations to extrude the continuous forming parison and blow the product can completely finish certain containers in the machine. Molds

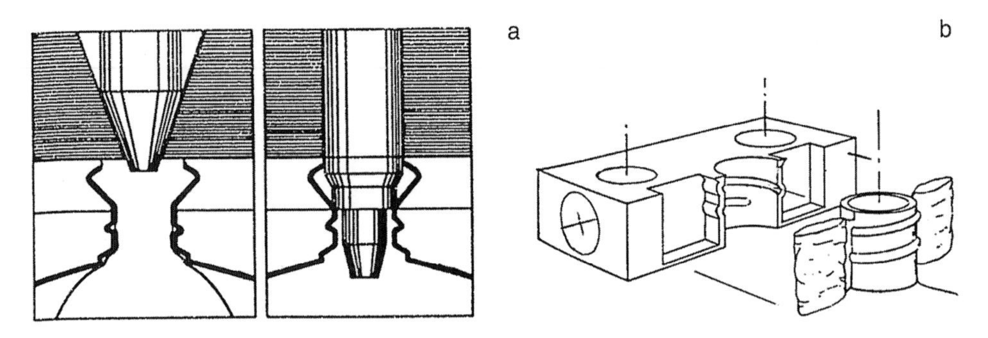

Fig. 6.24 Molded neck prefinished in mold. (a) Left: Parison extrusion station; right: blow molding station. The container neck is partially blown at parison stations. The molds then shift laterally to the blow pin station, where the container neck is finished and the containers are blown; (b) replaceable neck insert

close on the parison can partially blow the neck. The molds then shift laterally to the blow pin station, where the container is blown and the neck is finished. Tail flash is removed during transfer. This type of finish could be limited to center neck containers with neck sizes up to 48 mm and where the container is blown inside the neck diameter. There is no outside shoulder flash (Fig. 6.24).

The neck ring and the blow pin work together to form the container's finish. The finish, the area of the container where the closure will be applied, is the relationship of the threads to the top sealing surface and the shoulder of the container.

The neck ring can be an aluminum block mounted to the top of the mold body. For proper cooling, water lines in the neck ring are interconnected with water lines in the mold body. In the prefinish systems the blow pin has an important role in molding the finish. In the dome systems, the blow pin is not as critical, and in some applications it is not required at all. With the exception of cam and ratchet locks used on some specialty closures, most of the finish in the neck ring is cut on a lathe. The shrinkage applied to the neck is carefully calculated on the finishing technique used and the product wall thickness. The shrinkage amounts used for the neck are often different from that applied to the cavity.

With the dome system, the neck ring forms only the threads of the finish. The blow pin, if required, is used to seal and channel air inside the parison. After molding, the dome top is cut off and the container sized and trimmed to specification.

Three basic types of dome neck ring are used on molds. The one chosen depends on the type of container and the type of trimming equipment. The first type is a

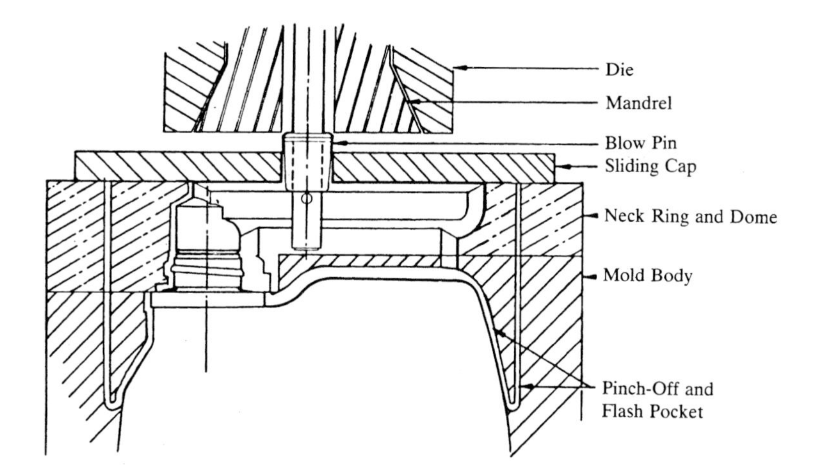

Fig. 6.25 *Neck ring with dome for "guillotine and face" method*

"guillotine and face method", which requires a blow pin and is used for handled and flat oval containers requiring the parison to flash outside the neck (Fig. 6.25). After molding, the trimmer guillotines the dome top from the container and machines the pour lip and inside diameter of the finish with a facing cutter [18].

The second type is a "fly cut" method used for wide-mouth containers where the parison will fall inside the neck (Fig. 6.26). A blow pin is not required; the parison is sealed with the mold against the die of the heat tooling. The dome is removed when the fly cutter sizes the inside diameter of the finish. The third type is the "spin-off" method (Fig. 6.27), used for any container in which the parison will fall inside the neck. A blow pin is not necessary. A special dome, shaped like a pulley, is molded above the finish. When trimmed, a V-belt fits into the dome, spinning the container against a stationary knife blade, which cuts through the dome top in the sharp groove at the top of the finish.

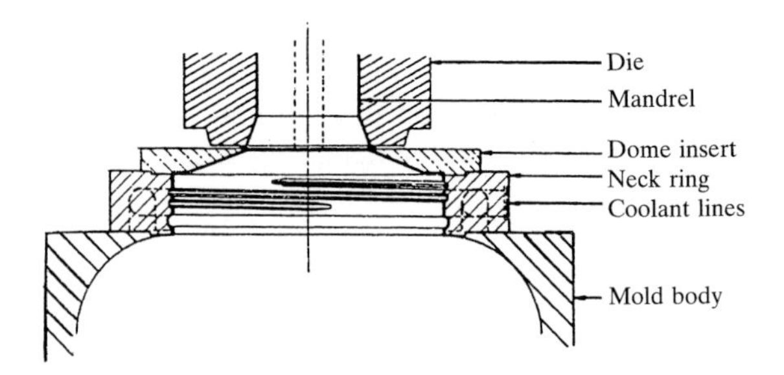

Fig. 6.26 *Neck ring and dome for "fly cut" method*

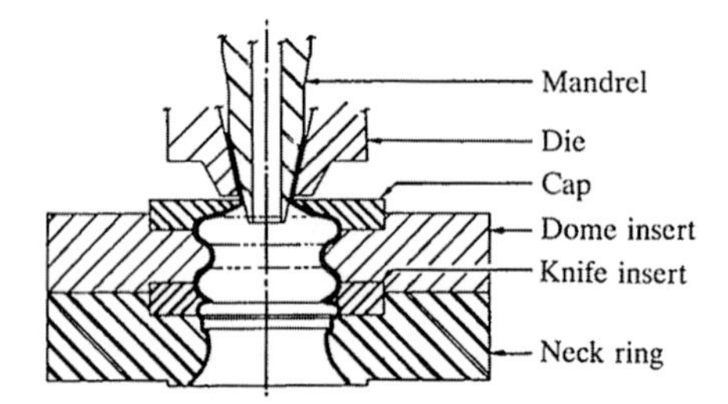

Fig. 6.27 *Neck ring and dome for "spin-off" method*

With the prefinish system, the neck ring and blow pin work together to mold and size the finish to specification before the container is removed from the mold cavity. Two basic types of prefinish neck ring are used on molds; both require that the parison be extruded over the blow pin and that flash is outside the neck. The one chosen depends on the type of closure used. The first type is the "pull-up" method, used primarily on dairy containers (Fig. 6.28). The closure used on these containers usually seals on the inside diameter of the finish. With the patented pull-up system, the blow pin moves upward just before the mold opening, shearing plastic material to form the finish

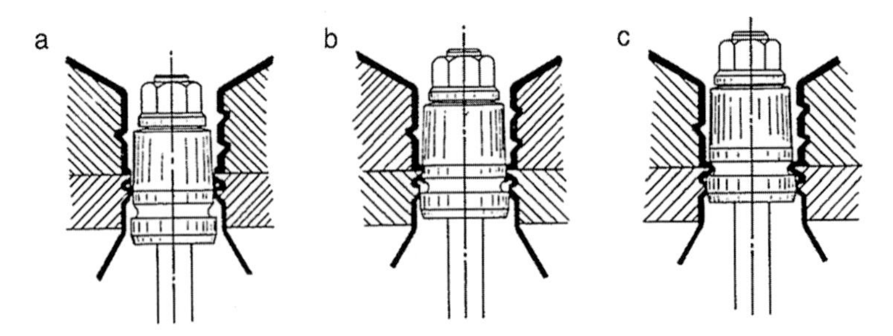

Fig. 6.28 *Pull-up prefinish system for center neck containers: (a) mold closed, (b) blown pin halfway up, (c) blown pin up*

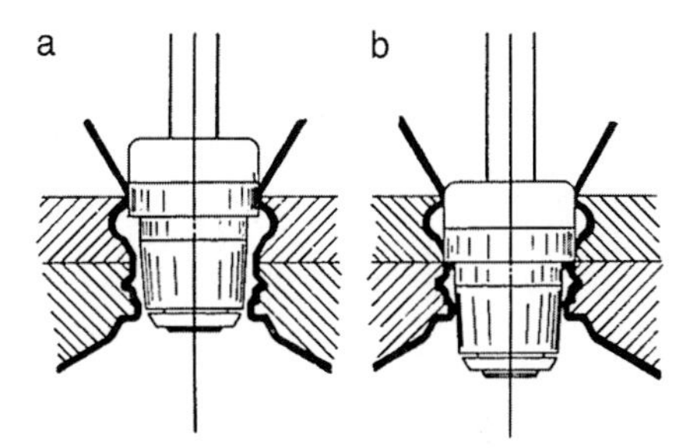

Fig. 6.29 Ram-down prefinish system for center neck containers: (a) mold close position, (b) blow pin down

inside diameter. The diameter of the hardened shear steel mounted to the top of the neck ring and the hardened shear ring mounted to the blow pin are held to a very close tolerance. This ensures that the finish inside diameter will be accurate, smooth, and free of burrs.

The second type of prefinish is the "ram-down" patented method (Fig. 6.29). Containers designed for a closure requiring a flat, smooth top-scaling face on the finish will generally use this method of prefinish. Immediately after the mold closing the blow pin moves down, pushing material into the finish. The stroke of the blow pin is adjusted so that the shear ring mounted on the blow pin contacts the shear steel mounted in the neck ring without interference, pressure, or stress. The diameters of the hardened shear steel and shear ring are also held to very close tolerances, which ensures that the container finish is held within specifications. This prefinish method can also be used for containers with an off-center neck or with the finish offset from the center line. The sequence of operation is the same except the mold side-shifts after the molds close and before the blow pin downstroke (Fig. 6.30).

Special features that can be added to molds to deal with unusual problems include side cores for deep undercuts, needle blow pins for container/cover combinations, and "sealed blow" for sterile applications. In "sealed blow" the container, prior to opening the molds, is sealed to maintain the special sterile conditions inside [18].

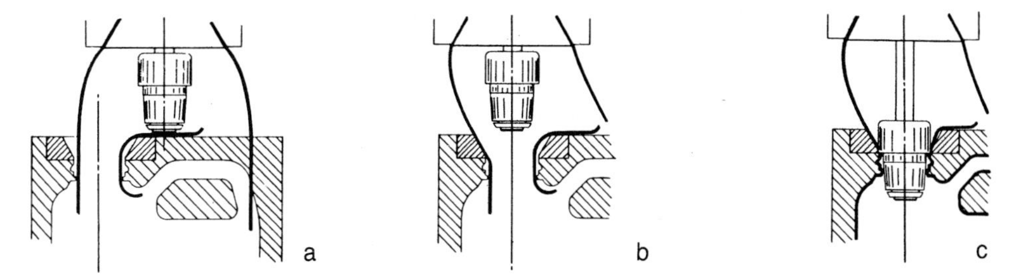

Fig. 6.30 Side-shift prefinish system for off-center neck: (a) mold close position; (b) side shift position; (c) blown pin down

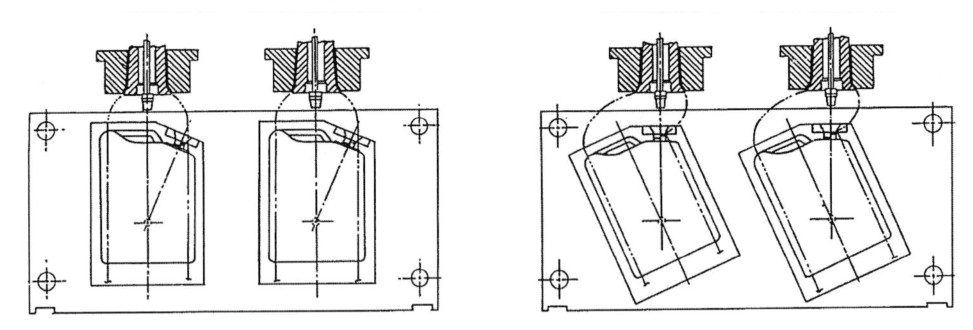

Fig. 6.31 Container is partially blown in the erect position, then rotated to align the blow pin with the container neck to complete the blow molding process

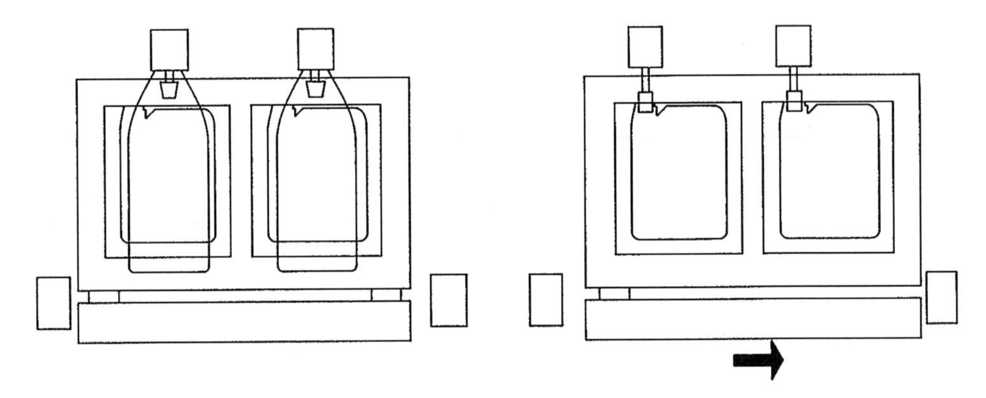

Fig. 6.32 BM off-center-neck containers with handles; the mold's side action follows blowing action

Two other special patented features developed are the "rotating mold" and the "sliding bottom." The "rotating mold" was developed for container designs that have a finish centerline at some angle to the container centerline (Figs. 6.31 and 6.32). The system permits the use of the ram-down prefinish method. The "sliding bottom" is a feature that prepinches and seals the bottom of the parison, permitting it to be puffed up like a balloon. This device will allow a smaller diameter parison to be used than normally expected. The smaller parison pinch that results at the bottom is of value on certain containers filled with a stress crack product.

There are different methods for molding necks. Figure 6.33 shows a method in which the filling is integrated during the BM process. Another method is displayed in Fig. 6.34, which shows a calibrated blowing mandrel and neck insert.

The straight-sided calibrating mandrel is usually introduced into the parison or preform before the mold is clamped. Unwanted melt can then be squeezed

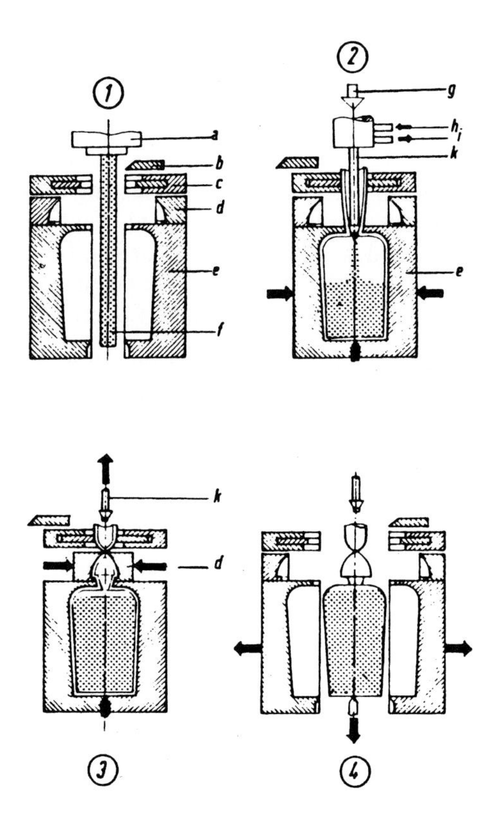

Fig. 6.33 *Integrated filling during blow molding*

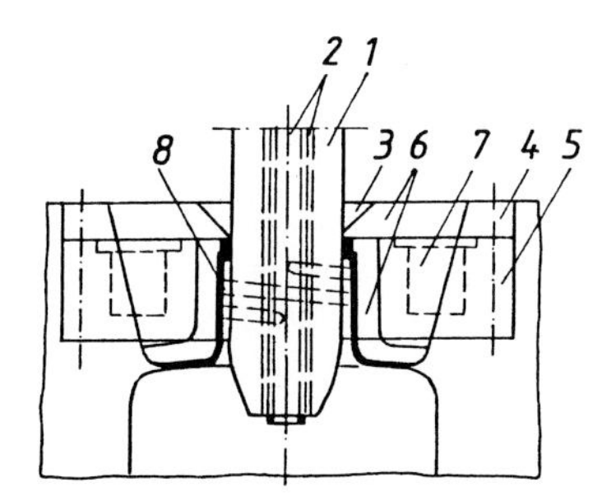

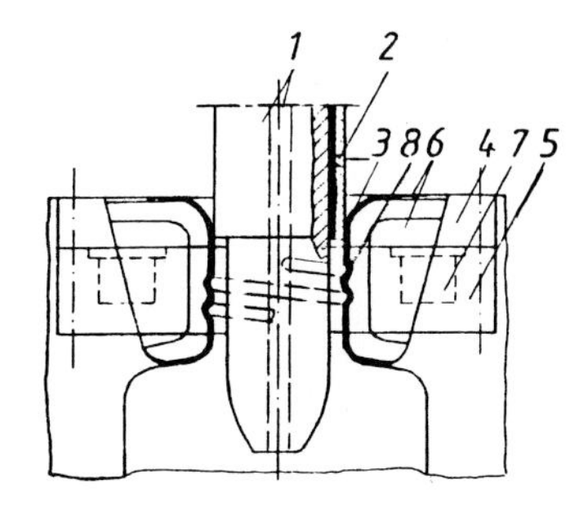

Fig. 6.34 *Calibrated blowing mandrel and neck insert.* ***Left: 1, straight-sided calibrating blowing*** *mandrel; 2, blowing air channel; 3, ring-shaped pinch relief; 4, neck knife; 5, neck insert with interrupted thread in the parting line; 6, pinch relief in the neck-shoulder area; 7, neck cooling; 8, thread pitch; the neck is stepped at the top to auxiliary air from escaping even with very thin walls.* ***Right: Stepped mandrel.*** *1, blowing air channel; 2, mandrel shoulder and cutting sleeve with venting gap (0.05 mm) and exhaust channels; 3, neck flash; 4, hardened cover plate; 5, neck insert with continuous screw threads; 6, pinch-relief; 7, neck cooling; 8, thread*

out into the relief. Further characteristics of this type of calibration are the weld around the inner edge and the radial weld on the horizontal sealing surface of the neck. The calibrating mandrel should be roughened slightly to facilitate venting.

The stepped mandrel results in a neck opening possessing only an external weld line. The horizontal sealing surface can be flat or profiled. Venting of the mandrel step is very important, and is obtained with a small annular gap [up to 0.05 mm (0.002 in.)] that blends into radial venting channels. The actual step is subject to wear and therefore is designed as a replaceable sleeve that is press fit on the mandrel.

A cup-shaped annular gap facilitates the insertion but makes calibration more difficult. A trumpet shape, however, makes setup and mandrel guidance easier but more readily permits material to be pushed through into the mold cavity. The conical annular gap is an acceptable compromise. These two mandrel designs are compared with each other in Fig. 6.34.

To standardize fabrication and help reduce the cost of blow pins, a simplified neck finish separation can be used. Machines can be supplied with basically a three-step calibration (Fig. 6.35). The calibrating blow pin is equipped, above its normal cutting sleeve, with a supplementary sleeve that has an undercut for better adhesion and removing the neck flash. This ensures that the finished bottle will always be first to be evacuated from the mold or be stripped from the blow pin, before the neck flash is removed from that pin. This latter will only be carried out when the mold has moved from the calibration station to the die head. Thus, sticking together of bottle and neck flash is avoided. Moreover, special channels for neck or bottom flash evacuation may be installed to collect flash in bins or transfer it to conveyor belts.

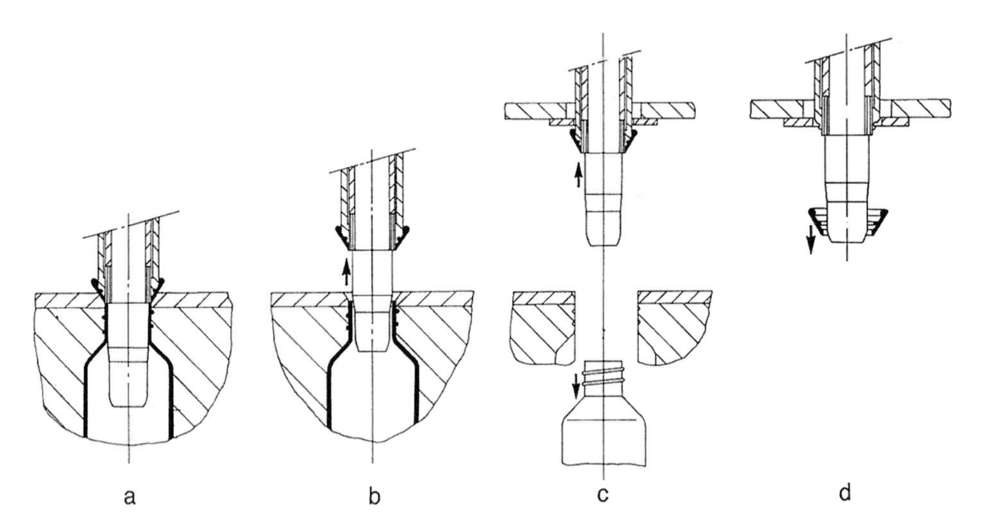

Fig. 6.35 Stripping neck flash from blow pin. (a) Situation during blow molding; (b) tearing off neck flash; (c) removing mold from container; (d) stripping neck flash from blow pin

When threads are included with neck, for ease of molding and simplifying flash control, the thread should start one pitch below the rise of the neck and finish similarly above the shoulder. The ends of the thread are "cleanly" molded if situated outside the mold parting line. To facilitate trimming of the neck flash the thread is interrupted along the mold parting. The calibrating mandrel must be unscrewed when dealing with internal threads.

6.3.4 Flash Trimming

By proper design of the mold, blow pin, and associated parts, it is possible to produce containers on an extrusion BM machine that are completely finished and ready for filling. However, in many cases, the pinch-off and the flash at the neck must be removed by a separate trimming operation. In addition, in many products the neck must be reamed and sized to produce a finished product.

Some of the equipment used for these operations includes:

- guillotines for fast trimming where tolerances are large;
- fly cutters used to square up a guillotine cut;
- reamers used to size the interior of the neck;
- trim dies, routers, punches, and so forth used to remove flash, excess material, or to separate products.

In most cases the trimming equipment is located immediately adjacent to the BM machine, if it is not a part of the blow molder. When the trimming is done in a second machine, a conveyor belt usually feeds the container automatically to the trimming machine.

When the parison is larger than the neck diameter of the container or when the container has a handle, a secondary finishing operation is required to remove neck, handle, and tail flash. Normally, containers are removed from the molds by a takeaway mechanism and deposited on a cooling conveyor that transfers them to the trimmer. Here neck, handle, and tail flash are removed by the punch of a blade. The container then is oriented for takeaway to a filling or packing station. Containers without handles can utilize the spin-off trimmer system when the parison is smaller than the parison neck.

Although all types of products have been trimmed in the blow molder, this method is generally used to remove the tail or bottom flash from small handle-less bottles. A sliding or pivoting plate is fitted into the flash pocket of the mold, and a cylinder moving the plate pulls the tail flash free from the bottle before mold opening and product removal.

The flash on bottles with handles is usually removed in a downstream trim press where the bottle is held in a contoured nest. A punch, also contoured, cuts off the tail, handle, and top flash. The flash cut from the bottle by the mold pinch-off is held in place by a very thin membrane.

Wide-mouth bottles are also trimmed downstream of the blow molder. In general, a fly cutter removes a dome (lost blowing volume) and trims the inside dimension of the finish; another method is to mold a special pulley-shaped dome above the product. Using a spin-off trimmer the dome engages a belt drive, which rotates the bottle against a stationary knife blade fitted into a sharp groove directly above the finish; this blade cuts the dome from the bottle as it rotates.

When producing wide-neck containers, the lost blowing volume has to be cut off after blowing. Different types are used. Bekum offers a unit for single and double cutting (Fig. 6.36) in which the pneumatically operated trimmer is positioned next to the

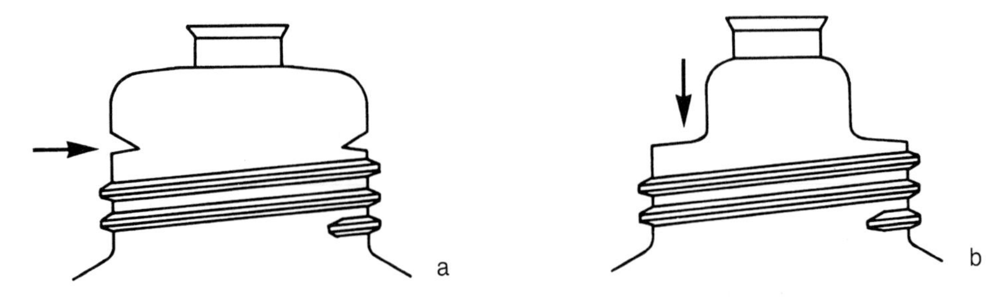

Fig. 6.36 Horizontal and vertical cutting at the neck

platens. The cutting process takes place while the container is held in the take-off mask.

The cutting unit is designed for twin cutting (dual container production), but can be converted to a single cutter for single container production. The cut can be made either horizontally or vertically without modification, to provide the desired neck finish. For a wide neck container suitable for a screw cap or snap-on lid, it is advisable to trim the neck section horizontally. For a wide neck container involving a press-fit lid, a precise concentric circular cut is necessary. In this case, the lost volume should be removed using the vertical cut principle. This will also be necessary if the container is to be sealed with a heat-seal film.

During and after cutting, the severed flash must be prevented from dropping into the container. If the container is made with relatively thick walls or from a stiff material (such as PVC) a claw is adequate for gripping the lost volume. However, if the container has relatively thin walls and is made from a soft material, a centering and ejector "bell" is necessary to ensure precise cutting. This bell holds the lost volume in position during the cutting process and prevents it from falling into the container. The dimensions of the bell depend on the product being cut.

Mold design predetermines a classification into bottom, handle, neck, and side flash, as these are trimmed by different methods. Tear-off jaws tear off the bottom flash by deflashing hooks, deflashing plates, or rolls while the mold is still closed. Cutting off along the pinch-off has also been suggested. The closed mold prevents the bottom from turning inside out. Handle flash and other flashes are bent in various directions by offset pushpins during mold opening and knocked out. Pushout plates with side punches are also employed for flash removal.

In the case of neck flash, the choice of trimming device depends on the calibration method. Only the ring of flash above the mouth of the neck has to be torn off or severed when using the stepped mandrel. When calibrating with a smooth mandrel, the side flash must also be removed from the neck using pushpins or tear-off jaws. Swiveling the molded product with the aid of the blowing mandrel toward a knife has also been employed. Side flash can either be torn off or knocked off sideways. Success depends on the best possible design and adjustment of the pinch-off edge.

Basically in extrusion BM, parting off the parison is a basic feature of the process except where parisons are fed tangentially into a rotary or chain machine, or drawn off by the "hand over hand" method. With all the various methods of calibration it is essential for the parison (before or after mold closure) to remain open on at least one side, and for the cross section of the separated parison to be minimally distorted. For economic reasons the parison should be as short as possible. The parison can be parted off below the die by pulling off, pinching off, cutting off, and squaring with a knife or shearing arrangement, and cutting off with a heated blade.

All methods have their advantages and disadvantages, and the choice of method must be determined by the particular circumstances. Both the blowing method used and the characteristics of the plastic must be taken into account.

The pulling-off method is used in the case of molds arranged on a vertically rotating wheel in parison transfer systems, and with semi-automatic demolding of larger containers when the parison is extruded intermittently into the open mold below the die. For this method to operate successfully, it is essential to ensure a sufficient pull-off rate (difference between rate of mold travel and average extrusion rate). The pull-off length, (the distance between the mold and die) should be as short as possible. With low viscosity melts, there is a danger of film formation due to plastic extension of the clamped length of parison. In the case of partially crystalline plastics with high viscosity and marked swelling, on the other hand, pulling off can cause fibrillation and necking down of the parison. It is, therefore, necessary to test whether the inner width of parison that remains is suitable for insertion of the calibrating mandrel.

Pinching off is used in the case of parison transfer by means of a jaw or gripper arrangement, or when mold closure takes place immediately below the die.

The parison can be separated from the die by means of a knife. In this operation only low knife speeds are required (peripheral speed $\approx 1\,\mathrm{m/s}$, $3.3\,\mathrm{ft/s}$). By correct positioning of the knife and suitable choice of the material from which it is made, it is possible to ensure that the die edges are not damaged. Cross sectional distortion of the cut parison can be countered by the use of support air. The ensuing parison emerges with an open, nondistorted, free end. Because this method requires a flush fit for the core with the die orifice, it is used only in special cases. In cutting with a cold knife it is essential that the knife rotates at sufficient speed (peripheral speed $\approx$ 5 m/s; 16.4 ft/s). Cutting should preferably involve a "drawing" action. Tensioning of the parison (e.g., by a rise-and-fall movement of the extruder or by movement of the mold or transfer arm) helps to obtain a clean cut.

In cutting with a shearing arrangement, although the parison is not pulled to one side, it is nevertheless pressed flat. For this reason, shear cutters are used in extrusion BM only when the parison is cut immediately below the die (while blowing with support air) and the shears then move aside for the next length of parison.

Heated blade cutters are made of nonscaling metal and usually have a sword-shaped cross section. They are heated to a dark red or cherry red by an electric current. The blade carrier must be insulated from the device that moves it, and grounded from possible leak voltage. For heating, the main voltage is stepped down to a 24- or 6-V supply. Heated blade cutters may need resharpening or replacing after about a week. When heated blade cutters are used to cut parisons containing inorganic pigments, plateout or crust formation frequently develops on the blade. To avoid these problems, it is necessary to adjust the blade temperature upwards. For this reason it is advisable to provide an infinitely variable or graduated temperature control.

In the processing of unplasticized PVC, the heated blade cutting causes undesirable thermal degradation in the cutting region. It is therefore necessary to optimize processing conditions by varying blade speed and temperature. In acetal processing a low blade temperature should be set to avoid igniting the melt. Acetals burn with a visible feathery flame.

6.3.5 Negative Pressure and Double Wall Blow Molds

The use of venting channels in the pressure (blow) molding of hollow containers is common practice. Underlying such systems is the basic principle that as the volume/pressure of air relative to the parison inner diameter (ID) increases, the entrapped air between the parison and the mold wall must be evacuated. Failure to do so results in an air entrapment between parison and mold, preventing the product from being fully expanded to the mold wall.

To ensure the free escapement of entrapped air, mold surfaces and ventilation ports provide passage for the air to exit the system. In the 1950s and 1960s, several companies fitted their blow molding machines with vacuum platen systems (see Fig. 6.37). The main use of vacuum on these machines was for the production of insulating double-walled containers and finely defined pinseat and Morocco leather embossings on ethylene–vinyl alcohol (EVA) and PVC components such as automotive visors.

A cross section of a 20-oz, double-walled, insulated flask is shown in Fig. 6.38. In the production sequence (Fig. 6.39) the inner jar was blow molded and retained for a

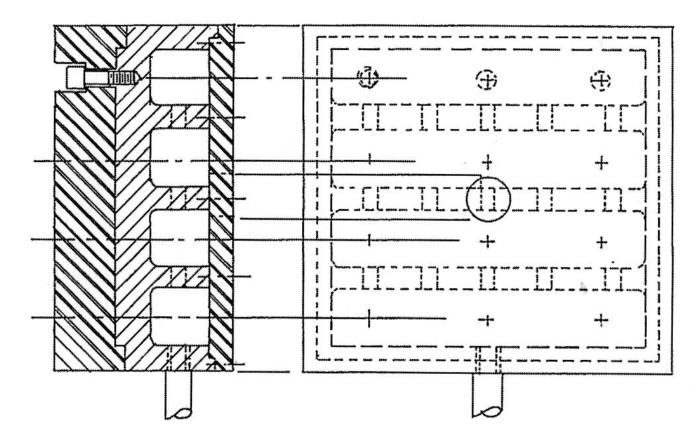

Fig. 6.37 BM with vacuum platen system

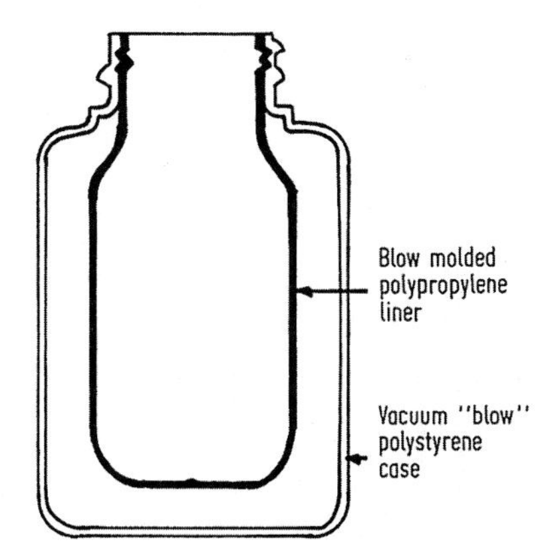

Fig. 6.38 Cross section of 20-oz double wall insulated flask

period of 48 h. This inner container was drawn from "stock," placed over a locator pin, the secondary parison extruded over the "insert," the mold closed, and the parison vacuum formed to the cooled mold wall. In the case of some double-walled container forms, the modulus of stiffness of the outer "vacuum formed" container portion must be substantially high to prevent implosion or collapse due to the negative pressure often created between the inner and outer walls of the external and internal containers.

The effectiveness of the mechanical seal achieved at the neck section was enhanced by (1) compression forming of the parison over the neck section and (2) the natural thermal shrink of the outer case at the neck area relative to the "preshrunk" outer diameter (OD) of the inner container neck.

The basic energy force exerted by a vacuum assist is greater than that obtained by a positive pressure of 2 atm or 29.4 psi. Further, the parison comes in intimate contact

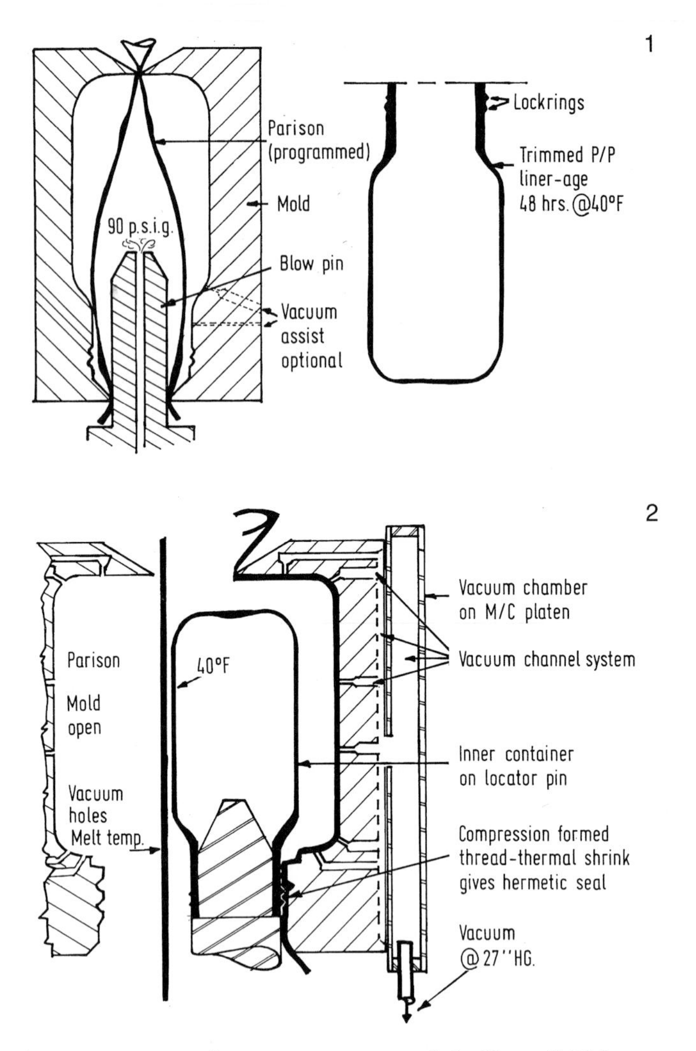

Fig. 6.39 *Production sequence of negative pressure and double wall BM*

with the cool mold surface, being restrained in that position by the positive/negative "clamping" action. The clamping force is further reflected in restrainment of shrinkage.

The viability of vacuum sheet forming technology applies equally to the practice of BM. Both disciplines are in effect forms of stretch or stress forming a sheet or parison to mold shape conformity. In a vacuum forming operation, the atmospheric air pressure between sheet and mold is reduced to 27 in. (650 mm) or better, of mercury, the opposite face being exposed to 1 atm of 14.9 in. As in BM, it is the pressure and elimination of restrictive resistance to pressure that determines the effectiveness of the process.

6.4 Injection Blow Molding Mold

There are two basic operations in IBM. The first is to mold the preform, using the conventional injection molding operation that requires the melt in the mold cavity to be at a high pressure of 4,000 psi (27.6 MPa) and higher depending on the plastic melted [6]. The second operation takes the hot preform via the injection mold core and transfers it to the blow mold that requires low pressure of 100 to 150 psi (0.7 to 1.0 MPa).

Thus, IBM requires two molds: one for molding the preform or parison, and the other for blow molding the bottle (Fig. 6.40). The preform mold consists of the preform cavity, injection nozzle, neck ring insert, and core-rod assembly (Figs. 6.41 and 6.42).

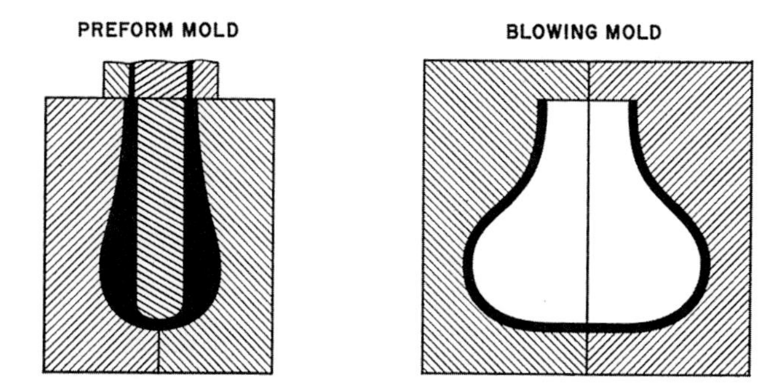

Fig. 6.40 *Example of a preform mold that is designed to obtain a product with a uniform wall thickness when blow molded*

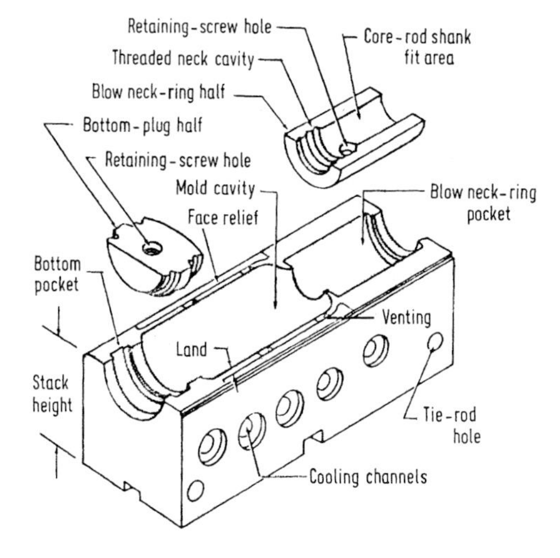

Fig. 6.41 *Exploded view of one half of a single preform injection blow mold*

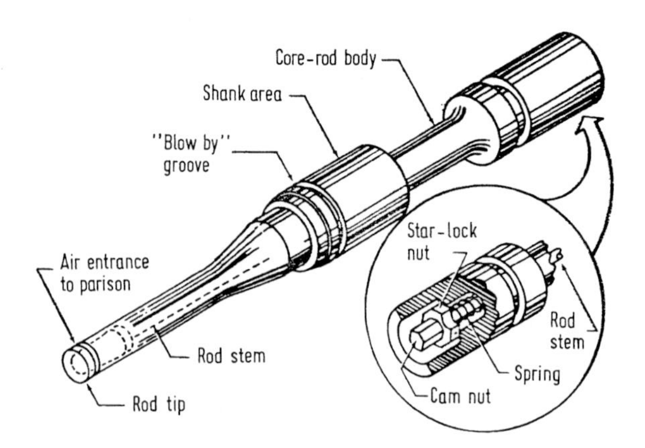

Fig. 6.42 *Example of bottom blow core pin and its principal elements*

Different type runners to direct the melt from the injection machine to multicavity preform mold can be used (cold, insulated, and hot [6]) but the typical type is hot runner (Fig. 6.43) where hot melt flow is controlled through a hot manifold. The two-part multicavity mold is clamped between upper and lower platens/plates.

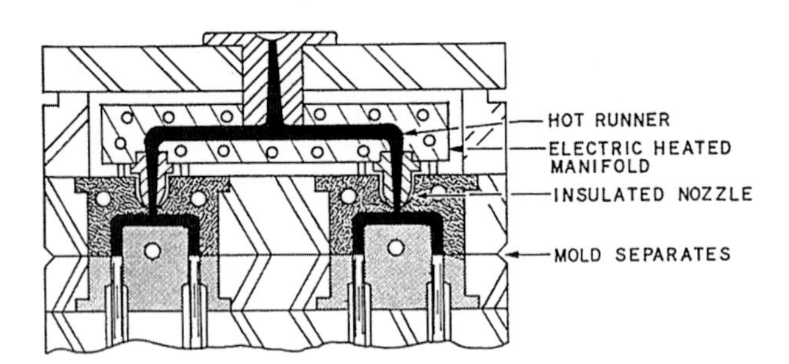

Fig. 6.43 *Hot runner mold [6]*

6.4.1 Preform

From the general concept of the preform molding operation it is important to design a mold that will safely absorb the forces of clamping, injection, and ejection. Furthermore, the flow conditions of the plastic path must be adequately balanced or proportioned to obtain uniformity of product quality in cycle after cycle. Finally, effective heat absorption from the plastic by the mold has to be incorporated for a controlled rate prior to removal from the molds [6].

Basically with flash free injection molding of the preform, the ratio of flow path length to flow path height is less critical than with conventional injection molding which relates to the ratio of cavity length to wall thickness. The higher temperatures of core and cavity and the considerable wall thickness reduction of the preform during the subsequent BM into a usually self-supporting hollow article are the decisive factors here. The core requires a certain draft; undercuts are not acceptable; and considerable reduced cross sections can be a technical problem. Only two basic geometric designs remain for the construction of the core: the bell and the truncated cone.

Pinpoint gating is preferred. In multicavity molds, hot runners and needle shutoffs are employed. The sprue gate can be used in special cases; it must be pinched off in the mold similar to the bottom flash in EBM. So that the gate and its root are not torn out of the preform, it must be tapered toward the nozzle. A lens-like thickening of the bottom opposite the gate can also be considered for eliminating this weak point on the preform [6].

The face of the core acts as a buffer during injection and needs to ensure advancing flow is uniformly distributed in the cavity. This predetermines the most central positioning; effective stability of the core; melt having the correct profile time, pressure, and rate of flow; proper clamping action; and proper mold temperature.

Heating the core and cavity accurately is important so that the individual preform temperature profile can be set properly and be repeatable particularly in gate and threaded areas. With inaccurate temperature differences on the preform difficulties occur in blowing the preform.

Production molds are usually made from steel for pressure molding that requires heating or cooling channels, strength to resist the forming forces, and/or wear resistance to withstand the wear due to plastic melts. However, most blow molds are casted or machined from aluminum because of its fast heat transfer characteristics. Also used are beryllium–copper, zinc, or kirksite. However, where extra and long term performance is required, steel is used.

6.4.2 Mold Type/Layout

There are many different molds designed to meet many different product requirements. The industry generally identifies six basic IM types in this multibillion-dollar thermoplastics market [6]:

- the cold runner two-plate mold;
- the cold runner three-plate mold;
- the hot runner mold;
- the insulated hot runner mold;
- the hot manifold mold;
- the stacked mold.

Hot runner molds, also known as “runnerless” molds, are typical in the design used for BM parisons. The runners are kept hot to keep the molten plastic in a fluid state at all times. In runnerless molds, the runner is contained in a plate/manifold of its own.

Runnerless molding has several advantages over conventional cold runner molding. There are no molded side products (gates, runners, or sprues) to be disposed of or reused, and there is no separating of the gate from the product. The cycle time is only as long as is required for the molded product to be cooled and ejected from the mold. In this system, a uniform melt temperature can be attained from the injection cylinder to the mold cavities. Shot size capacity and clamp tonnage required in the IMM are decreased by the size of the sprue and runners which also leads to an average 20% energy saving during plastication and a decrease in chiller size as sprues and runners are eliminated.

The size of the IMM relates to the maximum amount of plastics required to fill the mold cavities. The amount is usually 50% of machine shot capacity, or at least less than 60% to 70% to ensure proper plasticating action. The actual percentage use is based on experience to ensure providing the required amount to fill the mold, with a safety factor in case the extra melt is required.

The clamping force required to keep the mold closed during injection must exceed the force given by the product of the live cavity pressure and the total projected area of all impressions and runners. The projected area can be defined as the area of the shadow cast by the molded part(s) when it is held under a light source, with the shadow falling on a plane surface parallel to the parting line [6].

With the hot runner system used for thermoplastics (TPs), the force to move the melt in the runner does not open or push apart the molds at the parting line. As an example, if the total projected area is 132 in.^{2} and a pressure of 5,000 psi (pressure is dependent on plastic used) is required in the cavity(ies), the minimum clamping force required equals projected area $\times$ plastic pressure cavity or 132 in.^{2} $\times$ 5,000 psi $=$ 660,000 lb or 660,000 lb/2,000 lb $=$ 330 tons.

Consider adding a safety factor of about 10% to 20% to ensure sufficient clamping pressure, particularly when one is not familiar with the operation and/or plastic melt behavior. Thus, the standard IMM maximum clamping force could be 330, 375, or 400 tons. A guide to the clamping pressure requirement of PE, PP, and PS based on flow path length is given in Table 6.4.

It is important to avoid applying too much clamping force to the mold. If a small mold is installed in a large machine and closed under full clamp, the mold can actually sink into the machine platens; if this situation is liable to occur, steel plates can sandwich the mold to reinforce the platens. Also, if the area of mold steel in contact at the parting line is insufficient, the mold may be crushed under the excessively high clamping force. Steel molds will begin to crush when the unit clamp pressure exceeds 10-tons/in.^{2} of contact area. In less severe cases, the mold components may be distorted, or may fracture prematurely from fatigue.

Whereas the projected area determines the clamping force required, the weight or volume of the molded shot determines the capacity of the injection machine in which the mold must be operated. Capacities of IMMs are commonly rated in ounces of PS that can be injected by one full stroke of the injection plunger. Calculated against the mold clamping force, the total projected surface area of the moldings must not exceed

Table 6.4 *Clamping Pressure Required in PE, PP, and PS Based on the Melt Flow Path Length and Section Thickness*

Average component section thickness		Clamping pressure required (psi or kgf/cm^2 of projected area)				
		Ratio of flow path length to section thickness				
in.	mm	200 : 1	150 : 1	125 : 1	100 : 1	50 : 1
0.04		—	9,960	9,000	7,200	4,500
	1.02	—	706	633	506	316
0.06		12,000	8,500	6,000	4,500	3,000
	1.52	844	598	422	316	211
0.08		9,000	6,000	4,500	3,800	2,500
	2.03	633	422	316	267	176
0.10		7,000	4,500	3,500	3,000	2,500
	2.54	492	316	246	211	176
0.12		5,000	4,000	3,100	3,000	2,500
	3.05	352	281	218	211	176
0.14		4,500	3,500	3,100	3,000	2,500
	3.56	316	246	218	211	176

the machine's maximum permissible molding area, when subjected to injection pressure. Machinery manufacturers usually provide this information.

When heated during IM plastic expands and contracts on cooling. The total percent expansion/contraction becomes the average shrinkage. Predicting this average shrinkage is a major challenge to the mold builder. Shrinkage differentials may result in warpage; shrinkage is more predictable with amorphous plastics and much less predictable with crystalline plastics.

6.5 Multilayer Manufacturing

Various techniques are available for multiple component molding for parisons and preforms as reviewed. Schematics for coextrusion dies are shown in Figs. 6.44 and 6.45 that can be related to forming parisons. As reviewed, product performances gained are a combination of what each individual plastic provides. The layup can include recycled material and/or low-cost materials that will reinforce more expensive material(s) in terms of strength, tear resistance, and so on. Certain melt processing factors have to be considered to eliminate any problems, for example: (1) different melt temperatures of adjacent layers; (2) plastic viscosity differentials should not be greater than 2 to 4 : 1; and (3) minimum thickness of a cap (top) layer, because it is subjected to a high shear stress, is usually limited to 5% to 10% of the total thickness. Some of these factors can be compensated by the available extruder

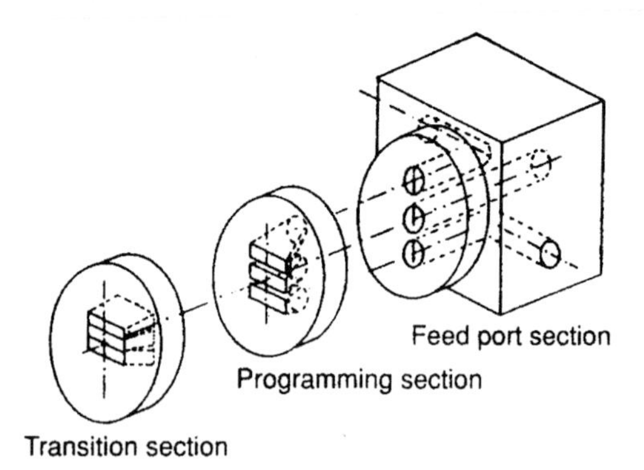

Fig. 6.44 Schematic of a three-layer feedblock die

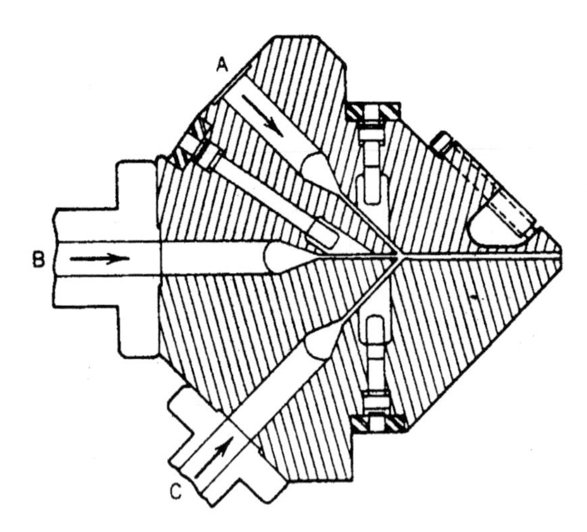

Fig. 6.45 Schematic of a multimanifold three-layer die

and die adjustments [5]. An unsteady balance of shear forces causes interfacial instability. There is a tendency for the less viscous plastic to migrate to the region of high shear stress in the flow channel, causing an interface deformation. With a great difference in viscosities existing between adjoining layers, the less viscous tends to surround or encapsulate the oper plastic. The result could be a fuzzy interface, orange peel, and so on.

The multilayer fabrication system is the simultaneous processing of two to at least seven plastic melt streams meeting in a die/mold to produce laminated plastic products. Product performances gained are a combination of what each individual plastic provides with a potential of obtaining synergistic gains. The layup can include recycled material and/or a low-cost material that will "reinforce" the strength, tear resistance, and so on of more expensive material(s). It produces products in which two or more different or similar plastics go through a single die/mold. Two or more orifices from each extruder or each injection molding machine processing the different plastics are arranged so that the extrudates merge and "weld" together into a single structure prior to cooling. Each ply of the "laminated" structure imparts a desired property such as impermeability or resistance to some environment, heat stability, toughness, tear resistance, lower cost, and so on.

6.6 Mold Design

A mold is a controllable, complex mechanical device, that must also be an efficient heat exchanger. Hot melt, under pressure, moves rapidly through the mold. Air is released from the mold cavity(ies) to prevent the melt from burning, to prevent voids in the product, and/or to prevent other defects. To solidify the hot melt, water or some other medium circulates in the mold to remove heat from thermoplastics; with thermoset plastics, heat is used to solidify the final product. There are various techniques to operate the mold such as sliders and unscrewing mechanisms. Computer-aided Design (CAD) and computer-aided engineering (CAE) programs are available that can assist in mold design and in setting up the complete fabricating process. These programs include melt flow to product solidification and the meeting of performance requirements.

The function of an EBM mold is twofold: imparting the desired shape to the melted plastic and cooling the molded product [81]. In terms of both performance and cost, it is necessary for the mold cavity to be at the required initial temperature to blow mold the desirable product (Table 6.5).

Table 6.5 *Examples of Temperatures for Mold Cavities in Blow Molds*

	Recommended temperature	
Plastic	(°C)	(°F)
Polyacetates	80–100	176–212
Polyamides	20–40	68–104
Polyethylenes and PVCs	15–30	59–86
Polycarbonates	50–70	122–158
Polymethyl methacrylates	40–60	104–140
Polypropylenes	30–60	86–140
Polystyrenes	40–65	104–149

In its simplest form, a two-part tool consists of two female cavities that close around a parison in an extrusion system or a preform in an injection system. An entrance into the formed melt is required for the blown air.

To the inexperienced eye, molds for BM may not look particularly different from conventional molds for injection molding [6]. The basic differences are few in number but they far outweigh the similarities. These differences are: (1) BMs use cavities only; (2) BMs do not require hardening, except for long production runs; (3) material movement within a blow mold is a stretching rather than a flowing action; and (4) deeper undercuts on hollow blow molded items, particularly PE, can be stripped from the mold. The first difference, use of cavities only, eliminates at the mold construction stage the complexities of matching forces with cavities as well as the problems of wall thickness usually associated with it. The wall thickness problem is not eliminated entirely from the overall problem; the bulk of this problem is merely shifted from the moldmaking to the design and processing stages.

Molding by a stretching rather than a flowing action reduces but does not eliminate the problems of weld and flow lines, mold erosion, trapped gas, and so forth. The problems of BM more closely parallel those of vacuum or thermoforming than either injection or compression molding. In turn, the construction details of blow molding molds show many similarities to those of molds used for thermoforming [2].

The possibility of stripping (such as snap fit) certain amounts of undercuts from the mold, because the inner substance is air rather than steel, eliminates the need for expensive sliding inserts, levers, cams, used extensively in injection molding for undercuts. Sliding inserts are used in BM to create completely enclosed shapes such as trash cans, toilet floats, bowling balls, Christmas tree ornaments, 55-gallon drums, etc.

The important first step of mold design is product or container design. The highest quality molds cannot and will not make up for poor product design. The basic requirements involved in product design that have to be defined and "compromised" before mold design starts are material selection, blow molding process strengths and limitations, product performance requirements, size (both volume and palletizing needs), type of closure and finish, and marketing style and appeal (Table 6.6).

6.6.1 Basic Design

On many BM machines, the mold closes in two steps in conjunction with expanding the parison by the blowing air. On these systems, the mold closes rapidly until generally $\frac{1}{3}$ to $\frac{1}{2}$ in. (5 to 13 mm) of daylight remains. The final clamping is done at reduced speed, but with increased clamping force. This two-speed operation protects the mold surface from being marred by objects that may fall between the mold halves and reduces mold surface wear.

The mold does not have to be positioned vertically between the platen area. It also does not have to be positioned so that there is equal resin all around the blow pin.

Table 6.6 *Mold Design Checklist*

- Was latest issue part drawing used?
- Will mold fit press for which intended?
- Is daylight of press sufficient for travel and ejection?
- Do guide pins enter before any part of mold?
- Can mold be assembled and dis-assembled easily?
- Has allowable draft been indicated?
- Is plastic material and shrinkage factor specified?
- Are mold plates heavy enough?
- Are waterlines, air lines, thermo-couple holes or cartridge holes shown and specified?
- Is ejector travel sufficient?
- Is the material type for mold parts specified?
- Do loose mold parts fit one way only? (Make fool proof.)
- Will molded part stay on ejector side of mold?
- Can molded part be ejected properly?
- Have trade marks and cavity numbers been specified?
- Has engraving been specified?
- Has mold identification been specified?
- Has plating or special finish been specified?
- Is there provision for clamping mold in press?
- Are runners, gates, and vents shown and specified?
- And others...

“Offsetting” of the molds is commonly done when containers with large handles on one side (such as jugs for milk, bleach, or detergent) are blow molded from PE. Offsetting of the molds makes it especially easy to “catch” the hollow handle of such containers.

Molds must have pins that perfectly align the mold when closed. These guide pins and bushings can be arranged to suit the opening and closing operation required for a given set of equipment. Unlike injection blow molds, which are mounted onto a die set, extrusion blow molds are fitted with hardened steel guide pins and bushings to ensure that the two mold halves are perfectly matched. Dies, mandrels, blow pin cutting sleeves, and neck ring striker plates are made from tool steel hardened to 56–58 HRC.

Standardized or preengineered mold components, such as ejector pins, guide pins, guide pin bushings, sprue bushings, bolts, and so forth and complete standardized mold assemblies are commercially available, principally for small molds. In these assemblies only the cavities and cooling channels need to be machined and the ejectors fitted. Such standardized mold assemblies are available in various types. Advantages include low cost, quick delivery, interchangeability, and promotion of industrial standardization.

The mold wearing surfaces should align and mate perfectly, or else thin parting lines may occur. Misaligned or poorly machined parting lines prevent intimate mold contact and cooling of the parison on blowing. The shrinkage of the faster cooling and shrinking adjacent parison areas pulls the hotter material from the parting line, making it excessively thin by the time it does cool. Polishing of the mating surfaces, however, is not recommended to avoid entrapped air between the parison and the mold.

Once the product is designed, a model or pattern is produced slightly oversize to compensate for final product molding shrinkage. This shrinkage is carefully calculated for the material being used and is often greater for horizontal than for vertical dimensions. For example, a cherry or mahogany wood pattern, held to dimensional tolerances of 0.005 in. (0.13 mm), is used to cast a so-called "stone" master cavity (no dimensional change of cavity occurs) which in turn is used to cut duplicate, virtually identical, mold cavities.

Most container designs require at least two "stone" masters, one for the side and handle details, the other for the bottom detail. (A tracer simultaneously cuts both cavity halves, the right and left mirror images, from the same "stone" master.) Even the most simple round container molds, which can be turned on a lathe, should have a model and stone made for the bottom detail. Shrinkage from molding will often distract the container contours, producing an unstable "rocker" bottom. By using a stone, these bottom contours can be adjusted to compensate for the shrinkage and to be duplicated identically in each cavity.

Good mold design is the combination of two interrelated sections (the mold cavity and the neck ring and blow pin) that form a system within the BM machine. The container or product design will determine many of the alternatives available in these areas.

An example of the makeup of an aluminum blow mold to produce HDPE milk bottles is summarized in Fig. 6.46.

The mold consists of:

- *Backing plates* that allow for easy mounting of cavities to moveable platens. These plates incorporate slots in the top and bottom for slight adjustments vertically and horizontally.
- *Cavity*, one half of the mold, consisting of three major parts: neck ring, body, and bottom insert.
- *Neck ring*, an aluminum block mounted to the top of the mold body. The neck ring is the portion of the mold that forms the threads of the finish. Within the container design, any of several styles of finish can be interchanged.
- *Mold body*, the portion of the mold that forms the body of the bottle.
- *Bottom insert*, an aluminum mounted to the bottom of the mold body that forms the bottom of the bottle. The bottom insert, like the neck ring, serves two purposes. First, it includes areas that are likely to wear and can therefore be replaced without scrapping the entire mold. Second, it provides an opening to the cavity that permits the moldmaker to accurately match up the cavity halves at the parting line.

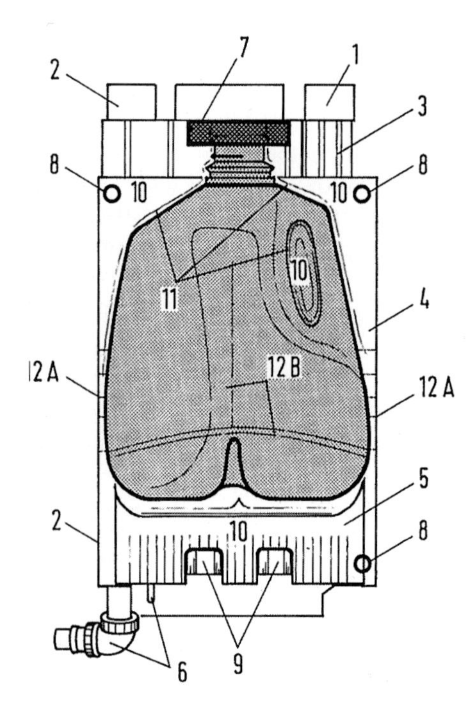

Fig. 6.46 *HDPE milk bottle*

- *Coolant inlet and outlet*, means by which chilled water is supplied to the molds. The inlet and outlet fittings are threaded into cooling channels machined into each cavity. These cooling channels are generally in the form of holes drilled from the top of the cavity to the bottom and are joined by crosswise channels. These channels are drilled and plugged in such a manner that the cooling water can be introduced into one of the channels in the bottom where it will flow in a zigzag pattern throughout the cavity and exit on the other side of the cavity at the bottom.
- *Shear steel*, in conjunction with the blow pin, forms the inside diameter of the finish. The shear steel is incorporated in the neck ring just above the finish, and is in the form of a hardened insert that may be replaced when worn. With the pull-up system, the blow pin moves up just prior to mold opening, shearing plastic material to form an accurate, smooth, and burr-free ID.
- *Locating pins and bushings* are located in the cavities to allow a means of alignment when closed. There must be at least two sets of hardened pins and bushings to accomplish this task, with the pins located in one cavity and the bushings in the other. For mold materials such as aluminum, a shoulder-type pin and bushing is desirable to prevent them from pulling out of the mold.
- *Tail grabber slots* are incorporated in the bottom insert to facilitate automatic unloading of the bottle from the mold to a conveyor system. The unloading system is comprised of fingers attached to a swinging arm. During unloading, the swing arm moves up and the fingers enter the tail grabber slots. The fingers then close on

the tail of the bottle. When the molds open, the swing arm moves down and places the bottle on a conveyor. This operation repeats for every cycle.

- *Flash pockets* are reliefs milled into the mating cavity surfaces to permit displacement of excess plastic material. The depth of the relief is critical for proper molding and trimming. If the relief is too shallow, the mold will not close completely, resulting in poor pinch-offs and parting lines. If the relief is too deep, the flash will not contact the mold surface for proper cooling, resulting in trimming problems.
- *Pinch-offs* are areas in the cavity where the parison is squeezed very thin and welded together. The pinch-off must be structurally sound to withstand the pressure of the plastic material and repeated closing cycles of the mold. It must push a small amount of material into the interior of the part to slightly thicken the weld area. It must also cut through the parison to provide a clean breakpoint for flash removal.
- *Vents* are used to allow trapped air between the outside walls of the expanding parison and the inside of the cavity to escape.

Mounting the mold into the machine can be facilitated considerably by laterally protruding base plates or strips. The mold is fastened to the platens of the clamping unit by bolts or clamps. Mold damage or accidents, caused by involuntary mold opening, can be prevented by bolting on mold straps. Heavier blow molds should be provided with tapped holes for eyebolts to permit lifting or handling the mold via over-head crane, swinging arm fitters, and so forth. The base plates are of particular importance for the setup and guidance of multicavity molds if these do not consist of one solid block. On molds provided with cooling channels the base plates also serve as a rear cover.

Eyebolts should be provided on both mold halves opposite each other; they should be placed in areas where balanced lifting of the mold base is possible. Holes should also be tapped on surfaces perpendicular to the slots. The forged steel eyebolts have a safe load carrying capacity. Only forged steel eyebolts should be specified for safe reasons, preferably those with a shoulder for better stability.

The blow mold is usually centrally aligned with the parison. In special cases inclined to laterally offset mounting can be technically advantageous.

Sinking the guide pins and bushings below the surface by approximately 1 to 2 mm (0.039 to 0.079 in.) ensures flush clamping of the mold halves. It is also essential not to press the bushings into blind holes but to drill out the mold block toward the rear, matched to the pin diameter, and keep it accessible from the outside by side channels. In this way accidentally pushed-in pieces of plastic can easily be removed and cracking of the mold by pressed-in plastic is avoided.

The guide components must be positioned at sensible distances from the mold cavity and channels, particularly in molds incorporating support air and the preblowing of the parison.

Mold opening on the workbench is greatly facilitated by milling 3 to 4 mm (0.1 to 0.16 in.) wide deep slots into the parting line.

6.6.2 Venting and Mold Surface

Well designed molds are vented, as entrapped air in the mold prevents good contact between the parison and mold cavity surface. When air entrapment occurs, the surface of the blown part is rough and pitted in appearance. A rough surface can be undesirable because it can interfere with the quality of decoration and can detract from the overall appearance. Examples of mold venting are illustrated in Fig. 6.47.

As shown in Fig. 6.47, to facilitate venting in the parting line, venting channels can be provided at proper intervals in the mold edge, with additional funnel-shaped inlets from the corner areas. Additional venting must be provided in the corner and edge areas and at changes in cross section, threads, and engravings. In these instances, narrow channels, slots, and fitted joints with enlarged outlets have proved useful. In some instances, the interface of the inserts have the added advantage of providing good venting conditions. In special cases, evacuating the mold cavity can assist venting, particularly when the parison is clamped all around by the mold. Sintered metals have also been recommended for venting, but it must be taken into account

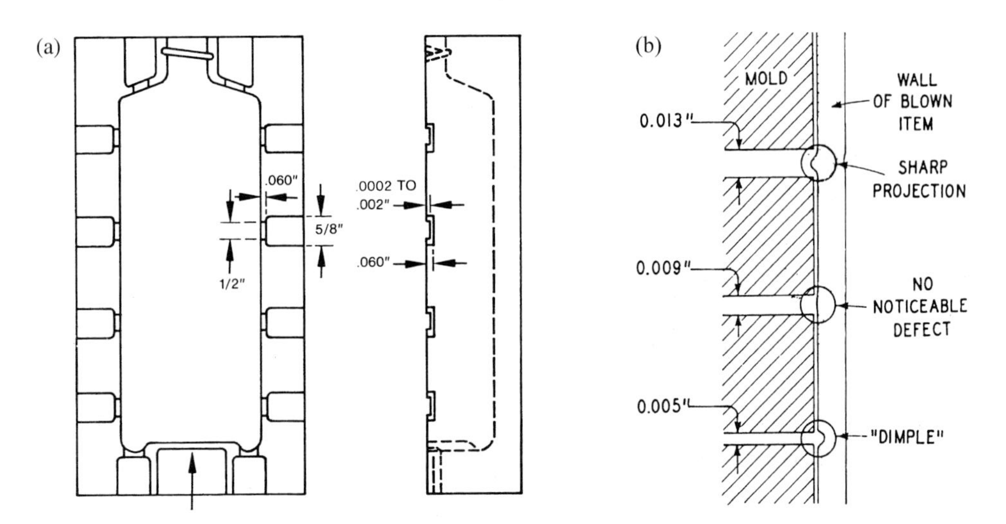

Fig. 6.47 Venting and mold surface designs: (a) pinch-off block vented; (b) effect of vent hole on surface; (c) parting line venting; (d) core venting; (e) locations of vents in baffles; (f) slotted or screen type vent plugs; (g) venting channels, joints, and slots [here the lettering denotes: (a) 0.1 to 0.3 mm, (b) 0.5 to 1.5 mm, (c) blind hole plug, (d, e, f, g) fitted hexagon with free circular section 0.1 to 0.2 mm. Depth of slot 1.5 mm, (h, i) sintered metal plug with vent channels, (k) venting in the fitted joint of the bottom insert]

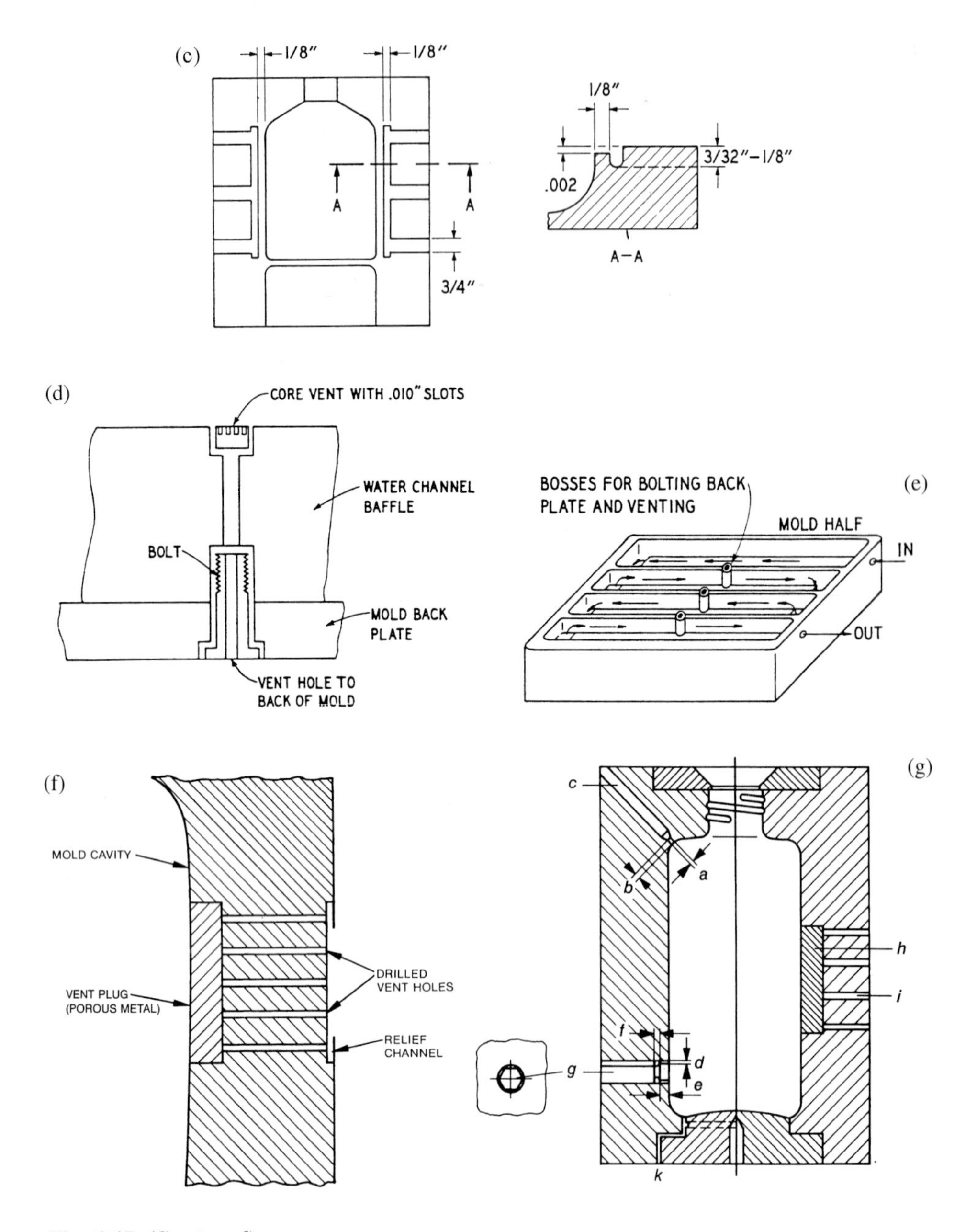

Fig. 6.47 (Continued)

that these cause visible surface markings that are imprinted by the blowing pressure. If the vent is too large or if the blowing pressure is too high, a protrusion is formed, and too small a vent can result in a dimple (Fig. 6.47b).

The air between parison (preform) and mold must be expelled as completely as possible during blowing so that the molded product receives the correct shape and can cool down by contact with the mold walls. To ensure minimum cooling time, it is important that the mold be adequately vented. Entrapped air or air that is not removed rapidly enough from the mold can hold the plastic away from the mold wall and thus prevent or delay efficient cooling relieved in areas to form a vent slot when assembled. To ensure proper venting, the bottom-split point is located at or near the widest or deepest point of the cavity. Because this is the last point filled by the parison, trapped air is generally pushed toward this vent.

For additional venting, some molds are made with "split" construction. The bottom and neck ring split points are a part of the mold for construction conveniences as well as for venting. With split construction, the mold is generally split again at a point perpendicular to the parting line to provide a vent slot. This slot, as others, is usually 0.003 in. (0.08 mm) wide and runs vertically between the bottom split and the neck ring split points. Vent slots, generally more expensive, are preferred over other methods (e.g., vent plugs, drilled holes, or knockout pins). Large vent openings can be provided with slots without unsightly marks to distract and spoil the surface and lines of the molded product.

When using HDPE, a somewhat rough mold surface (e.g., generated by sandblast) can improve venting. Because of the relatively cold mold temperature (2–18 °C, 35–65 °F) and relatively low mold pressure (max 150 psi), the plastic will lie and bridge across the high points, providing minute passageways for trapped air to move.

Special venting problems can occur with materials other than HDPE. Polypropylene (PP) and LDPE, for example, will duplicate the rough sandblasted finish. As a result, the cavity finish must be a smoothed sandblast to provide a semigloss surface finish on the container. As expected, these molds do not vent as well. In some cases, the problem can be corrected by processing techniques when the mold is sampled; however, if trapped air is still found, a fine vent hole is drilled in the mold at the point the trapping occurs. These holes will leave small prick points on the container; however, because they are located only where needed, they are relatively unnoticeable.

The mold surface should be finely finished and scratch free. Polishing makes sense only if glass-clear material is being processed and/or particular surface brilliance is required. Chrome plating is usually not necessary as it could flake off around the pressurized edges.

When sandblasting to roughen the surface, care must be taken that the grains do not lift the metal and that the edges are not removed. The grain size depends on the required surface condition to be obtained. For surfaces to be printed by the offset process, a fine grain below 0.1 mm is required. In most other cases grain sizes of 0.2

to 0.3 mm (0.008 to 0.01 in.) are recommended. Carborundum, quartz, and steel grit are used. Too rough a mold surface and excessive mold temperatures cause dull surfaces on the molded product.

The mold cavity surface has an important bearing on mold venting and on the surface of the molded product. With PEs and PPs (as reviewed) a mild roughened mold cavity surface is necessary for the smoothest surface. Grit blasting with 60- to 80-mesh grit for bottle molds and 30- to 40-mesh grit for larger molds is a common practice. The clear plastics such as PVC and styrene require a polished mold cavity for the best surface. A grit blasted surface will reproduce on the clear plastics and this is not normally desirable. When processing plastic materials that cause corrosion or plate out, roughing can be a disadvantage as it makes cleaning of the cavities more difficult and encourages corrosion. Apart from venting the mold cavity, venting of the calibrating blowing mandrel is essential during calibration for the internal walls as well as the horizontal seating surface of the neck. The calibrating blowing mandrel introduced into the parison should also be slightly roughened. A slight ring-shaped relief should be provided in the shoulder of the stepped mandrel, with a vent gap and outlets.

In selecting a texture for a particular application, the type of molding process that will be used should be considered. In the case of BM, a larger design with a pattern to it that allows air to escape rather than be trapped in the small etched surfaces usually will yield a better, more uniform appearance. The product should also be large enough to allow for deeper than normal etching, generally in the area of at least 0.008 to 0.014 in. (0.20 to 0.36 mm).

In the case of a BM product with a 0.125 in. (3.175 mm) thick wall section, the texture depth of the mold will need to be 0.015 to 0.020 in. (0.38 to 0.51 mm) deep to obtain an even 0.010 to 0.012 in. (0.254 to 0.304 mm) depth on the product. In this example, there is approximately a 25% to 30% texture loss in processing of heavy walled products. Most molds, however, fall into the 95% to 100% texture reproduction range. The actual effect is directly related to the type of plastic used.

The various types of materials used in mold construction can affect texture appearance. The most commonly used material is a prehardened CSM-2 or P-20 tool steel. Most texture-standard plaque molds have been made from this tool steel.

Hardened tool steels will accept a texture pattern slightly different because of a tighter grain structure. High-chrome and nickel-content tool steels can also change the appearance of a texture. Stainless steels require different etching reagents than carbon steels, as do aluminum or beryllium–copper molds. The type of metal used to build a mold often dictates the type of acid and methods used to produce a pattern on the surface of the tool. As an example, if plaques are molded using the same patterns on molds of aluminum and P-20 steel, the molded products will have a visual difference. In addition, processing of molds is limited by the quality of the mold being made. Poor quality materials or alloys usually produce substandard texturing results.

Different alloys can cause some depth variations in texturing. An aluminum mold that is etched 0.015 to 0.020 in. (0.381 to 0.508 mm) deep will have a tolerance range of just that. The mean depth will be 0.0175 in. (0.445 mm). Processing of aluminum molds tends to be less controllable than of steel molds. A mold that is made of top grade tool steel can produce a top quality etch with texture depth tolerances of within ±0.0003 in. (0.008 mm).

What sort of finish is best or most suitable on a mold before the texture is etched in? Opinions seem to vary among the major mold texturers. One company claims that a 320 stone finish (SPI-SPE No. 3) is sufficient, whereas another says it is not necessary to put a high polish on a mold prior to texturing. Normally, finer patterns require a slightly more finished surface. However, as long as the cutter, file, and disk marks are removed, the equivalent of a 240 to 320 emery finish is satisfactory. Scale that is the result of heat treating, bobbing, or EDM must be removed prior to texturing.

All chrome plating must be stripped before the mold can be processed. If a mold is to include highly polished surfaces, it is recommended that the polishing be done after the mold is textured, as solutions used in the texturing process might discolor the polished surfaces.

6.6.3 Mold Cooling

Cooling is particularly important in mold design because it consumes much of the cycle time (up to 80%) and therefore bears on product economics. Cooling can take two thirds or more of the entire mold-closed time in a cycle. Best economics require cooling as quickly as possible. Cooling action occurs principally through mold cooling (cool the outside surface of blown part) and internal cooling of the blown part.

Controlled cooling channels in the mold for TPs are an essential mold feature and require special attention in mold design [6]. One of the most effective and low cost coolants is water, which should be in turbulent, rather than laminar flow, to transfer heat out of the molded product at a much faster rate (Fig. 6.48). The coolant is any liquid and/or gas (air, etc.) having the property of absorbing heat and transferring it efficiently away from its source. Coolants are used in molds, chillers, and so forth. Coolants operating below freezing incorporate antifreeze such as ethylene glycol.

Channels or passageways are located within the body of the mold through which the cooling and/or heating medium (of water) can be circulated to control temperature on the cavity surface, providing required cooling action to solidify the TP melt into a molded product (Fig. 6.49). A few TPs may require the higher heat to complete their cooling/solidification.

A laminar, nonturbulent flow is not desirable in a coolant system because it flows in parallel stream lines with a stagnant layer at the wall that prevents heat transfer [76]. Required is a nonlaminar flow in which a fluid moves in all different directions in the mold cooling channels. With turbulence, more heat will be removed because as the

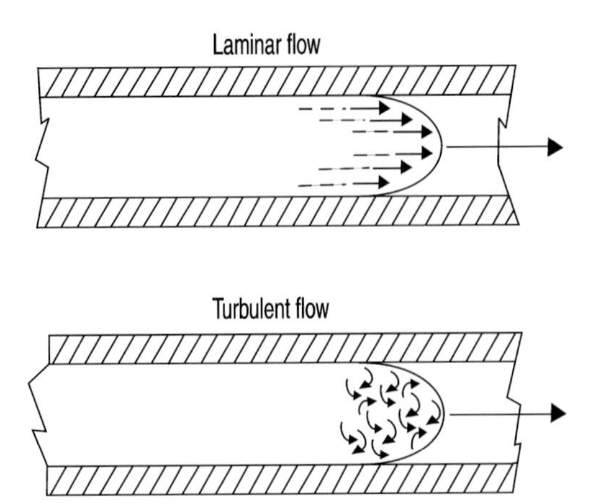

Fig. 6.48 *Laminar and turbulent flow*

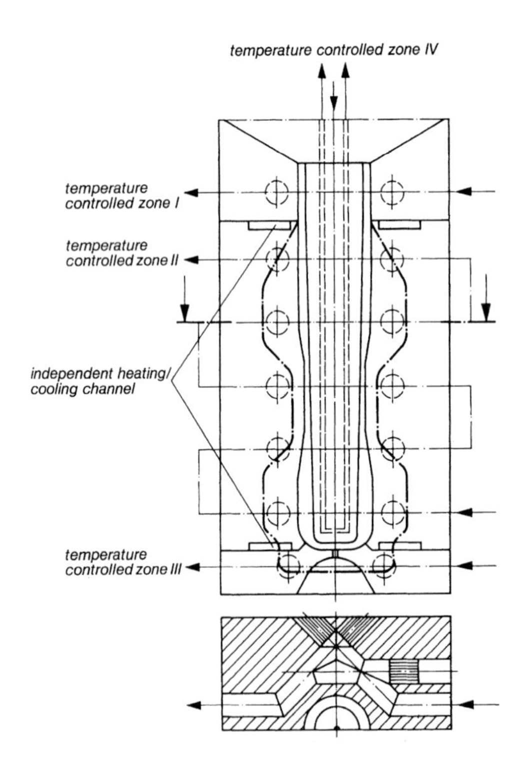

Fig. 6.49 *Example of cooling line arrangements for temperature control of the mold cavity (Chapter 8)*

fluid on the inside surface of the channel is heated, it moves away to be replaced by cooler fluid to rapidly remove more heat. With laminar flow the fluid heat built up on the wall would act as an insulator so that inside laminar flow would be an inefficient heat remover.

When good turbulence cannot be achieved (such as in very small diameter passages) baffles designed to divert or restrict the flow to a desired path can be used in the mold's water channels. Baffles channel the fluid to the wall and then away again to circulate the fluid from the wall center of passage and back to remove heat from the plastic at the required uniform rate and in the shortest time period. These cooling baffles (ribs, plugs, etc.) are used to provide more uniform cooling action within the mold. Other devices are used such as bubblers and baffles. When inserted into a mold cavity they allow water to flow deep inside the cavity into which they are inserted and also discharged.

With other molding factors being equal, the greater the velocity of coolant flow, the greater the heat transfer from the mold to the coolant. But this principle runs into the law of diminishing returns. After a certain point, increasing cooling velocity does not appreciably improve heat transfer from the mold.

At low coolant velocities, coolant flow is laminar. In laminar flow, the coolant moves in layers parallel to the wall of the coolant passage, each thermal layer acting as an insulator, impeding heat transfer from the mold to the stream of coolant. By contrast, turbulent flow, the desired condition, creates random movement of coolant and substantially increases heat transfer. However, once turbulent flow conditions are reached, higher coolant velocities bring diminishing returns in improved heat transfer.

6.6.3.1 Reynolds Number

Turbulent conditions in a passage can be calculated based on the Reynolds number equation:

$$\mathrm{Re} = (7{,}740\ VD)/v \quad \text{or} \quad (3{,}160Q)/Dv$$

where Re = Reynolds number, V = fluid velocity, ft/s, D = diameter (in.) of the cooling channel, v = kinetic viscosity in centistokes, and Q = coolant flow rate in gpm.

The kinetic velocity is defined as viscosity divided by density. In the preceding formula the proper units must be used. The coefficients 7,740 or 3,160 take care of the conversion factors to make each expression come out dimensionless. The Reynolds number should preferably be approximately 10,000. Above 10,000 no gain in heat transfer is realized. $\mathrm{Re} \leq 2{,}100$ is laminar flow and $\mathrm{Re} < 3{,}000$ is transition flow. Increased flow up to a Reynolds number of 10,000 removes additional convective heat. Above $\mathrm{Re} > 10{,}000$ one is wasting electricity pumping the coolant fluid. Except for very small or very large molds, the cooling passages are usually 0.5 to 0.75 in. (12 to 19 mm) in diameter. A 0.5-in. (12-mm) passage is rated 2 gpm under normal pressure drop conditions.

Cooling passages above 0.75 in. (19 mm) are not economical. They require a tremendous volumetric flow rate, as the cross sectional area increases with the square of the diameter; for an increased area the velocity decreases for a constant flow that leads to laminar flow. Thus the flow rates for large diameters burn up a great deal of electricity to pump the fluid and maintain the Reynolds number at 5,000 to 10,000. The size of the cooling passage also determines the cavity and core plate thickness.

As a rule of thumb, the center of the cooling passage should be located 1.5 diameters below the cavity surface for mechanical strength of the plate. Hence a cooling passage with a 0.5 in. (12 mm) diameter requires a plate of 3 diameters or 1.5 in. plus the cavity depth. Similarly, to maintain a uniform temperature along the cavity surface for uniform cooling, the pitch or spacing between passages should be 3 to 5 diameters.

6.6.3.2 Pressure Drop

One of the major mistakes in mold construction relates to pressure drop in the cooling channel. A 0.5 in. (12 mm) diameter cooling passage is rated at 2 gpm with a pressure drop of 10 ft of head of water. Note that 34 ft of head is 1 atm or 14.7 psi. In most molds, the cooling passage is no more than 2 or 3 ft long; hence, the pressure drop should be less than 1 psi. The internal diameter of the passages should be the same diameter as the diameter of the cooling passage.

The most common passage diameter is $\frac{11}{16}$ in. (18 mm). This is the diameter that is perfect for a $\frac{3}{8}$ in. (9 mm) pipe thread tap that chokes the flow and causes a high-pressure drop. A well-designed cooling passage with elbows or turns should have a passage drop of less than 5 psi (0.03 MPa). Commercial plumbing fittings such as manifolds and water bridges are available as off-the-shelf items that make it easy to keep the pressure drop low.

6.6.3.3 Heat Control

Based on the production rate of plastic used in pounds per hour, one can calculate the total amount of heat to be removed. The chiller load is expressed in tons of refrigeration, where 1 ton is 12,000 Btu/h of heat removed. A simple formula relates the heat load to the flow rate for a mold temperature controller or chiller in tons of refrigeration $=$ (gmp)(ΔT)/24. The ΔT is the difference in inlet and outlet temperature of the coolant. Preferably it is 3 °F (1.5 °C) and it should not exceed 5 °F (1.8 °C). The ΔT increases as the cooling circuit adds additional loops. Using the above formula one can solve for the gpm. Using the rule of thumb that a 0.5-in. (12-mm) channel can carry 2 gpm, one can estimate how many separate passes are required.

In case a cooling tower is used, the formula is changed slightly, as the efficiency is not as great so tower tons of refrigeration $=$ (gmp)(ΔT)/30.

Based on the heat content of various plastics and their processing temperatures the following chiller guidelines can be used: HDPE at 30 lb/h/ton, LDPE at 35, PP at 35, PS at 40, PVC at 45, and PET at 40.

In a cooling tower, a small part of the water evaporates into the air to cool the rest. The system must be provided with makeup water and one must purge about 1.5% of the flow with treated water to keep the chemical balance in the system and prevent mineral buildup that will reduce water flow in the mold.

A mold temperature controller only heats the water; thus, it can only control the mold temperature from ambient conditions to the boiling point of the pressurized system. For water with a minimum pressure of 30 psi (0.18 MPa) at the suction side of the pump, the attainable temperature is about 250 °F (130 °C). In case a mold has cooled below ambient temperature conditions, a chiller must be used. Here a combination of antifreeze/ethylene glycol and water is circulated. The lowest recommended temperatures for the coolant/temperature (°F) are water/40, 10% antifreeze/25, 20% antifreeze/15 antifreeze, 30% antifreeze/0, 40% antifreeze/15, and 50% antifreeze/40.

No mixture higher than 50/50 should be used. High antifreeze concentrations give increased viscosities, lower Reynolds numbers, and less heat transfer than pure water [76]. One should remember that for mold cooling it is not how cold the mold gets but how fast the coolant flows that is important. The cavity/core surface cannot be cooled below the dew point of ambient air because moisture will condense on the surface. The moisture will flash to steam when the hot plastic melt hits it; this will cause a cosmetic defect known as the "orange peel" effect. Because humidity is high in the summer, the mold cannot be run as cold and the cooling capacity is less compared to the winter months.

One of the plastics with the most stringent cooling requirements is PET in the molding of preforms for later stretch BM. The mold has to be kept below 40 °F (4 °C) or the plastic will cool too slowly, crystallize, and turn opaque. Also, the blown products will stick to the mold cavity.

A new mold will have the highest heat transfer efficiency when it is first installed. The stagnant fluid layer next to the channel wall only limits convective heat transfer. As the mold operates, additional resistance to heat transfer develops in the form of buildup of rust, minerals (scales), and algae. To minimize this action the water has to be treated with chemicals and the mold has to be maintained including descaling of the cooling passages periodically.

With laminar flow conditions, the overall heat transfer coefficient in mold cooling passages might be in the range of 10 to 50 $Btu/h/ft^2/°F$. With turbulent flow, the overall heat transfer coefficient might be 300 to 1000 $Btu/h/ft^2/°F$.

The most common mistake made by molders is to run their coolant at too low a temperature, which may actually reduce cooling effectiveness instead of increasing it. In a large number of cases, they would be better off running at a higher temperature and a higher flow rate. This also illustrates why a careful cooling analysis is needed. If a molder simply goes out and buys a bigger pump to increase the flow rate, one may not achieve the desired results. Either the mold may already be operating at turbulent flow conditions, or the larger pump may not be sufficient to boost coolant velocity into the turbulent flow region; thus the improvement will be marginal.

Different techniques are used to develop different degrees of the cooling action. In pulse cooling, rather than having the coolant continually running, it flows only after the melt fills the cavity(ies). Solenoid-controlled valves placed in the incoming coolant are used to open or shut off the coolant. Thus when the melt enters the mold, it is not subjected to a fast cooling shock. Flood cooling used in BM involves an internal flooding action in a confined open chest that surrounds the mold cavity rather than using drilled holes. However, drilled holes can also be used or combined with the flooding action.

6.6.3.4 Cooling Factors

Mold cooling is a function of three factors:

- The coolant circulating rate,
- The coolant temperature,
- The efficiency of the overall heat transfer system.

The most common mold coolant is tap water. Hard water should be softened before use. The overall efficiency of a mold cooling system would be greatly reduced if water scale were allowed to plate out on, or partially block cooling channels. Generally, cooling water circulates through a mold cooling system that lowers the mold's temperature to between 40 and 70 °F (4 to 21 °C). If the temperature in the molding shop is high, cooling water can cause moisture to condense on mold surfaces. This will cause defects in finished parts, or small blotches or pock marks on the surface. Condensation problems can be eliminated by raising the mold coolant temperature, or by air conditioning the molding shop. This solution merits careful economic consideration: overall production increases resulting from shorter cycles can pay for the initial capital investment in air conditioning.

It is often useful to cool the neck rings and the blow pin independently. Here, as with mold cooling, basic rules of heat transfer optimization must be applied.

To obtain good quality with maximum stress resistance, refrigerated coolant should be passed through the neck ring and bottom pinch-off. Coolant in the range of 35 to 70 °F (2 to 24 °C) should be used for the mold body. Too low a temperature in the pinch-off area can have a detrimental effect on the blowing of the parison tube. When the pinch-off plates squeeze the parison, the material in contact with the cold surface of the pinch-off crystallizes and cannot blow outward to the extremities of the molding. This causes thin walls at the extremities. Too hot a pinch-off is also detrimental, as it produces the opposite effect (thin sections at the pinch-off and heavier walls at the extremity of the product).

Cold mold temperatures require high blowing pressures to produce good surface finishes. The exact temperatures for each section of a blow mold depend on the product design. Properly balanced zone cooling will result in faster cycles with lower levels of strain in the finished product. Mold temperature and temperature control must also take into consideration the properties of the plastic being blown, required properties of the product, and shrinkage of the molded part. In general, the lower the mold temperature, the less shrinkage takes place, other conditions being equal.

Different techniques to cool molds are used in very limited applications. An air spray to such an extent that folded over flash will not stick on ejection can cool overhanging flash. Mist (spray) cooling is used mainly on large blow molds to save mold weight and water ballast. The chambers in the mold cavity are sprayed with a system of jet pipes. The outgoing water film carries the absorbed heat away with it. Other techniques are also used such as heat pipes.

A heat pipe is a means of heat transfer that is capable of transmitting thermal energy at near isothermal conditions and at near sonic velocity. The heat pipe consists of a tubular structure closed at both ends and containing a working fluid. For heat to be transferred from one end of the structure to the other, the working liquid is vaporized; the vapors travel to and condense at the opposite end, and the condensate returns to the working liquid at the other end of the pipe. The heat transfer ability of saturated vapor is many times greater than that of solid metallic material.

6.6.3.5 Internal Cooling

In addition to the external systems that circulate liquid coolants through the molds to cool both mold and part simultaneously, there are also internal cooling systems. These use circulating air, a mixture of air and water, carbon dioxide, and other injections to cool the inside of the parts while they are in the mold. The following are typical of the commercial methods currently in use: (1) liquid carbon dioxide is injected into the blown part, followed by vaporization and exhausting of the coolant as hot gas through the blow pin exhaust; (2) highly pressurized moist air is injected into the part where it expands to normal blow pressures, producing a cooling effect by lowering the air temperature and freezing the moisture present into ice crystals (crystals and cold air strike the hot walls of the part where they melt and vaporize); (3) air is passed through a refrigeration system and into the hot parison; (4) normal plant air is cycled into and out of the blown parts by a series of timers and valves; (5) liquid nitrogen, and others. These types of systems can reduce many BM cycle times up to 50%.

As an example, the use of atomized water or liquid carbon dioxide mixed with the blowing air for blowing parisons has resulted in reduction of molding cycles by up to 50% through the rapid cooling produced. However, in certain applications the sudden shock of these methods of cooling can create strains in the plastic and can spot or pock mark the inside surface of the blown parisons.

The liquid nitrogen process has been developed and proven in the BM industry. The Extrublas process utilizes liquid nitrogen as both a cooling and pressurizing fluid to blow and mold a parison. The cooling effect has been found to decrease the total cycle time by as much as 50%. Cycle reduction depends on a number of properties of the blown part such as material, weight, wall thickness, and shape. Blown parts with hold times greater than 20 s have been shown to realize significant economic benefits from the Extrublas process.

6.6.4 Computer-Aided Mold and Product Design

Mold designers and builders can benefit from the use of CAD, CAM, and CAE techniques. Computer programs permit analysis of the flow of plastics into cavities, mold cooling systems design, mechanical stress in molded product, and so forth. These programs can simplify the design of molds with lower stress levels, less warpage, and shorter cycle time (Chapter 10).

6.7 Mold Construction

The design and construction of a mold plays a significant role in the dimensional integrity of the final product. The cavity that forms the final product can be shaped using a variety of steel removal/machining methods from jig grinding to wire-feed electrical discharge machining. Each removal method has a corresponding range of variations. Although the tolerances of these processes are geometry-dependent, some processes are more accurate than others. Studies can be performed to determine an average range of variation for each method. Among other important tool design and construction considerations for determining tool performance for a given dimension are the mold's construction details, such as its main parting lines. All these factors add to the overall variations in related dimensions of the final products.

The proper choice of materials of construction for die openings or mold cavities is paramount to quality, performance, and longevity (number and/or length of products to be processed). Desirable properties are good machinability of component metal parts, material that will accept the desired finish (polished, textured, etc.), ability with most molds or dies to transfer heat rapidly and evenly, capability of sustained production without constant maintenance, and so on.

Choice of construction material ranges from computer-generated plastic to specialty alloys or even pure carbide tooling, with the usual being aluminum and steels. Because blow molds do not have to withstand the high pressures of injection molding, a wide selection of materials is available.

6.7.1 Materials

Mold materials can be classified according to the size and the numbers of articles required, such as (1) blow molds for continuous production and large batches and (2) blow molds for small batches and prototype production. Only metal molds can be considered for continuous production and long runs, whereas for small batches and prototypes, casting resins in water-permeable backed shell construction are also suitable.

It is necessary to further subdivide metals into tool steels, aluminum, and alloys. Table 6.7 shows the materials normally used for blow molds along with their hardness, tensile strength, and thermal conductivity. Fastenings and guide components and, if possible, also the areas of high edge loading, should be made of steel in all the molds. In addition, guide components and edge inserts should be hardened.

The advantages of blow molds produced in aluminum, beryllium, and zinc alloys include excellent thermal conductivity, low weight, and economical manufacture by precision casting. These materials should be considered mainly for BM in the range from a 10-liter volume and up, but they can also be used for mass production of smaller hollow articles if they are equipped with steel or beryllium–copper inserts. Plating may be considered where processing of plastic materials containing corrosive

Table 6.7 *(a) BM Tool Materials and (b) Properties of BM Materials*

(a)

Material	Hardness[a]	Tensile strength (MPa[b])
Aluminum		
No. A356	BHN-80	255
No. 6061	BHN-95	275
No. 7075	BHN-150	460
Beryllium–copper		
No. 25	HRC-30	930
No. 165	BHN-285	
Steel		
No. O-1	HRC-52-60	2000
No. A-2	BHN-530-650	
No. P-20	HRC-32 BHN-298	1000

[a] HRC = Rockwell hardness (C scale); BHN = Brinell.
[b] To convert MPa to psi, multiply by 145.

(b)

Material	Thermal conductivity[a]	Wear resistance	Ability to be cast	Ability to be repaired	Density lb/in.^{3} (g/cm^3)	Ability to be machined and polished
Aluminum	0.53	Poor	Fair	Good	0.097 (2.699)	Excellent
Beryllium–copper	0.15–0.51	Excellent	Good	Good	0.129–0.316 (3.589–8.793)	Fair
Cast iron	0.08	Good	Good	Good	0.24 (7.6)	Good

[a] Calories per square centimeter per second per degree centigrade.

volatiles is a problem, in which case even gold plating is used. Owing to the reduced strength and hardness, damage to edges and surfaces can easily be caused. With proper care, life spans of 10^5 to 10^6 cycles can be expected.

Although beryllium–copper and Kirksite are better conductors of heat, aluminum is by far the most popular material for blow molds because of the high cost of beryllium–copper and the short life of the Kirksite (soft zinc alloy) molds. Aluminum is light in weight, a relatively good conductor of heat, very easy to machine (but also easy to damage if the mold is abused), and low in cost. Its potential porosity may easily be eliminated by coating the inside of the mold halves with a sealer, such as automotive radiator sealant.

Aluminum is used for single molds, molds for prototypes, and large numbers of identical molds, as might be used on wheel type blowing equipment or equipment with multiple die arrangements. Aluminum may tend to distort somewhat after prolonged use. Thin areas, as at pinch-off regions, can wear in aluminum. With the exception of large industrial blow molds made from aluminum (typically no. A356), most extrusion blow molds today are cut from no. 7075 or no. 6061 aluminum or from no. 165 or no. 25 beryllium–copper. The latter is corrosion resistant and very hard, making it the choice for PVC BM. However, compared with aluminum, it weighs about three times as much, costs about six times as much per cubic centimeter, and requires about one third more time to machine. In addition, thermal conductivity is slightly lower.

Compared with injection molds, those for BM can be considerably less rugged in construction. Clamping pressures applied to blow molds range from 100 to 300 psi (0.7 to 2 MPa) and blowing pressures from 30 to 150 psi (0.18 to 1 MPa). This contrasts with the pressure of 2,000 to 25,000 psi (14 to 175 MPa) used in injection molding applications.

The use of unhardened cavities permits many lower cost procedures with a resulting lower overall cost for the completed mold. Unhardened cavities are feasible in BM because of the very low internal molding pressures required. Hardening, where required in blow molds, is usually limited to the use of hardened inserts.

6.7.1.1 Steel Dies and Molds

Tool steels should be considered mainly for long production runs in the volume range of up to 10 liters (2.64 US gallons). Machining produces the blow mold. Any of the steel types generally used in plastics processing can be chosen. If blow molded plastics contain corrosive volatiles, it is necessary to employ corrosion- or acid-resistant steels. Apart from the exceptions already mentioned, hardening is not necessary and chrome plating is not customary. For large-volume articles with a content of 3 m^3 (4 yd^3) and up, welded machined plate construction can also be considered. Cast steel is of no value for mold bases.

Given the right care, the life span of steel blow molds is the longest, at least 10^7 parts per blow mold. Inserts subject to wear and tear either have to be refurbished or exchanged at intervals of 10^5 or 10^6 cycles.

For dies and molds P-20 steel, a high grade of forged tool steel relatively free of defects and available in a prehardened steel, is commonly used. It can be textured or polished to almost any desired finish and it is a tough mold material. H-13 is usually the next most popular mold steel used. Stainless steel, such as 420 SS, is the best choice for optimum polishing and corrosion resistance. Other steels and materials are used to meet specific requirements such as copper alloys (fast cooling), aluminum (low cost), etc. They are used to meet specific requirements to meet mold life and cost. The choice of steel is often limited particularly by the available sizes of blocks or plates that are required for the large molds.

As a guide to life expectancy consider P-20 steel for long runs (one million molded products, etc.), QC-7 aluminum for medium runs (250,000 molded products), sintered metal for short runs (100,000 molded products, etc.), and for relatively shorter runs filled epoxy plastic (50 to 200 molded products, etc.).

The flow surfaces of the tool usually have protective coatings such as chrome plating to provide corrosion resistance. With proper chrome-plated surfaces, microcracks that may exist on the steels are usually covered. The exterior of the die is usually flash chrome plated to prevent rusting. Where chemical attack can be a severe problem (processing PVC, etc.), various grades of stainless steels are used with special coatings. Coatings will eventually wear, so it is important that a reliable plater properly recoat the tool, usually the original tool manufacturer.

Fortunately, the needs for the vast majority of materials, particularly steel, can be satisfied with a relatively small number of these materials. The most widely used steels have been given identifying numbers by the American Iron and Steel Institute (AISI). The properties of the tool material usually are as follows:

- wear resistance to provide a long life;
- toughness to withstand processing and particularly factory handling;
- high modulus of elasticity so that the die channels do not deform under melt operating pressure and the die's weight;
- high uniform thermal conductivity;
- machinability so that good surface finish can be applied particularly near the die exit.

Different characteristics and performances identify steels. As an example, higher hardness of steel improves wear, dent and scratch resistance, and polishability but lowers machinability and weldability. High sulfur content degrades the stainless qualities and polishability of the steel. Hardness, as a measure of the internal state of stress of the steel, has an adverse effect on weldability, fracture toughness, and dimensional stability. A low carbon steel undergoing a surface hardening process is used to resist wear and abrasion. It is used in molds, dies, and other machine parts. The steel is heat treated in a box packed with carburizing material, such as wood charcoal, to 2,000 °F (1,093 °C) for several hours.

The most important requirements for tools are high compression strength at the processing temperatures of the plastics, wear resistance, adequate toughness, possibly corrosion resistance, and good thermal conductivity. In addition, so that the tools may be manufactured economically, good machinability is expected and, in certain cases, also cold hobbing is of less importance as wire and die sink EDM has taken over most of these applications that require hobbing. Dimensional stability during heat treatment generally is necessary.

Machining rolled or forged steels makes most of the tools used. Extensive machining is required, particularly for large-size tools. For this reason, consideration is given to using blanks with a premachined cavity for constructing large-size tools. The disadvantages of this technique, such as high allowances, local impurities, shrink holes, or subcutaneous blowholes, are a thing of the past. Steel foundries are able to produce tools with close tolerances and the smallest allowances for machining.

There are grades that offer a slight allowance for grinding. The tolerances for these tools, whose patterns are made from metal or plastic, are stated to be +0.25%, based on the nominal size. With little effort for making the pattern and the mold it is possible to cast tools for processing plastics with only a slight allowance for machining of about 1 to 1.5 mm (0.04 to 0.06 in.) per surface, depending on the cavity. Another grade are tool blanks with a machining allowance of at least 5 mm (0.2 in.) per surface. Even this relatively rough finish can be of interest economically for the manufacture of large-size tools.

Before deciding in favor of any of the forms in which the tools are supplied, consideration should be given to the cost of the material and to the machining required for the tool in question. These should be compared with the costs of the material of a cast tool and the costs of the pattern required. If the tool will be manufactured repeatedly, the cost of the pattern can be in proportion to the anticipated number of tools to be manufactured. It is, however, possible that the costs of the pattern can be amortized already by a single tool.

An additional economic advantage of the cast tool is the possibility of casting the cooling core in place, where required, simultaneously with the tool. In smaller or medium size tools, ducts can be introduced during the casting operation. In large tools it has proved to be advantageous to cast cooling chambers, which are subsequently closed off by plates. But the effectiveness of this cooling is slight, and it must be decided whether this cooling effect is adequate for the fabricated plastic product.

Gray cast iron or castings are used sometimes for large size tools. Cast tools normally consist of steel castings and, moreover, of those grades of steel whose chemical composition largely corresponds to that of forged or rolled grades.

Casehardened steels still have a tough core as well as a wear-resistant surface after any heat treatment. The high surface hardness and modern deoxidation methods offer the best conditions for polishing. In special cases this property of steels can be improved even further with the help of remelting processes (electroslag refining).

These steels are molten with carbon contents of less than 0.25%. Apart from the alloying elements, the carbon content has an effect on the core strength available after the heat treatment. To attain a high hardness at the surface it is necessary to increase the carbon content in the case by adding carburizing agents during the heat treatment. Usually the aim is a carburizing depth of about 0.6 to 1 mm (0.024 to 0.04 in.) and, after quenching and tempering, a case hardness of 58 to 62 HRC is achieved. Because of the extensive heat treatment cycle certain dimensional changes cannot be avoided; these require an additional expenditure in the finishing.

Because of the low carburizing depth subsequent changes in the engraving are no longer possible. Should they be necessary, it is recommended that the mold be refinished in all the regions of the engraving area. Those areas of the surface of the tool that have already been subjected to carburizing in the first heat treatment will be supercarburized by the additional heat treatment. Nitriding steels are preferred for plasticating tools. Particularly high surface hardnesses are reached with the aluminum alloyed steels Cr–Al–Mo and Cr–Al–Ni. Because of the lower toughness of the nitrided layer the depth of the nitrided case is limited. These steels have a lesser surface hardness after nitriding, but it is possible to increase the nitriding depth. The total service life of the plasticating tools of aluminum-free nitriding steels is consequently longer than that of the aluminum alloyed steels.

In general nitriding steels are supplied in the already quenched and tempered condition. The aluminum-alloyed types have higher core strength than aluminum-free grades. They are used in special cases for tool construction such as for tools with very thin sections. But since the core strength of these steels is low, the hot-work steels, which are also known from their use in light-metal die-casting tool construction, are preferred. The high retention of tempering of these hot-work steels, which are normally premachined in the soft-annealed condition, permits tools to be heat treated to a higher core strength without a decrease in strength in the subsequent nitriding treatment.

Depending on the process, nitriding is carried out at a temperature between about 450 and 590 °C (842 and 1,094 °F) in gas, powder, or salt baths. Because of the low treatment temperature, which causes no structural transformations, nitriding is associated with only very slight dimensional changes.

Because of the problems with heat treating other steels, quenched and tempered tool steels were originally used only for constructing large molds. Since then they have gained acceptance in plastics processing for tools of all dimensions. Quenched and tempered tool steels are normally supplied and used.

Because a heat treatment is not required there are no problems due to unavoidable dimensional changes and associated reworking. The time required for manufacturing a mold is reduced. The availability of standard molds and premachined bars and plates in these steel grades contributes to this reduction in manufacturing time.

To improve the machinability of quenched and tempered mold steels, grades with a higher sulfur content of about 0.05% to 0.08% are available. Because the sulfide inclusions in the steel have a low hardness they impair the polishability only slightly. On photoetching, however, the sulfide can lead to a nonuniform structure. For this reason also, quenched and tempered tool steels with a low sulfur content are required in spite of the disadvantages with respect to machinability (Table 6.8).

The process of photochemical machining (PCM) is recognized by the metalworking industry as one of several effective methods for metal parts fabrication. The technique, also called photoetching, chemical etching, and chemical blanking, competes with stamping, laser cutting, and EDM. It uses chemicals, rather than mechanical or electrical power or heat, to cut and blank metal.

Some of the advantages of photochemical machining over these other processes include low tooling costs associated with the photographic process, quick turnaround times, and the intricacy of the designs that can be achieved by the process, as well as high productivity and the ability to manufacture burr-free and stress-free parts. Of paramount importance in using this process are the cost savings associated with generating prototypes; a test prototype can be made for as little as $250. Because it is chemical rather than mechanical, the process does not alter the physical characteristics of the metal. However, many engineers do not completely appreciate its capabilities. It has been the mission of the Photo Chemical Machining Institute (Littleton, MA, USA) to promote the process, to educate engineers throughout industry as to the advantages of PCM, and to develop and enhance the technology via its technical publications and biannual technical conferences. For more information go to www.pcmi.org.

The advantages of using fully hardening tool steels rather than case-hardening steels for the manufacture of tools are primarily the simpler heat treatment and the possibility of making corrections to the cavity at a later time without a new heat treatment. However, the greater risk of cracking is a disadvantage, particularly for tools with a greater cavity depth, because tools from these steels do not have a tough core. Moreover, the tougher steels with a carbon content of about 0.4% do not attain the high surface hardness of about 60 HRC, which is desirable with respect to wear and polish.

Maraging steels for plastic tools are special steels that were developed originally as high-strength steels for aviation and space travel. They are suitable for tools with particularly complicated cavities. These special steels are supplied in the solution-annealed condition. In spite of their higher tensile strength the soft nickel-martensite of these steels can be machined readily. After a subsequent aging treatment at a temperature of from 480 to 490 °C (896 to 914 °F), these steels reach a higher tensile strength. When the machining operation is performed prior to the heat treatment, consideration has to be given that a uniform reduction in volume of about 0.05% occurs. If plastics with high filler content are to be processed, heavy tool wear may result.

It is recommended that maraging steels be nitrided, which can be combined with the process of precipitation hardening. It is a low temperature treatment that is possible in an appropriate gas atmosphere.

Stainless steels have their place. Their cost is not excessive but the benefit toward eliminating corrosion inside water channels, at O-ring face seals, and so forth is great

Table 6.8 *Comparing Properties of Tool Steels*

AISI designation description	Typical Rc	Hardening temp (°F)	Tempering temp (°F)	Wear resistance	Toughness	Compressive strength	Corrosion resistance	Thermal conductivity	Hobbability	Machinability	Polishability	Heat treatability	Weldability
4140	30–36	1,500	1,200	2	8	4	1	5	1	6	5	10	4
P20	30–36	1,600	1,100	2	9	4	2	5	1	6	8	10	4
414SS	30–34	1,550	750	3	9	4	7	2	1	4	9	10	4
420SS	35–40	1,885	1,050	3	9	4	6	2	1	4	9	10	4
P5	59–61	1,575	450	8	6	6	2	3	9	10	7	6	9
P6	58–60	1,475	425	8	7	6	3	3	8	10	7	6	8
O1	58–62	1,475	475	8	3	9	1	5	5	8	8	7	2
O6	58–60	1,475	500	8	4	8	1	5	7	10	5	6	2
H13	50–52	1,875	1,000	6	7	7	3	4	6	9	5	6	2
S7	54–56	1,725	550	7	5	8	3	4	6	9	8	8	3
A2	56–58	1,750	1,000	9	3	9	3	4	4	8	7	9	2
A6	56–58	1,600	450	8	4	8	2	5	5	10	7	7	4
A10	58–60	1,475		9	5	9	2	5	5	8	6	7	2
D2	56–58	1,850	950	10	3	8	4	2	4	4	6	9	1
420SS	50–52	1,885	480	6	6	6	7	2	4	7	10	8	6
440SS	56–58	1,900	425	8	3	8	8	2	3	6	9	7	4
M2	60–62	2,225	1,125	10	2	10	3	3	2	4	6	8	2
ASP23	61–63			10	5	10	4	3	1	4	7	8	2
BECU (2%)	36–42	625	NR	1	1	2	6	9	10	10	9	7	7
BECU (2%)	26–30			1	1	2	6	9.5	10	10	9	7	7
BECU (0.5%)	20–24	900	NR	1	1	1	7	10	10	10	9	7	9

Note: Hardening and tempering temperatures are approximate and not intended for heat treatment, but only to reference when hardness may be altered (e.g., surface treatments may use elevated temperatures.)

if the mold is expected to run for a long time or for many cycles. Conversely, if a mold runs only 250,000 pieces and at the end of year the mold will be retired, then 4130/4140 plates (no. 2 steel) will be sufficient. If the molding material is corrosive, such as PVC, then stainless steel may be needed for molding surfaces. If the molding plastic has certain fillers then abrasive wear resistance is more of a concern.

Sometimes the mechanical action of the mold may require certain steel selections so as to permit steel on steel sliding without galling. Molding surfaces of precision optics will need steel that can be polished to a mirror finish. If the inserts will receive coatings to further enhance performance, then steel characteristics to receive coating or endure a coating process must be considered (coating application temperature vs. tempering temperature). Hot runner mold components often use hot work steel because of their superior properties at elevated temperatures. Very large molds and/or short run molds may use prehardened steel (270 to 350 Brinell) to eliminate the need for additional heat treatment.

When mold steels of high hardness are used they are supplied in the soft annealed condition (hardened mold inserts for cores, cavities, other molding surfaces and gibs, wedge locks, etc. are typically hardened to a range of 48 to 62 Rc). They are then rough machined, stress relieved, finish machined, and heat treated for hardening and tempering to desired hardness. The core or cavity typically must then be finish ground and/or polished. In some applications, there will be additional coatings or textures to further treat the tool surfaces.

While steels typically have increased wear resistance, they should be selected when appropriate, as those same steels may be more difficult to machine. A D2 or A2 insert will have higher hardness and excellent wear resistance, but will be much more difficult to size than a H13 or S7 insert. Inserts such as those with titanium carbide alloy are even more difficult to grind, but afford outstanding wear resistance.

When processing particularly highly abrasive plastics, the wear can still be too high even when using high-carbon, high-chromium steels. Metallurgical melting cannot produce steels with even higher amounts of carbides. In such cases hard material alloys, produced by powder metallurgy are available as a tool material. These alloys contain about 33 wt% of titanium carbide, which offers high wear resistance because of its very high hardness.

Like other tool steels, hard material alloys are supplied in the soft-annealed condition where they can be machined. After the subsequent heat treatment, which should if possible be carried out in vacuum-hardening furnaces, the hard materials attain a hardness of about 70 HRC. Because of the high carbide content, dimensional changes after the heat treatment are only about half as great as those in steels produced by the metallurgical melting processes.

In machining as well as in noncutting shaping processes stresses develop chiefly as a result of the solidification of surface layers near the edge. These stresses may already exceed the yield point of the respective material at room temperature and consequently lead to metallic plastic deformations. As the yield point decreases with increasing temperature, additional stresses can be relieved by plastic deformation during the subsequent heat treatment. To avoid unnecessary, expensive remachining it is advisable to eliminate these stresses by stress-relief annealing.

6.7.1.2 Aluminum

Molds for forming hollow products are generally made from aluminum (Al). The mold can have water jackets, flood cooling, cast-in tubing, and/or drilled cooling lines. The commonly employed materials for these small parts are aluminum and aluminum alloys, steel, beryllium–copper (Be/Cu), and cast iron and cast zinc alloys.

Aluminum is used for single molds, molds for prototypes, and large numbers of identical molds, as might be used on wheel type blowing equipment or equipment with multiple die arrangements. Aluminum tends to distort somewhat after prolonged use. Thin areas, as at pinch-off regions, can wear in aluminum. With the exception of large industrial blow molds cast from aluminum (typically No. A356), most extrusion blow molds today are cut from No. 7075 or No. 6061 aluminum or from No. 165 or No. 25 Be/Cu. The latter is corrosion resistant and very hard, making it the choice for polyvinyl chloride (PVC) blow molding. However, compared with aluminum, it weighs about three times as much, costs about six times as much per cubic centimeter, and requires about one-third more time to machine. In addition, thermal conductivity is slightly lower (Table 6.7).

Aluminum molds are excellent heat conductors, easy to machine, can be cast, and are reasonably durable, particularly when fitted with harder pinch blades and neck inserts. Aluminum also provides faster heat transfer than steel. However, steel is also used to provide improved wear resistance, handling, and longer life cycles for certain type products and operations. Pressure cast beryllium–copper also has excellent heat transfer properties.

Isolated areas, such as a thread or pinch-off, can use alternative material inserted in Al molds to extend the Al longevity. An aluminum mold body is quite adequate for the pinch-off. If properly set up and maintained, molds with aluminum pinch-off areas could last for 1 million to 2 million cycles. When poorer heat transfer metals or less durable metals are used, it is advisable to have pinch-off inserts made of Be/Cu or steel. The pinch-off in these molds is more durable and can easily be repaired. Be/Cu is preferred because of its high thermal conductivity. Neck inserts of a highly conductive metal are also recommended. The thickest portion of the blown part will determine the cooling cycle. The pinch and the neck are the thickest portions of a bottle, and attention paid to these mold areas will result in shorter cycles. Water lines are placed throughout the mold for proper cooling and the neck area or neck ring is generally a separate insert since these can be damaged and then easily replaced at a low investment.

When compared to steel, fabricators may consider aluminum too soft with limited durability and shortcomings relative to compressive strength, reparability, and surface finish. In recent years, however, current grades have overcome these concerns. In many cases good design practices can compensate for these limitations. With the advent of the automobile industry driving the implementation of large molds, production expectations have dramatically been upgraded. It is common to see aluminum plate molds that have run anywhere from 500,000 to 1 million products. With better thermal conductivity of aluminum compared to steel, cycle time is reduced. From a handling and operating approach, the lighter weight aluminum provides about a 3 : 1 weight advantage. Other advantages include lower cost with machining that is much faster than tool steel and very good machinability.

6.7.1.3 Other Materials

Beryllium–copper is used in molds to provide relatively fast heat transfer. There are two basic families of beryllium–copper alloys—those with high heat conductivity and those with high strength. Heat conductivity is about 10 times greater than those of stainless steels and tool steels. Conductivity is double that of aluminum, such as alloy 7075, and higher than that of others. They have higher hardness and strength than aluminum. When heat treated, they are the strongest of all copper-based alloys (Table 6.9). Copper alloys such as beryllium–copper and bronze are sometimes used for the high copper content that yields outstanding thermal conductivity in the tools. They can be used just as inserts in the tool to help direct heat transfers in critical areas of the tools.

An alloy of copper and zinc is used in the manufacture of molds, dies, instruments, and so on. One of its desired and excellent properties is good heat transfer (for fast cooling). Care has to be taken with these tools in structural foam molding. Some chemical blowing agents have decomposition products that corrode the copper alloy.

Kirksite is a popularly used alloy of aluminum and zinc to make prototype molds or for short production runs. A low melting alloy with a high degree of heat conductivity makes it easy to produce a mold. Pouring temperatures are as low as 800 °F (440 °C) as compared to 3,000 °F (1,640 °C) for steel and 2,000 °F (1,093 °C) for beryllium–copper castings.

For quickly made low-cost tooling metal spraying with low melting alloys is used. Metal spraying is performed in a way similar to paint spraying, with the important difference that the metal has to be liquefied and is therefore hot. Although metal leaves the gun at a temperature in the range of 200 to 300 °C (400 to 600 °F), the very small particles produced in the nozzle cool so rapidly in flight that they solidify almost instantaneously on the pattern. Each tiny droplet is roughly spherical while it is in the air, but on impact it spreads out into a liquid film. Succeeding droplets do the same and a well-bonded structure can be produced as long as the correct temperature is maintained.

If the droplets have cooled too much that they are virtually solid by the time they contact the pattern, they neither flatten out nor fuse together and make only point contact with their neighbors. This results in a weak and brittle deposit with a good

Table 6.9 *Properties of the Major Mold Steels and Beryllium–Copper*

Type	AISI designation	Recommended hardness, Rockwell C	Point ratings, 1 to 10 (10 is highest)											
			Wear resistance	Tough-ness	Com-pressive strength	Hot hardness	Corrosion resistance	Thermal conductivity	Hobb-ability	Machine-ability	Polish-ability	Heat treatability	Weldability	Nitriding ability
Prehardened	4140	30–36	2	8	4	3	1	5	1	5	5	10	4	4
	P20	30–36	2	9	4	3	2	5	1	5	8	10	4	5
Prehardened stainless	414SS	30–35	3	9	4	3	7	2	1	4	9	10	4	6
	420SS	30–35	3	9	4	3	6	2	1	4	9	10	4	7
Carburizing	P5	59–61	8	6	6	5	2	3	9	10	7	6	9	8
	P6	58–60	8	7	6	5	3	3	8	10	7	6	8	8
Oil-hardening	O1	58–62	8	3	9	5	1	5	5	8	8	7	2	3
	O6	58–60	8	4	8	5	1	5	7	10	5	6	2	3
Air-hardening	H13	50–52	6	7	7	8	3	4	6	9	8	8	5	10
	S7	54–56	7	5	8	8	3	4	6	9	8	8	3	8
	A2	56–60	9	3	9	7	3	4	4	8	7	9	2	8
	A5	56–60	8	4	8	7	2	5	5	10	7	7	4	7
	A10	58–60	9	5	9	7	2	5	5	8	6	7	2	NA
	D2	56–58	10	3	8	8	4	2	4	4	6	9	1	10
Stainless	420SS	50–52	6	6	6	8	7	2	4	7	10	8	6	8
	440C SS	56–58	8	3	8	7	8	2	3	6	9	7	4	NA
Maraging	250	50–52	5	10	6	7	4	3	4	4	7	9	5	9
	350	52–54	6	10	7	7	4	3	4	4	7	9	5	9
Maraging stainless	455M	46–48	5	10	5	7	10	2	3	4	8	9	5	NA
High-speed	M2	60–62	10	2	10	10	3	3	2	4	6	8	2	10
	ASP30	64–66	10	4	10	10	4	3	1	4	7	8	2	NA
Beryllium–copper	Be Cu	28–32	1–2	1	2	4	6	10	10	10	8–9	7	7	NA

NA, not available.

deal of entrapped air. On the other hand, if the spray is too hot, a puddle of metal is formed, which is then blown away by the air stream. In practice it is not difficult to work at the right temperature, providing it is understood that the secret of success is to keep a proper distance between the nozzle and the pattern. Adjustment of this distance is the most effective way of controlling temperature.

Metal spraying can be either a high-temperature or a low-temperature form. In the former, solid metal is fed slowly into an oxyacetylene flame; in the latter, a charge of solid metal is liquefied inside the spray gun by means of electric heating elements that will melt metals up to a maximum melting point of about 200 °C (400 °F). This means that only fusible (low melting point) alloys can be handled.

Examples of the alloys used include bismuth, lead, tin, cadmium and, where very low melting points are required, indium. Different expansion properties for the different alloys exist. The extreme is bismuth, which expands 3.3 vol% when it solidifies. By adjusting the composition it is possible to obtain alloys with dimensional stability on casting or a significant degree of growth. When more than 25% lead is added, this growth persists after solidification and can be detected as much as a month later.

These are relatively soft, heavy metals that are brittle to shock yet perform well when subjected to heated plastic melt flow under sustained load. Their mechanical properties improve with age. Conductivity of heat and electricity is rather poor. Being stable metals, they have the great advantage that they can be remelted and repeatedly used. They are normally applied by simple gravity casting, but they also lend themselves to pressure die-casting and can be sprayed. This technique is unrivaled for high-fidelity reproduction and makes possible the application of low melting point alloys to the pattern surface at well below their casting temperature. Wood, plaster, or wax forms can consequently be sprayed, with a fineness of definition that is capable of reproducing the lettering from a printed page.

From the mold making aspect the use of fusible alloys has two important advantages where exact reproduction is concerned. First, because so little heat is involved, no thermal stresses are left in the mold and the tendency toward warping is negligible. Second, because fusible alloys containing bismuth have no casting contraction, the sprayed metal shell is dimensionally an exact replica of the pattern. Spraying with alloys is comparable in several respects to electroforming.

Porous metals provide another construction material used in different processes, particularly in thermoforming. They are made up of bonded or fused aggregates (powdered metal, coarse pellets, etc.) such that the resulting mass contains numerous open interstices of regular or irregular size allowing air or liquid to pass through the mass of the mold, thus permitting removal of air and other gases from the mold cavities.

A special material category of so called soft tooling covers materials such as elastic or stretchable tools made of rubber or plastic elastomer (silicone, etc.), so that complex shaped products can be removed without tool side actions, etc. They can be stretched to remove cured products having undercuts, and so on. In general soft tooling can be anything other than the usual steels used in the production of tools. It includes materials such as cast or machined aluminum grades, cast plastics (epoxy, silicone, etc.), cast rubbers, and cast zincs. By definition they are the least expensive, more flexible, usually faster to fabricate products, and have limited lives compared to steel molds. Today's choices range from computer-generated plastic molds to specialty alloys or even pure carbide. However, each of them has limitations in durability and capabilities.

These molds can last a relatively limited time if they are properly prepared and maintained. Steel wear resistant edge plates can be used to expand life expectancy. Preventative maintenance, such as cleaning, is very important. To clean, use a mold cleaner designed to loosen normal parting line and vent residue. The cleaning fluid should be compatible with the tooling material.

6.7.2 Cooling Design

Coolants can be circulated through the mold in several ways. One of the simplest is to "hog" out the back of each mold half, leaving an open area, and enclose this cavity with a gasketed back plate. Then with simple entrance and exit openings, the coolant can be circulated in a fairly large volume. Another method (much more efficient) is to drill holes through the mold block from top to bottom and from side to side so that they intersect. Excess openings are plugged and hose connections inserted for the exit and entrance lines. This type of cooling arrangement can provide extra cooling where needed in the mold, such as at the neck finish and bottom pinch-off areas where the plastic material is thicker and requires more cooling [6].

As the thickest portion of the part controls the cooling cycle, multizone cooling is desirable in blow molds. The coring can be more closely spaced in mold areas adjacent to the thickest parts, or the coring made larger with each zone having its own inlet and outlet. Mold temperature controllers can also be used to cool the pinch and neck zones more than the body. Coring should be as close to the parting line as possible to prevent thin parting lines, and can be brought to within $\frac{3}{16}$ in. (4.8 mm) of the mold surface. Coring diameters of $\frac{7}{16}$ to $\frac{9}{16}$ in. (11 to 14 mm) are recommended. Generally the passageways run parallel approximately $1\frac{1}{2}$ to 2 in. (38 to 50 mm) apart and $\frac{1}{2}$ in. (13 mm) from the cavity surface. Double water lines are often used in larger molds, which have duplicate water systems, each with an inlet and outlet. More frequent smaller diameter corings are preferable to fewer large diameter corings. Baffling of cores with twisted copper strips to increase the surface area and therefore heat transfer capacity has been found useful. Coolant flow rates should be at least 5 gallons (23 liters) per minute per mold. Many molders find city or well water satisfactory, but the use of controlled coolant temperatures and multizone coring is increasing.

Casting-in of cooling coils is applied mainly with precision castings of blanks in aluminum, beryllium, and zinc alloys. The coil-to-coil distances correspond to those of the cooling bores. Copper pipes are ideal for fast heat transfer. The advantage over cooling bores is that the cooling coils can be designed and led so that the required distance from the mold cavity is observed even though it has a complex shape. Cooling channels with metered water flow caused by numerous diversions can be constructed as steps and surround the mold cavity. The diversion baffles also serve as supports for the base plate. The width of the baffles should amount to approximately one third of the channel width. Cooling chambers allow the heat to permeate uniformly if they surround the mold cavity as much as possible. They permit retention of a uniform shell thickness or its gradation to the mold cavity.

Results are best when uniform temperatures are maintained throughout the mold. The problem most frequently encountered is inadequate mold cooling. It can be improved by increasing the rate of coolant flow through the mold or by making the mold of material with better heat transfer. Flow of coolant can also be improved by increasing the size of cooling channels (including the adjacent piping) and by increasing the coolant flow rate. Heat transfer can be enhanced by lowering the temperature of the coolant or by using a mold material with better heat transfer properties.

Mold materials such as aluminum and beryllium–copper are often porous in some areas, and even with careful mold making, coolant can penetrate through to the inner mold surface and cause problems. However, applying sealants can stop this leakage.

Extrusion blow mold cavities are generally cooled in parallel from a manifold system. When possible, the neck ring insert, bottom insert, and mold body are cooled with an individual circuit, permitting each section to be set differently; this, in turn, promotes more even and uniform cooling of the part. Water lines are straight holes drilled parallel to the cavity axis with several right-angle turns for greater turbulent flow. The lines are located approximately 1 to 2 diameters below the cavity surface. Spacing between the lines is between 3 and 5 diameters. Large cast-aluminum industrial molds follow the same rules; however, the water lines are often cast-in-place stainless steel tubing.

Cooling channels should be installed as close as possible to the mold cavity, and should be baffled to increase the surface area with which cooling water is in contact. A low-pressure [30 psi, (0.18 MPa)] coolant supply system is common. Higher pressure systems are generally difficult to maintain owing to the possibility of leakage through porous portions of the mold metal or through joints in segmented molds.

If a soft mold material is used, steel inserts should be used for the neck; both the bottom pinch-off and the neck insert should be separately cored for cooling because of the greater amount of heat that must be removed owing to the greater thickness of the material.

Heat pipes can be used either to remove or to add heat. The smaller heat pipes, which can be used to operate against gravity, are equipped with thick homogeneous wicks and have higher thermal resistance, so that the heat transfer will not be quite as fast as in the case of gravity-positioned pipes. Even with the higher thermal resistance, these heat pipes still have a very high heat transfer rate in comparison with solid metals.

Both extrusion and injection blow molders have increased productivity from 17% to as much as 32% with sealed copper tube heat pipes. Thermal pins have been used extensively in EBM to speed cooling of the neck portion of the bottle. Because the neck is usually the thickest area of plastic and also the most critical in terms of dimensions, its cooling rate limits the overall mold cycle. The problem is compounded by the fact that it is usually difficult or impossible to water cool inside the portion of the blow pin in contact with the neck ID. Stainless steel inserts in the aluminum mold, which have relatively poor heat transfer characteristics, often form the neck OD.

Both difficulties can be overcome with the use of thermal pins. First a "hollow" pin with an opening down its center can be inserted in a standard blow pin. The ID of the thermal pin provides an adequate air passage for blowing the bottle, while the pin efficiently conducts heat away from the blow pin. For maximum cooling efficiency, it would be desirable to place a water line in close proximity to the upper end of the heat conductor where it extends beyond the upper end of the blow pin. This would allow the heat conductor to draw heat from the blow pin and conduct it into the cooling water. If this is not possible, significant improvement can nonetheless be realized in many cases, because the thermal pin effectively diffuses the heat concentrated in the blow pin by distributing it into the mass of the machine (Fig. 6.50).

In most cases, addition of a hollow thermal pin to provide internal blow pin cooling is sufficient. But if necessary, cooling of the bottleneck from the outside can be enhanced by inserting several standard pins (no hole down the center). They extend from within the steel neck ring insert out into the main body of the mold to use the mold as a heat sink or to move the heat into close proximity to a water line. This provides much faster heat transfer than by unaided conduction through all that metal. Not only does this speed cycles, but it can also help improve product quality and reduce rejects by providing more uniform cooling around the neck circumference, thus preventing warping and unwanted ovalization of the neck.

Two other bottle areas where it is often helpful to insert a thermal pin in the mold are (1) in the handle pinch-off area, if there is one, which tends to concentrate heat because it is surrounded by plastic on two sides and (2) at the bottom of the mold, where the mold itself is commonly thinner and thereby less able to conduct heat by itself. The approach of placing a pin as a fast "escape route" for heat from any slow cooling area of a mold to the vicinity of a water line can be used in any blow molded shape. An example is a mold for truck wheels in which a heat conductor is placed in the hub section of the mold, near where the plastic is thickest. The heat conductor has both increased productivity by 50% and prevented wrinkling caused by differential cooling rates in different areas of the part.

In one unusual application, a thermal pin was designed to permit continuous flushing of a bottle with air during cooling, without necessitating a "huff and puff" approach. A hollow pin was inserted in the blow pin, and a small tube threaded through the pin. A lateral hole was drilled through the wall of the blow pin, and the pin inside it had been built with a matching hole. This permitted cooling air to be continuously

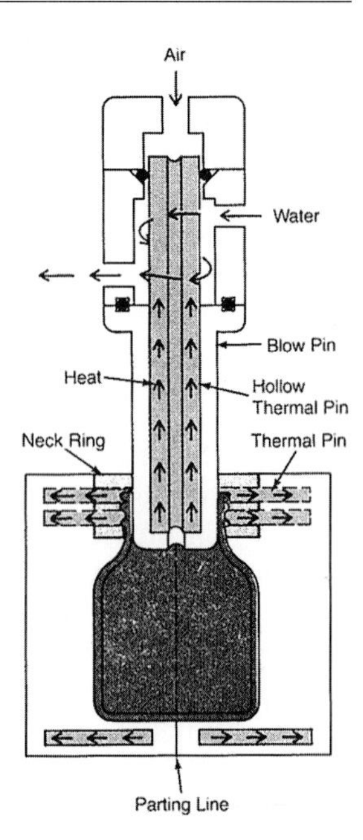

Fig. 6.50 *Hollow thermal pin cools neck from inside. Other pins can add cooling to the neck bottom of the BM container*

blown through the small tube and exit via the space left in the ID of the thermal pin.

6.7.3 Tool Surface

Different surface treatments are used to protect the processing tool (die, mold, etc.) against conditions such as abrasion and corrosion owing to their contact with the melt. When fabricating corrosive reacting plastics such as CPVC (chlorinated PVC) products, the metal die and mold can be subjected to corrosion and pitting. Certain steels such as stainless steel can provide a degree of protection. There are surface platings and hard coatings that can provide excellent protection such as abrasion resistance and also enhance molded product release. Different coating systems are used to protect the metal tool's polished surface and/or extend its useful life based on the plastic being processed. They boost lubricity while avoiding melt adhesion problems.

Prevalent are chromium-based materials that can be applied at rather low temperatures to provide resistance to corrosion, abrasion, and/or erosion if needed. Used materials include pure Cr, CrN (chromium nitride), and CrC (chromium carbide). Some coatings such as physical vapor deposition (PVD) and chemical vapor deposition (CVD) subject the mold steels to excessive high temperatures that reduce steel hardness. Systems have been developed with PVD and CVD (plasma CVD (PCVD)) that operate at lower temperatures of 200 to 400 °C (400 to 750 °F). Popular is the so-called nitrided coating; it is actually a hardened nitride casing (nitrogen is absorbed into the surface of the steel).

Fabricators can be faced with sweating and moisture condensation on their die and mold surfaces, particularly during the summer months. This can lead to corrosion and rust, and in turn to poor finishes and inferior quality parts. In addition, rust on guide pins can cause damage to the mold. By keeping the plant air around the mold dry, one can improve part quality and increase production rate.

Surface treatments used on tools are generally plated, coated, and/or heat treated to resist wear, corrosion, and release problems. Treatments such as those reviewed in Table 6.10 reduce wear. Other treatments resist the corrosion damage inflicted by chemicals such as hydrochloric acid when processing PVC, formic acid or formaldehyde with acetals, and oxidation caused by interaction between dies/molds and moisture in the plant atmosphere. Release problems require treatments that decrease friction and increase lubricity in mold cavities.

No single treatment is ideal for solving all these problems. The fabricator must determine which problems are causing the greatest loss of productivity (or could cause loss) and then select the treatments that will be most effective in solving the problems.

Plating and coating affect only the surface of a tool or component, while heat treating generally will affect the physical properties of the entire tool. Treatments such as carburizing and nitriding are considered to be surface treatments; although heat is applied in these processes, they are not considered to be heat treatments. Heat treating is more often the province of the steel manufacturer and tool maker than the fabricator. However, stress relieving is a heat treatment that the fabricator can perform.

Some mold wear cannot be prevented. This wear should be observed, acknowledged, and dealt with at intervals in the tool's useful life; otherwise, the tool could be allowed to wear past the point of economical repair [6]. Periodic checks of how platings and coatings are holding up will allow the fabricator to have a tool resurfaced before damage is done to the substrate.

While a poorly finished tool is being used for the first time, heat, pressure, and plastic are actually reworking its surface. Fragmented metal is pulled out of the metal fissures, and plastic is forced into them. While the fissures are plugged with plastic, the fabricator may actually be processing plastic against plastic.

Breaking in a poorly finished tool can be haphazard without proper presurfacing. If the underlying tool surface is unsound (and no prior treatment was used although required), a thin layer of metal plating, particularly chrome plating, will not make it correct. A poorly prepared surface makes for poor adhesion between treatment and the base metal.

Most fabricators and tool makers will not find it practical to do very much surface treatment themselves. Few of the plating, coating, and lubricating treatments can be done in a treatment fabricating shop, because, with few exceptions, treatments involve processes and chemicals that should not be used anywhere near a fabricating machine (because of corrosiveness). Treatments are best handled by custom plating and treating shops specializing in their use.

Table 6.10 *Surface Plating and Coating Treatments*

Process	Material	Applied to	Purpose
Coating by impingement—molecularly bound	Tungsten disulfide	Any metal	Reduce friction and metal-to-metal wear with dry film—nonmigrating
Coating by impingement—organically bound	Graphite	Any metal	Reduce sticking of plastics to mold surface—can migrate
Electrolyte plating	Hard chrome	Steel, nickel, copper alloys	Protect polish, reduce wear and corrosion (except for chlorine or fluorine plastics)
	Gold	Nickel, brass	Corrosion only
	Nickel	Steel and copper alloys	Resist corrosion except sulfur-bearing compounds, improve bond under chrome, build up and repair worn or undersized molds
Electroless plating	Nickel	Steel	Protect nonmolding surfaces from rusting
	Phosphor nickel	Steel and copper	Resist wear and corrosion
Nitriding	Nitrogen gas or ammonia	Certain steel alloys	Improve corrosion resistance, reduce wear and galling; alternative to chrome and nickel plating
Liquid nitriding	Patented bath	All ferrous alloys	Improve lubricity and minimize galling
Anodizing	Electrolytic oxidizing	Aluminum	Harden surface, improve wear and corrosion resistance

The distinction between platings and coatings is not entirely clear. Generally, thin layers of metals applied to the surface of tool components are considered platings. The application of alloys, fluorocarbons, or fluoropolymers (such as tetrafluoroethylene (TFE)) or dry lubricants is considered a coating. The effectiveness of a surface treatment depends not only on the material being applied, but also on the process by which it is applied. For any plating or coating to stick to the surface of a tool component, it has to bond to the surface in some way. The bonding may be relatively superficial, or a chemical/molecular bond may accomplish it. The nature and strength of the bond directly affect the endurance and wear characteristics of the plating or coating. The experience of the plater is an important factor in applications where cut-and-dried or standard procedures have not been developed.

6.7.3.1 Plating

Various plating methods can be used to meet different performance requirements. Electroless nickel plating provides a good protective treatment for tool components, including the holes for the heating media, etc. It prevents corrosion; also, any steel surface that is exposed to water, PVC, or other corrosive materials or fumes benefits from this minimal plating protection. The term electroless is used to describe a chemical composition/deposition process that does not use electrodes to accomplish the plating. Plating by the use of electrodes is called electrodeposition.

Electroless nickel deposits up to 0.001 in. (0.0025 cm) of nickel uniformly on all properly prepared surfaces. It has the characteristic of depositing to the same depth on all surfaces, which eliminates many of the problems associated with other metallic platings. Grooves, slots, and blind holes will receive the same thickness of plating as the rest of the tool. This allows close tolerances to be maintained. The surface hardness of electroless nickel is 48 Rockwell C, and the hardness can be increased by baking to 68 Rockwell C. Plated components will withstand temperatures of 700 °F (371 °C).

Chrome plating is very popular. Chromium is a hard, brittle, tensile-stressed metal that has good corrosion resistance to most materials. As it becomes thicker, it develops a pattern of tiny cracks because the stresses become greater than the strength of the coating. These cracks form a pattern that interlaces and sometimes extends to the base metal. A corrosive liquid or gas could penetrate to the base metal. This action can be prevented in three ways: 1) A nickel undercoat can be applied to provide a corrosion-resistant barrier; 2) the chrome plating can be applied to a maximum thickness; 3) or thin dense chrome can be substituted.

Hard chrome can be deposited in a rather broad range of hardnesses, depending on plating bath parameters. The average hardness is in the range of 66 to 70 Rockwell C. A deposit of over 0.001 in. (0.0025 cm) thickness is essential before chromium will assume its true hardness characteristics when used over unhardened base metals. Over a hardened base, however, this thickness is not necessary because of the substantial strength/backing provided. Precise control of thickness tolerances can be achieved in a particular type of chrome plating generally referred to as thin, dense chrome. This kind of plating provides excellent resistance to abrasion, erosion, galling, cavitation, and corrosion wear.

An atomic bond achieves adhesion between a chrome layer and the base metal. The bond strength is less upon highly alloyed steels and on some nonferrous metals; however, bonds in excess of 35,000 psi (245 MPa) tensile strengths are common. It is often said that the chromium layer will reproduce faithfully every defect in the base material; actually, as the chrome deposit thickens, it will level imperfections. For smooth chrome plating, the base material should be at least as smooth as the expected chrome. Imperfections can be ground or polished after plating to erase them.

In some instances, cracks in the base material that are not visible through normal inspection techniques may become apparent only after plating. This phenomenon is attributed to the fact that grinding of steel often causes a surface flow of material, which spreads over cracks and flaws. However, this cover is dissolved during the preplating treatment, and the cracks become apparent as the coating thickness increases. The deposited layer of chromium, although extremely thick, will not bridge a large crack.

With electrodeposition, the current distribution over different areas of a component greatly varies, depending on its geometrical shape. Elevations and peaks, as well as areas directly facing the anodes, receive a higher current density than depressions, recesses, and areas away from the anodes that do not directly face the anodes. The variation in current density over different areas produces a corresponding variation in the thickness of the deposited metal.

Hard chrome plating is recommended to protect polished surfaces against scuffing and to provide a smooth release surface that will minimize sticking of the parts in the tool. Some precautions are necessary. Hydrochloric acid created in the processing of PVC will attack chrome. Chrome that is stressed and cracked under adverse conditions will permit erosion from water and gas penetration into the steel. To deal with hydrogen embrittlement created when hydrogen is absorbed by steel during the plating process, chrome-plated components should be stress relieved within a half-hour of completion of the plating. To protect against galling, chrome should not be permitted to rub against chrome or nickel.

Although it is not uncommon for a tool to be sampled in the machine under pressure before plating, this is an undesirable practice. The effects of moisture on the steel can cause chrome plating to strip later. Carelessness can also result in scratching an unplated tool. Dimensional checks can readily be made outside the press using wax or other sampling materials. Chrome can be stripped from a tool after sampling in order to make essential changes. Periodic checks after a chrome-plated tool is in full production are desirable to find evidence of wear, which will show up first in the tool corners and high flow areas. Swabbing a copper sulfate solution in the mold areas can serve as a simple check. If the copper starts to foam on the surface plating, the chrome is gone and must be replaced.

Numerous metals other than nickel and chromium have been used at one time or another to plate the components of plastics tools. As an example, gold plating can be used to create a protective surface for PVC and some fluorocarbon materials. It prevents tarnishing, discoloration, and oxidation of the tool surface. Gold will protect the original finish and provide a hydrogen barrier. Fifty millionths of an inch of gold plating [0.00005 in. (0.0013 cm)] is adequate for the purpose. Gold can also be used as a primer under polished chrome. Platinum and silver have also been used to plate tools, but they share with gold the notable drawback of high cost.

As discussed previously, steels also are nitrided and carburized to improve the surface hardness, thus making the surface more wear resistant. Nitriding will penetrate the surface from 0.003 to 0.020 in. (0.008 to 0.051 cm), depending on the steel, and can result in hardnesses of 65 to 70 Rockwell C. Another process for imparting a surface hardness is carburizing, in which carbon is introduced into the surface of the cavity or core steel, and the inserts are heated to above the steel's transformation temperature range while in contact with a carbonaceous material. This process frequently is followed by a quenching operation to impart the hardened case. Hardness as high as 64C and a depth of 0.030 in. (0.08 cm) are possible with this process.

6.7.3.2 Coating

Ultrahard titanium carbides distributed throughout a steel or alloy matrix are used very effectively for coating. The carbides are very fine and smooth, and the coatings are applied by sintering, a process in which the component to be coated is preheated to sintering temperature, then immersed in the coating powder, withdrawn, and heated to a higher temperature to fuse the sintered coating to the component.

Flexibility in selecting and controlling the composition of the matrix alloy makes it possible to tailor the qualities of the alloy according to the requirements of the application. When heat-treatable-matrix alloys are used, the composite can be annealed and heat treated, permitting conventional machining before hardening to 55 to 70 Rockwell C. Standard alloy matrices have been developed with quench-hardenable tool steel, high-chromium stainless steels, high-nickel alloys, and age-hardenable alloys. Special alloys can be formulated for even the most corrosive conditions.

These types of coatings are effective in combating the severe wear that occurs with abrasive plastic compounds. Ceramic/metal composites provide the hardness and abrasion resistance to withstand wear by the most damaging glass-fiber-, mineral-, or ferrite-filled compounds. With the right metal matrix selection, resistance to corrosion and heat is also obtained.

Treatments are available via a chemical process that utilizes thermal expansion and contraction to lock polytetrafluoroethylene (PTFE) particles into a hard electro-deposited surface such as chromium. Surfaces treated by the process have the sliding, low-friction, nonstick properties of PTFE, along with the hardness, thermal conductivity, and damage resistance of chrome. Core rods and other internal tool components benefit from this surface treatment. The process builds a thickness of 0.002 in. (0.0051 cm) of electrodeposited chromium on component surfaces and is available for ferrous, copper, and aluminum alloy parts (Table 6.10).

Anodizing is an electrochemical process in which a thin, protective deposit of aluminum oxide is applied to aluminum that imparts different benefits such as providing a hard coating. One of the advantages of anodizing is that it significantly improves appearance, leaving parts looking cleaner and more finished. The process also fights corrosion and enhances the areas of abrasion resistance, heat absorption, and heat radiation. The anodizing process is complicated by the fact that every alloy anodizes differently. Each job may call for knowledge and experience totally unlike that required by the job before it.

Differences may take the form of new mixtures in chemical compounds, changes in the preparation of parts, and adjustments in the regulation of electrical current and temperature. Sulfuric acid is, by far, the most popular form of anodizing. Sulfuric anodizing is widely favored because it adds dielectric strength to aluminum, and staves off corrosion and abrasion. All forms of anodizing have the advantage of being both cost effective and relatively inexpensive.

Titanium carbonitride is the material most commonly applied to provide wear resistance to tools and wear parts made from tool steels. The coating inhibits galling and results in a favorable coefficient of friction. The process protects against wear from abrasive fillers and corrosion from unreacted plastics that release acids. The coating is said to be 99% dense, that is, to have virtually no porosity, and is inert to acids. Note that the processes used for hardening screws and barrels are not necessarily the same as those applicable to tool components.

The PVD coating is used to optimize the surface properties in a layer up to 10 mm (0.4 in.) deep that has little effect on the contour of product. PVD coatings lead to an insignificant and usually discernible roughening of the surface. Tools that have been machine finished can therefore be improved. No expensive post-treatment is necessary. In its vacuum chamber (10^{-2} to 10^{-4} mbar (1 to 0.01 Pa)), metals are converted to a gaseous state by the introduction of thermal (electron beam or arc) or kinetic energy (atomization) and condense on the surface being coated.

6.7.3.3 Polishing

A large part of tool cost is polishing cost, which can represent from 5% to 30% of the mold cost. Tools usually require a high polish. An experienced polisher could polish from 2 to 5 in.^{2}/h (5.1 to 12.7 cm^2/h). Certain shops can at least double this rate if they have the proper equipment. Polishing is rarely done for appearance alone, but rather to obtain a desired surface effect on the fabricated product, facilitate the ejection of the product from the mold, or prepare the tool for another operation such as etching or plating. If a plastics product is to be plated, it is important to remember that plating does not hide any flaws; it accentuates them. Therefore, on critical plated-part jobs, the tool polish must be better than for nonplated parts.

Another purpose of polishing is to remove the top layer of metal that may be weak from the stresses induced by machining (conventional or EDM) or from the annealing effect of the heat generated in cutting. When it is not removed, this layer very often breaks down, showing a pitted surface that looks corroded.

Although the operation is seemingly gentle, polishing can damage the steel unless it is properly done. A common defect is the so-called orange peel, a wavy effect that results when the metal is stretched beyond its yield point by overpolishing and takes a permanent set. Attempts to improve the situation by further vigorous polishing only make matters worse; eventually, the small particles will break away from the surface. The more complicated the tool, the greater the problem. Hard carburized or nitrided surfaces are much less prone to the problem. The harder the steel, the higher the yield point and therefore the less chance of orange peel. The surest way to avoid orange peel is to polish the tool by hand, as it is easier to exceed the yield point of the metal when using powered polishing equipment. If power polishing is done, use light passes to avoid overstressing.

Orange peel surfaces usually can be salvaged: Remove the defective surface with a very fine grit stone, stress-relieve the mold, restone, and diamond polish. If orange peel recurs after this treatment, increase the surface hardness by nitriding with a case depth of no more than 0.005 in. (0.0125 cm) and repolish the surface.

The techniques used to achieve a good and fast polish are basically simple, but they must be followed carefully to avoid problems. The first rule is to make sure the part is as smooth as possible before polishing. If EDM is used, the final pass should be made with a new electrode at the lowest amperage. If the part is cast or hobbed, the master should have a finish with half the roughness of the desired mold finish.

When the tool is machined, the last cut should be made at twice the normal speed, the slowest automatic feed, and a depth of 0.001 in. (0.0025 cm). No lubricant should be used in this last machining, but the cutting tool should be freshly sharpened, and the edges honed after sharpening. The clearance angle of the tool is usually from 6 to 9°, and if a milling cutter or reamer is used, it should have a minimum of four flutes. A steady stream of dry air must be aimed at the cutting tool to move the chips away from the cutting edge.

Polishing a tool begins when the designer puts the finishing information on the drawing. Such terms as “mirror finish” and “high polish” are ambiguous. The only meaningful way to describe what has to be polished and to what level is to use an accepted standard. It is also important that the designer specify a level of polish no higher than is actually needed for the job, because going from just one level of average roughness to another greatly increases the cost of the tool. Figure 6.51 provides an estimate on the cost of polishing as a function of the roughness of the finish.

Roughness is given as the arithmetical average in microinches. A microinch is one-millionth of an in. (10^{-6} in.) and is the standard term used in the United States. Sometimes written as MU in. or μin., it is equal to 0.0254 μm. A micron, one-millionth of a meter (10^{-6} m), equal to 39.37 μin, is the standard term in countries that use the metric system.

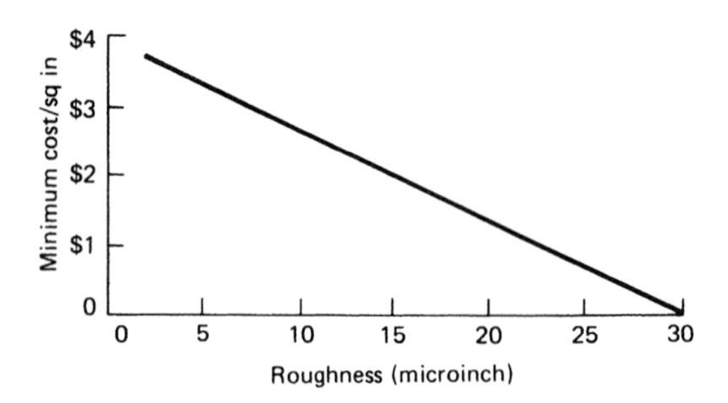

Fig. 6.51 *Cost of polishing tool steels*

There are several ways to specify a certain level of roughness. One common standard is the SPI Mold Finish Comparison Kit. Both SPI and SPE sponsored the old standard. The revised standard adopted in the 1990s is solely the responsibility of SPI. It consists of six steel disks finished to various polish levels and covered with protective plastics caps processed in molds with those finishes. One disk has a roughness of 0 to 3 μin., another a roughness of 15 μin., and all the others are coarser. Because 15 μin. roughness is acceptable in fewer than 10% of all jobs, the result is that the disk with zero to 3 μin. roughness, or the highest level of polish, must be selected in almost all other cases.

The no. 6 finish is equivalent to 24 grit, dry blast at 100 psi (0.7 MPa) from a distance of 3 in. (75 mm). A no. 5 finish is achieved with 240 grit, dry blast at 100 psi (0.7 MPa) at 5 in. (127 mm). The no. 4 finish is the equivalent of that achieved with a 280-grit abrasive stone. The no. 3 finish is installed by finishing with a 320-grit abrasive cloth. The no. 2 finish is the surface finish obtained when the final phase of polishing is completed using a 1,200-grit diamond (up to 15 diamond μm). The highest finish is the SPI no. 1 finish, the equivalent of a 8,000-grit diamond (0 to 3 μm range).

Other standards have come into general use. An example is specification of a finish as produced by a polishing compound containing diamond particles within a certain microinch range. It is not a perfect system but works in most cases. A near-perfect system is the American Standard Association's standard ASA B 46.1, which corresponds to the Canadian standard CSA B 95 and British standard BS 1134. The use of this standard in specifying finishes leaves no room for disputes about what is called for, and because all quotes apply to the same standard, the polishing costs tend to be more uniform. The biggest drawback to its use is that many tool shops do not have the necessary test equipment.

6.7.3.4 Polishing Technique

Hand benching is the first step in mold finishing. It normally involves the use of both hand tools and power-assisted grinders to prepare the surfaces to be polished. Depending on the last machining operation and roughness of the remaining stock, the tool maker will select the method of preparing the surface in the quickest manner possible. Typical hand tools include files and die maker rifflers. The files range from 20 to 80 teeth/in. and will assist in removing stock quickly and accurately. Rifflers are available in a wide range of shapes to fit into any conceivable contour found on a tool and have from 20 to 220 teeth/in.

Power-assisted tools include electric, pneumatic, and ultrasonic machines, which can be equipped with a wide variety of tool holders. Rotary pneumatic and electric grinders, with the tool mounted either directly or in a flexible shaft, are extremely popular with the tool maker for rapidly removing large amounts of stock. With the grinding or cutting medium mounted in a 90° or straight tool, depending on the surface, the tool maker will start with the coarser medium and work the surface progressively to the finer medium. The finishing process can best be described as one in which each preceding operation imparts finer and finer scratches on the surface. The initial stages are perhaps the most important in the finishing state. Spending too little time in this phase normally will be detrimental to the final surface finish.

A variety of cutting tools, from carbide rotary cutters, abrasive drum, band, cartridge, tapered cone, and disk to flap wheels, are used to reduce the roughness of the surface in the quest for the desired tool surface finish. The use of abrasive stones, either worked by hand or in conjunction with reciprocating power tools, normally is the next phase in tool finishing. These stones are a combination of grit particles suspended in a bonding agent. For general purpose stoning, silicon carbide is available in type A stones for steel hardness under 40 R, and type B stones for a higher degree of hardness. Most stones are available in square, rectangular, triangular, and round shapes in grits of 150, 240, 320, 400, and 600. Other stones consist of type E aluminum oxide for working EDM surfaces, type M that has oil impregnated for additional lubrication, and type F for added flexibility or reduced breakage of thin stones.

The procedure normally consists of using abrasive sheets or disks to continue smoothing of the tool surface. All the finished work is carried out in the direction of the scratches installed in the line of draw or ejection. Machine, file, stone, or abrasive marks installed perpendicular to the direction of draw are detrimental to removal of the part from either the cavity or core, and must be avoided. At this point, the desired tool surface finish may have been achieved, or the piece parts grit blasted, and finishing is considered complete. These finishes may be acceptable for nonfunctional core/cavity surfaces in which the part does not pose a problem.

Some polishers just polish and others can engrave or bench by hand or bench any shape on any mold block. Most engravers used to utilize a collection of Swiss files that they developed throughout the years; different jobs demanded different shapes and so their collections grew. Seeing an engraver with a hundred or more different scrapers only attested to the person's skill and long years of patient work.

Different jobs demand different approaches. For example, a job has a core and it requires a 0.030-in (0.9 mm) radius on all molding corners. The veteran already knows that pulling a scraper toward himself will be 10 times faster than pushing one away from himself. This action occurs because for some reason the human hand has more natural "depth control" pressure when it pulls than when it pushes. Today many polishers simply grind or blitz radii. Some are quite good at maintaining perfect size and consistency (which is good), but the next time they get an electrode that cannot be ground or a contour that requires dimensions to be benched into the steel and those dimensions must be held, blitzing may not be the best option.

Carbide blanks have become an extremely useful tool in the benching and engraving tool box; blanks of $\frac{3}{16}$ in. (4.5 mm) and $\frac{1}{4}$ in. (6 mm) diameter are a perfect fit to most people's hands and the polisher can retain a good feel (the secret to successful engraving) without the fingers cramping or getting sore. In benching the tool is the determining factor and not the polisher. Take any piece of scrap steel, sharpen an old Swiss file into a three-pointed scraper, and practice different pushing and dragging moves. You will immediately find that the tool works better one way than the other. Now change the angle at which you are holding the file to the steel from perpendicular to 45°. Notice that when you push forward the movement of the scraper becomes much stronger and the scraper probably will bury itself into the steel and stop, which is a complete disaster and should always be avoided.

No contour will come out looking professional with uncontrolled nicks and gouges in the steel. Other approaches are used such as rotating the scraper. Only experimentation will demonstrate how a scraper will cut because each polisher naturally uses a different amount of speed in his stroke, a different amount of pressure with his hands, and holds the scraper at a different angle. However, the true way to learn which combination is best is to try the following: (1) scrape away from your body as fast as you can (exactly like peeling a potato) using a light pressure; (2) look at the size and type of the chips that you are producing; and (3) try to produce an exact pile of chips that are all the same. This is the beginning of control.

It is important to learn how your hand feels as you are doing the work. Go for a consistent feeling in your hand and you will get a consistent result every time. Next, after you have been "peeling the potato" for a while, roll right in and change something in your stroke. Change the angle from perpendicular to 10° forward or backward, but keep your speed, pressure, and stroke speed the same.

There is also the aid for finishing involving ultrasonic finishing and polishing systems. The ultrasonic controls deliver strokes that can be adjusted to between 10 and 30 μm at speeds up to 22,000 cycles/s. This action eliminates costly hand finishing and is extremely effective in polishing deep narrow ribs that have been EDM installed.

6.7.4 Conventional Manufacturing

Manufacturing components such as the cavity and core inserts are fabricated by a large variety of methods. However, conventional machining of the cavity and core inserts accounts for the most widely employed of these methods. The term machining is used here to denote milling, drilling, boring, turning, grinding, and cutting.

As greater sophistication in temperature control of molds is required to meet the challenges presented by today's design engineers, programs have been developed and are commercially available to accurately predict the amount and placement of temperature control channels. Generally, the core will be required to remove approximately 67% of the British thermal units (Btu's) generated in the injection molding process. This requirement presents one of the greatest challenges to the moldmaker, as less space usually is available in the core than in the cavity.

A generation ago, the standard of the industry was the reliable Bridgeport, with the skill of the journeyman moldmaker coaxing accuracy from the equipment. Today, the standards are the NC (numerical control), CNC (computer numerical control), or DNC (digital numerical control) milling, drilling, duplicating, electric-discharge machining (EDM), and grinding machines. Often, codes are generated by the mold designer, a practice made possible by the wide acceptance of CAD/CAM equipment in even the smallest of operations. The addition of CAD/CAM equipment has made the skilled craftsman more productive and more valuable.

Cavities are formed by methods other than machining such as electroforming, EDM, and casting. After machining, hobbing ranked as the next most common method employed to produce cavities, although it is no longer as important as it was 30 years ago. Many hobbing applications have been replaced by die sink EDM. EDM can be used on all steels and puts much less stress onto the metal. As compared to conventional machining, in EDM it is not important whether the steel is hard or soft. EDM does leave a damaged surface layer that must be removed before polishing or texturizing.

Cast cavities have certain advantages not obtainable with machining, in that large amounts of excess stock do not have to be removed by machining. A variety of cavity materials are suitable for casting, including steel, beryllium–copper, kirksite, aluminum, and others.

6.7.5 Preengineered Mold

Standardized mold (and die) components have been available at least since 1943. They provide for exceptional quality control on materials used, quick delivery, interchangeability, and lower cost. These preengineered molds and mold parts provide high-quality manufacturing techniques that result in consistent quality and reduced mold cost. The different manufacturers of these preengineered mold bases and components provide similar but also different products that can provide unique and different approaches to meeting complex product designs.

A major advantage to the molder is in saving time and money should a component ever need replacement, enabling one to meet quality demands and tight deadlines Preengineered molds reduce cost risks in most mold shops because they are bought at fixed prices and delivered from stock [6]. Most often these components serve the function of the mold and are not designed for use as plastic-forming mold members.

6.7.6 Quick Mold Change

Completely automated quick mold change (QMC) devices are used to handle molds (also dies/molds in EBM lines). The cost could be higher than that of just an injection molding machine; however, they pay for themselves "quickly" when required. QMC devices, with microprocessor controls, provides cost-effective approaches to plantwide automation. Different designs are used, such as overhead or side loading/unloading platforms, to move molds from an inventory that can have hundreds of molds.

The concept is best suited to processors with relatively short production runs and frequent mold changes. In such operations, the benefits of QMC devices are many: increased productivity, reduced inventory, increased scheduling flexibility, and more efficient processing.

Completely automated systems contain (1) mold conveyors that propel the molds in and out of position on motorized rollers and (2) mold carriers that index on a track parallel to the IMM to align the mold conveyors and clamp unit. They can also convey molds to and from the machines from a central mold storage area.

Basically, when the microprocessor-based control signals the end of a mold production run, the moveable platen indexes to a preset position for mold removal. Automatic mold clamps are deactivated, releasing the old mold. Mold clamps or straps connect to lock the mold halves together. The mold conveyor removes the old mold. Computer control resets platens for the new mold, and a carrier device aligns the preheated new mold with the clamp unit. As mold conveyors insert the new mold, automatic mold positioners center the mold within seconds. Automatic clamps are activated by the control system, and new molding cycle information stored in the microprocessor initiates the next production cycle. The estimated downtime between the old- and new-mold shot is 2 min with a highly sophisticated QMC device.

6.8 Buying a Mold

Mold making has been moving toward more specialization. Various machine operators on milling, lathe, EDM, or jig grinding machines manufacture parts of the finished mold separately. At the final stage, all these parts are brought together and assembled by the moldmaker.

Studies have shown that the total hours required for mold production involve 25% mold construction, 20% additional work on the mold, and 55% contour of parts. Using standard elements can save at least 25% of the required capacity. In addition special machining requirements can also be handled by the manufacturer of standard elements. In total, these add up to some 40% of the capacity required for the production of a mold.

It is estimated that the initial cost of mold construction is about 20% of the cost of the mold. Another 40% of the cost is scheduled maintenance, and the other 40% is repair after breakdowns. As part of its continuing effort to improve service to molders, the Moldmakers Division of SPI has a bulletin on mold making intended to assist buyers seeking guidance in mold procurement. It points out the various difficulties that can result, unless thorough understanding and communication are established between the mold buyer (molder) and moldmaker.

Also, the Moldmakers of the SPI provides industry an updated directory of its members and their special capabilities. The SPI Moldmaker members are in constant contact with the plastics industry and its ever-changing technology. The directory lists moldmakers as contract or custom services and in turn by type of process mold such as IM and BM.

There are also publications that provide buyer guides. As an example, *Moldmaking Technology* magazine issues an annual buyers' guide that features directories on:

- mold making equipment, supplies, and accessories;
- mold components;
- mold design and engineering equipment;
- mold material;
- machining equipment;
- electrical discharge machining equipment and supplies;
- machining tools and accessories;
- hot runner systems and supplies; and
- mold polishing and repair equipment and supplies.

6.9 Mold Safety

The Mold Safety Committee of the SPI, organized under the SPI Moldmakers Division, formed two subcommittees to set criteria to ensure that molds meet certain guidelines for safety and good electrical practices, as described in the following sections.

6.9.1 Electrical Subcommittee

Members of this group create guidelines for:

- the recommended wire size within the mold;
- the current capacity of the wire to be used in molds;
- the temperature rating of the wire to be used in molds;
- the name of the wire, for example, appliance wire, UL listed, CSA approval and other approved standards for wire (5107A);
- wire capacity of the wire ways (wire channels) in the molds for the thermocouples and the heater wires;
- assuring that the thermocouples are put in wire ways (channels) separate from the heater wires;
- the proper way to ground the mold;
- setting up the proper way to wire molds that would include such things as how wires are to be spliced within the mold, that is, by wire splices, butt splices, wire nuts, and so on.

(Soldering of wires within the mold is not recommended, and the guidelines will specify this. The use of connectors that require the wires to be soldered to them also is not recommended and the guidelines will specify this as well.)

6.9.2 Mold Interface Subcommittee

This group provides guidelines regarding:

- how the mold is to be wired;
- the location of the first zone (cavity);
- whether the manifolds will be wired first or the bushings or the sprue;
- how the cavities will be numbered from the parting line of the mold;
- the top nonoperator side looking at the cavities toward the stationary side of the molding machine;
- the numbering of the cavities from top to bottom, then left to right;
- the numbering of the cavities in columns. (Top nonoperator looking into the cavity top to bottom appears to be the best way, as there are some issues that have to do with the wiring channels within the mold. By numbering the cavities from top to bottom, the heater and the thermocouple wires will not have to be routed across or under the manifold to the wiring channels.);
- that manifold always will have bushing under it;
- that pin of the connector always will be wired to bushing;
- where connectors/junction boxes are to be mounted to the mold;
- the ratings junction boxes should have.

6.9.2.1 Universal Guidelines

Within the guidelines, there also will be definitions of terms such as connector, plug, receptacle, parting line of the mold, operator side of the mold, manifold, bushing, nozzle, drop, cavity, wiring channel, and junction box. It is the aim of the subcommittees to make the guidelines universal. That is, wherever a mold is built using SPI guidelines, users could be sure of the first zones being the bushing or manifolds; the location of the first zone, or cavity; the numbering of the cavity; and how the first zone is wired and to what pin on what connector.

7 Plastic Types and Processability

7.1 Overview

Worldwide annual plastic consumption is now about 340 billion lb of plastics for all processes (Table 7.1), with approximately one third of the total in the United States. At least 10% of the total plastics are used in BM. Practically any TP can be blow molded, particularly the commodity type plastics. Different properties can be obtained with the different available and useful plastics that are now used to blow mold different size and shaped products (Fig. 7.1).

Properties can include formability, light weight to strength, toughness, clarity, gas barriers, chemical resistance, heat resistance, color, and many others. Different combinations of properties can be achieved by combining two or more plastics. Parisons can be coextruded, coated, dipped, and coinjected to provide the combinations of different properties available in different plastics. Stretch-BM is also an important processing technique to provide improved properties and reduce the quantity of plastics to improve cost/performance characteristics. BM also permits consolidation of parts.

Examples of properties that can be obtained with the different plastics are as follows: clarity (crystal polystyrene (PS), acrylonitrile (AN), styrene acrylonitrile (SAN), polyvinyl chloride (PVC), stretched polyethylene terephthalate (PET), polysulfone (PSF)), moisture barrier (polypropylene (PP), high-density polypropylene (HDPE), transparent nylon), oxygen barrier (PET, ethylene vinyl alcohol copolymer (EVOH), polyvinylidene dichloride (PVDC), PVC, AN, PSF, transparent nylon), and so on.

Extrusion BM (EBM) represents the biggest outlet for BM containers, with HDPE being the major plastic used. There are different grades of HDPE that include those with changes in density, stiffness, strength, hardness, toughness, stress crack resistance, permeability, creep resistance, softening temperature, and gloss.

In poundage, PVC is the second largest plastic used in the United States for EBM due to its low oxygen transmission, water clear clarity, and relative ease of processing, low cost. It is being challenged by stretch blown PP and stretch blown PET. PP and PET copolymers are used in containers such as packaging cosmetics; drugs; some food containers (e.g., syrup), especially where hot filling is required; and for medical-surgical fluid [79].

Table 7.1 *World Plastic Consumption (million lb) in 2000*

Region / Plastic	United States	Canada	Mexico	Brazil	Other Latin America[a]	Western Europe[b]	Eastern Europe[c]	Japan	China	Other Asia-Pacific[d]	Africa and Middle East	Rest of World	Grand Total
LLDPE	8468	610	480	610	900	4093	1023	1439	2661	6121	580	367	27,352
LDPE	7748	1281	1176	1590	619	10,254	2563	1804	1938	3391	1210	457	34,031
HDPE	14,065	1136	952	1388	417	9178	2294	2158	1218	4967	1325	532	39,630
Urethane	5265	475	380	410	390	5481	1808	1512	1825	4850	390	310	23,096
PVC	14,698	1394	605	1456	1837	12,388	2477	3761	5338	11,633	1250	773	57,610
Polystyrene	6589	725	405	280	495	6180	1236	2175	2299	5275	190	352	26,201
Polypropylene	13,739	796	695	1366	1307	13,566	2713	5001	2667	11,176	1275	738	55,039
ABS	1409	145	175	365	290	1410	282	948	460	950	305	92	6831
Acrylic	613	75	80	165	140	576	115	302	155	325	140	37	2723
Unsaturated Polyester	1681	180	125	250	210	1036	207	1285	645	1175	215	95	7104
Nylon	1267	110	90	170	145	1210	242	339	170	425	155	59	4382
PET	4330	410	192	482	386	2464	493	1178	715	5441	390	224	16,705
Polycarbonate	857	90	70	155	140	607	121	443	215	475	145	45	3363
Thermoplastic Polyester	346	45	30	65	50	243	48	164	85	322	42	20	1460
Acetal	389	40	30	70	45	317	63	212	106	235	63	21	1591
Recycled Plastics	1800	166	121	195	162	1625	220	502	453	1254	165	88	6751
Other Plastics[e]	8150	763	412	875	74	7250	1750	255	198	5310	740	344	26,121
Total	91,414	8441	6018	9892	7607	77,878	17,655	23,478	21,148	63,325	8580	4554	339,990

[a] Argentina, Chile, Columbia, Venezuela and all other.
[b] European Union plus Norway and Switzerland.
[c] Includes Russia and the Balkans.
[d] Australia, India, Indonesia, Malaysia, North Korea, Pakistan, South Korea, Taiwan, Thailand, Philippines, Singapore, Vietnam.
[e] High performance, other thermosets, specialty elastomers, tailored blends and alloys.

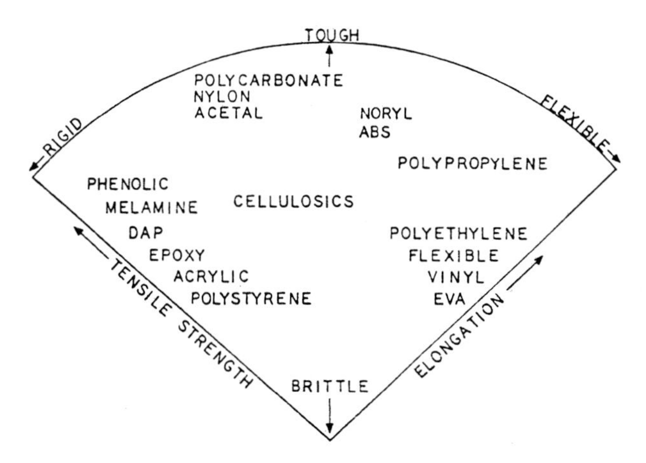

Figure 7.1 *Example of range of mechanical properties*

This brief review on the different plastics emphasizes the importance of using the correct plastic to meet specific requirements. The real test on any material is when it is blown, either unoriented or stretched oriented. Unfortunately, in the past, many simulated "fast" tests to evaluate plastic bottle performance have had "bad" records. But now "fast" tests have been developed to simulate actual and reliable performance. Detailed information on plastic material processes and product performances is available from material suppliers and the literature.

7.2 Basic Blow Molding Material

The term *plastic* comes from the Greek word "to form." It identifies many different plastic materials. Practically all plastics at some stage in their manufacture or fabrication can be formed into various simple to extremely complex shapes ranging from extremely flexible (rubbery/elastomeric) to extremely hard (high performance properties). The use of a virtually endless array of additives, fillers, and reinforcements permits compounding from the raw material suppliers to the fabricators, imparting specific qualities to the basic materials (polymers) and expanding opportunities for plastics. These many combinations are endless so that new materials are always on the horizon to meet new industry requirements.

Polymers, the basic ingredients in plastics, can be defined as high molecular weight organic chemical compounds, synthetic or natural substances consisting of molecules characterized by the repetition (neglecting ends, branch junctions, and other minor irregularities) of one or more types of monomeric units. A repeated small unit, the mer, such as ethylene, butadiene, or glucose, can represent its structure. Practically all of these polymers (base materials) use certain types of additives to perform properly during fabrication and/or in service.

The terms plastic, polymer, resin, elastomer, and reinforced plastic (RP) are somewhat synonymous. Worldwide, the term preferred is plastics. However, polymer and resin usually denote the basic material, whereas plastic pertains to polymers or resins (as well as elastomers and RP) containing additives, fillers, and/or reinforcements. An elastomer is a rubberlike material (natural or synthetic). Reinforced plastics (also called plastic composites) are plastics with reinforcing additives such as fibers and whiskers, added principally to increase the product's mechanical properties.

There are terms that overlap and also interfere with each other. An example is stating that TPs are cured during processing. Cure occurs only with thermoset (TS) plastics or when a TP is converted to a TS plastic. The term, curing TPs, has been in use since the beginning of the 20th century rather than just the correct term *melting*. TSs represented practically all the plastic used worldwide. Thus, the incorrect term *curing* was applied to TPs even though there is no chemical reaction or curing action.

Plastic also refers to a material that has a physical characteristic such as plasticity and toughness. The general term engineering plastics, advanced plastic, advanced reinforced plastic, or advanced plastic composite is used to indicate higher performance materials.

The term plastic is not a definitive one. Metals, for instance, are also permanently deformable and therefore have a plastic behavior. How else could roll aluminum be made into foil for kitchen use, or tungsten wire be drawn into a filament for an incandescent light bulb, or a 90-ton ingot of steel be forged into a rotor for a generator? Likewise the different glasses, which contain compounds of metals and nonmetals, can be permanently shaped at high temperatures. These cousins to polymers and plastics are not considered plastics within the plastic industry.

The plastics industry provides a wide variety of many different plastic materials with their different processing methods. There are well over 35,000 different types of plastic materials worldwide [50]. However, most of them are not used in large quantities; they have specific performance and/or cost capabilities generally for specific products by specific processes that principally include many thousands of products.

Petroleum is currently the major source of raw materials for most high-volume plastics. Also used are natural gas and coal. Other sources to produce plastics include vegetation and biotechnology. Figure 7.2 shows the manufacturing stages for plastics from raw materials to products.

Natural gas, crude oil, and coal can be starting points for a variety of plastics; over a half century ago the main source was coal. Natural gas liquids, captured during the production of natural gas, can be moved directly into the process stream.

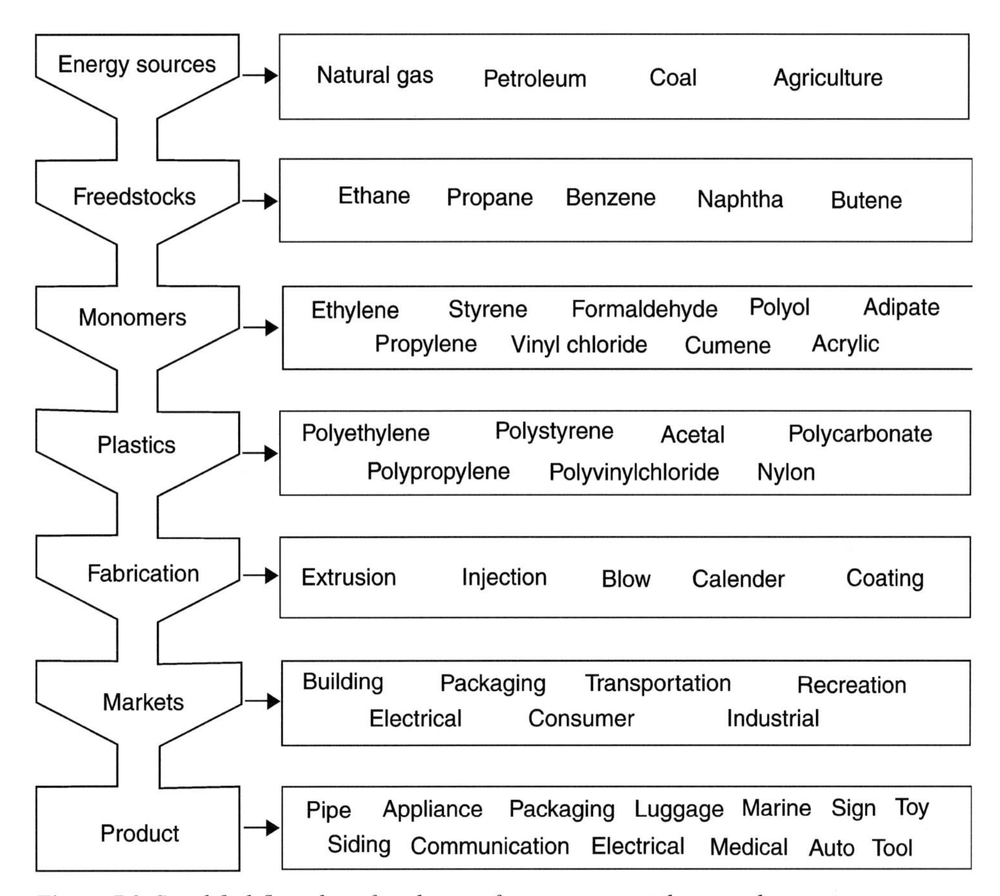

***Figure** 7.2 Simplified flow chart for plastics, from raw materials to products*

Crude oil and coal must undergo some primary processing such as distillation, cracking, or solvent extraction to produce ethylene, C_2H_4; propylene, C_3H_6; or benzene, C_6H_6.

The chemistry of plastics is an important aspect to understand, as it permits an endless growth of new plastics with improved performances [2]. Chemistry is the science that deals with the composition, structure, and properties of substances and of the transformations that they undergo. The majority of chemical compositions are basically organic polymers that are very large molecules composed of chains of carbon (C) atoms generally connected to hydrogen (H) atoms and often also oxygen (O), nitrogen (N), chlorine (Cl), fluorine (F), and sulfur (S). While polymers thus form the structural backbone of plastics, they are rarely used in pure form. In almost all plastics, other useful and important ingredients are added to modify and optimize their properties for each desired process and application; additives affect both processability and product performance.

7.3 Polymer to Plastic

Chemical and physical characteristics of plastics are derived from the chemical structure, form, arrangement, and size of the polymer. As an example, the chemical structure, that is the types of atoms and the way in which they are joined to one another, influences density. The form of the molecules, their size, and their disposition within the material influence the mechanical behavior. It is possible deliberately to vary the crystal state in order to vary the hardness or softness, toughness or brittleness, resistance to temperature, and so forth. The chemical structure and nature of plastics have a significant relationship not only to the properties (mechanical, physical, barrier, density, etc.) of the plastics but also to the ways in which they can be processed, designed, or otherwise translated into a finished product.

Understanding the basic plastic characteristics and behaviors provides a better understanding as to why plastic material variables exist during processing such as BM. Even though equipment operations and plastic compounded materials have understandable, but controllable variables that influence processing, the usual most uncontrollable variable in the process can be the plastic material. The degree of properly compounding or blending by the plastic manufacturer, converter, or in-house by the fabricator is important.

In chemistry, a polymer is described by drawing its physical structure of atoms. An example is PE with its hydrogen (H) and carbon (C) atom bonds. The bonds on the carbon atoms indicate that atoms can form bonds with like atoms to form a polymer chain. Many polymers can be made from the same carbon atom backbone by substituting another element or chemical compound for one or more of the H atoms.

Polymer molecules are distinguished from other chemical compositions primarily by their tremendous size; most of their properties result from this unusual size. Many of the properties and product applications of plastics are greatly influenced by the individual atoms and functional groups within the polymer molecule.

The size and flexibility of molecules explain how individual molecules would behave if they were completely isolated from their neighbors. Such isolated molecules are encountered only in theoretical studies of dilute solutions. In actual practice, different molecules always occur in a mass, and the behavior of each individual molecule is very greatly affected by its intermolecular relationships to adjacent molecules in the mass. Three basic molecular properties affect processing performances (flow conditions, etc.) that in turn affect product performances (strength, dimensional stability, etc.):

- the mass or density (*d*),
- average molecular weight (AMW), and
- molecular weight distribution (MWD).

7.3.1 Density/Specific Gravity

The mass of any substance per unit volume of a material represents the absolute density (*d*). It is usually expressed in grams per cubic centimeter (g/cm^3) or pounds per cubic inch ($lb/in.^3$). Specific gravity (sp. gr.) is the ratio of the mass in air of a given volume compared to the mass of the same volume of water, both *d* and sp. gr. being measured at room temperature (23 °C (73.4 °F)). Because sp. gr. is a dimensionless quantity, it is convenient for comparing different materials. Like density, specific gravity is used extensively in determining product cost vs. average product thickness, product weight, quality control, etc. It is frequently used as a means of setting plastic specifications and following product consistency.

In crystalline plastics, such as PE, density has a direct effect on properties such as stiffness and permeability to gases and liquids. Changes in density may also affect some mechanical properties. For greatest usefulness, density needs to be measured to an accuracy of at least $\pm 0.001\ g/cm^3$.

The apparent density is the weight in air of a unit volume of material including voids usually inherent in the material. Also used is the term bulk density that is commonly used for compounds/materials such as molding powders, pellets, flakes, etc. Bulk density is the ratio of weight of the compound to the volume of a solid material including voids.

7.3.2 Molecular Weight

Molecular weight (MW) is the sum of the atomic weights of all the atoms in a molecule. It represents a measure of the chain length for the molecules that make up the polymer. Atomic weight is the relative mass of an atom of any element based on a scale in which a specific carbon atom (carbon-12) is assigned a mass value of 12 [2].

The MW of plastics influences their properties. As an example, with increasing MW properties increase for abrasion resistance, brittleness, chemical resistance, elongation, hardness, melt viscosity, tensile strength, modulus, toughness, and yield strength. Decreases occur for adhesion, melt index, and solubility.

Adequate MW is a fundamental requirement to meet desired properties of plastics. With MW differences of incoming material, the fabricated product performance can be altered; the more the difference, the more dramatic change occurs in the product. *Melt flow rate* (MFR) tests are used to detect degradation in products where comparisons, as an example, are made of the MFR of pellets to the MFR of products. MFR has a reciprocal relationship to melt viscosity. This relationship of MW to MFR is an inverse one: as the MW drops, the MFR increases. The MW and melt viscosity is directly related: as one increases the other increases.

The average molecular weight (AMW) is the sum of the atomic masses of the elements forming the molecule, indicating the relative typical chain length of the

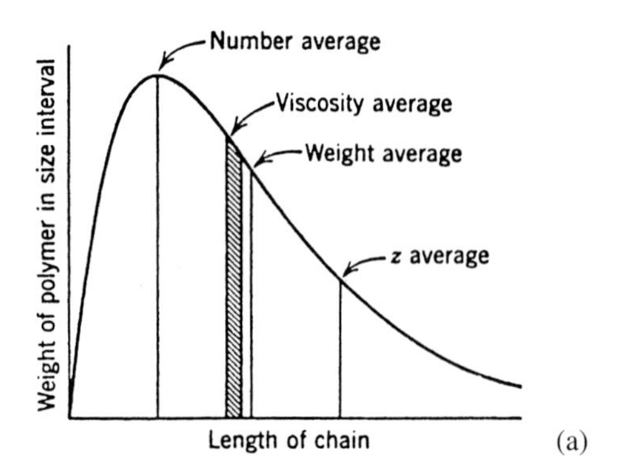

Figure 7.3 *(a) Typical molecular weight distribution and averages*

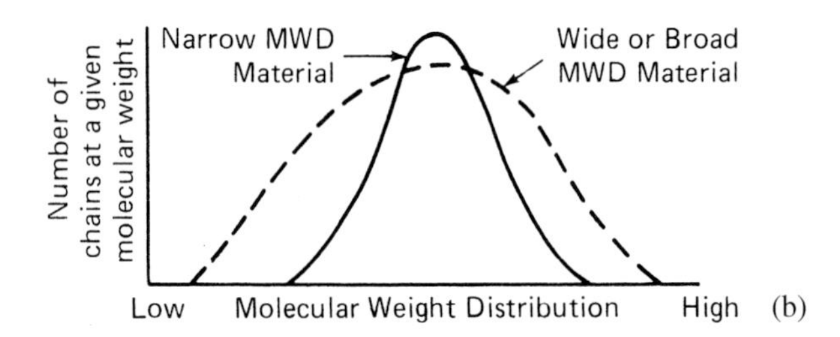

Figure 7.3 *(b) Examples of molecular weight distributions*

polymer molecule (Fig. 7.3a). Many techniques are available for its determination. The choice of method is often complicated by limitations of the technique as well as by the nature of the polymer because most techniques require a sample in solution.

The molecular weight distribution (MWD) is basically the amounts of component polymers that go to make up a polymer (Fig. 7.3b). Component polymers, in contrast, are a convenient term that recognizes the fact that all polymeric materials comprise a mixture of different polymers of differing molecular weights. The ratio of the weight average molecular weight to the number average molecular weight gives an indication of the MWD. Average molecular weight information is useful; however, characterization of the breadth of the distribution is usually more valuable. For example, two plastics may have exactly the same or similar AMWs but very different MWDs. There are several ways to measure MWD such as fractionation of a polymer with broad MWD into narrower MWD fractions.

7.3.3 Rheology

Rheology is the science that deals with the deformation and flow of matter under various conditions. The rheology of plastics is complex but understandable and manageable. These materials exhibit properties that combine those of an ideal viscous liquid (with pure shear deformation) with those of ideal elastic solid (with pure elastic deformation). Thus, plastics are said to be viscoelastic. The mechanical behavior of these viscoelastic plastics is dominated by such phenomena as tensile strength, elongation at break, stiffness, and rupture energy, which are often the controlling factors in a design [2, 64].

7.3.3.1 Newtonian and Non-Newtonian

The rheological processed melt flow viscosity are also important considerations in the fabrication of BM plastic product. As shown in Fig. 7.4, the volume of a so-called Newtonian fluid, such as water, is forced through an opening (in a tool) resulting in a direct proportion to the applied pressure (straight dash line). The flow rates of a non-Newtonian fluid/melt such, as plastics are nonproportional. The solid curve represents a non-Newtonian material in which the flow rate increases more rapidly than the pressure exerted. Plastics can have different non-Newtonian curves that can be directly related to their melt flow during the BM process.

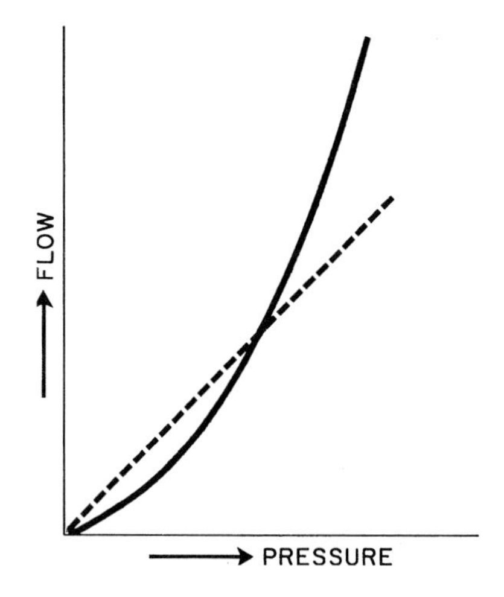

***Figure** 7.4 Rheology of a Newtonian fluid (dashed line) and non-Newtonian fluid (solid line)*

7.3.4 Rheology and Mechanical Analysis

Rheology and mechanical analysis are usually familiar techniques, yet the exact tools and the far-reaching capabilities may not be so familiar. Rheology is the study of how materials flow and deform, or when testing solids, it is called dynamic mechanical thermal analysis (DMTA). Rheometers and dynamic mechanical analyzers impose a deformation on a material and measure the material's response, which in millimeters gives a wealth of information about structure and behavior of the basic polymer.

During dynamic testing, the elastic modulus (relating to energy storage), and loss modulus (relative measure of a material's damping ability) are determined. Steady testing provides information about creep and recovery, viscosity, rate dependence, and more. Stress rheometers are used for testing melts or liquids with a moderate temperature range. Strain controlled rheology is the ultimate in material characterization, with the ability to handle anything from light fluids to solid bars, films, and fibers.

7.3.5 Viscoelasticity

A material having this property is considered to combine the features of a perfectly elastic solid and a perfect fluid, representing the combination of elastic and viscous behavior of plastics. A phenomenon of time-dependent, in addition to elastic, deformation (or recovery) is response to load. This property, possessed by all plastics to some degree, dictates that while plastics have solidlike characteristics such as elasticity, strength, and form stability, they also have liquidlike characteristics such as flow depending on time, temperature, rate, and amount of loading.

Viscoelasticity is perhaps better viewed more broadly as mechanical behavior in which the relationships between stress and strain are time-dependent, as opposed to the classical elastic behavior in which deformation and recovery both occur instantaneously on application and removal of stress, respectively. On stressing a viscoelastic material, three deformation responses may be observed. They are an initial elastic response, followed by a time-dependent delayed elasticity (also fully recoverable), and finally a viscous, nonrecoverable, flow component. Most polymer-containing systems (solid polymers, polymer melts, gels, dilute and concentrated solutions) exhibit viscoelastic behavior due to the long-chain nature of the constituent molecules.

For a high molecular weight glassy polymer [an amorphous polymer well below its glass transition temperature (T_g) value], very few chain motions are possible and hence the material largely behaves elastically, with a very low value for the creep compliance of about 10^{-9} Pa^{-1}. On the other hand, well above the T_g value (for a rubbery polymer), the creep compliance is about 10^{-4} Pa^{-1}, as considerable segmental rotation can occur. Sometimes the intermediate temperature region, which corresponds to the region of the T_g value, is referred to as the viscoelastic region, the leathery region, or the transition zone. Well above the T_g value is the region of rubbery flow followed by the region of viscous flow. In this last region, flow occurs due to the possibility of slippage of whole molecular chains occurring by means of coordinated segmental jumps. These five temperature regions give rise to the five regions of viscoelastic behavior. Light crosslinking of a polymer will have little effect on the glassy and transition zones, but will considerably modify the flow regions.

A constitutive relationship between stress and strain describing viscoelastic behavior will have terms involving strain rate as well as stress and strain. If there is direct proportionality between the terms, then the behavior is that of linear viscoelasticity described by a linear differential equation. Polymers may exhibit linearity but usually only at low strains. More commonly, complex nonlinear viscoelastic behavior is observed.

7.3.5.1 Viscous/Elastic Basics

The novel properties of polymer melts have stimulated considerable interest in rheology. This interest is vitally important to the plastic conversion industry, because it concerns the techniques that are used and were mostly developed for other, older materials. There are two classes of deformation: viscous, in which the energy causing the deformation is dissipated; and elastic, in which that energy is stored. The result is viscoelastic plastics.

The resistance to viscous deformation is called viscosity, the basic unit of measurement being the poise (dyne/s/cm^2). Typical viscosity levels include glass at 10^{22} poise, treacle at 10^4 poise, and water at 10^{-2} poise. Polymer melts have viscosities in the range 10^2 to 10^8 poise, which may be termed the treacle level.

The resistance to elastic deformation is called modulus and is measured in dyne/cm^2 (psi). Typical values of modulus for solids are steel at 10^{12} dynes/cm^2 and rubber at 10^6 dyne/cm^2. The modulus of polymer melts is commonly in the range 10^4 to 10^7 dyne/cm^2 rubbery range. The consistency of a polymer melt is that of rubbery treacle.

As well as two classes of deformation, there are two modes in which deformation can be produced: simple shear and simple tension (Fig. 7.5). In simple shear, the deforming force is applied tangentially to opposite faces of a body. In simple tension, the direction of application is at right angles to the opposite faces. These two modes of deformation are geometrically related and it may be shown that simple shear is an extensional deformation plus a rotation. The rotational component produces qualitative differences in response.

In both simple shear and simple tension, the stress is defined as the force applied per unit area. Deformation is measured by considering the relative movement of points in two parallel planes. For small deformations, the strain is the relative movement divided by the perpendicular distance between the planes. The rate of deformation (stress rate) is the first differential of strain with respect to time. From the above we

Figure 7.5 Simple shear and simple tension (extension) modes

define viscosity as stress/strain rate and modulus as stress/recoverable strain, a basic engineering principle.

Practical deformations are nearly always complex in form—part shear and part extension. They are also complex in nature, having both viscous and elastic components.

The viscosity under simple shear is the most commonly used rheological measurement. Together with engineering design, it determines the pumping efficiency of a screw extruder and controls such features as the relationship between output rate and pressure drop through a die system.

Polymer molecules are long flexible chains. These chains entangle with each other and it is these entanglements that are largely responsible for the high viscosity of melts. Shear may be envisioned as planes sliding over each other (Fig. 7.6). The molecules in these planes are rotated, which causes the molecules to disentangle.

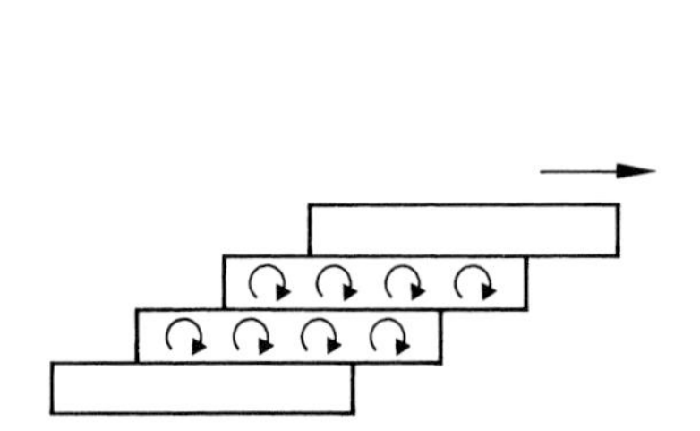

Figure 7.6 Shear of plastic melt envisioned as sliding planes

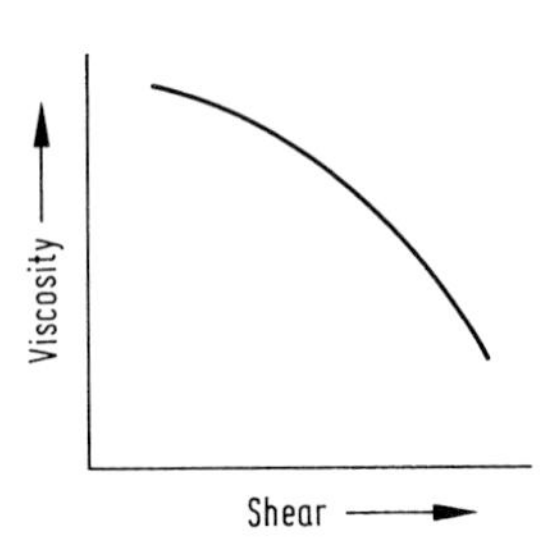

Figure 7.7 Melt flow curve that shows viscosity reduction as shear increases

At low shear rates the molecules will also reentangle. However, as the shear rate is increased, the reentanglement rate drops behind that of disentanglement and the total number of entanglements present is reduced. This results in a corresponding reduction in viscosity (Fig. 7.7) that is called the flow curve. In the shear rate range of practical interest (10 to 1,000,000 s^{-1}), the viscosity may be several orders of magnitude lower than its original value.

7.3.5.2 Molecular Weight Distribution

Processability is highly dependent on molecular weight (chain length). Any increase in chain length increases the number of entanglements and therefore the viscosity. The molecules in any polymer sample are not all the same lengths. Their length may vary from hundreds to millions of monomer units. This variation gives rise to the concept of MWD.

A narrow distribution implies that all the chains are of similar length, a wide distribution that the lengths vary widely. If a material with a wide distribution is deformed, then the long molecules absorb a high proportion of the work done, the shorter molecules a lesser amount. Thus, on a molecular scale, different parts of the body respond differently. As a result, the flow curve for a material of wide distribution is qualitatively different from that for one of narrow distribution (Fig. 7.8).

As for all materials, an increase in temperature reduces viscosity while an increase in pressure increases viscosity. Typical values that will produce an order of magnitude change are 50 °C (160 °F) and 15,000 psi (1000 kg/cm^2) but these figures vary with chemical structure. The major factors influencing viscosity are summarized in Fig. 7.9.

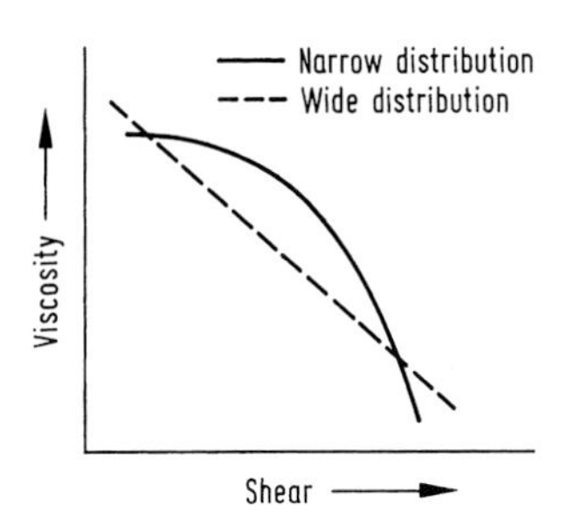

Figure 7.8 *Viscosity vs. shear rate as related to molecular weight distribution*

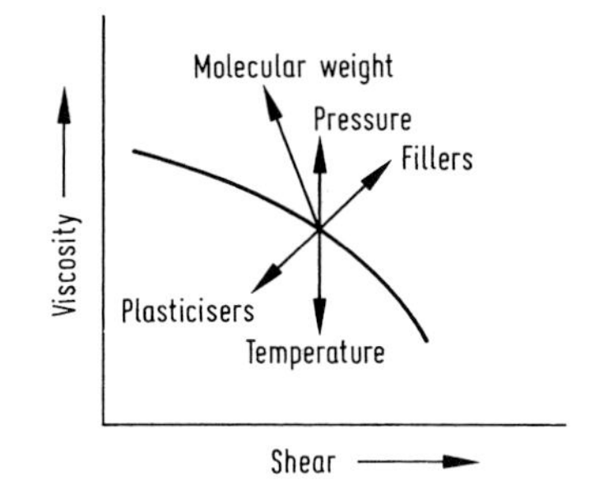

Figure 7.9 *Factors influencing viscosity*

7.3.5.3 Elasticity

If a polymer melt is subjected to a fixed stress, the deformation/time curve will have two components. An initial rapid deformation is followed by a steady continuous flow (Fig. 7.10). The former is elastic, the latter viscous. If the stress is then removed the elastic component is recovered. The relative importance of elasticity and viscosity depends on the time scale of the deformation.

At short times elasticity dominates, while at long times the flow becomes purely viscous. A simple way of estimating the relative importance of elasticity and viscosity in a process is to anneal the end product and see how it changes in shape. One of the

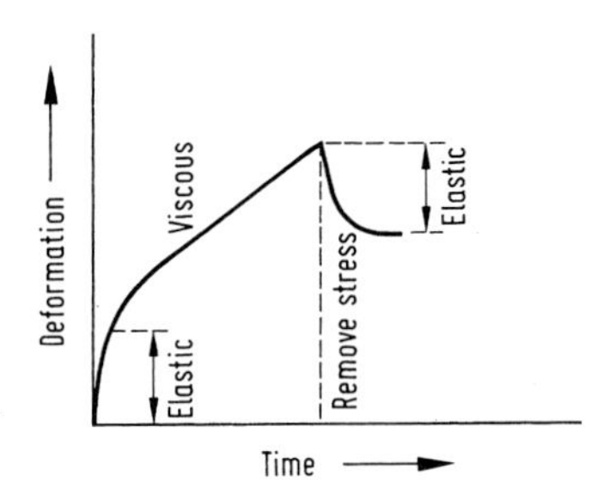

Figure 7.10 *Effect of melt flow rate on molecular weight distribution*

most obvious manifestations of elasticity is post-extrusion swelling. Elasticity also contributes significantly to "flow defects."

Take an elastic band and stretch it. For small deformations, deformation and force are proportional. However, there is a limit to the amount of elastic deformation that can be introduced. If we try to exceed this limit the band breaks. Polymer melts respond in much the same way. For small deformations, stress and strain are proportional, but as the stress is increased the recoverable strain tends to reach some limiting value. It is in the high stress range, near to the elastic limit, that conversion processes operate.

Molecular weight, temperature and pressure have but a small influence on elasticity. The main controlling factor is MWD. Discussion of MWD suggests that, on a molecular scale, there are stress variations within a deforming body. In particular the long molecules are under a high stress and therefore contain a large recoverable component. Materials with a wide MWD, and particularly those with a tail of high molecular weight material, have a lower modulus than those of narrow distribution. For a given deformation they will have a larger elastic component.

Practical elasticity phenomena are often concerned less with the actual value of modulus than with the balance between modulus and viscosity. Thus, although modulus is but little influenced by molecular weight and temperature, these parameters have great effect on viscosity, and can therefore alter the balance.

7.3.5.4 Shear Sensitivity

As an example in reviewing shear sensitivity, melt index (MI) and density are two useful numbers for describing a polyethylene such as that used to blow mold milk bottles (HDPE). But, there should be a third value for shear sensitivity. To see how this can be important, examine how melt index differences alone affect production rates and costs.

Shear sensitivity relates to moldability and performance characteristics. The viscosity of HDPE melt appears to change depending on how rapidly it is undergoing shear and how hard it is being worked. In the extruder, under moderate rate of shear, it is moderately viscous. Forced at high speed through a narrow die gap, the melt is sheared at a fast rate, and appears to have a low viscosity. Then, hanging from the die as a parison, under practically no shear, it appears to be very viscous.

All HDPEs show shear sensitivity where the faster the shear rate, the lower the apparent viscosity. All HDPE's exhibit shear sensitivity, but some are more shear sensitive than others—much more.

Extrusion data for three HDPEs are given in Tables 7.2 and 7.3, which show the effect of MI on the best performance possible with the three plastics run on the same machine at the same extrusion pressure. It will be no surprise that the higher MI resins is easier to run. Between the resins with the highest and lowest MIs, cycle time is increased 9%, yet 8% more power is needed. Put in other terms, there will be 200

Table 7.2 *Effect of Melt Index on Extrusion*

Resin	Melt index	Cycle time (s)	Extrusion time (s)	Power (amp)	Temperature	
					(°F)	(°C)
A	0.3	9.6	1.8	92	385	196
B	0.6	9.3	1.5	87	375	191
C	0.8	9.0	1.2	85	365	185

Table 7.3 *Effect of Shear Sensitivity on Extrusion*

Resin	Shear sensitivity	Extrusion pressure (psi)	Extrusion rate		Temperature	
			(lb/h)	(lb/h/hp)	(°F)	(°C)
D	low	2300	324	3.7	435	224
E	medium	1800	390	4.8	412	211
F	high	1500	406	5.8	390	199

fewer gallon bottles molded in each 8-h shift with the low MI resin, and each bottle will cost more to produce.

Therefore, it would seem preferable to choose the fast cycle time; however, the higher MI would be undesirable for quality reasons. This is where shear sensitivity is involved. The three resins in Table 7.2 have different MIs, but the same shear sensitivity. Now, in Table 7.3 are three more HDPEs, the same MIs this time, but different shear sensitivity. Compare the performance of the resins with the highest and lowest degrees of shear sensitivity. Over 50% more extrusion pressure is needed to extrude 20% less melt. There is about 26% poorer efficiency in terms of melt extruded per horsepower. And to get even this performance from the low shear sensitivity resin, melt temperature had to be increased by 45 °F (25 °C): More heat in, more taken out.

It becomes evident that MI is not all that great a tool for controlling output. Shear sensitivity is a far more powerful one, and the reason becomes clear when one considers the viscosity characteristics of these resins. The HDPE with the poorest shear sensitivity is about 15 times less viscous at high shear rates than at low. This information initially can appear desirable until one learns that the HDPE with high shear sensitivity is 200 times less viscous at high shear than at low. Obviously, this HDPE will require less power, less heat input, and will yield faster extrusion rates. It has the same MI as the other resin and will exhibit even higher melt strength in the parison.

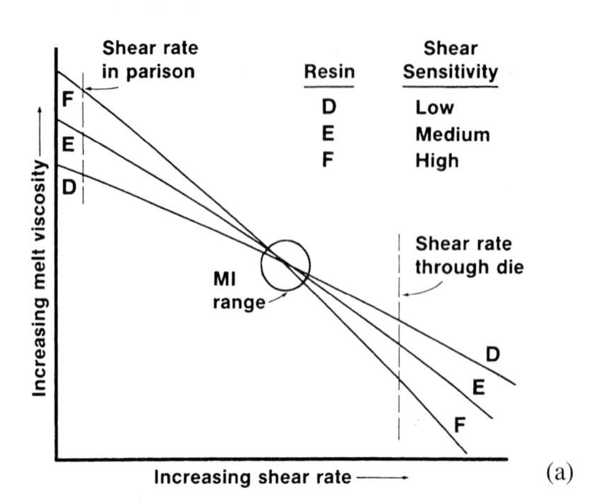

Figure 7.11 *(a) Data show the importance of good shear sensitivity resulting in high viscosity in the parison and low viscosity through the die*

The effect of shear sensitivity can be shown in another way, as in Fig. 7.11a. Here, for the three resins of Table 7.3, shear rate is plotted against melt viscosity. Thus, the resin with poor shear sensitivity provides undesirable results—lowest viscosity of the three in the parison instead of high melt strength; and highest viscosity of the three through the die instead of low viscosity for efficient use of power.

In examining the possibility of obtaining desirable results from a low shear sensitivity HDPE, the result is nonprofitable technically and economically. Poor cooling/cure process performance occurs with low shear sensitivity having low MI; it shows that one cannot overcome and compensate for the problem.

The first step is to increase the melt temperature to reduce viscosity and avoid long cycle times. But, the higher melt temperature will need longer cooling to avoid sticking problems; so the cycle time increases. Plus, the higher melt temperature will result in a parison with poor melt strength, drawdown will be high, and a poor wall thickness distribution in the bottle (uneven, etc.). The low melt strength will also make the parison increasingly likely to pleat, and upper and lower handle webbing will be frequent.

7.3.5.5 Behavior of Plastics

To obtain the best melt, start with the plastics manufacturer's heat profiles and/or ones experience. Or use the profile used on another machine, but it is important to continue changing the barrel temperature profile to obtain the best heat profile that is repeatable during the plasticizing process for each particular machine.

An amorphous material such as PVC, SAN, acrylonitrile–butadiene–styrene (ABS), PS, acrylic, etc. usually requires a fairly low feed end heat; the purpose is to preheat the material, but not really melt it before reaching the compression zone of the screw. The crystalline materials such as PE, PP, nylon, and acetals require a higher heat load in the rear to ensure melting by the time they reach this compression zone.

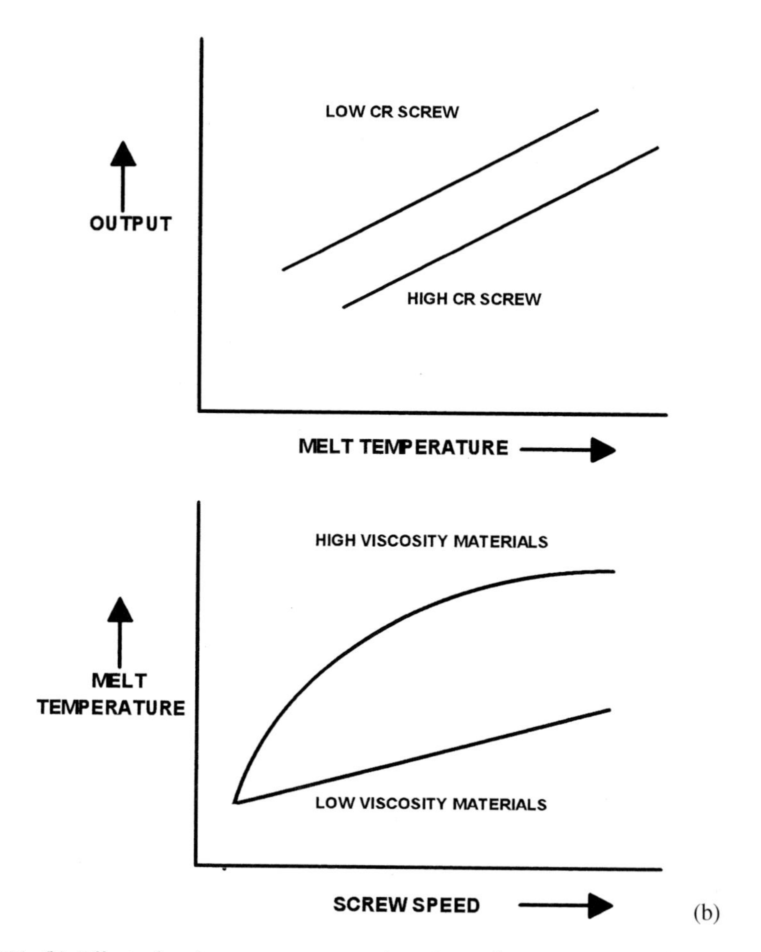

Figure 7.11 *(b) Effect of melt temperature on viscosity and screw compression ratio (C/R)*

Molded part properties, machine cycle times, cooling rates, and melt output rates (Fig. 7.11b) are very strongly influenced by the plasticating process. The success of molding as a production technique, is due mainly to the efficiency of the plasticating units in both melting the plastic feedstock and, simultaneously, providing sufficient mixing to ensure good melt uniformity.

7.4 Plastic Type

Plastic materials to be processed are in the form of pellets, granules, flakes, powders, flocks, liquids, and so on. Of the 35,000 types available worldwide, there are about

200 basic types or families that are commercially recognized with less than 20 that are popularly used. Examples of these plastics are shown in Table 7.4.

Within the fewer than 20 popular plastics, there are five major TP types that consume about two thirds of all thermoplastics. Approximately 18 wt% are LDPEs, 17% PVCs, 12% HDPEs, 16% PPs, 8% PSs. Each have literally thousands of different formulated compounds. The relatively new generation of high performance metallocene and elastomeric plastics fall in this group [62]. The basic types, with their many modifications of different additives/fillers/reinforcements, grafting, alloying, and so on, provide different processing capabilities and/or product performances.

With so many plastics available, it is best to recognize that all materials fit into systematic group categories or classifications. Examples of various plastic groupings are basically TPs and TSs. In addition to the broad categories of TPs and TSs, TPs can be further classified in terms of their structure as either crystalline, amorphous, or liquid crystalline. Other classes include NEAT (*N*othing *E*lse *A*dded *T*o it) plastics, commodity plastics, or engineering plastics. Additives, fillers, and reinforcements are other classifications that relate directly to plastics' properties and performance. In turn there are elastomers, reinforced plastics, foams, and so on, classified by performance (high strength, high stiffness, electrically conductive or nonconductive, etc.), environments (weather resistant, etc.), applications, and so on.

7.4.1 Thermoplastics

TPs are plastics that after processing via heat (they soften) and cooling (they harden) into products that are capable of being repeatedly softened by reheating due to their crystalline or amorphous morphology (molecular structure). Many are soluble in specific solvents and burn to some degree. Their softening temperatures vary (Fig. 7.12b). During the heating cycle care must be taken with these materials to avoid degrading, decomposing, or ignition. Generally, no chemical changes take place during processing. An analogy would be a block of ice that can be softened (turned back to a liquid), poured into any shape mold or die, then cooled to become a solid again. TPs generally offer easier processing and better adaptability to complex designs than do TS plastics.

Most TP molecular chains can be thought of as independent, intertwined strings resembling spaghetti. When heated, the individual chains slip, causing plastic flow. On cooling, the chains of atoms and molecules are once again held firmly. With subsequent heating the slippage again takes place. There are practical limitations to the number of heating and cooling cycles before appearance and/or mechanical properties are drastically affected. Certain TPs have no immediate changes while others have immediate changes.

7.4.1.1 Crystalline and Amorphous Polymers

In crystalline plastics (polymers) the molecules tend to be arranged in a relatively regular repeating structure. They are usually translucent or opaque and generally have

Table 7.4 *Types of Plastics*

- Acetal (POM)
- Acrylics
  - Polyacrylonitrile (PAN)
  - Polymethylmethacrylate (PMMA)
- Acrylonitrile butadiene styrene (ABS)
- Alkyd
- Allyl diglycol carbonate (CR-39)
- Allyls
  - Diallyl isophthalate (DAIP)
  - Diallyl phthalate (DAP)
- Aminos
  - Melamine formaldehyde (MF)
  - Urea formaldehyde (UF)
- Cellulosics
  - Cellulose acetate (CA)
  - Cellulose acetate butyrate (CAB)
  - Cellulose acetate propionate (CAP)
  - Cellulose nitrate
  - Ethyl cellulose (EC)
- Chlorinated polyether
- Epoxy (EP)
- Ethylene–vinyl acetate (EVA)
- Ethylene–vinyl alcohol (EVOH)
- Fluorocarbons
  - Fluorinated ethylene propylene (FEP)
  - Polytetrafluoroethylene (PTFE)
  - Polyvinyl fluoride (PVF)
  - Polyvinylidene fluoride (PVDF)
- Furan
- Ionomer
- Ketone
- Liquid crystal polymer (LCP)
  - Aromatic copolyester (TP polyester)
- Melamine formaldehyde (MF)
- Nylon (Polyamide) (PA)
- Parylene
- Phenolic
  - Phenol formaldehyde (PF)
- Phenoxy
- Polyallomer
- Polyamide (nylon) (PA)
- Polyamide-imide (PAI)
- Polyarylethers
  - Polyaryletherketone (PAEK)
  - Polyaryl sulfone (PAS)
- Polyarylate (PAR)
- Polybenzimidazole (PBI)
- Polycarbonate (PC)
- Polyesters
  - Aromatic polyester (TS polyester)
  - Thermoplastic polyesters
    - Crystallized PET (CPET)
    - Polybutylene terephthalate (PBT)
    - Polyethylene terephathalate (PET)
  - Unsaturated polyester (TS polyester)
- Polyetherketone (PEK)
- Polyetheretherketone (PEEK)
- Polyetherimide (PEI)
- Polyimide (PI)
  - Thermoplastic PI
  - Thermoset PI
- Polymethylmethacrylate (acrylic) (PMMA)
- Polymethylpentene
- Polyolefins (PO)
  - Chlorinated PE (CPE)
  - Crosslinked PE (XLPE)
  - High-density PE (HDPE)
  - Ionomer
  - Linear LDPE (LLDPE)
  - Low-density PE (LDPE)
  - Polyallomer
  - Polybutylene (PB)
  - Polyethylene (PE)
  - Polypropylene (PP)
  - Ultrahigh molecular weight PE (UHMWPE)
- Polyoxymethylene (POM)
- Polyphenylene ether (PPE)
- Polyphenylene oxide (PPO)
- Polyphenylene sulfide (PPS)
- Polyurethane (PUR)
- Silicone (SI)
- Styrenes
  - Acrylic styrene acrylonitrile (ASA)
  - Acrylonitrile butadiene styrene (ABS)
  - General-purpose PS (GPPS)
  - High-impact PS (HIPS)
  - Polystyrene (PS)
  - Styrene acrylonitrile (SAN)
  - Styrene butadiene (SB)
- Sulfones
  - Polyether sulfone (PES)
  - Polyphenyl sulfone (PPS)
  - Polysulfone (PSU)
- Urea formaldehyde (UF)

Table 7.4 (continued)

Vinyls
Chlorinated PVC (CPVC)
Polyvinyl acetate (PVAc)
Polyvinyl alcohol (PVA)
Polyvinyl butyrate (PVB)
Polyvinyl chloride (PVC)
Polyvinylidene chloride (PVDC)
Polyvinylidene fluoride (PVF)

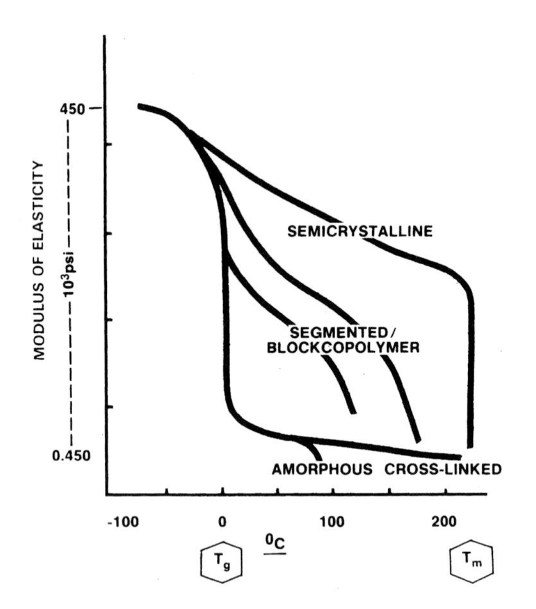

Figure 7.12 (a) Example of the dynamic and mechanical properties of TPs and TSs in relation to their T_g (glass transition temperature) and T_m (melting temperature)

higher softening points than the amorphous plastics (Fig. 7.12b). They can be made transparent with chemical modification. Polymer molecules that can be packed closer together can more easily form crystalline structures in which the molecules align themselves in some orderly pattern. Since commercially perfect crystalline polymers are not produced, they are identified technically as semicrystalline TPs (normally up to 80% crystalline and the rest amorphous).

The amorphous plastic is the term used for a TP having no crystalline plastic structure. They form no pattern instead their structure tends to form like spaghetti; their molecules go in all different directions. These TPs have no sharp melting point and are usually glassy and transparent such as PS and polymethyl methacrylate

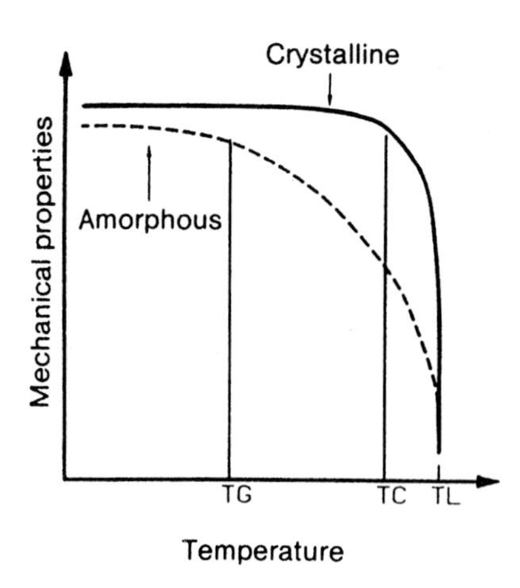

TC: Crystalline melting point
TG: Amorphous glass transition temp
TL: Temperature material completely liquid

***Figure** 7.12 (b) Example of differences in processing temperature of crystalline and amorphous TPs*

(PMMA, acrylic). Amorphous plastics soften gradually as they are heated. If they are rigid, they may be brittle unless modified with certain additives.

During processing, plastics are normally in the amorphous state with no definite order of molecular chains. For TPs that normally crystallize (crystalline plastics), they may not be properly quenched (when hot melt is cooled to solidify the plastic) and the result is an amorphous or partially amorphous solid state usually resulting in inferior properties. Compared to crystalline types, amorphous polymers undergo only small volumetric changes when melting or solidifying during processing.

As symmetrical molecules approach within a critical distance during melt processing, crystals begin to form in the areas where they are the most densely packed. A crystallized area is stiffer and stronger, a noncrystallized (amorphous) area is tougher and more flexible. With increased crystallinity, other effects occur (Fig. 7.12c). As an example, with PE (crystalline) there is increased resistance to creep and heat as well as increased mold shrinkage.

In general, crystalline types of plastics are more difficult (but controllable) to process, requiring more precise control during fabrication, have higher melting temperatures, and tend to shrink and warp more than amorphous types. They have a relatively sharp melting point. That is, they do not soften gradually with increasing temperature but remain hard until a given quantity of heat has been absorbed, then change rapidly into a low-viscosity liquid. If the amount of heat is not applied properly during processing, product performance can be drastically reduced and/or an increase in processing cost occurs.

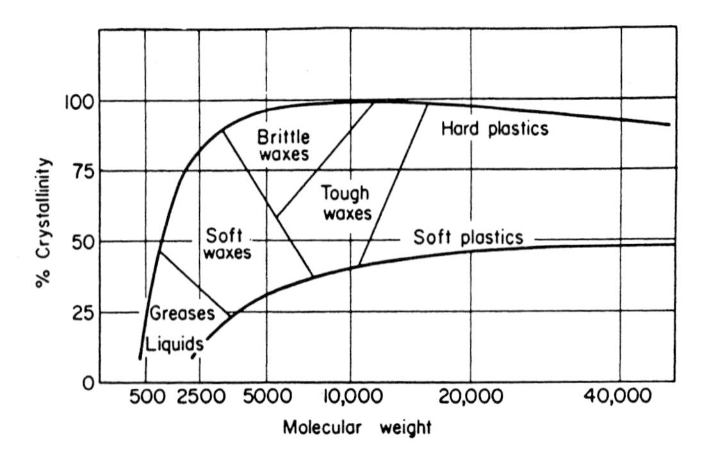

Figure 7.12 (c) Effect of crystallinity and MW on the physical properties of PE

Different processing conditions influence the performance of plastics. For example, the effects of time are similar to those of temperature in the sense that any given plastic has a preferred or equilibrium structure in which it would prefer to arrange itself timewise. However, it is prevented from doing so instantaneously or at least on short notice. If given enough time, the molecules will rearrange themselves into their preferred pattern. Proper heating time causes this action to occur sooner. Otherwise, with a fast action, severe shrinkage and property changes could occur in all directions in the processed plastic products. This characteristic morphology of plastics can be identified by tests. It provides excellent control as soon as material is received in the plant, during processing, and after fabrication.

7.4.2 Thermoset Plastic

Thermoset plastics (TS) are rarely used in BM. Basically after final processing into products, TSs are substantially infusible and insoluble. They undergo a crosslinking chemical reaction by the action of heat and pressure (exothermic reaction), oxidation, radiation, and/or other means often in the presence of curing agents and catalysts. Cured TSs cannot be resoftened by heat (Fig. 7.13a). However, they can be granulated and used as filler in TSs as well as TPs. TSs generally cannot be used alone in structural applications and must be filled with additives and/or reinforced with materials such as glass fibers, wood fibers, and so on.

The TS molecules are crosslinked, that is, chemical links occur between the molecular chains of polymers. It is the principal difference between TSs and TPs. During curing or hardening of TSs the crosslinks are formed between adjacent molecules, producing a complex, interconnected network that can be related to its viscosity and performance. These cross bonds prevent the slippage of individual chains, thus preventing plastic flow of products under the addition of heat. If excessive heat is applied, degradation rather than melting will occur.

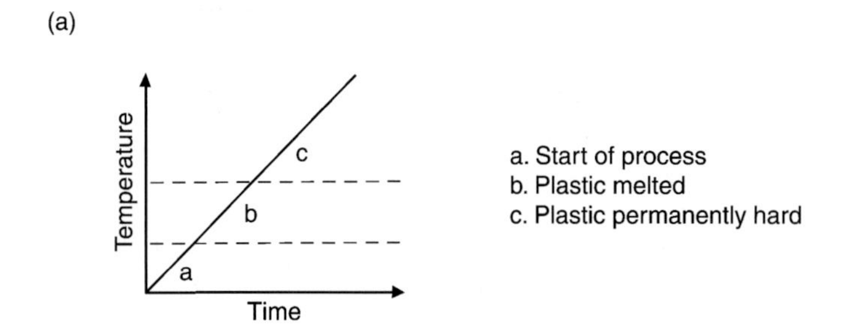

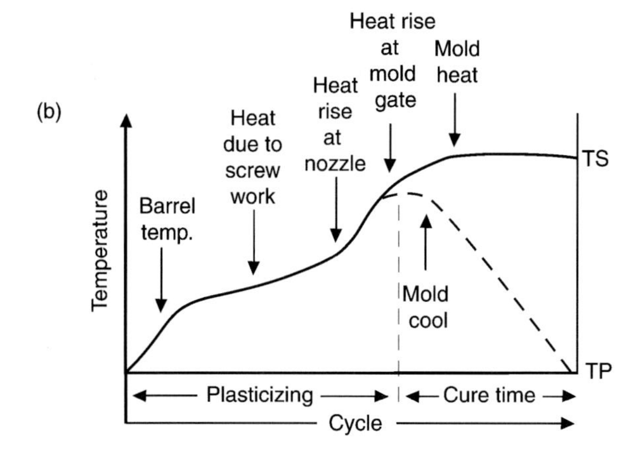

Figure 7.13 *(a) Example of heat-time processing profile cycle for a thermoset. (b) Processing temperatures for TS and TP during IBM*

Within the TS family there are natural and synthetic rubbers (elastomers) such as styrene-butadiene, nitrile rubber, millable polyurethanes, silicone, butyl, and neoprene, which attain their properties through the process of vulcanization. The best analogy of all TSs is that of a hard-boiled egg that has turned from a liquid to a solid and cannot be converted back to a liquid. In general, with their tightly crosslinked structure TSs resist higher temperatures and provide greater dimensional stability and strength than do most TPs (Fig. 7.13b).

7.4.3 Crosslinked Thermoplastics

Certain TPs can be converted to TSs, providing improved properties. They can be crosslinked by different processes such as chemical and irradiation. PE is a popular plastic that is crosslinked. Crosslinking is an irreversible change that goes through a chemical reaction, that is, by condensation, ring closure, or addition. Cure is usually accomplished by the addition of curing (crosslinking) agents with or without heat and pressure.

The accurate characterization of crosslinked polymer networks is among the more difficult tasks of polymer analyses. Yet crosslinks have a major effect on physical and mechanical properties of the fabricated plastic products. Crosslinked materials are relatively difficult to handle by solution techniques due to limited solubility. Solution techniques can be used after a degradative process has been applied. Many of the characterization tests are based on mechanical properties such as tensile, compression, hardness, and other properties.

7.4.4 Liquid Crystalline Polymer

Liquid crystalline polymers (LCPs) are best thought of as being a separate, unique class of TPs. Their molecules are stiff, rodlike structures organized in large parallel arrays or domains in both the melted and solid states. These large, ordered domains provide LCPs with characteristics that are unique compared to those of the basic crystalline or amorphous plastics. They are called self-reinforcing plastics because of their densely packed fibrous polymer chains.

These LCPs provide unparalleled combinations of properties, such as resisting most solvents and heat. Unlike many high-temperature plastics, LCPs have a low melt viscosity and are thus more easily processed, resulting in faster cycle times than those with a high melt viscosity, thus reducing processing costs. They have the lowest warpage and shrinkage of all the TPs. When they are injection molded or extruded, their molecules align into long, rigid chains that in turn align in the direction of flow and thus act like reinforcing fibers giving LCPs both very high strength and stiffness. As the melt solidifies during cooling, the molecular orientation "freezes" (solidifies) into place. The volume changes only minutely with virtually no frozen-in stresses.

In service, products experience very little shrinkage or warpage. They have high resistance to creep. Their fiberlike molecular chains tend to orient near the surface, resulting in products that are anisotropic, meaning that they have greater strength and modulus in the flow direction, typically on the order of three to six times of the modulus in the transverse direction. However, adding fillers or reinforcing fibers to LCPs significantly reduces their anisotropy, more evenly distributing strength and modulus and even boosting them. Most fillers and reinforcements also reduce overall cost and place mold shrinkage to near zero. Consequently, products can be molded to tight tolerances. These low-melt-viscosity LCPs thus permit the design of products with long or complex flow paths and thin sections.

They have outstanding strength at extreme temperatures, excellent mechanical property retention after exposure to weathering and radiation, good dielectric strength as well as arc resistance and dimensional stability, low coefficient of thermal expansion, excellent flame resistance, and easy processability. Their UL continuous-use rating for electrical properties is as high as 240 °C (464 °F), and for mechanical properties it is 220 °C (428 °F). The high heat deflection value of LCPs permits LCP molded products to be exposed to intermittent temperatures as high as 315 °C (600 °F) without affecting their properties. Their resistance to high-temperature flexural creep is excellent, as are their fracture-toughness characteristics. This family of different LCPs resists most chemicals and weathers oxidation and flame, making them excellent replacements for metals, ceramics, and other plastics.

LPCs with their high strength-to-weight ratios are particularly useful for weight-sensitive products. They are exceptionally inert and resist stress cracking in the presence of most chemicals at elevated temperatures, including the aromatic and halogenated hydrocarbons as well as strong acids, bases, ketones, and other aggressive industrial products [89]. Their hydrolytic stability in boiling water is excellent, but high-temperature steam, concentrated sulfuric acid, and boiling caustic materials will deteriorate LCPs. In regard to flammability, LCPs have an oxygen index ranging from 35% to 50%. When exposed to open flame they form an intumescent char that prevents dripping.

7.4.5 NEAT Plastic

A NEAT plastic is a true virgin polymer, as it does not contain additives, fillers, and so on. This type is rarely used.

7.4.6 Virgin Plastic

Plastic materials in the form of pellets, granules, powders, flakes, liquids, and so on that have not been subjected to any fabricating method or recycled.

7.4.7 Copolymer

Polymer properties can be varied during polymerization. The basic chemical process is carried during their manufacture; the polymer is formed under the influence of heat, pressure, catalyst, and/or combination inside vessels or tubular systems called reactors. One special form of property variation involves the use of two or more different monomers as comonomers, forming them to produce copolymers (two comonomers) or terpolymers (three monomers). Their properties are usually intermediate between those of homopolymers, which may be made from the individual monomers, and sometimes superior or inferior to them. (A polymer such as PE is formed from its monomer ethylene, PVC polymer from its vinyl chloride monomer, and so on.)

7.4.8 Compounding and Alloying

Basically compounds are intimate compositions of a polymer alloyed with all the additional materials necessary, such as additives, fillers, and/or reinforcements required to fabricate a product.

Alloys are combinations of polymers that are mechanically blended. They do not depend on chemical bonds, but do often require special compatibilizers. Plastic alloys are usually designed to retain the best characteristics of each constituent.

The classic objective of alloying and blending is to find two or more polymers whose mixture will have synergistic property improvements beyond those that are purely additive in effect.

Alloys can be classified as either homogeneous or heterogeneous. The former can be depicted as a solution with a single phase or single glass transition temperature (T_g). A heterogeneous alloy has both continuous and dispersed phases, each retaining its own distinctive T_g. Until recently, blending and alloying were either restricted to polymers that had an inherent physical affinity for each other or else a third component, called a compatibilizer, was employed. These constraints severely limited the types of polymers that could be blended without sacrificing their good physical properties. As a rule, incompatible polymers produce a heterogeneous alloy with poor physical properties.

7.4.9 Commodity and Engineering Plastic

About 90 wt% of plastics can be classified as commodity plastics, the others being engineering plastics. The five families of commodities LDPE, HDPE, PP, PVC, and PS account for about two thirds of all the plastics consumed. The engineering plastics such as nylon, PC, acetal, and so on are characterized by improved performance in higher mechanical properties, better heat resistance, higher impact strength, and so forth [50] (Table 7.5). Thus, they demand a higher price. About a half century ago the price per pound difference was at 20¢; now it is about $1.00. There are commodity plastics with certain reinforcements and/or alloys with other plastics that put them into the engineering category. Many TSs and RPs are engineering plastics.

7.4.10 Elastomer

An elastomer is a rubberlike material (natural or synthetic) generally identified as a material that, at room temperature, stretches under low stress to at least twice its length and snaps back to approximately its original length on release of the stress (pull) within a specified time period. The term elastomer is often used interchangeably with the term plastic or rubber; however, certain industries use only one or the other.

Table 7.5 Examples of Key Properties for Engineering TPs

Crystalline	Amorphous
Acetal	*Polycarbonate*
Best property balance	Good impact resistance
Stiffest unreinforced thermoplastic	Transparent
Low friction	Good electrical properties
Nylon	*Modified PPO*
High melting point	Hydrolytic stability
High elongation	Good impact resistance
Toughest thermoplastic	Good electrical properties
Absorbs moisture	
Glass reinforced	
High strength	
Stiffness at elevated temperatures	
Mineral reinforced	
Most economical	
Low warpage	
Polyester (glass reinforced)	
High stiffness	
Lowest creep	
Excellent electrical properties	

Although rubber originally referred to a natural thermoset elastomeric (TSE) material obtained from a rubber tree (*Hevea braziliensis*), it identifies a TSE or TPE material. They can be differentiated by reversible vs. irreversible links or on the basis of how long a material deformed requires to return to approximately its original size after the deforming force is removed and by its extent of recovery. Different properties also identify the elastomers such as strength and stiffness, abrasion resistance, solvent resistance, shock and vibration control, electrical and thermal insulation, waterproofing, tear resistance, cost-to-performance, and so on.

The natural rubber materials have been around for over a century, and will always be required to meet certain desired properties in specific products. TPEs principally continue to replace traditional TS natural and synthetic rubbers (elastomers). TPEs are also widely used to modify the properties of rigid TPs usually by improving their impact strength. Both synthetic TSE and TPE have made major inroads to product markets previously held by natural rubber and also expanded into new markets (Table 7.6). The three basic processing types are conventional (vulcanizable) elastomer, reactive type, and TPE.

They present creative designers with endless new and unusual product opportunities. More than hundreds of different major groups of TPEs are produced worldwide with

Table 7.6 *Growth of the Worldwide Natural and Synthetic Rubber/Elastomer Market*

	1984	1989	1995	Ann. growth %
Rubber demand	13,136.9	15,064	17,800	2.8
natural	4278.2	4929	5830	2.9
synthetic	8858.7	10,135	11,970	2.7
tire	5921.5	6214	6696	1.3
nontire	7315.4	8850	11,104	3.9
United States	2811.8	2995	3290	1.3
USSR	2156.0	2445	2760	2.6
Japan	1440.0	1620	1880	2.4
Germany	588.0	635	685	1.6

new grades continually being introduced to meet different performance requirements (Table 7.7).

Quite large elastic strains are possible with minimal stress in TPEs. TPEs have two specific characteristics: their glass transition temperature (T_g) is below that at which they are commonly used, and their molecules are highly coiled as in natural TS rubber (isoprene). When a stress is applied, the molecular chain uncoils and the end-to-end length can be extended several hundred percent with minimum stress. Some TPEs have an initial modulus of elasticity of less than 10 MPa (1,500 psi); once the molecules are extended, the modulus increases.

The modulus of metals decreases with an increase in temperature. However, in stretched TPEs and particularly conventional elastomers, the opposite is true, because at higher temperatures there is increasingly vigorous thermal agitation in their molecules. Therefore, the molecules resist more strongly the tension forces attempting to uncoil them. To resist requires greater stress per unit of strain, so that the modulus increases with temperature. When stretched into molecular alignment, many elastomers can form crystals, an impossibility when they are relaxed and kinked. TPEs can be fine-tuned to meet coefficient of linear thermal expansion required in product performance product requirements.

TPEs range in hardness from as low as 25 Shore A up to 82 Shore D (ASTM test). They span a temperature range of −34 to 177 °C (−29 to 350 °F), dampen vibration, reduce noise, and absorb shock. However, designing with TPEs requires care, because unlike TS rubber that is isotropic, TPEs tend to be anisotropic during processing with injection molding. Tensile strengths in TPEs can vary as much as 30% to 40% with direction.

Overall, an elastomer may be defined as a natural or synthetic material that exhibits the rubberlike properties of high extensibility and flexibility. It identifies any TSE or TPE material. Such synthetics as neoprene, nitrile, styrene butadiene, and polybutadiene are grouped with natural rubber that are TSEs. The term's "rubber" and "elastomer" are usually used interchangeably.

Table 7.7 *Guide to Data for Elastomers (1 = Excellent, 2 = Good, 3 = Fair, 4 = Poor)*

	Elastomers													
Properties	Natural Rubber	Synthetic Polyisoprene	SBR	Chloroprene	NBR	Butyl	Polybutadiene	EPDM	Polysulfide	Polyurethane	Polyacrylate	Silicone	Fluorocarbon	Chlorosulfonated Polyethylene
Physical properties:														
Tensile strength	1	2	2	2	2	2	2	2	4	1+	3	4	3	3
Elongation	2	1	2	1	2	1	2	2	2	1	3	1	3	3
Compression set	2	3	2	2	2	3	3	3	4	1	2	1	1	2
Resilience	1+	1+	2	1	2	3	1+	2	3	1	2	2	3	2
Electrical resistivity	1	1	1	3	4	1	1	1	3	2	3	1	2	2
Mechanical resistance:														
Tear	1	2	3	2	2	2	2	4	3	1	3	3	3	2
Abrasion	1	1	1	1	1	2	1	2	3	1	2	3	2	2
Cut growth	1	1	2	2	2	1	3	2	4	2	2	3	3	2
Temperature resistance:														
Heat	2	3	2	2	2	1	3	1	2	2	1	1	1+	1
Low temperature	2	2	3	3	3	3	2	2	2	3	4	1	4	2
Enrivonmental resistance:														
Water	1	1	1	2	1	1	1	1	2	2	3	2	2	2
Acid	2	2	2	2	2	1	2	1	3	3	3	3	1	1
Alkali	2	2	2	2	2	1	2	1	2	3	4	3	3	1
Aliphatic hydrocarbons	4	4	4	2	1	4	4	4	1	1	1	2	2	2
Aromatic hydrocarbons	4	4	4	4	2	2	4	3	1	3	4	3	1	1
Chlorinated solvents	4	4	4	4	4	4	4	4	2	2	4	3	3	1
Ketones	2	2	2	3	4	2	4	1	2	4	4	3	4	3
Alcohol	2	2	3	3	2	2	2	4	2	2	4	2	1	2
Lubricating oils	4	4	4	2	1	4	4	4	1	2	1	4	2	4
Synthetic oils	3	3	4	4	2	3	3	3	2	4	2	3	2	4
Hydraulic oils	4	4	3	2	3	3	4	2	3	4	2	4	2	2
Fuels	4	4	4	2	1	4	4	4	1	1	1	3	1	3
Weather	3	3	3	1	2	1	3	1	1	1	1	1	1	1
Oxidation	2	1	2	2	2	1	1	2	1	1	1	1	1+	1
Ozone	4	4	4	1	4	1	4	1	1	1	1	1	1+	1+

7.4.11 Foamed Plastic

Practically all plastics can be made into foams with some being BM. When compared to solid plastics, density reduction can go from near solid to almost a weightless plastic material. There are so called plastic structural foams that have up to 40% to 50% density reduction. The actual density reduction obtained will depend on the product's thickness, the product shape, and the melt flow distance during processing such as how much plastic occupies the mold cavity.

The microcellular foam processing techniques by Trexel, Inc. have been used for both extrusion (including wheel type BM) and injection BMs [82]. There has been weight reduction in extrusion of up to 25% with increases in productivity, while maintaining excellent physical properties. In injection molding, the same basic processing techniques have provided molders with the ability to produce parts with reduced warpage, greater dimensional accuracy, at lower molding pressures and at faster cycle times—all this combined with weight reduction in these parts.

Attempts to gain weight and cost reduction through standard foaming techniques have not been particularly successful. Generally speaking, bottle weight is as low as it can be using design and wall thinning without the loss of the bottle performance. Microcellular foam technology is being developed to further reduce the bottle weight while maintaining key performance characteristics. The application of this technology to BM requires the development of equipment and process techniques using the basic microcellular approach but with specific adaptations required for the BM process.

The practice of microcellular foaming in commercial plastics processing requires

- use of a precise amount of blowing agent in the form of a supercritical fluid into the plasticator and formation of a single phase solution of the blowing agent (nitrogen or carbon dioxide) and the polymer melt,
- nucleation of a large number of cells as a result of a rapid drop in pressure,
- expansion of the cells using process conditions to control size, and
- the shaping of the part to final size and dimensions.

A screw was designed specifically for allowing the uniform introduction of supercritical fluid blowing agent and its homogeneous mixing into the melt. The molding end of the equipment remains virtually unchanged. Small changes in mold dimensions may be required to maintain the volume capacity of the bottles being molded; these are accomplished through shims and other simple mold modification techniques.

BM variables are similar to those in normal BM, and are used to control the dimensions and integrity of the bottles. Selection of appropriate conditions is done with appropriate consideration to the microcellular foaming technique. Mold temperature and blow pressure is important to maintain the desired bottle size and appearance and to maintain the low weight of the microcellular-foamed bottle. The results are products that have

- uniform foam structure in the foamed layer;
- no delamination of foam and nonfoam layers; and
- inner, middle, and outer layer integrity, excellent surface quality and appearance, and lower weight than products of solid construction.

7.5 Fundamentals of Melt Processing

Fabricating products involves conversion processes that may be described as an art. Like all arts, they have a basis in science and one of the short routes to processing improvement is a study of the relevant sciences. This review presents a general analysis of conversion processes that relates to BM and continues to discuss the sciences relevant to them including rheology, thermodynamic (heating), and chemistry previously reviewed. It provides some of the scientific background of plastic/polymer processing, analyzes conversion processing techniques, and describes some of the factors that influence them.

The target of plastic processing is to take the plastic in the form of pellets, powders, granules, and so on and convert them into a more useful form. Usually, the change involved is largely physical, although in some cases chemical reactions are an essential part of the conversion process. The common features of processing may be organized by considering them under the four headings of:

- **Mixing, Melting, and Homogenization** The aim of this stage is to take the plastic and turn it into a homogeneous melt. This is often carried out in a screw plasticator, where melting takes place in the feed and compression zones as a result of heat conducted through the barrel wall and heat generated in the plastic by the action of shear via the screw. Homogeneity is called for at the end of this stage, not only in terms of material but also in respect to temperature.
- **Melt Transport and Shaping** Heading is the concern of a melt uniform in composition and temperature. In a screw plasticator, the next step would be to build up an adequate pressure in the plasticator so that it will produce the desired shape to be fabricated. In an extruder, the die (which initiates the shape) can vary from a simple cylindrical shape to a complex crosshead shape. In an injection molding process, pressure is applied to force the melt into a mold that defines the product shape in three dimensions.
- **Drawing and Blowing** A common technique is to stretch the melt to produce orientation, as a desired shape, as in BM.
- **Finishing** The final stage after a process fabricated a solid product usually does not require secondary operations. However, there are materials that may require annealing, sintering, coating or other decoration, and so on.

7.5.1 Melt Deformation

The most obvious feature of any conversion process is deformation of the melt. The study of deformation and flow refers to rheology. Also prominent is heat exchange. The relevant science here is thermodynamics. Lastly, chemical changes may take place. Thus the arts of melting, shaping, drawing, and finishing are furthered by study of the sciences of rheology, thermodynamics, and chemistry [47].

The novel properties of plastic melts have stimulated considerable interest in rheology. This study is vital to the fabricating/conversion industry because it concerns the techniques that are used and that were mostly developed for other, older materials. The classes of deformation are (1) viscous in which the energy causing the deformation is dissipated and (2) elastic in which that energy is stored. The result is a viscoelastic plastic.

As well as two classes of deformation, there are two modes in which deformation can be produced: simple shear and simple tension. In simple shear, the deforming force is applied tangentially to opposite faces of a body. In simple tension, the direction of application is at right angles to the opposite faces. These two modes of deformation are geometrically related and it may be shown that simple shear is an extensional deformation plus a rotation. The rotational component produces qualitative differences in response.

In both simple shear and simple tension, the stress is defined as the force applied per unit area. Deformation is measured by considering the relative movement of points in two parallel planes. For small deformations, the strain is the relative movement divided by the perpendicular distance between the planes. The rate of deformation (stress rate) is the first differential of strain with respect to time. From this summation the basic engineering principals such as stress–strain rate and modulus as stress recoverable strain define viscosity.

7.5.2 Viscosity

Practical deformations are nearly always complex in form because they are part shear and part extension. They are also complex in nature, having viscous and elastic components. The viscosity under simple shear is the most commonly used rheological measurement. Together with engineering design, it determines the pumping efficiency of a screw in a plasticator and controls such features as the relation between output rate and pressure drop through a die system and so on with the other processes.

Basically, viscosity is a measure of the internal friction resulting when one layer of fluid is caused to move in relationship to another layer. Viscosity is measured in terms of flow in Pascal second (Pa·s) with water as the base standard value of 1.0. The higher the number, the less flow: it relates viscosity and shear rates to the speed of moving a plastic melt in a processing machine.

When the ratio of shearing stress to the rate of shear is constant, as in the case with water and thin motor oils, the material/fluid is called Newtonian fluid. In the case of non-Newtonian fluids, such as plastics, the ratio varies with the shearing stress; viscosities of such fluids/materials are called apparent viscosities.

The relationships between viscosity, shear and molecular weight distribution were discussed in Section 7.3.5 (Viscoelasticity). Particularly, Fig. 7.7 depicts a melt flow curve that shows viscosity reduction as shear increases. Figure 7.14 shows the viscosity behavior of plastics and other materials with respect to shear rate.

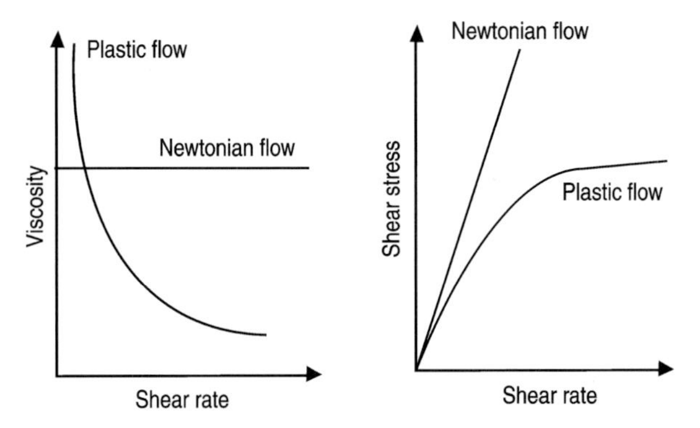

***Figure** 7.14 Viscosity behavior of plastics and other materials*

If a plastic melt is subjected to a fixed stress, the deformation/time curve will have two components. An initial rapid deformation is followed by a steady continuous flow. The former is elastic, the latter viscous. If the stress is then removed, the elastic component is recovered.

The relative importance of elasticity and viscosity depends on the time scale of the deformation. At short times elasticity dominates, while at long times the flow becomes purely viscous. A simple way of estimating the relative importance of elasticity and viscosity in a process is to anneal the end product and see how it changes in shape. As an example, one of the most obvious manifestations of elasticity is post-extrusion swelling. Elasticity also contributes significantly to "flow defects."

Take an elastic band and stretch it. For small deformations, deformation and force are proportional. However, there is a limit to the amount of elastic deformation that can be introduced. If one tries to exceed this limit the band breaks. Plastic melts respond in much the same way. For small deformations, stress and strain are proportional, but as the stress is increased the recoverable strain tends to reach a limiting value. It is in the high stress range, near to the elastic limit, that conversion processes operate.

7.5.3 Molecular Weight Distribution

Molecular weight, temperature, and pressure have but a small influence on elasticity. The main controlling factor is molecular weight distribution (MWD). Molecular weight suggests that, on a molecular scale, there are stress variations within a deforming body. In particular, the long molecules are under a high stress and so contain a recoverable component. Materials with a wide MWD, and particularly those with a tail of high molecular weight material, have a lower modulus than those of narrow distribution. For a given deformation, they will have a larger elastic component. Figure 7.15 relates melt flow to MWD where A = wide or broad MWD plastics and B = narrow MWD plastics. With MWD differences of incoming

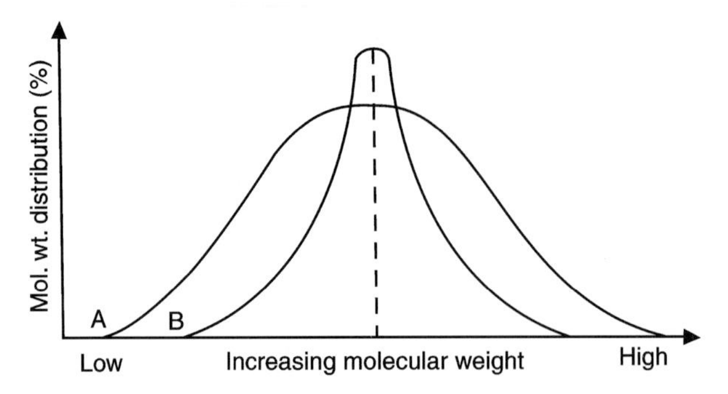

Figure 7.15 *Melt flow rates as a function of molecular weight distribution*

material, the fabricated performances can be altered; the more the difference, the more dramatic change occurs in the products.

Practical elasticity phenomena are often concerned less with the actual value of modulus than with the balance between modulus and viscosity. Thus, although modulus is not highly influenced by MW and temperature, these parameters have great effect on viscosity, and so can alter the balance.

A review can be made concerning the difference between shear and extensional flows. Simple shear flow involves a rotation that causes disentanglements and a reduction in viscosity. Under simple tension, the couple causing rotation is absent. Thus, extensional flows are not subject to disentanglements and the viscosities under tension are largely independent of stress. In fact, there is a small dependence on stress that may be related to such molecular features as branching. This dependence may result in either increasing or decreasing viscosity as the stress is raised. Small though this dependence is, it has important practical significance.

In any practical deformation there are local stress concentrations. Should viscosity increase with stress, the deformation at the stress concentration will be less rapid than in the surrounding material. The stress concentration will be smoothed out and deformation will be stable. However, when viscosity decreases with increased stress, then any stress concentrations will be exploited and catastrophic failure will result.

There are two classes of practical extensional flows. The first, or obvious class, varies from the sagging of a melt such as an extrudate under its own weight, a slow deformation, to the blowing of a bottle, an explosively fast deformation. This class includes the drawing of fibers and films. The second, concealed class is the extensional deformation that takes place when a flowing melt converges at a change in cross section. Typical of such flows is that occurring as melt enters an extrusion die.

7.5.4 Melt Defect

Flow defects, especially insofar as they affect the appearance of a product, play an important part in many processes. Flow defects are not always undesirable. Some processes take advantage of them to produce matte finishes. The following six important types of defects can be distinguished that are applied to extrusion, as it provides less complications and/or restrictions. These flow analyses can be related to the more complex flow during injection molding.

Laminar flow is desired. The plastic melt flows in a steady streamline pattern. Thus, when it leaves the die it recovers evenly and a smooth extrudate is obtained. If the streamline pattern breaks down, nonlaminar flow occurs, and adjacent parts of the melt have quite different shear histories.

In the sharkskin defect, during flow through a die the material next to the die wall is not moving, while that in the center flows rapidly. This is the same effect as the current being fastest at the center of a river.

The deformation of nonhomogeneous material produces an inhomogeneous stress distribution. After extrusion, this stress distribution will produce uneven recovery that may cause lumps. The lumps may be gross or appear only as a fine matte finish. Because materials of wide MWD are nonhomogeneous on a molecular scale, some lack of gloss will be encountered with such materials.

Some plastics contain small quantities of volatile material that boil at processing temperatures. Water, which may be absorbed from the atmosphere, is one example. As the volatiles are driven off they may cause a pitted or scarred surface; alternatively, bubbles may be formed. Some materials depolymerize at high temperatures and the monomer so formed is volatile. Bubbles form only as a plastic melt leaves the die because volatiles are kept in solution by hydrostatic pressure.

In going from room temperature to the maximum temperature during processing, plastic density may decrease by as much as 25%, causing shrinkage. Local variations in cooling will produce stresses that may cause surface distortion or voiding. The surface distortion varies from a fine haze to sink marks covering a large area. Such defects may be overcome during ejection by cooling the material under pressure. These defects are most pronounced in products of thick cross section.

Melt structure can cause defects. High shear at a temperature not far above the plastics melting point may result in the melt taking some ordered or semicrystalline form. This inevitably results in a distortion of the rheology of the system.

Wherever there is heat exchange, the rules of thermodynamics are relevant to processing. It is the high heat content of melts (about 100 calories/g) combined with the low rate of thermal diffusion (10^{-3} cm^2/s) that limits the rate of output or cycle time of many processes. Also prominent are density changes that, for semicrystalline materials, may exceed 25% as melts cool. Plastic melts are highly compressible; 10% volume change for 10,000 psi (700 kg/cm^2) is typical.

When viscous flow takes place, energy is dissipated as heat. This principle is used to achieve homogeneity in the plastifying action of the plasticator. The rate of heat generation is proportional to the square of the deformation rate; at high rates shear heating increases rapidly. The highest rates are found in extrusion dies or injection molding nozzles.

The chemical changes that can occur during a melt process include:

- polymerization and crosslinking that increase viscosity,
- depolymerization or breaking of molecules that reduce viscosity, and
- sometimes complete changes in the chemical structure that may result in a color change.

The most potent agents of chemical change include heat, shear, oxygen, and water. Already degraded material may catalyze further degradation. In some cases, chemical reaction may be an integral part of a process while at other times it may cause nonhomogeneity. If chemical change is initiated by heat or shear, then the design of the equipment may need modification.

Plastic melts have many different properties. Combined with the complexity of conversion processes, this has made detailed and factual prediction difficult. Thus, during recent years, research has been mainly directed at explaining the anomalous behavior of materials as observed during processing. Modern techniques (in terms of equipment and instruments or controls) are overcoming the difficulties of prediction and one may look forward to the time when more conversion processes will be designed to take advantage of the novel properties of plastics rather than to overcome them.

7.6 Influence of Processing Behavior

Different plastic characteristics that influence processing and properties of plastics have been reviewed. This section highlights glass transition and melt temperatures (Table 7.8).

7.6.1 Melt Temperature

The T_m, also called melting point, is the temperature at which a plastic melts or softens and begins to have flow tendency. It refers to processing temperature and melt strength. Basically, the melt strength of the plastic occurs while in the molten state. It is an engineering measure of the extensional viscosity and is defined as the maximum tension that can be applied to the melt without breaking. Linear plastics, such as crystalline LLDPE, HDPE, and PP, generally have poor melt strength. The reason is that in extension the linear chains slide by without getting entangled. However, branched plastics such as LDPE exhibit higher melt strength and elongation viscosity because the long branches get entangled. There is also the absolute melt temperature. In a hypothetical characterization, it is a perfectly free or relative free melt in consistency with regard to temperature, pressure, mixture, and so on.

The T_m occurs at a relatively sharp point for crystalline materials (Fig. 7.16). They have a true melting point with a latent heat of fusion associated with the melting and freezing process, and a relatively large volume change during fabrication, the transition from melt to solid. Crystalline plastics have considerable order of the molecules in the solid state, indicating that many of the atoms are regularly spaced.

Amorphous plastics do not have a T_m, but rather a softening range and undergo only small volume changes when solidified from a melt, or when the solid softens and becomes a fluid melt. They start softening as soon as the heat cycle begins. It is often taken at the peak of the differential scanning calorimeter thermal analysis test equipment.

The melt temperature is dependent on the processing pressure and the time under heat, particularly during a slow temperature change for relatively thick melts during processing. Also, if the melt temperature is too low, the melt's viscosity will be high and more power is required to process it. If the viscosity is too high, degradation will occur. There is the right processing window used for the different melting plastics. Table 7.8 provides a guide on decomposition temperature (T_d).

Trying to measure the melt temperature could be deceiving. As an example, an extrudate with a room temperature pyrometer probe will often give a false reading because when the cold probe is inserted it becomes sheathed with the plastic that has been cooled by the probe. A more effective method is by using what some call the 30/30 method. One simply raises the temperature of the probe about 30 °F (15 °C) above the melt temperature and then keep the probe surrounded with hot melt for 30 s. The easiest way to preheat the probe is to place the probe on, near, or in a hole in the die. Preheating above the anticipated temperature, just prior to inserting it into the melt, requires the probe to actually be cooled by the melt. The lowest temperature reached will be the stock temperature. It also helps to move the probe around in the melt to have the probe more quickly reach a state of equilibrium. To be more accurate, repeat the procedure.

7.6.2 Glass Transition Temperature

The glass transition temperature, T_g, is the temperature at which plastic changes in phase from a viscous or rubbery state to a brittle glassy state. T_g, also called glass-rubber transition, is the point below which plastic behaves like glass but is very strong and rigid. Above this temperature, it is not as strong or rigid as glass, but neither is it brittle. At and above T_g, the plastic's volume or length increases more rapidly and rigidity and strength decrease (Fig. 7.17). The amorphous TPs have a more definite T_g when compared to crystalline plastics. It is usually reported as a single value, although it occurs over a temperature range and is kinetic in nature.

A plastic's thermal properties, particularly its T_g, influence its processability in many different ways. The selection of a plastic should take these properties into account.

Table 7.8 *Examples of Plastics Melting Temperatures, Glass Transition Temperatures, and Decomposition Temperature*

Class	Polymer	Melting temperature (°C)	Glass transition (°C)	Decomposition temperature (°C)
Model	Polymethylene	137	−118	390
Polyolefins	Polyethylene, linear (HDPE)	135	−123	360
	Polyethylene, branched (LDPE)	110	−60	335
	Polypropylene (PP)	170	−10	330
	PPcoE	128		350
	Polyisobutene (PIB)	45	−73	290
	Poly-4-methyl pentene 1 (TPX)	235	30	335
	PE cosodium acrylate (Ionomer)	120	−110	340
	P4,4′-methylene diphenylene (PPX)	375	280	420
Vinyls	Polystyrene (PS)	240	82	325
	Polyvinyl acetate (PVA)	240	71	250
	Polyvinyl chloride (PVC)	210	80	160
	Polyvinylidene chloride (PVDC)	190	−20	200
	PVDC co-VC etc.	150	−17	150
	Polyvinyl fluoride (PVF)	200	40	372
	Polyvinylidene fluoride (PVDF)	168	−40	400
	Polychlorotrifluoroethylene (PCTFE)	220	−35	290
	Polytetrafluoroethylene (PTFE)	330	130	475
	PTFEcoHFP (FEP)	270	127	373
	Polyacrylonitrile (PAN)	320	87	220
	Polymethyl methacrylate (PMMA)	160	65	170
Ethers	Polymethylene oxide (POM)	181	−85	100
	Polyphenylene oxide (PPO)	261	83	270
	Polyphenylene sulphone	300	245	310
	Poly-β-glucoside (cellulose)	450	230	270
	Trinitrocellulose (NC)	700	60	160
	Triacetyl cellulose (CA)	306	105	250
Esters	Polyethylene terephthalate (PET)	265	81	283
	Polycarbonate (PC)	267	147	330
Amides	Polycaprolactam (nylon 6)	220	40	200
	Polyhexamethylene adipamide (nylon 6.6)	265	50	200
	Polyundecanoamide (nylon 11)	180	46	200
	Polyaromatic amide[a]	635	546	635
	Polypyromellitimide	900	390	406

[a] *m*-Phenylene isophthalamide.

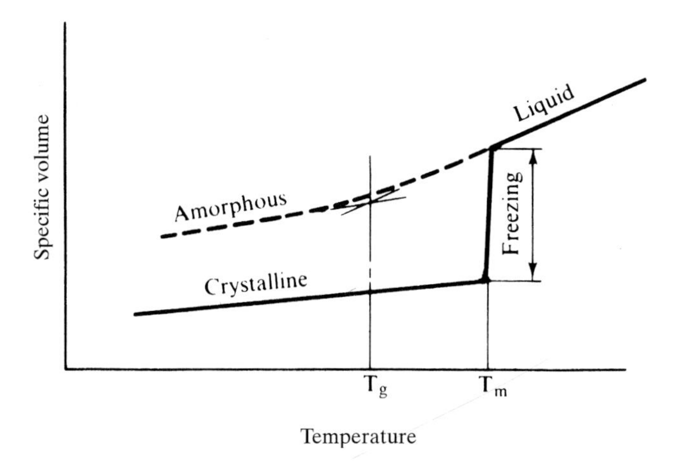

Figure 7.16 *Volume of crystalline and amorphous plastics vs. temperature at melt temperature (T_m) and glass transition temperature (T_g)*

The operating temperature of an amorphous TP is usually limited to below its T_g. Note that a more expensive plastic could cost less to process because of its T_g location that results in a shorter processing time, requiring less energy for a particular weight, and so on.

The glass transition generally occurs over a relatively narrow temperature span and is similar to the solidification of a liquid to a glassy state. Not only do hardness and brittleness undergo rapid changes in this temperature region, but other properties such as the coefficient of thermal expansion and specific heat also change rapidly. This phenomenon has been called second-order transition, rubber transition, or rubbery transition. The word transformation has also been used instead of transition. When more than one amorphous transition occurs in a plastic, the one associated with segmental motions of the plastic backbone chain, or accompanied by the largest change in properties, is usually considered to be the glass transition.

Designers and fabricators should know that above T_g, many mechanical properties of TPs are reduced. Most noticeable is a reduction in stiffness by a factor that may be as high as 1,000.

The T_g can be determined readily only by observing the temperature at which a significant change takes place in a specific electrical, mechanical, or physical property. Moreover, the observed temperature can vary significantly, depending on the specific property chosen for observation and on details of the experimental technique (e.g., the rate of heating or frequency). Therefore, the observed T_g should be considered to be only an estimate. The most reliable estimates are normally obtained from the loss peak observed in dynamic mechanical tests or from dilatometric data (ASTM D-20).

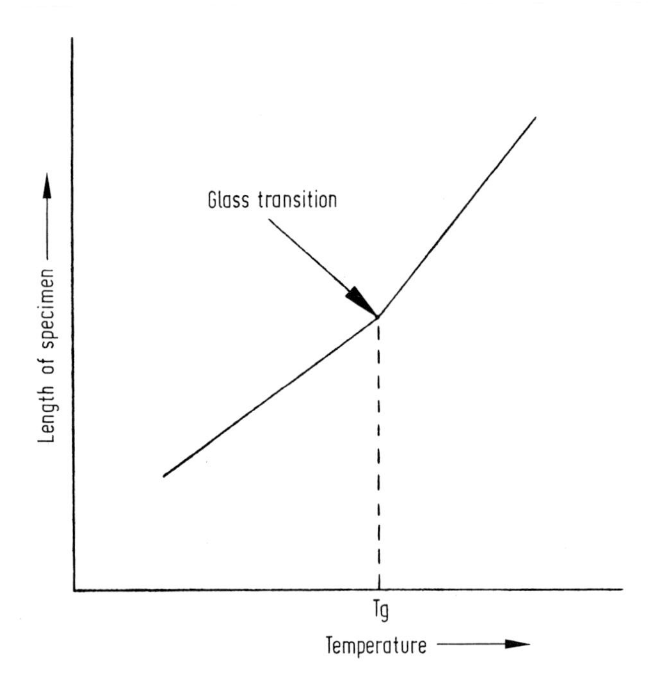

Figure 7.17 Effect of T_g on the volume or length of TPs

7.6.2.1 Mechanical Property and T_g

Figures 7.18 and 7.19 provide examples of modulus vs. T_g for amorphous and crystalline plastics. Temperature can help explain some of the differences observed in plastics. For example, at room temperature polystyrene and acrylic are below their respective T_g values; we observe these materials in their glassy state. In contrast, at room temperature natural rubber is above its T_g ($T_g = -75\,°C$ ($-103\,°F$); $T_m = 30\,°C$ ($86\,°F$)), with the result that it is very flexible. When it is cooled below its T_g natural rubber becomes hard and brittle.

7.7 Barrier Container Material

There are barrier requirements for BM products that range from packaging ketchup to gasoline. The barrier requirements to meet package needs vary depending on the product being packaged. Properties that may need to be retained or excluded can include gases, water vapor, aroma, taste, solvents, and so on (Tables 7.9 and 7.10).

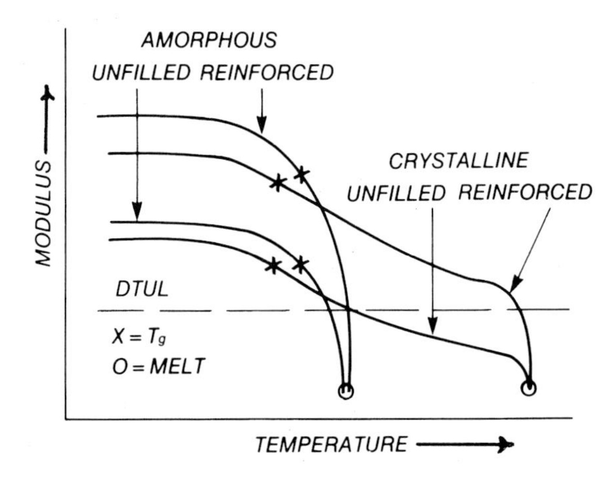

Figure 7.18 *Modulus behavior of amorphous and crystalline plastics showing T_g and melt temperatures*

It should also be noted that many oxygen-sensitive food products could be either hot filled or retorted. In these cases the container must have requisite thermal as well as barrier properties (Table 7.11). Further, if the container is to be subjected to elevated temperatures during filling or processing it must be designed to either resist or anticipate paneling. Both of these factors can considerably increase the complexity of developing satisfactory rigid barrier containers.

Table 7.12 gives an overview of the permeability of water vapor, gas, and oils/grease for various plastic materials.

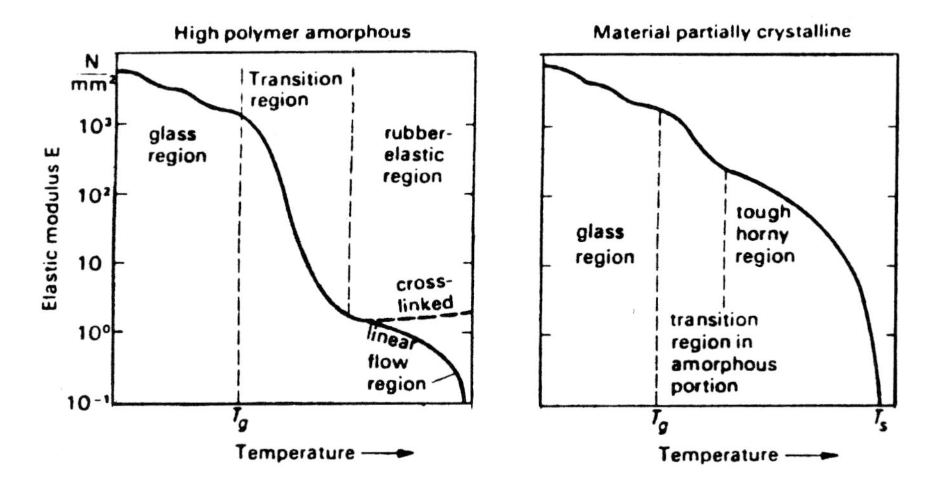

Figure 7.19 *Modulus vs. temperature dependence going through different processing stages*

Table 7.9 Aroma Retention

		Days to aroma leakage			
Construction	Thickness (mils)	Vanillin (vanilla)	Menthol (peppermint)	Piperonal (heliotropin)	Camphor
PET/EVOH/PE	0.5/0.6/2.0	15	25	27	> 30
OPP/EVOH/PE	0.7/0.6/2.0	30	> 30	27	> 30
PET/EVOH	0.5/0.6	> 30	> 30	30	> 30
PET/PE	0.5/2.0	2	16	5	> 30
OPP/PE	0.7/2.0	6	2	1	13

An example is the success of packaging carbonated beverages using PET plastic. The plastic bottle industry continues its research into expanding plastic bottles in other large volume applications to replace glass and metal with BM plastic containers. There are still many more high-volume markets that could be expanded and/or switched into a plastic container, such as beer, aseptic juice packages, salad dressing (both pourable and spoonable), toppings such as catsup and barbecue sauces, baby foods, cosmetics and pharmaceuticals, chemicals, and insecticides.

7.7.1 Monolayer

There are three basic ways in which to increase the barrier properties of a rigid plastic container. The first is to improve the barrier properties of the plastic being used in a

Table 7.10 Multilayer Combinations of Outside, Middle, and Inside Layers

Purpose	Inside and outside layers	Middle layer	Applications
Barrier packaging	PET	EVAL Special nylon Special polyester	For carbonated beverage, beer, etc. For foodstuff (juice, ketchup, mayonnaise, etc.)
High heat beverage packaging High strength packaging	PET	PC Polyarylate PET with glass fiber	For sterilization For high filling temp. up to 85 °C (185 °F) For high strength
Bottle made with recycled PET	PET	Recycled PET	Bottle with recycled PET middle layer For cost reduction
Others	PET	Colored PET Ultraviolet absorbing PET	

Table 7.11 Example of Performance Requirements (Qualitative) of High-Barrier Plastics

Foods	Prime barrier requirements	Shelf life requirements (yrs)	Container	
			Environment	Clarity
Low acid	Gas/flavor/odor moisture	2+	Retort (250 °F)	Opaque–clear
High acid	Gas/flavor/odor	0.5–2	Hot fill/pasteurization (180–210 °F)	Contact clear/clear
Jams, jellies, desserts	Gas/flavor/odor	0.5–2	Hot fill/pasteurization (180–210 °F)	Contact clear/clear
Peanut butter	Gas/flavor/odor	0.5–1	Hot fill (85–95 °F)	Contact clear/clear
Dressings	Gas/flavor/odor	0.5–1	Hot fill/pasteurization (180–210 °F)	Contact clear/clear
Pickles	Gas/flavor/odor		Hot fill/pasteurization (180–210 °F)	Clear

monolayer container. The second is to place a barrier coating on either the inside or the outside of the monoplastic layer. The third is to employ several distinct plastics in a multilayer coextruded structure with each providing a different specific barrier or performance quality (Table 7.10).

There are chemical reactive systems that are used to change the barrier behavior of certain plastics. As an example fluorination, a secondary process used on monolayer HDPE containers, involves the additional step of treating the container with fluorine gas. This treatment adds side chains of fluorine to the molecular structure of the HDPE, resulting in improved barrier properties.

Examples of fluorination processes used include Air Products & Chemicals' process that injects fluorine gas during the blowing cycle. The Union Carbide/Linde process is a post-molding treatment in which molded bottles are placed in a chamber and exposed to fluorine gas at elevated temperatures. Both processes are accepted by the industry and fluorinated HDPE bottles are being used for packaging pesticides, paint thinners, charcoal lighter fluid, kerosene, and so on.

Linde's Surface Modified Plastics (SMP) process was a spinoff from their aerosol container research. The process is unique among current fluorination methods for PE because it allows precise control over the degree of treatment. In effect, the user is able to custom-build a barrier that is specifically suited for the application at hand.

The development of barrier plastics that can be used in monolayer containers to pack oxygen-sensitive food products has not, to date, been commercially successful. As an example, plastic producers have developed barrier PET copolymers; but in all cases they have had to trade off some other critical performance characteristics such as impact strength or creep resistance. Further, barrier plastics, even when they do achieve an acceptable balance of characteristics, are too expensive, running at least between $1 to $3 a pound. Very little development work is being done in this area and the likelihood of an inexpensive barrier PET is several years off at best.

Table 7.12 *Overview Showing Permeability of Plastics*

Type of polymer	Specific gravity (ASTM D792)	Water vapor barrier	Gas barrier	Resistance to grease and oils
ABS (acrylonitrile butadiene styrene)	101–1.10	Fair	Good	Fair to good
Acetal—homopolymer and copolymer	1.41	Fair	Good	Good
Acrylic and modified acrylic	1.1–1.2	Fair	—	Good
Cellulosics acetate	1.26–1.31	Fair	Fair	Good
Butyrate	1.15–1.22	Fair	Fair	Good
Propionate	1.16–1.23	Fair	Fair	Good
Ethylene vinyl alcohol copolymer	1.14–1.21	Fair	Very good	Very good
Ionomers	0.93–0.96	Good	Fair	Good
Nitrile polymers	1.12–1.17	Good	Very good	Good
Nylon	1.13–1.16	Varies	Varies	Good
Polybutylene	0.91–0.93	Good	Fair	Good
Polycarbonate	1.2	Fair	Fair	Good
Polyester (PET)	1.38–1.41	Good	Good	Good
Polyethylene				
Low density	0.910–0.925	Good	Fair	Good
Linear low density	0.900–0.940	Good	Fair	Good
Medium density	0.926–0.940	Good	Fair	Good
High density	0.941–0.965	Good	Fair	Good
Polypropylene	0.900–0.915	Very good	Fair	Good
Polystyrene				
General purpose	1.04–1.08	Fair	Fair	Fair to good
Impact	1.03–1.10	Fair	Fair	Fair to good
SAN (styrene acrylonitrile)	1.07–1.08	Fair	Good	Fair to good
Polyvinyl chloride				
Plasticized	1.16–1.35	Varies	Good	Good
Unplasticized	1.35–1.45	Varies	Good	Good
Polyvinylidene chloride	1.60–1.70	Very good	Very good	Good
Styrene copolymer				
(SMA) Crystal	1.08–1.10	Fair	Good	Fair
Impact	1.05–1.08	Fair	Good	Fair

7.7.2 Coating

The concept of coating a plastic to improve its barrier performance is not new. Some of the first and still most widely used flexible barrier packages are PVDC coated polyester or oriented PP. Coating is perceived by many as the most reasonable low cost method to improve the barrier properties of rigid monolayer containers.

Coating has several drawbacks. The primary functional problem is adhesion. Another is the cost of high-speed continuous coating lines. There is also the problem of achieving an adequate barrier with a single pass. Assuming one is coating the outside of a container, there are three basic types of coating application systems: roll coating, dip coating (with or without a wipe step), and spray coating.

Roll and dip coating are used in Europe to apply a layer of PVDC directly on oriented PET bottles for beer. In the United States, these application methods have been used to coat both preforms and finished PET bottles. Coating the preform appeared originally to have some advantages because the subsequent stretch blowing step oriented the PVDC and increased its barrier properties. During 1980, wine bottles used coated preform. There were, however, resultant adhesion problems and an overthinning of the PVDC layer during blowing that effectively reduced the net barrier effectiveness in spite of the biaxial orientation achieved. In addition, the lines were too slow to be an effective commercial process.

The remaining target is to achieve an adequate barrier with a single pass. Efforts have been made to develop dual pass systems or coating systems where a superior oxygen barrier [ethylene–vinyl alcohol copolymer (EVOH or EVA)] is then covered with a water barrier but the cost has outweighed the benefits. The critical point to make is that although coating does not produce a totally successful barrier container, it is a commercially available technology that offers a modest degree of barrier improvement.

Some of the other emerging coating technologies that may generate interest in the future include the introduction of other sprayable barrier materials including EVOH, and the development of sinter coating techniques which would permit the use of certain plastics that can be powdered but not sprayed.

7.7.3 Multilayer

Most of the development efforts to produce high quality rigid barrier containers involve the use of multilayer structures (see Fig. 2.37). There are two basic approaches to the production of multilayer structures: coextrusion and coinjection. The basic issues to consider when developing a multilayer barrier container are which barrier material to use and which manufacturing technology to employ. BM multilayer hollow products is currently experiencing a phase of worldwide interest and a rapid upsurge. It utilizes the material-specific properties of various plastics for the production of high-grade multilayer containers. Tables 3.5 (page 106) and 4.1 (page 127) provide information on the plastic bonding capability in coextrusion and coinjection. When bonds are poor or do not exist, an adhesive or so-called tie-layer is used.

The introduction of the five-layer to seven-layer technique in the United States brought the breakthrough in 1984 for commercial production of these containers, particularly in the foods industry (ketchup, mayonnaise, soup concentrates, baby food, and the like). A crucial factor was the development of a completely new multilayer head extrusion technology, permitting wall thicknesses down to less than 0.03 mm (0.001 in.) with excellent, constant coaxial wall thickness distribution. This was the key to using expensive barrier and bonding agent materials at commercially viable production costs.

The advantages of these bottles are many, some of which, related to different material combinations, include:

- excellent barrier properties against gases such as nitrogen, carbon dioxide, and air as well as against water vapor (foodstuffs, pharmaceutical industry);
- resistance to aroma transmission (cosmetics);
- protection against UV radiation (mild);
- higher strength (greater product stackability);
- light weight (lower transport costs);
- scratch-resistant, glossy surfaces (cosmetics);
- resistance to aggressive media (insecticides, herbicides);
- barrier to solvents (paint industry);
- easy printability, clear printed impression (cosmetics); and
- prevention of static charging (explosion protection).

Figure 7.20 shows the relationship between comonomer concentration, the moisture transmission, and gas transmission properties of these copolymers. It is these exceptional barrier properties that allow the use of extremely thin layers of these materials in combination with other inexpensive plastics to BM economical, high performance products. Table 7.13 indicates the range of commercially available EVOH plastics. Within this range, conventional melt BM processes readily process EVOH plastics.

The method that has the greatest application potential is to arrange the barrier layer in the middle between the inner and outer body layers. Whether reclaimed material is used in the outer layer is a quality question that the container filler has to answer based on product performance requirements. The seven-layer combination with reclaim in the penultimate layer from the outside, without any further bonding agent to the body material is the technically/economically most satisfactory solution to date.

With the coextruded technology well established and the hardware in position, the variety of achievable properties is readily extendable through the correct combination of different materials. The potential for BM packaging products to be much more

***Table** 7.13 EVOH Plastic Range of Properties*

Melt index (g/10 min)	Density (g/cc)	Ethylene content (mol%)	Melting point (°C)
0.7–20	1.13–1.21	29–48	158–189

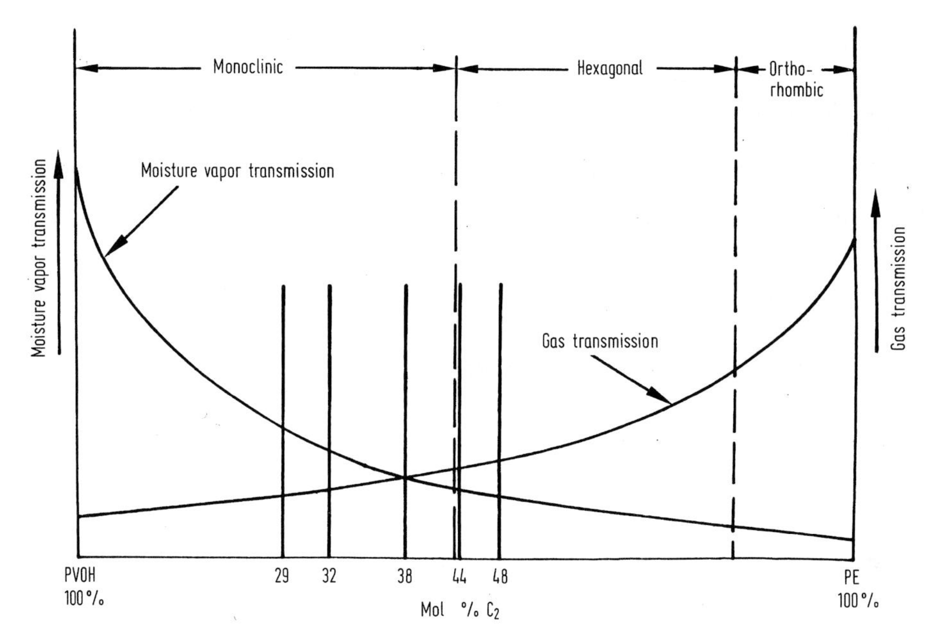

Figure 7.20 *Comonomer concentration vs. barrier properties of crystalline structures*

than simple bottles has been known for some years. Now the know-how and economics are such that many previously futuristic ideas are much closer to reality.

7.7.4 Barrier Material

There are four popular basic categories of barrier materials currently being used by producers of multilayer containers: EVOH, PVDC, EVA, and nylon. Most attention is

Table 7.14 *Examples of Barrier Properties of Commercially Available Plastics*

Polymer	O_2 transmission rate (25 °C, 65% RH cc.mil/100 in.^{2}/24 h)	Moisture vapor trans. rate (40 °C, 90% RH g.mil/100 in.^{2}/24 h)
Ethylene vinyl alcohol	0.05–0.18	1.4–5.4
Nitrile barrier resin	0.80	5.0
High barrier PVDC	0.15	0.1
Oriented PET	2.60	1.2
Oriented nylon	2.10	10.2
Low density polyethylene	420	1.0–1.5
High density polyethylene	150	0.69
Rigid PVC	5–20	0.9–5.1

focused on the first two of these, with EVOH having the most continual significant current prospects because of its greater processing window and compatibility. PVDC remains the preferred material for flexible packaging, and has the advantage of being unaffected by moisture. PVDC is, however, basically difficult to process and recycle due to its thermal sensitivity. EVA is an excellent barrier, but extremely hygroscopic, which severely limits its use (Table 7.14).

7.8 Orientation and Crystallinity

Intermolecular order refers to the geometric arrangement of adjacent polymer molecules in the solid mass. There are three distinct types of intermolecular order:

- Amorphous random coils are isotropic and entangled, and their properties depend primarily on molecular flexibility. Polystyrene, acrylic plastics, polyphenylene oxide, polysulfone, and polycarbonate are typical examples.
- Crystalline polymers have their molecules arranged in a very regular repeating lattice structure, so precise that every atom of the polymer molecule must recur at very specific points in the repeat structure. While no polymer is completely crystalline, very regular polymers can be mostly crystalline with only small amorphous areas remaining between the crystallites. Typical highly crystalline polymers include high-density HDPE, PP, polyacrylonitrile, polyformaldehyde (acetal resin), thermoplastic polyesters such as PET, and polyamides (nylons). Less regular structural polymers may be partly crystalline (LDPE) or only slightly crystalline (PVC).
- Oriented polymers have some of their molecules lying fairly parallel to each other. When crystalline polymers are oriented, their crystallites lie fairly parallel to each other. Such orientation produces anisotropic properties, very different when tested in the direction of orientation than when tested perpendicular to it [18].

7.8.1 Orientation in Blow Molding

The primary effect of blowing a rubbery-leathery molten parison is naturally to fill the mold and make the product conform to the walls of the mold. But close behind this in importance is the way that stretching the parison orients polymer molecules in the direction(s) of stretch so that they become fairly parallel to each other in the final product. The precise conditions of the stretching process determine the degree of orientation, and this in turn determines the improvement in many end-use properties. In crystallizable polymers, orientation also encourages crystallization, faster and further than it would occur in an isotropic system, and this in turn also affects end-use properties.

7.8.1.1 Processing

For stretching and orientation to take place, fairly large segments of the polymer molecule (5 to 30 atoms in length) must be free to move and align in the direction of the stretching force; that is, the polymer must be somewhat above its glass transition temperature. Obviously, the higher the temperature, the more readily the polymer can align in this way. On the other hand, increasing temperature means increasing vibration and Brownian motion of atoms and segments of molecules, a randomizing effect that destroys orientation in favor of the higher entropy of the completely random amorphous state (Fig. 7.21). The only reason that overall orientation is possible is that the mechanical stress during rapid stretching favors orientation temporarily over randomization. As soon as orientation is accomplished, it must be frozen in by cooling and/or crystallization, before randomization can occur.

Understanding and techniques of orientation were developed first for synthetic fibers, and applied later to plastic films. More recently they have been applied to the blow molding process specifically, with growing sophistication and improvement in production and in end-use properties. Increasing molecular weight produces a stronger melt and permits higher degrees of stretch orientation as the long molecules slide over each other. While wide latitude in processing temperature range would be desirable, higher temperatures generally accelerate randomization and premature crystallization, so practical orientation temperature ranges are generally narrow and closely specified (Table 7.15), for example T_g to 130 °C for polyacrylonitrile and $T_g + 23$ °C for PET [38]. Faster rate of stretching favors orientation over randomization. Temperature and rate of stretching correlate directly with each other, both because higher temperature produces higher molecular mobility which *permits* faster stretching, and also because higher molecular mobility produces faster randomization and thus *requires* faster stretching to retain orientation; properly combined, higher temperature and faster stretching can produce maximum orientation [13].

The pressure needed to blow the parison depends on molecular flexibility, melt strength, and melt elasticity, which in turn also depend on processing temperature; it

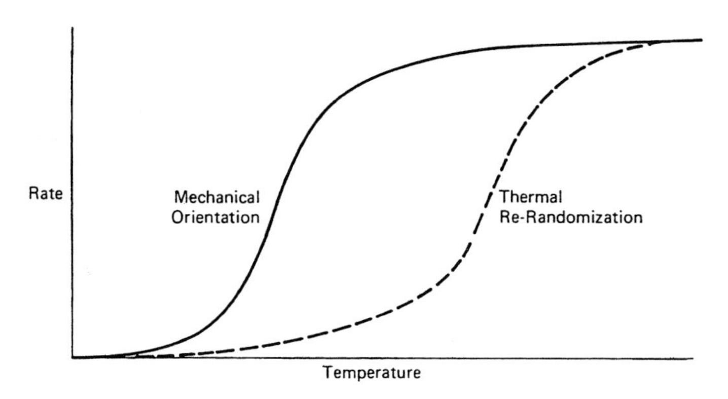

Figure 7.21 *Mechanical orientation vs. thermal randomization in warm stretching*

Table 7.15 Optimum BM Temperatures

Material	Melt (°C)	Stretch (°C)
PET	280	107
PVC	180	120
PAN	210	120
PP	240	160

Table 7.16 Typical Radial Stretch in BM

Material	Stretch ratio
PET	4 : 1
PVC	2.65 : 1
PAN	3 : 1
PP	2.45 : 1

is generally provided by air pressure, typically 300 to 650 psi (2.1 to 4.5 MPa) for PET and 100 psi (0.7 MPa) for acetal resin. Since air pressure produces mainly radial stretching, a vertical stretch rod may be needed to produce sufficient stretching in the longitudinal direction. The extent of stretch orientation, commonly expressed as the stretch ratio or draw ratio, is typically 2.5 to 4 x (Table 7.16); for example in PET radial stretch may be 4.7 x inside and 3.7 x outside while axial stretch is 2.8 x inside and 2.6 x outside. It has been noted that the "natural draw ratio" of PET ranges from 1.5 to 4.5 at 80 to 100 °C (176 to 212 °F) and is inverse to molecular weight (Fig. 7.22). Stretching beyond this range to produce strain-hardening can produce superior properties. Finally, design of the product and the blowing process to produce maximum optimum orientation remains a major challenge for the future.

7.8.1.2 Effects on Properties

While stretching is used primarily to make the parison conform to the walls of the mold, the resulting orientation produces major improvements in many properties which are particularly important, especially in bottles and other types of packaging. These improvements include:

- modulus (Table 7.17) and creep resistance to withstand internal pressure from carbon dioxide and vertical pressure from stacking in the warehouse,

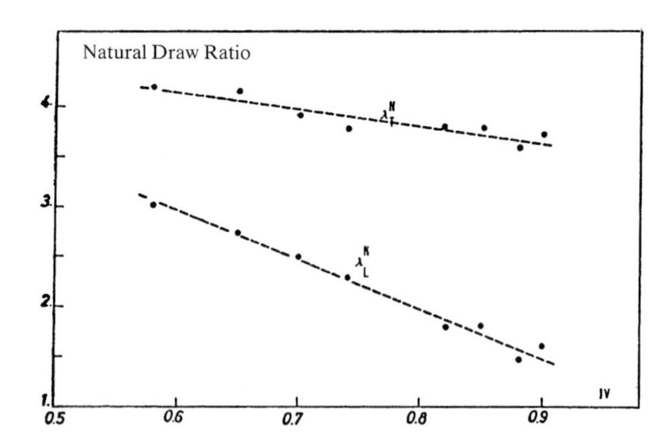

Figure 7.22 Effect of molecular weight on natural draw ratio of PET

Table 7.17 Young's Modulus Ratio of Highly Oriented Polymers

Polymer	Modulus parallel to orientation/ Modulus of unoriented polymer
Polyethyleneterephthalate	17.8
Nylon 66	19.2
Viscose rayon	9.3
Polyacrylonitrile	5.8

- tensile/hoop/burst strength to withstand internal pressure (Table 7.18), ultimate elongation [4],
- impact strength far superior to glass (Fig. 7.23), relaxation temperature at which the oriented shape retracts [29],
- transparency (Fig. 7.24), solvent resistance [28],
- impermeability for long shelf life (Fig. 7.25).

Table 7.18 Effect of Orientation on Mechanical Properties

	Polystyrene		Polymethyl methacrylate	
Property	Unoriented	Biaxially oriented	Unoriented	Biaxially oriented
UTS (psi/MPa)	7000 (48.3)	9500 (65.5)	8750 (60.3)	9500 (65.5)
UE (%)	2.3	13	7.5	37.5
IS (FPI)	0.375	> 3	4	15

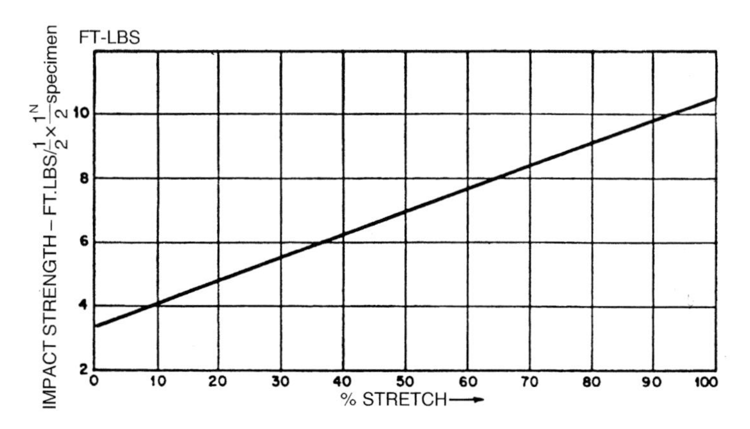

Figure 7.23 Charpy impact strength of stretched acrylic sheet

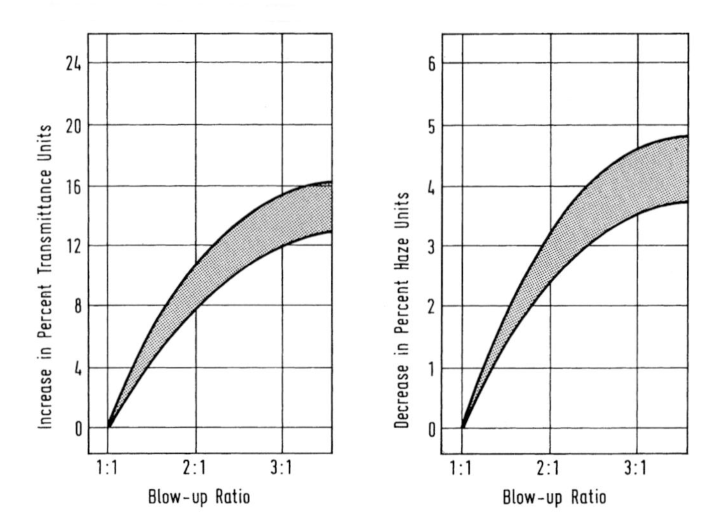

Figure 7.24 *Effect of orientation on clarity of PE film*

Many of these consequently permit design of thinner wall sections of lighter weight and therefore lower cost. Thus, while orientation is simply a consequence of stretching during blow molding, it provides major improvements in properties, product performance, and economics.

7.8.2 Crystallization in Blow Molding

In a molten polymer, the molecules exist in the form of random coils intertwined with each other. On cooling, one of two distinctly different events can occur. Irregular molecules in structures gradually lose their mobility and gradually freeze into an amorphous glassy state, in which the molecules are still intertwined random coils. Very regular molecular structures tend to fit into the imaginary points of a very regular lattice structure and form sub-microscopic crystallites which are densely packed and therefore contribute to rigidity, strength, heat deflection temperature, chemical resistance, and impermeability. During processing, the major difference between amorphous and crystalline polymers is that amorphous polymers gradually lose their molecular mobility as the temperature cools, whereas crystallizing polymers change suddenly from mobile liquids to crystalline solids at a sharply-defined melting/freezing point (Table 7.19).

Thus amorphous polymers generally offer a broader temperature range in which blowing and stretching can be optimized to produce maximum orientation. Regular crystallizable polymers such as PE, PP, polyacrylonitrile, and nylon crystallize rapidly in a narrow temperature range, above which the melt is too weak to stretch and the mobility is too high to retain any stretch orientation, and below which the crystallites

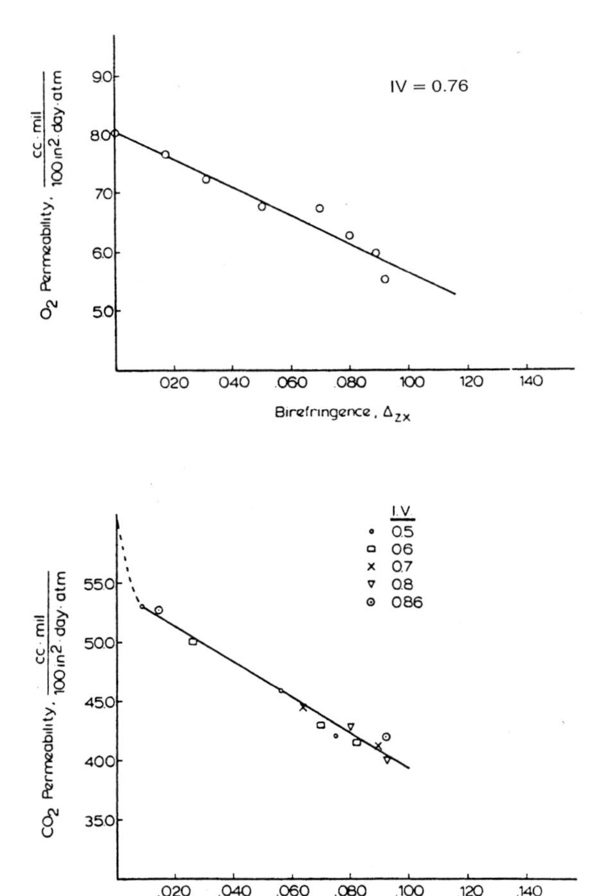

Figure 7.25 *Effect of orientation on permeability of PET*

are too rigid to permit blowing, stretching, and orientation; thus, they are more difficult to blow mold successfully.

PET occupies an intermediate position: its molecular structure is regular enough to fit into a crystal lattice structure, but its molecular stiffness hinders its ability to align itself into the precise points required to form the crystal lattice structure. It can be quenched from the melt to a solid amorphous form, which is very convenient for rewarming to blow mold and stretch orient it later. When it is warmed and stretched, the balance of molecular mobility and orientation brings the molecules into parallel alignment and initiates the crystallization process. And it can then be annealed at temperatures between glass transition and melting point a provide sufficient molecular mobility to permit it to crystallize at any desired rate and to any desired degree (Figs. 7.26 and 7.27).

Table 7.19 Molecular Stiffness, Polarity, and Melting Point

Polymer	Crystalline melting point (°C)
Polytetrafluoroethylene	327
Polyacrylonitrile	317
Cellulose triacetate	306
Polyethylene terephthalate	267
Nylon 66	265
Isotactic polystyrene	240
Polybutylene terephthalate	232
Nylon 610	227
Nylon 6	225
Polycarbonate	220
Polychlorotrifluoroethylene	220
Polyvinyl chloride	212
Polyvinyl fluoride	200
Polyvinylidene chloride	198
Nylon 11	194
Polyoxymethylene (acetal)	181
Polypropylene	176
Isotactic polymethyl methacrylate	160
Polyethylene	137

When regularity of polymer structure causes crystallization too rapidly to permit convenient BM and stretch orientation, the rate of crystallization can be controlled somewhat by the cooling process. Ultimate crystallinity may also be controlled somewhat by quenching; thus, blow molding PP into a cold mold (60 to 80 °F) reduces final crystallinity and improves impact strength.

A more effective way to retard and reduce crystallization is to copolymerize with a second comonomer, producing a more random polymer structure which is less able to fit into a regular crystal lattice. Copolymerization of PET with 15% or more of comonomers such as cyclohexane dimethanol, diethylene glycol, neopentyl glycol, or isophthalic acid produces amorphous copolymers which are much easier to blow mold and stretch orient. Recent development of amorphous nylons permits easier blow molding at lower temperatures and over wider temperature range. And copolymerization of vinyl chloride with propylene or with vinyl alkyl ethers similarly lowers crystallinity, and thus lowers processing temperature and minimizes thermal discoloration (Table 7.20).

7.8.2.1 Effects on Properties

Crystallization is useful in blow molding because (1) it freezes in the stretch orientation and thus gives the oriented structure permanence; and (2) it improves

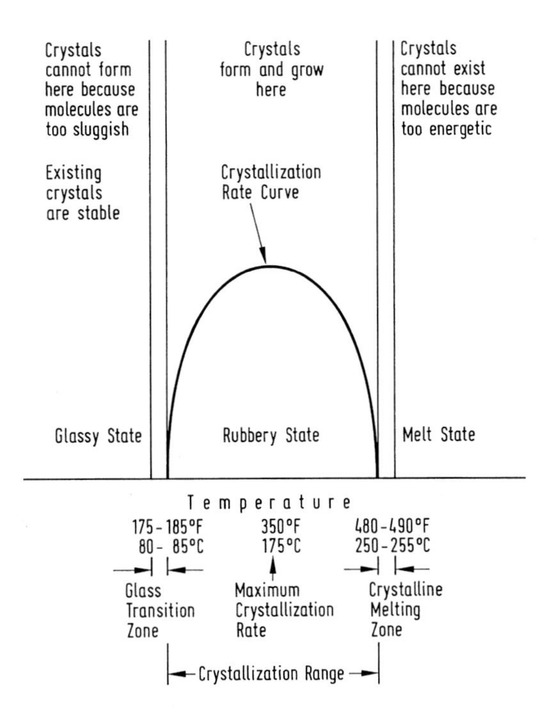

Figure 7.26 *PET crystallization vs. temperature*

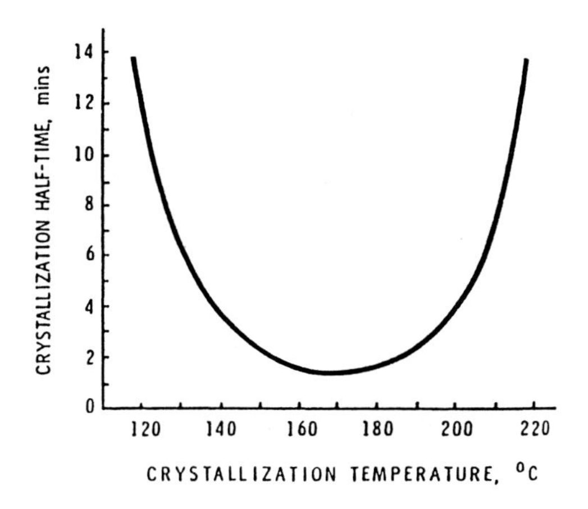

Figure 7.27 *PET crystallization time*

Table 7.20 *Dynamic Mill Stability of PVC Blow Molding Compounds at Equiviscous Temperatures*

	Equiviscous temperature		Time in minutes	
Polymer	(°C)	(°F)	Absorbance 0.1/476 μ	Yellowness index = 20
Homopolymer	188	370	20	8
Copolymers				
Vinyl alkyl ether	182	360	40+	20
Propylene	179	354	40+	30

many end-use properties of particular importance in bottles for food packaging, including rigidity, dimensional stability on reheating, and impermeability.

On the other hand, crystallinity tends to harm some useful properties such as ultimate elongation, impact strength, transparency, and environmental stress crack resistance. When these properties are critical, generally copolymerization is used to reduce molecular regularity and crystallinity, often going all the way to amorphous polymers which are held together simply by the stiffness of the entangled random coils.

Many other effects of crystallinity on melt processing in general include heat transfer, heat capacity, nucleation, and crystal growth; these have been discussed in detail elsewhere, and do not require further space here.

7.9 Intermolecular Bonding

Entanglement of amorphous random coils, parallel bundles of oriented molecules, and even neat tight packing in crystallites, are not enough to explain the remarkable mechanical and thermal properties of plastics. It is the attractive forces between polymer molecules, which hold them firmly together and make them able to resist mechanical and thermal stress, that give them their useful properties. On the other hand, when the blow molder wants to melt these polymers and make the liquids flow into parisons and stretch to the walls of the mold, these same attractive forces can make the molding process much more difficult.

There are a variety of such intermolecular attractive forces. They may be arranged from weakest to strongest in the following order: London dispersion forces, polarity, hydrogen bonding, orientation and crystallinity, ionic bonding, and permanent primary covalent crosslinking.

7.9.1 London Dispersion Forces

Linear hydrocarbon polymers have no polarity and there is no obvious reason why their molecules should be attracted to each other. Nevertheless, their mechanical strength and resistance to melt flow show that even here some attractive forces must exist. They are vaguely ascribed to dynamic fluidity and mobility of the valence electron cloud in the polymer molecule, producing transient unsymmetrical states of electrical imbalance and thus momentary polarity (Fig. 7.28). These weak fugitive polar attractions draw polymer molecules to within 3 to 5 Å of each other with a force of 1 to 2 kcal/mol. While one such attraction would be negligible, the large number of such attractions between large polymer molecules accumulates to a large total effect.

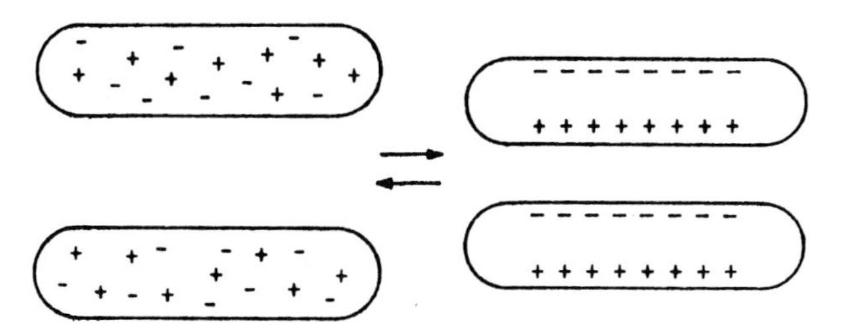

Figure 7.28 *Schematic model for London dispersion forces*

For amorphous flexible molecules such as atactic polypropylene, polyisobutylene, and unvulcanized rubber, such attractions produce viscous liquids to tacky gums to soft rubbery solids; when heated, they liquify and flow easily. For amorphous stiff molecules such as polystyrene, these attractions produce rigidity and strength up to 80 to 90 °C (176 to 194 °F), but they melt out easily to permit melt processing. In crystalline polymers such as polyethylene and isotactic PP, the same London dispersion forces recur so frequently in the tightly-packed crystal lattice that they add up to a very significant total, producing rigidity and strength up to heat deflection temperatures as high as 120 °C (248 °F) and requiring the molder to go to melting points as high as 176 °C (349 °F) or more in order to melt out the attractions and produce liquid flow.

These are polymers in which there are no intermolecular attractions other than London dispersion forces. In all other polymers, where there are also polar and hydrogen bonding attractions, there are still these same London dispersion attractions, and they still form a considerable portion of the total intermolecular attraction in all of them.

7.9.2 Polarity

When polymers contain other atoms in addition to carbon and hydrogen, these are generally more negative (Table 7.21), and the resulting polarity produces electrical

Table 7.21 *Electronegativities of Atoms in Polymers*

Sodium	0.9	Bromine	2.8
Silicon	1.8	Nitrogen	3.0
Hydrogen	2.1	Chlorine	3.0
Phosphorus	2.1	Oxygen	3.5
Carbon	2.5	Fluorine	4.0
Sulfur	2.5		

attractions between the polymer molecules. These attractions pull and hold polymer molecules closer together. They are most often measured by solubility in different solvents, quantified as the “solubility parameter”.

These polar attractions improve many end-use properties. They generally require more heat to overcome them and permit melt processing, so melting points and processing temperatures are generally higher. The over-enthusiastic polymer chemist, keeping his eye only on end-use properties, can easily synthesize polymers with so much polar attraction that they will not melt and flow even when they finally reach the decomposition temperature. Thus processability vs. properties generally requires some compromise level of polarity.

7.10 Property

An extensive amount of information on plastics material properties are available from suppliers and/or software recognizing that a specific plastic usually has many modifications to meet different properties and/or processing requirements. Also new developments in plastic materials are always on the horizon, requiring updates. It is important to ensure that the fabricating process to be used to produce a product provides the properties desired. Much of the market success or failure of a plastic product can basically be attributed to the initial choices of material, process, and cost.

For many materials (plastics, metals, etc.) it can be a highly complex process if not properly approached, particularly when using recycled plastics. As an example, its methodology ranges from a high degree of subjective intuition in some areas to a high degree of sophistication in other areas. It runs the gamut from highly systematic value engineering or failure analysis such as in aerospace to a telephone call for advice from a material supplier in the decorative houseware business. Different helpful publications, seminars, and software programs are available.

Guides to selecting materials can be provided by using examples of properties vs. plastics and plastics vs. properties (Table 7.22). Worldwide there are over 35,000

Table 7.22 *Thermoplastic and Thermoset Plastic Processing Guide*

	Melt density specific gravity g/cm³ at processing temperature	Density specific gravity g/cm³ at 70 °F (solid)	Density lb/ft³	Specific volume in.³/lb	Specific volume cm³/g	Extrusion temperature, °F	Injection temperature, °F	Linear mold shrinkage in./in.	Specific heat Btu/lb/°F	Water absorption % in 24 h	Maximum water content allowable for molding
ABS—extrusion	0.88	1.02	64.0	27.0	0.980	435		0.005	0.34	0.25	
ABS—injection	0.97	1.05	65.0	26.6	0.952		500	0.005	0.40	0.40	0.20
Acetal—injection	1.17	1.41	88.0	19.7	0.709		390	0.020	0.35	0.25	
Acrylic—extrusion	1.11	1.19	74.3	23.3	0.839	375		0.004	0.35	0.30	
Acrylic—injection	1.04	1.16	72.0	24.1	0.868		450	0.005	0.35	0.20	0.08
CAB	1.07	1.20	74.6	23.1	0.833	380	440	0.004	0.35	1.50	0.15
Cellulose acetate—extrusion	1.15	1.28	80.2	21.6	0.781	380		0.005	0.40	2.50	
Cellulose acetate—injection	1.13	1.26	79.0	21.9	0.794		450	0.005	0.36	2.40	0.20
Cellulose proprionate—extrusion	1.10	1.22	76.1	22.7	0.821	380		0.004	0.40	1.70	
Cellulose proprionate—injection	1.10	1.22	75.5	22.9	0.828		425	0.004	0.40	2.00	0.25
CTFE	1.49	2.11	134.0	13.1	0.473		550	0.008	0.22	0.01	
FEP	1.49	2.11	134.0	12.9	0.465	600	600	0.010	0.28	< 0.01	
Ionomer—extrusion	0.73	0.95	59.6	29.0	1.050	500		0.007	0.54	0.07	
Ionomer—injection	0.73	0.95	59.1	29.2	1.060		420	0.007	0.54	0.20	
Nylon 6	0.97	1.13	70.5	24.5	0.886	520	550	0.013	0.40	1.60	0.15
Nylon 6/6	0.97	1.14	71.2	24.3	0.878	510	510	0.015	0.40	1.50	0.15

(continued)

Table 7.22 *(continued)*

	Melt density specific gravity g/cm^3 at processing temperature	Density specific gravity g/cm^3 at 70 °F (solid)	Density lb/ft^3	Specific volume in.^{3}/lb	Specific volume cm^3/g	Extrusion temperature, °F	Injection temperature, °F	Linear mold shrinkage in./in.	Specific heat Btu/lb/°F	Water absorption % in 24 h	Maximum water content allowable for molding
Nylon 6/10	0.97	1.08	67.4	25.6	0.927		450	0.011	0.40	0.40	0.15
Nylon 6/12	0.97	1.07	66.8	25.9	0.935	475	500	0.011	0.40	0.40	0.20
Nylon 11	0.97	1.04	64.9	26.6	0.962	460	450	0.005	0.47	0.30	0.10
Nylon 12	0.97	1.02	63.7	27.1	0.980	450	445	0.003		0.25	0.10
Phenylene oxide based	0.90	1.08	67.5	25.6	0.926	480	525	0.006	0.32	0.07	
Polyallomer	0.86	0.90	56.2	30.7	1.110	405	405	0.015	0.50	0.01	
Polyarylene ether	1.04	1.06	66.2	30.7	0.940	460	535	0.006		0.10	
Polycarbonate	1.02	1.20	74.9	23.1	0.832	550	575	0.006	0.30	0.20	0.02
Polyester PBT	1.11	1.34	83.6	20.7	0.746		460	0.020		0.08	0.04
Polyester PET	1.10	1.31	81.8	21.1	0.746	480	490	0.002	0.40	0.10	0.005
HDPE—extrusion	0.72	0.96	59.9	28.8	1.040	410		0.025	0.55	< 0.01	
HDPE—injection	0.72	0.95	59.3	29.1	1.050		480	0.025	0.55	< 0.01	
HDPE—blow molding	0.73	0.95	56.9	28.8	1.040	410		0.025	0.55	< 0.01	
LDPE—film	0.77	0.92	57.4	30.1	1.090	350		0.032	0.55	< 0.01	
LDPE—injection	0.76	0.92	57.4	30.1	1.090		400	0.032	0.55	< 0.01	
LDPE—wire	0.76	0.92	57.4	30.1	1.090	400		0.025	0.55	< 0.01	
LDPE—extrusion coating	0.68	0.92	57.1	30.0	1.090	600		0.025	0.55	< 0.01	
LLDPE—extrusion	0.75	0.92	57.4	30.1	1.087	500		0.022		< 0.01	
LLDPE—injection	0.070	0.93	58.0	29.8	1.075		425	0.022		< 0.01	
Polypropylene—extrusion	0.75	0.91	56.8	30.4	1.100	450		0.005	0.50	0.03	

(continued)

Table 7.22 (continued)

	Melt density specific gravity g/cm^3 at processing temperature	Density specific gravity g/cm^3 at 70 °F (solid)	Density lb/ft^3	Specific volume in.^{3}/lb	Specific volume cm^3/g	Extrusion temperature, °F	Injection temperature, °F	Linear mold shrinkage in./in.	Specific heat Btu/lb/°F	Water absorption % in 24 h	Maximum water content allowable for molding
Polypropylene—injection	0.75	0.90	56.2	30.7	1.110		490	0.018	0.50	< 0.01	
Polystyrene—impact/sheet	0.96	1.04	64.9	26.6	0.963	450		0.005	0.34	0.10	
Polystyrene—G.P. crystal	0.97	1.05	65.5	26.2	0.943	410	425	0.004	0.32	0.03	
Polystyrene—injection impact	0.96	1.04	64.9	26.6	0.968		440	0.006	0.34	0.10	
Polysulfone	1.16	1.25	77.4	22.3	0.807	650	680	0.007	0.28	0.30	0.05
Polyurethane (non-elastomer)	1.13	1.20	74.9	23.1	0.834	400	400	0.020	0.40	0.10	0.03
PVC—rigid profiles	1.30	1.39	86.6	19.9	0.720	365		0.025	0.25	0.01	
PVC—pipe	1.32	1.44	87.5	19.7	0.714	380		0.025	0.25	0.10	
PVC—rigid injection	1.20	1.29	83.6	21.0	0.756		380	0.025	0.25	0.10	0.07
PVC—flexible wire	1.27	1.37	85.5	20.2	0.731	365		0.025	0.25	0.02	
PVC—flexible extruded shapes	1.14	1.23	76.8	22.5	0.814	350		0.025	0.25	0.02	
PVC—flexible injection	1.20	1.29	80.5	21.4	0.776		300	0.025	0.25	0.10	
PTFE	1.50	2.16	134.8	12.9	0.464		650	0.006	0.25	< 0.01	
SAN	1.00	1.08	67.4	25.6	0.927	420	470	0.005	0.31	0.03	0.02
TFE	1.50	1.70	106.1	16.3	0.589		610	0.040	0.46	0.01	
Urethane elastomers (TPE)	0.82	0.83	51.6	33.5	1.210	390	400	0.001	0.46	0.07	0.03

different plastics that provide all types of property [2]. In turn a certain process such as BM can only manufacture certain plastics. Of all these plastics, only a few dozen are used in large quantity.

An endless amount of data is available for many plastic materials. Unfortunately, as with other materials, there does not exist only one plastic material that will meet all performance requirements. However, it can be stated that for practically any product requirement(s), particularly when not including cost for a few products, more so than with other materials, there is a plastic that can be used. Plastics provide more property variations than any other material.

Plastics are families of materials each with their own special advantages. An example is polyethylene (PE) with its many types that include low density PE (LDPE), high density PE (HDPE), high molecular weight PE (HMWPE), etc. The major consideration for a designer and/or fabricator is to analyze what is required as regards to performances and develop a logical selection procedure from what is available [12, 19].

Recognize that most of the plastic products produced only have to meet the usual requirements humans have to endure such as the environment (temperature, etc.). Also, recognize that most plastics in use do not have a high modulus of elasticity or long creep and fatigue behaviors because they are not required in their respective designs. However, there are plastics with extremely high modulus and very long creep and fatigue behaviors. These type products have performed in service for long periods of time with some performing well over a half century. For certain plastic products there are definite properties (modulus of elasticity, temperature, chemical resistance, load, etc.) that have far better performance than steels and other materials.

The ranges of properties in plastics encompass all types of environmental and load conditions, each with its own individual, yet broad, range of properties (Table 7.23). These properties can take into consideration wear resistance, integral color, impact resistance, transparency, energy absorption, ductility, thermal and sound insulation, weight, and so forth.

Different plastics can be combined producing a product meeting different properties. They can just be stacked together, but with available processes they can also be fabricated so that each material retains its individuality, yet, has a bond with the adjoining plastics. These processes include coinjection and coextrusion. Each of the individual plastics can provide such characteristics as wear resistance, water barrier, electrical conductor, and adding strength. Low-cost, recycled (solid waste) plastics can be "sandwiched" between other expensive, high-performance plastics so they only act as a filler, increase strength, and so on.

7.10.1 Selecting the Plastic

Even though the range of plastics has become large and the levels of their properties so varied, in any proposed application only a few of the many plastics will be suitable.

Table 7.23 *Examples of a Few Properties of Thermoplastic Bottles*

A.

Properties	Acrylic multipolymer	Nitrile	Polycarbonate	Polyester (oriented) (PET)[b]	PETG copolyester	Polyethylene	
						Low density	High density
Resin density	1.09–1.41	1.15	1.2	1.35–1.40	127	0.91–0.925	0.94–0.965
Clarity[a]	Clear	Clear	Clear	Clear	Clear	Hazy transparent	Hazy translucent
Permeability to water vapor	High	Moderate	High	Moderate	Moderate	Low	Very low
Oxygen	Low	Very low	Moderate to high	Low	Low	Very high	High
CO_2	Moderate	Very low	Moderate to high	Low	Low	Very high	High
Resistance to acids	Poor to good	Poor to good	Fair	Fair to good	Fair	Fair to very good	Fair to very good
Alcohol	Good	Fair	Fair	Good	Good	Fair to very good	Good
Alkalis	Poor to fair	Good	Poor to fair	Poor to fair	Poor to fair	Good to very good	Good to very good
Mineral oil	Good	Very good	Good	Good	Good	Poor	Fair
Solvents	Poor	Good	Poor to fair	Good	Poor to good	Poor to fair	Poor to good
Heat	Fair	Poor to fair	Very good	Poor to fair	Poor to fair	Fair	Fair to good
Cold	Poor	Fair	Good	Good	Good	Very good	Very good
Sunlight	Good	Fair	Good	Good	Fair	Fair	Fair
Temperature at which finished product distorts[d]	180 °F to 195 °F	140 °F to 150 °F	260 °F to 280 °F	100 °F to 160 °F	160 °F to 220 °F	160 °F to 220 °F	160 °F to 250 °F
Stiffness	Moderate to high	Moderate to high	High	Moderate to high	Moderate to high	Low	Moderate

(continued)

Table 7.23 *(continued)*

A.

Properties	Acrylic multipolymer	Nitrile	Polycarbonate	Polyester (oriented) (PET)[b]	PETG copolyester	Polyethylene	
						Low density	High density
Resistance to impact	Poor to good	Poor to good	Excellent	Good to excellent	Poor to fair	Excellent	Good to very good
Unit cost	Moderate to high	High	Very high	Moderate	Moderate to high	Low	Low
Typical uses	Foods, drugs, cosmetics	Cosmetics, household chemicals	Mineral oil, baby nurser bottles, water, milk	Carbonated beverages, mouthwash, liquor, edible oil, drugs, cosmetics	Cosmetics, foods, mineral oil, personal care	Cosmetics, personal products, mustard, drugs	Detergents, bleaches, milk, chocolate, syrup, industrial cleansing powders, drugs, cosmetics, lube oil, edible oil

Table 7.23 *Continued*

B.

Properties	Polypropylene		Polystyrene	Polysulfone	SAN (styrene acrylonitrile)	TPX[c]	PVC
	Regular	Oriented					
Resin density	0.89–0.91	0.90	1.0–1.1	1.24	1.07–1.08	0.83	1.35
Clarity[a]	Moderate haze	Clear	Clear	Clear	Clear	Moderate haze	Clear
Permeability to water vapor	Very low	Very low	High	High	High	Very low	Moderate
Oxygen	High	High	High	Low to moderate	High	Very high	Low
CO_2	Moderate to high	Moderate to high	High	Low to moderate	High	High	Low
Resistance to acids	Fair to very good	Fair to very good	Fair to good	Fair to very good	Fair to good	Fair to good	Good to very good
Alcohol	Good	Good	Fair	Fair to good	Poor	Good	Good to very good
Alkalis	Very good	Very good	Good	Very good	Good	Very good	Good to very good
Mineral oil	Fair	Fair	Fair	Very good	Fair	Fair to good	Good
Solvents	Poor to good	Poor to good	Poor	Fair to good	Poor	Fair to good	Poor to good[e]
Heat	Good	Good	Fair	Very good	Fair	Very good	Poor to fair
Cold	Poor to fair	Very good	Poor	Very good	Poor	Poor to fair	Fair
Sunlight	Fair to good	Fair to good	Fair to poor	Fair	Fair to poor	Fair	Poor to good
Temperature at which finished product distorts[d]	250 °F to 260 °F	250 °F to 260 °F	200 °F to 220 °F	330 °F to 365 °F	170 °F to 190 °F	230 °F to 260 °F	140 °F to 150 °F
Stiffness	Moderate to high	Moderate to high	Moderate to high	High	Moderate to high	Moderate to high	Moderate to high

(*continued*)

Table 7.23 Continued

B.

Properties	Polypropylene Regular	Polypropylene Oriented	Polystyrene	Polysulfone	SAN (styrene acrylonitrile)	TPX[c]	PVC
Resistance to impact	Poor to good	Very good	Poor to good	Good to very good	Poor to good	Poor to good	Fair to good
Unit cost	Moderate	Moderate to high	Moderate	Very high	Moderate to high	High	Moderate to high
Typical uses	Drugs, cosmetics, syrups, juices, detergents, mouthwash, shampoo, IV solutions	Detergents, drugs, mouthwash, shampoo, house hold chemicals, IV solutions, liquid soaps	Dry drugs, petroleum, jellies, vitamins, spices	Laboratory ware, microwave cooking	Dry drugs	Laboratory ware, microwave cooking	Cosmetics, personal care, household chemicals, edible oils, vinegar

[a] Bottles made from any of these resins are available in opaque colors
[b] Polyethylene terephthalate
[c] Polymethylpentane
[d] Range given here is limiting temperature in normal use
[e] Ketones, esters, chlorinated and aromatic solvents: poor; aliphatic solvents: good

A compromise among properties, cost, and manufacturing process generally determines the material of construction. Selecting a plastic is very similar to selecting a metal. Even within one class, plastics differ because of varying formulations, just as steel compositions vary (tool steel, stainless steel, etc.).

As an initial step, the product designer must know and/or anticipate the conditions of use and the performance requirements of the product, considering such factors as life expectancy, size, condition of use, shape, color, strength, and stiffness. These end use requirements can be ascertained through market analysis, surveys, examinations of similar products, testing, and general experience. A clear definition of product requirements will often lead directly to choice of the material of construction. At times incomplete or improper product requirement analysis is the reason why a product fails.

Whether the product is a new model of an established commodity or a completely new development, a list can be made of the properties the material or group of materials to be employed must possess, and of those that are also most desirable. By reference to the relevant material properties and prices, an analysis can be made to determine the plastic most likely to be suitable from all requirements.

As reviewed within each one of the major classes of plastics (PE, PVC, PC, etc.), there are usually a very wide variety of specific formulations, each of which has slightly different properties and/or processing capabilities at various costs. Prices, too, will tend to vary depending on the supplier, the current state of the market, and the volume of plastic that the processor is prepared to purchase.

7.10.2 Preliminary Considerations

As a general rule, it is considered desirable to examine the properties of three or more materials before making a final choice. Material suppliers should be asked to participate in type and grade selection so that their experience is part of the input. The technology of manufacturing plastic materials, as with other materials (steel, wood, etc.), results in the same plastic compounds supplied from various sources generally not delivering the same results in a product. As a matter of record, even each individual supplier furnishes product under a batch number, so that any variation can be linked to the exact condition of the raw material production. Taking into account manufacturing tolerances of the plastics, plus variables of equipment and procedure, it becomes apparent that checking several types of materials from the same and/or from different sources is an important part of material selection.

Experience has proven that the so-called interchangeable grades of materials have to be evaluated carefully as to their effect on the quality of a product. An important consideration as far as equivalent grade of material is concerned is its processing characteristics. There can be large differences in properties of a product and test data if the processability features vary from grade to grade. It must always be remembered that test data have been obtained from simple and easy to process shapes and do not necessarily reflect results in complex product configurations.

7.10.3 Mechanical Properties

Most plastics are used to produce products because they have desirable mechanical properties at an economical cost. For this reason their mechanical properties may be considered the most important of all the physical, chemical, electrical, and other considerations for most applications. Thus, everyone designing with such materials needs at least some elementary knowledge of their mechanical behavior and how they can be modified by the numerous structural geometric shape factors that can exist in plastics.

Plastics have the widest variety and range of mechanical properties of all materials (Figs. 7.29 and 7.30). They vary from basically soft to hard, rigid solids. Many structural factors determine the nature of their mechanical behavior, such as whether a load occurs over the short term or the long term. As a rule, design is based on certain minimum strength or minimum deformation criteria.

Short-term testing is important in designs and for quality control of plastics to ensure the required properties of plastics used in production [9]. The short time data provides the designer information that permits comparisons of one material with another.

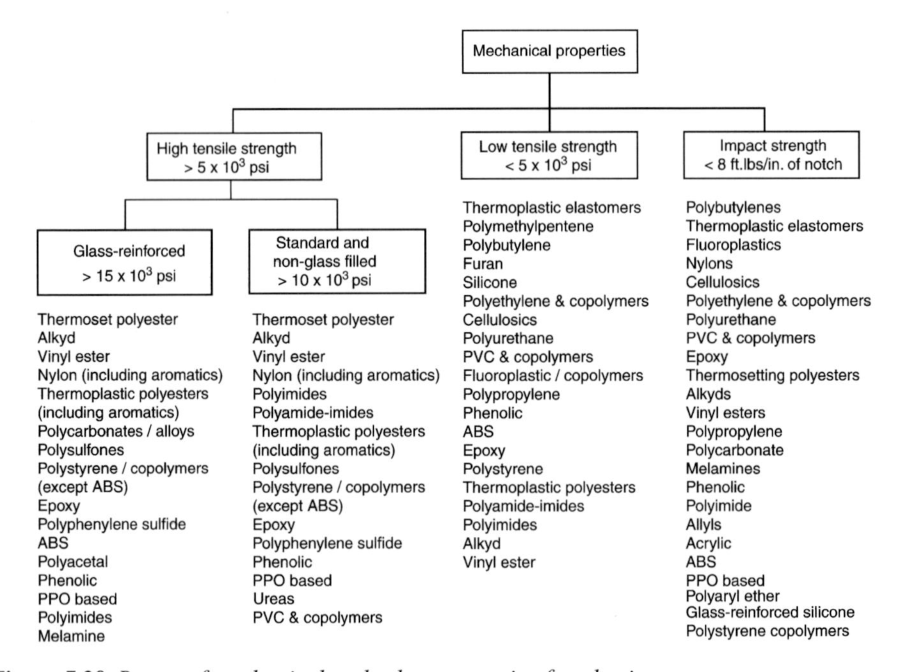

Figure 7.29 *Range of mechanical and other properties for plastics*

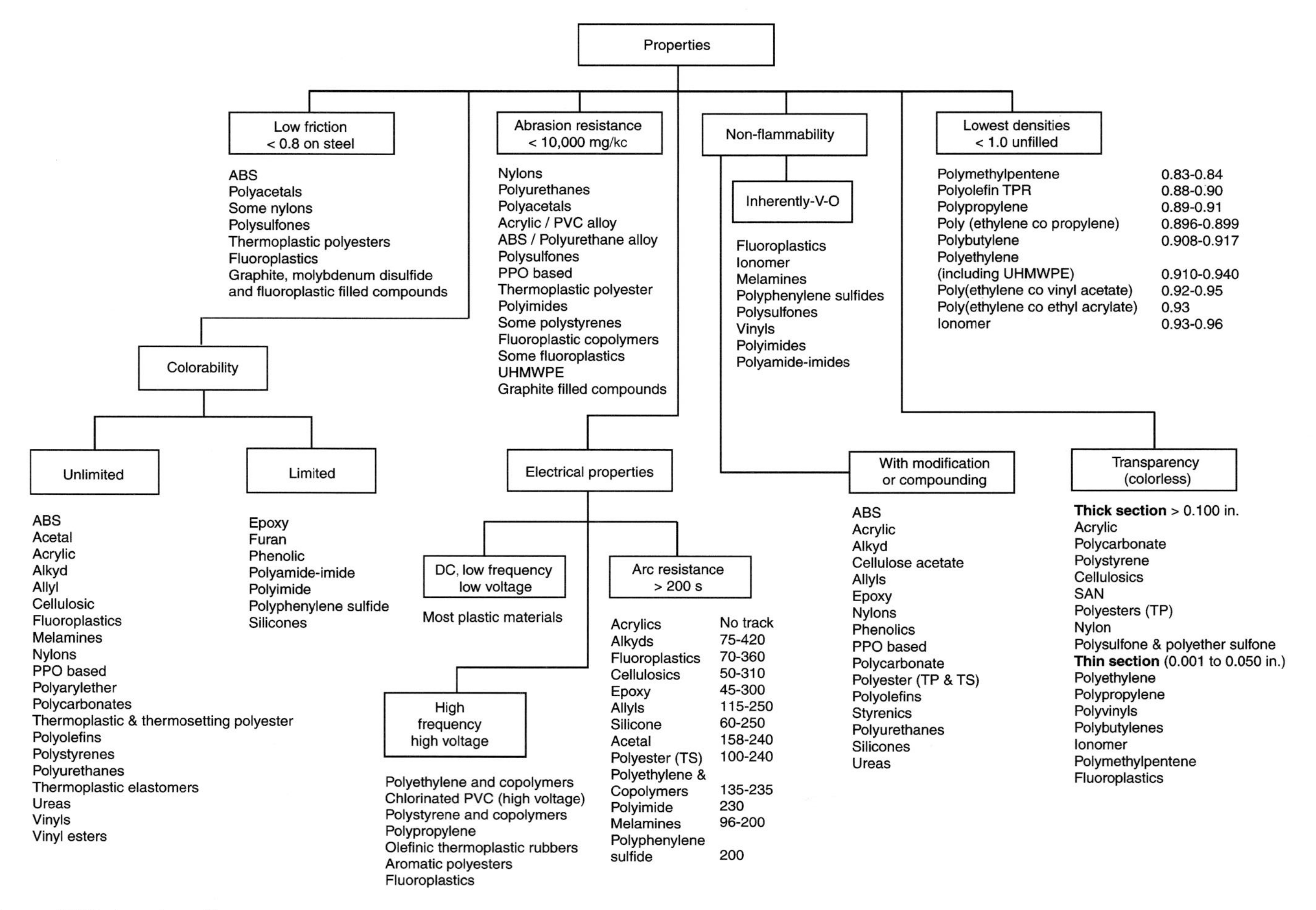

Figure 7.29 *(continued)*

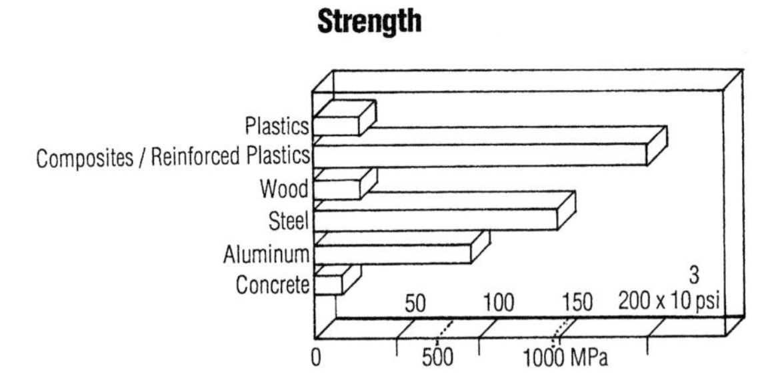

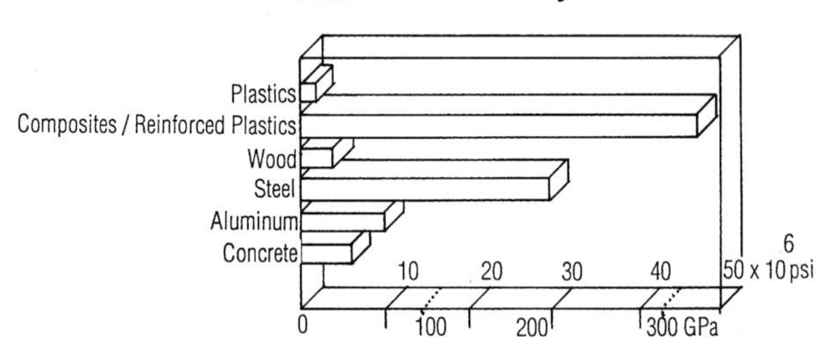

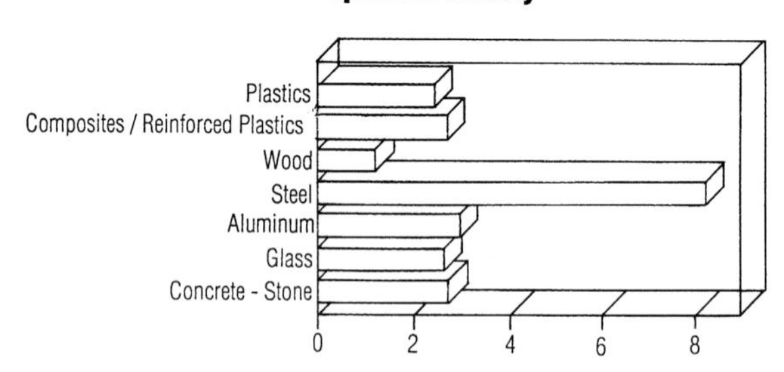

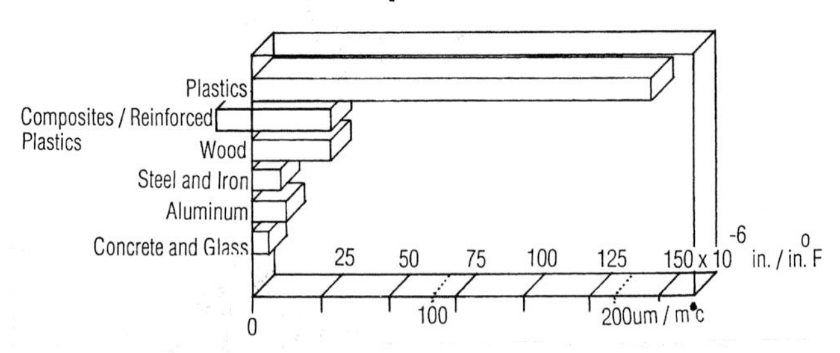

Figure 7.30 General comparison of different materials

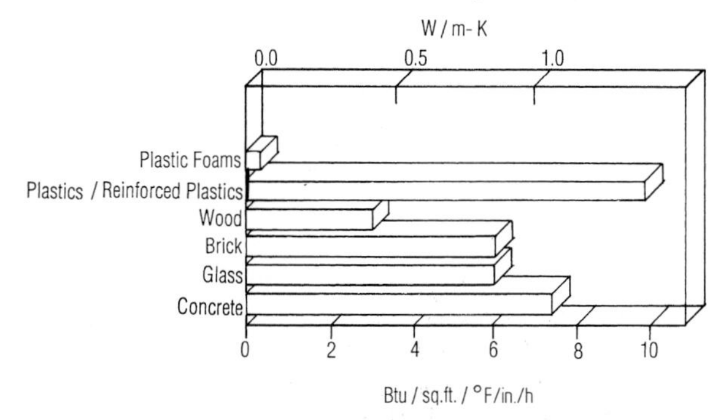

Maximum Continuous Service Temperature

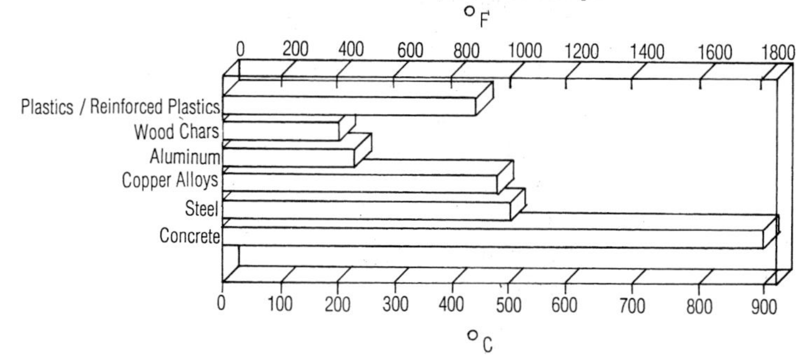

Figure 7.30 *(continued)*

However, a true comparison is possible only if both sets of data were determined in exactly the same way. For example, the speed of loading tensile test specimens influences performance factors such as deformation. Also, comparing the impact resistance of $\frac{1}{2}$ in. specimen with that of a $\frac{1}{8}$ in. specimen will result in a different analysis of the material's properties. Thus, it is necessary to describe the exact testing conditions along with each set of data sheets. The data from short-term testing give the user an important overall picture of the material.

The long-term testing of certain plastics allows their strength properties to be identified rapidly. Three of the major control test procedures for long-term testing and predicting product lifetime is creep, fatigue, and impact [9]. Figure 7.31 shows (a) long-term tensile creep curves at 20 °C (49 °F) for PP and nylon; numbers in parentheses refer to stress levels in MPa and (b) long-term tensile fatigue curves for dry nylon 6. Therefore, designing in this type product environment requires understanding all the variables so that the proper plastics can be used successfully with other materials that is 4.5 mm (0.18 in.) thick, acrylic (PMMA) that is 6.4 mm (0.25 in.) thick, and polytetrafluoroethylene (PTFE) that is 6.6 mm (0.26 in.) thick. The test frequency is at 1,800 cpm.

7.10.3.1 Tensile Property

It should be recognized that tensile properties would most likely vary with a change in speed of the testing machine pulling jaws and with variation in the atmospheric conditions. In a tensile test machine, the test specimen is held between the fixed and movable crossheads. When the speed of the pulling force is increased, the material reacts like brittle material; when the temperature is increased, the material reacts like ductile material.

The tensile data show the stress necessary to pull the specimen apart and the elongation prior to breaking. A moderate elongation (about 6%) of a test specimen generally implies that the material is capable of absorbing rapid impact and shock. The area under the stress–strain curve is indicative of overall toughness except for reinforced plastics [9, 10]. A material of very high strength, high rigidity, and little elongation would tend to be brittle in service. For applications where almost rubbery elasticity is desirable, a high ultimate (over 100%) elongation is an asset.

The tensile test and the calculated property data provide a most valuable source of information for the designer in determining product dimensions. The important consideration is that use conditions compare reasonably close to test conditions as far as speed of load, temperature, and moisture are concerned. Should use conditions differ appreciably, a test should be requested for data that are comparable to service requirements, which would thereby ensure that applicational needs will be based on more exacting data. It is to be noted that tensile data should be applied only to short-term stress conditions, such as operating a switch or shifting a clutch gear, and so on.

7.10.3.2 Impact Energy

Impact energy is a measure of toughness or impact strength of a material, for example, the energy needed to fracture a specimen in an impact test. It is the difference in kinetic energy of the striker before and after impact, expressed as total energy per inch of notch of the test specimen for plastic and electrical insulating material [in. lb (J/m)]. Higher energy absorption indicates a greater toughness. For notched specimens, energy absorption is an indication of the effect of internal multi-axial stress distribution on fracture behavior of the material. It is merely a qualitative index and cannot be used directly in design.

7.10.3.3 Hardness

Hardness is closely related to strength, stiffness, scratch resistance, wear resistance, and brittleness. The opposite characteristic, softness, is associated with ductility. The usual hardness tests are listed in three categories:

- to measure the resistance of a material to indentation by an indentor; some measure indentation with the load applied, and some measure the residual indentation after it is removed;

(a)

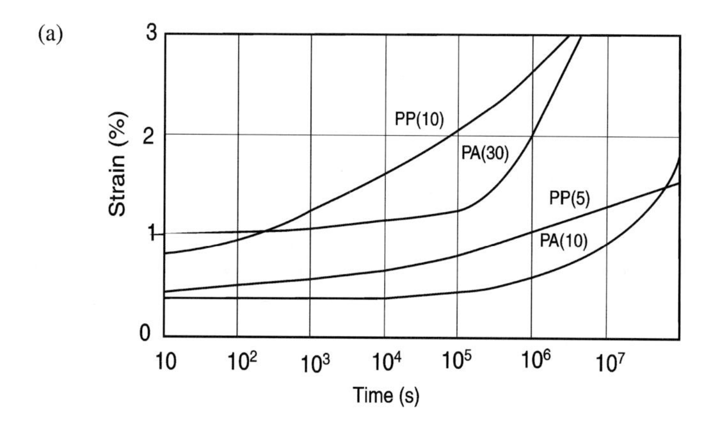

(b)

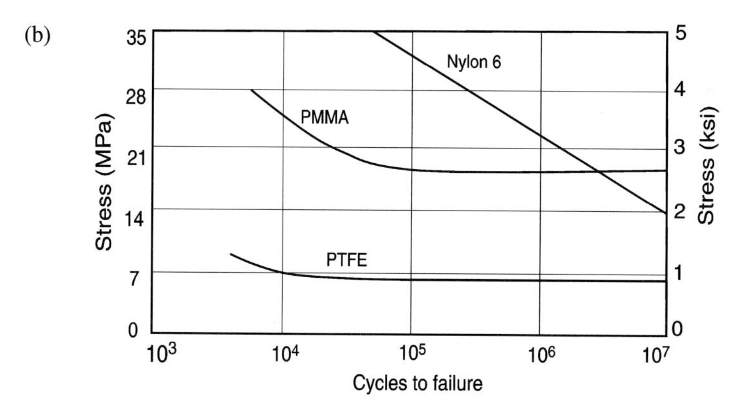

Figure 7.31 *(a) Example of long-term tensile creep curves and (b) long-term tensile fatigue curves*

- to measure the resistance of a material to scratching by another material or by a sharp point; and
- to measure rebound efficiency or resilience.

7.10.3.4 Fatigue Strength

The fatigue strength is defined as the stress level at which the test specimen will sustain N cycles prior to failure. The data are generated on a machine that runs usually at 1,800 cycles/min. This test is of value to material manufacturers in determining consistency of product and for products subjected to fatigue loads.

7.10.3.5 Long-Term Stress Relaxation/Creep

Tests such as those outlined by ASTM D2990 that describe in detail the specimen preparations and testing procedure are intended to produce consistency in observations and records by various manufacturers, so that they can be correlated to provide meaningful information to product designers.

The procedure under this heading is intended as a recommendation for uniformity of making setup conditions for the test, as well as recording the resulting data. The reason is the time consuming nature of the test (many years duration), which does not lend itself to routine testing. The test specimen can be round, square, or rectangular and manufactured in any suitable manner meeting certain dimensions. The test is conducted under controlled temperature and atmospheric conditions [9, 10].

7.10.4 Shrinkage

Shrinkage behavior of different TPs and the product geometry must be considered when fabricating BM products. Factors to consider are reviewed in Chapter 8.

The use of correct shrinkage information is very important, not only to ensure the desired proportions of a product, but also for functional purposes. The choice of shrinkage for a selected material and a specific product is the responsibility initially of the designer, but also, involves the mold or die designers and the fabricators. If experience with the selected grade of plastic is limited, the design should be submitted to the material supplier or someone knowledgeable for recommendations, and the data coordinated with the interested parties.

Where very close tolerances are involved, preparing a prototype of full-size product may be necessary to establish critical dimensions. If this step is not practical, it may be necessary to test a mold or die during various stages of cavity or die opening manufacture with allowances for correction to determine the exact shrinkage needed.

7.10.5 Thermal Property

To select materials that will maintain acceptable mechanical characteristics and dimensional stability, designers must be aware of both the normal and extreme thermal operating environments to which a product will be subjected. TS plastics have specific thermal conditions when compared to TPs that have various factors to consider that influence the product's performances and processing capabilities. The properties and processes of TPs are influenced by their thermal characteristics such as melt temperature (T_m), glass-transition temperature (T_g), dimensional stability, thermal conductivity, thermal diffusivity, heat capacity, coefficient of thermal expansion, and decomposition (T_d). Table 7.24 provide some of these data on different plastics.

The behaviors of plastics when subjected to heat and chemicals vary. There are those that are basically not affected when compared to other materials (steel, aluminum, wood, etc.). Table 7.25 provides information on the effects of chemicals when subjected to 77° and 200 °F.

Table 7.24 *Examples of Thermal Properties of TPs (Properties of Common Materials Included for Comparison)*

Plastics (morphology)	Density g/cm^3 (lb/ft^3)		Melt temperature T_m, °C (°F)		Glass transition temperature T_g °C (°F)		Thermal conductivity (10^{-4} cal/ s·cm °C) (BTU/lb °F)		Heat capacity cal/g °C (BTU/lb °F)		Thermal diffusivity 10^{-4} cm^2/s (10^{-3} ft^2/h)		Thermal expansion 10^{-6} cm/ cm °C (10^{-6} in./ in. °F)	
PP (C)	0.9	(56)	168	(334)	5	(41)	2.8	(0.068)	0.9	(0.004)	3.5	(1.36)	81	(45)
HDPE (C)	0.96	(60)	134	(273)	−110	(−166)	12	(0.290)	0.9	(0.004)	13.9	(5.4)	59	(33)
PTFE (C)	2.2	(137)	330	(626)	−115	(−175)	6	(0.145)	0.3	(0.001)	9.1	(3.53)	70	(39)
PA (C)	1.13	(71)	260	(500)	50	(122)	5.8	(0.140)	0.075	(0.003)	6.8	(2.64)	80	(44)
PET (C)	1.35	(84)	250	(490)	70	(158)	3.6	(0.087)	0.45	(0.002)	5.9	(2.29)	65	(36)
ABS (A)	1.05	(66)	105	(221)	102	(215)	3	(0.073)	0.5	(0.002)	3.8	(1.47)	60	(33)
PS (A)	1.05	(66)	100	(212)	90	(194)	3	(0.073)	0.5	(0.002)	5.7	(2.2)	50	(28)
PMMA (A)	1.20	(75)	95	(203)	100	(212)	6	(0.145)	0.56	(0.002)	8.9	(3.45)	50	(28)
PC (A)	1.20	(75)	266	(510)	150	(300)	4.7	(0.114)	0.5	(0.002)	7.8	(3.0)	68	(38)
PVC (A)	1.35	(84)	199	(390)	90	(194)	5	(0.121)	0.6	(0.002)	6.2	(2.4)	50	(128)
Aluminum	2.68	(167)	1000				3000	(72.5)	0.23		4900	(1900)	19	(10.6)
Copper/bronze	8.8	(549)	1800				4500	(109)	0.09		5700	(2200)	18	(10)
Steel	7.9	(493)	2750				800	(21.3)	0.11		1000	(338)	11	(6.1)
Maple wood	0.45	(28.1)	400	(burns)			3	(0.073)	0.25		27	(10.5)	60	(33)
Zinc alloy	6.7	(418)	800				2500	(60.4)	0.10		3700	(1430)	27	(15)

C = Crystalline resin, A = amorphous resin

Table 7.25 *Effects of Temperature and Chemical Agents on the Stability of Plastics (1 Equals Best)*

	Aromatic solvents		Aliphatic solvents		Chlorinated solvents		Weak bases and salts		Strong bases		Strong acids		Strong oxidants		Esters and ketones		24-h water absorption
	Temperature °F																
Plastic material	77	200	77	200	77	200	77	200	77	200	77	200	77	200	77	200	% change by weight
Acetals	1–4	2–4	1	2	1–2	4	1–3	2–5	1–5	2–5	5	5	5	5	1	2–3	0.22–0.25
Acrylics	5	5	2	3	5	5	1	3	2	5	4	4–5	5	5	5	5	0.2–0.4
Acrylonitrile–butadiene–styrenes (ABS)	4	5	2	3–5	3–5	5	1	2–4	1	2–4	1–4	5	1–5	5	3–5	5	0.1–0.4
Aramids (aromatic polyamide)	1	1	1	1	1	1	2	3	4	5	3	4	2	5	1	2	0.6
Cellulose acetates (CA)	2	3	2	3	3	4	2	3	3	5	3	5	3	5	5	5	2–7
Cellulose acetate butyrates (CAB)	4	5	1	3	3	4	2	4	3	5	3	5	3	5	5	5	0.9–2.0
Cellulose acetate propionates (CAP)	4	5	1	3	3	4	1	2	3	5	3	5	3	5	5	5	1.3–2.8
Diallyl phthalates (DAP, filled)	1–2	2–4	2	3	2	4	2	3	2	4	1–2	2–3	2	4	3–4	4–5	0.2–0.7
Epoxies	1	2	1	2	1–2	3–4	1	1–2	1	2	2–3	3–4	4	4–5	2	3–4	0.01–0.10
Ethylene copolymers (EVA) (ethylene-vinyl acetates)	5	5	5	5	5	5	1	2	1	5	1	5	1	5	2	5	0.05–0.13
Fluorocarbons:																	
Ethylene/tetrafluoroethylene copolymers (ETFE)	1	1	1	1	1	1	1	1	1	1	1	1	1	1	1	1	< 0.03
Fluorinated ethylene propylenes (FEP)	1	1	1	1	1	1	1	1	1	1	1	1	1	1	1	1	< 0.01
Perfluoroalkoxies (PFA)	1	1	1	1	1	1	1	1	1	1	1	1	1	1	1	1	< 0.03

(*continued*)

Table 7.25 (continued)

Plastic material	Aromatic solvents		Aliphatic solvents		Chlorinated solvents		Weak bases and salts		Strong bases		Strong acids		Strong oxidants		Esters and ketones		24-h water absorption
	Temperature °F																
	77	200	77	200	77	200	77	200	77	200	77	200	77	200	77	200	% change by weight
Polychlorotrifluoroethylenes (CTFE)	1	1	1	1	3	4	1	1	1	1	1	1	1	1	1	1	0.01–0.10
Polytetrafluoroethylenes (TFE)	1	1	1	1	1	1	1	1	1	1	1	1	1	1	1	1	0
Furans	1	1	1	1	1	1	2	2	2	2	1	1	5	5	1	1	0.01–0.20
Ionomers	2	4	1	4	4	4	1	4	1	4	2	4	1	5	1	4	0.1–1.4
Melamines (filled)	1	1	1	1	1	1	2	3	2	3	2	3	2	3	1	2	0.01–1.30
Nitriles (high barrier alloys of ABS or SAN)	1	4	1	2–4	1–4	2–5	1	2–4	1	2–4	2–5	5	3–5	5	1–5	5	0.2–0.5
Nylons	1	1	1	1	1	2	1	2	2	3	5	5	5	5	1	1	0.2–1.9
Phenolics (filled)	1	1	1	1	1	1	2	3	3	5	1	1	4	5	2	2	0.1–2.0
Polyallomers	2	4	2	4	4	5	1	1	1	1	1	3	1	4	1	3	< 0.01
Polyamide-imides	1	1	1	1	2	3	1	1	3	4	2	3	2	3	1	1	0.22–0.28
Polyarylsulfones (PAS)	4	5	2	3	4	5	1	2	2	2	1	1	2	4	3	4	1.2–1.8
Polybutylenes (PB)	3	5	1	5	4	5	1	2	1	3	1	3	1	4	1	3	< 0.01–0.3
Polycarbonates (PC)	5	5	1	1	5	5	1	5	5	5	1	1	1	1	5	5	0.15–0.35
Polyesters (thermoplastic)	2	5	1	3–5	3	5	1	3–4	2	5	3	4–5	2	3–5	2	3–4	0.06–0.09
Polyesters (thermoset-glass fiber filled)	1–3	3–5	2	3	2	4	2	3	3	5	2	3	2	4	3–4	4–5	0.01–2.50
Polyethylenes (LDPE-HDPE—low-density to high-density)	4	5	4	5	4	5	1	1	1	1	1–2	1–2	1–3	3–5	2	3	0.00–0.01

(continued)

Table 7.25 (continued)

Plastic material	Aromatic solvents		Aliphatic solvents		Chlorinated solvents		Weak bases and salts		Strong bases		Strong acids		Strong oxidants		Esters and ketones		24-h water absorption
	Temperature °F																
	77	200	77	200	77	200	77	200	77	200	77	200	77	200	77	200	% change by weight
Polyethylenes (UHMWPE—ultrahigh molecular weight)	3	4	3	4	3	4	1	1	1	1	1	1	1	1	3	4	< 0.01
Polyimides	1	1	1	1	1	1	2	3	4	5	3	4	2	5	1	1	0.3–0.4
Polyphenylene oxides (PPO) (modified)	4	5	2	3	4	5	1	1	1	1	1	2	1	2	2	3	0.06–0.07
Polyphenylene sulfides (PPS)	1	2	1	1	1	2	1	1	1	1	1	1	1	2	1	1	< 0.05
Polyphenylsulfones	4	4	1	1	5	5	1	1	1	1	1	1	1	1	3	4	0.5
Polypropylenes (PP)	2	4	2	4	2–3	4–5	1	1	1	1	1	2–3	2–3	4–5	2	4	0.01–0.03
Polystyrenes (PS)	4	5	4	5	5	5	1	5	1	5	4	5	4	5	4	5	0.03–0.60
Polysulfones	4	4	1	1	5	5	1	1	1	1	1	1	1	1	3	4	0.2–0.3
Polyurethanes (PUR)	3	4	2	3	4	5	2–3	3–4	2–3	3–4	2–3	3–4	4	4	4	5	0.02–1.50
Polyvinyl chlorides (PVC)	4	5	1	5	5	5	1	5	1	5	1	5	2	5	4	5	0.04–1.00
Polyvinyl chlorides-chlorinated (CPVC)	4	4	1	2	5	5	1	2	1	2	1	2	2	3	4	5	0.04–0.45
Polyvinylidene fluorides (PVDF)	1	1	1	1	1	1	1	1	1	2	1	2	1	2	3	5	0.04
Silicones	4	4	2	3	4	5	1	2	4	5	3	4	4	5	2	4	0.1–0.2
Styrene acrylonitriles (SAN)	4	5	3	4	3	5	1	3	1	3	1	3	3	4	4	5	0.20–0.35
Ureas (filled)	1	3	1	3	1	3	2	3	2	3	4	5	2	3	1	2	0.4–0.8
Vinyl esters (glass fiber filled)	1	3	1–2	2–4	1–2	4	1	3	1	3	1	2	2	3	3–4	4–5	0.01–2.50

All these thermal properties relate to determining the best useful processing conditions for meeting product performance requirements. There is a maximum temperature or, to be more precise, a maximum time-to-temperature relationship for all materials preceding loss of performance or decomposition. The effects of temperature on plastics are discussed throughout this book. There is also the effect of heat during recycling plastics [74].

An examination of the effect of temperature on the modulus of elasticity (also viscosity) of a typical TP is shown in Fig. 7.32. As the temperature is increased the plastic changes through different stages from a rigid solid to a liquid through the stages of being glassy, in transition, rubbery, and flow. Figure 7.33 shows the effect of the degree of crystallinity of TPs on modulus vs. temperature.

The process of heating and cooling TPs can be repeated indefinitely by granulating scrap, defective products, and so on. During the heating and cooling cycles of injection, extrusion, and so on, the material develops a "time at heat" history or residence time. With only limited repeating of the recycling, the properties of certain plastics are not significantly affected by residence time. However, some TPs can significantly lose certain properties. If incorrect methods were used in granulating recycled material, more degradation will occur.

Dimensional stability is an important thermal property for the majority of plastics. It is determined by the temperature above which plastics lose their dimensional stability. For most plastics, the main determinant of dimensional stability is their T_g. Only with highly crystalline plastics is T_g not a limitation.

Substantially crystalline plastics in the range between T_g and T_m are referred to as leathery, because they are made up of a combination of rubbery noncrystalline regions and stiff crystalline regions. The result is that such plastics as PE and PP are still useful at room temperature and nylon is useful to moderately elevated temperatures even though those temperatures may be above their respective T_g.

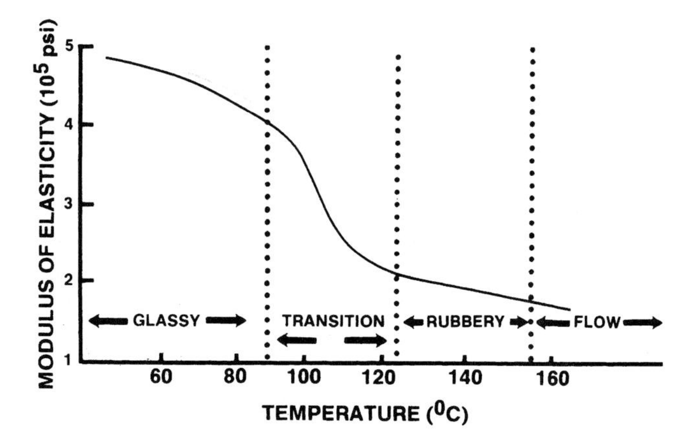

Figure 7.32 *Example of TP modulus during different temperature phases*

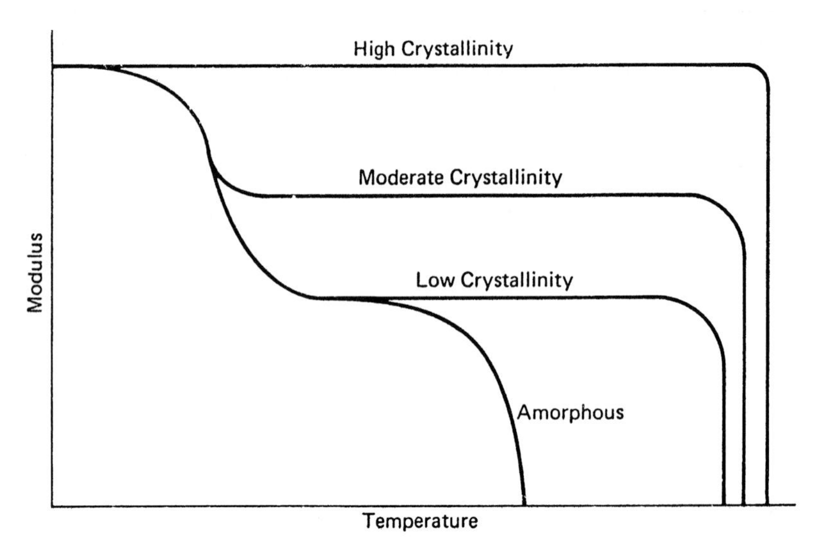

***Figure** 7.33 Effect of crystallinity on modulus vs. temperature*

7.10.5.1 Thermal Aging

Section UL 746B provides a basis for selecting high-temperature plastics and provides a long-term thermal aging index, the relative thermal index (RTI). The testing procedure calls for test specimens in selected thicknesses to be oven aged at certain elevated temperatures (usually higher than the expected operating temperature, to accelerate the test), then removed at various intervals and tested at room temperature.

Another reason for using higher temperatures is that, for an application requiring long-term exposure, a candidate plastic is often required to have an RTI value higher than the maximum application temperature. The properties tested can include mechanical strength, impact resistance, and electrical characteristics. The RTI of a plastic is based on the temperature at which it still retains 50% of its original properties. In general, these RTIs are very conservative and can be used as safe continuous-use temperatures for low-load mechanical products.

7.10.6 Plastic Memory

TPs can temporarily assume a deformed shape but always maintain internal stresses that tend to force the material back to its original shape. This tendency to change back is called plastic memory. Plastic memory is reviewed in Section 8.6.1.

7.10.7 Coefficient of Linear Thermal Expansion

Like metals, plastics generally expand when heated and contract when cooled. Usually, for a given temperature change, many TPs have a greater change than metals. The coefficient of linear thermal expansion (CLTE) is the ratio between the change of a linear dimension to the original dimension of the material per unit change in temperature (per ASTM standards). It is generally given as cm/cm/°C or in./in./°F.

The CLTE is an important consideration if dissimilar materials such as one plastic to another or a plastic to metal and so forth are to be assembled where material expansion or contraction is restricted. The CLTE is influenced by the type of plastic (liquid crystal, for example) and RP, particularly the glass fiber content and its orientation. It is especially important if the temperature range includes a thermal transition such as T_g. Normally, all this activity with dimensional changes is readily available from material suppliers to let the designer apply a logical approach and understand the results.

The design of products has to take into account the dimensional changes that can occur during fabrication and during its useful service life. With a mismatched CLTE there could be destruction of plastics from factors such as cracking or buckling.

Expansion and contraction can be controlled in plastic by its orientation, crosslinking, adding fillers or reinforcements, and so on. With certain additives, the CLTE value could be zero or near zero. For example, plastic with a graphite filler contracts rather than expands during a temperature rise. RPs with only glass fiber reinforcement can be used to match those of metal and other materials. In fact, TSs can be specifically compounded to have little or no change.

In a TS, the ease or difficulty of thermal expansion is dictated for the most part by the degree of crosslinking as well as the overall stiffness of the units between the crosslinks. The less flexible units are also more resistant to thermal expansion. Such influences as secondary bonds have much less effect on the thermal expansion of TSs. Any crosslinking of TPs has a substantial effect. With the amorphous type, expansion is reduced. In a crystalline TP, however, the decreased expansion as a result of crosslinking may be partially offset by a loss of crystallinity.

7.10.8 Thermal Stress

If a plastic product is free to expand and contract, its thermal expansion property will usually have little significance. However, if it is attached to another material, one having a lower CLTE, then the movement of the part will be restricted. A temperature change will then result in developing thermal stresses in the product. The magnitude of these stresses will depend on the temperature change, the method of attachment and relative expansion, and the modulus characteristics of the two materials at the point of the exposed heat.

The goal is to eliminate or significantly reduce all sources of thermal stress. This can be achieved by keeping the following factors in mind:

- when adding material for local reinforcement, select a material with the same or a similar CLTE;

- where plastic is to be attached to a more rigid material, use mechanical fasteners with slotted or oversized holes to permit expansion and contraction to occur;
- do not fasten dissimilar materials tightly; and
- adhesives that remain ductile, such as urethane and silicone, through the product's expected end-use temperature range can be used without causing stress cracking or other problems [2, 89].

7.10.9 Property and Moisture

In addition to dimensional changes from changes in temperature, other types of dimensional instability are possible in plastics as in other materials. Water-absorbing plastics, such as certain nylons, may expand and shrink as they gain or lose water, or even as the relative humidity changes. The migration or leaching of plasticizers, as in certain PVCs, can result in slight dimensional change.

The effect of having excess moisture can manifest itself in various ways depending on the plastic and process being employed. The common result is a loss in both mechanical and physical properties for hygroscopic or nonhygroscopic plastics (Fig. 7.34). During injection molding sprays, nozzle drool between controlled shot size, sinks, and other losses may occur. The effects during extrusion can include gels, trails of gas bubbles in the extrudate, arrowheads, wave forms, surging, lack of size control, and poor appearance.

Moisture or water absorption is an important design property. It is particularly significant for a product that is used in conjunction with other materials that call for fits and clearances along with other close tolerance dimensions.

The moisture content of a plastic affects such conditions as electrical insulation resistance, dielectric losses, mechanical properties, dimensions, and appearance. The effect of moisture content on the properties depends largely on the type of exposure (by immersion in water or by exposure to high humidity), the shape of the product, and the inherent behavioral properties of the plastic material. The ultimate proof for

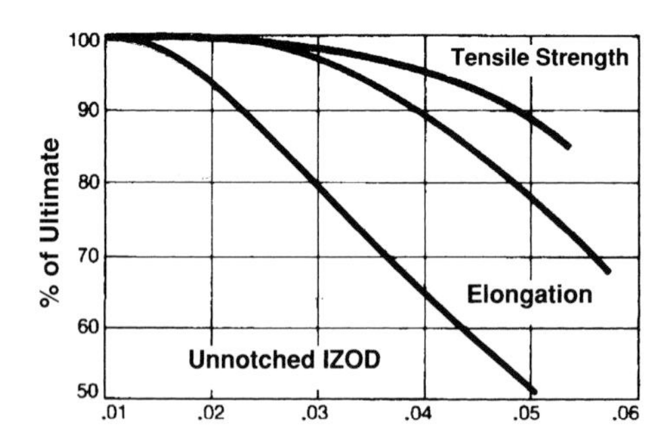

Figure 7.34 *Example of the effects of moisture on the mechanical properties of a hygroscopic PET plastic*

tolerance of moisture in a product has to be a product test under extreme conditions of usage in which critical dimensions and needed properties are verified. Plastics with very low water-moisture absorption rates tend to have better dimensional stability.

Water absorption data should indicate the temperature and time of immersion and the percentage of weight gain of a test specimen. The same applies to data at the saturation point of 73.4 °F (23 °C), and, if the material is usable at 212 °F (100 °C), also to saturation at this temperature.

7.10.9.1 Water Vapor Permeability

There are substantial differences in the rates at which water vapor and other gases can permeate different plastics. For instance, PE is a good barrier for moisture or water vapor, but other gases can permeate it rather readily. Nylon, on the other hand, is a poor barrier to water vapor but a good one to other vapors. The transmission rates are influenced by such different factors as pressure and temperature differentials on opposite sides of the film.

The effectiveness of a vapor barrier can be rated in a term such as perms. A rating of 1 perm means that one 1 ft^2 of the barrier is penetrated by 1 g of water vapor per hour under a pressure differential of 1 in. Hg. One inch of mercury equals virtually 0.5 psi; 1 g is 0.007 lb.

The material to be tested is fastened over the mouth of a dish that contains either water or a desiccant. This assembly is placed in an environment of constant humidity and temperature. The gain or loss in weight of the assembly is used to calculate the rate of water vapor movement through the specimen under prescribed conditions of humidity inside and outside of the dish. The results are reported in grams per 100 square inches over 24 h, or equivalent metric units.

Over time, all plastic materials allow a certain amount of water vapor, organic gas, or liquid to permeate the thickness of the material. It has been found that the permeability coefficient is a function of the solubility coefficient and diffusion coefficient. The process of permeation is explained as the solution of the vapor into the incoming surface of the barrier, followed by diffusion through the barrier thickness, and evaporation on the exit side.

Different factors influence permeation:

- The composition of the barrier, including additives, fillers, colorants, plasticizers, etc. Even when data on permeability are available for a specific industry grade of plastic, they cannot be used for evaluation because different commercial grades can contain ingredients that can change the values.
- Crystalline plastics are better vapor barriers than amorphous plastics. Also, TS plastics have better barrier properties than TPs, especially when the fillers do not absorb moisture.

- An increase in temperature brings about an increase in permeability. In addition, an increase in vapor pressure of the permeating agent also causes acceleration of transmission.
- Product thickness is inversely proportional to permeation, that is, with double the thickness, there is one half of the evaporation.
- Coatings such as epoxy-based finishes will improve resistance to permeation.
- In the case of organic vapors, the permeation will depend not only on the composition of the barrier, but also, on the molecular configuration of both the barrier material and the permeating agent.

7.10.10 Degradation Problem

Details on cause/solution are provided in Chapter 12. Degraded resin can result from a variety of different sources or conditions.

7.10.11 Checking Plastic Received

An important factor in the production of products is that the quality of the raw materials always conforms to specification. Certain properties must be checked when the goods are received. In view of the wide variety of applications for plastic products, a testing schedule of general validity should be set.

Testing yields basic information about material properties, its quality with reference to standards or material inspections, and can be applied to designing products with plastics.

Variations exist with the same basic types of plastic from different manufacturers and even in different batches from the same manufacturer. Taking into account manufacturing tolerances of the plastic, plus variables of equipment and procedure, it becomes apparent that checking several types of materials from the same or different sources is an important part of material selection.

The preliminary check must proceed without loss of time. For this reason, rapid tests with specific aims are frequently used. Since molding has been caught up in the automation trend, it is feasible that checking the goods received will become part and parcel of the actual production process. However, this entails that any deviations from standard must remain within narrow limits. This procedure is called adaptive process control.

7.10.11.1 Sampling

It is important to obtain a representative portion of the material or product to be tested. The number of samples required generally is specified for each test to obtain a reasonably reliable test value. Information on variation test values, sample-to-sample, and other sampling procedures are given in various documents such as ASTM D 2188 (American Standards for Testing Material).

7.10.11.2 Acceptable Quality Level

The acceptable quality level (AQL) is the most important part of the sampling standard (ISO-2859 (International Organization for Standardization)) because the AQL and the sample size code letter indexes the sampling plan. AQL is defined as the maximum percent defective, or the maximum number of defects per hundred units, that, for purposes of sampling inspection, can be considered satisfactory as a process average. The phrase "can be considered satisfactory" is interpreted as a producer's. When the standard is used for percent defective plan, AQLs from 0.010% to 10.0% are possible. For defective-per-unit plans, there are additional AQLs, so AQLs from 0.010 defect per 100 units to 1,000 defects per 100 units are possible. The AQLs are in a geometric progression, each being approximately 1.585 times the preceding one. AQLs are determined from: (1) historical data; (2) empirical judgment; (3) engineering information, such as function, safety, interchangeable manufacturing, etc.; and (4) experimentation.

7.10.11.3 Sampling Size

The lot size and the inspection level per sampling plan determine the sample size. The inspection level to be used for a particular requirement will be prescribed by the responsible authority. The decision on the inspection level is also a function of the type of product. For inexpensive products, for descriptive testing, or for harmful testing, different inspection levels are considered.

7.10.11.4 Testing

There are destructive and nondestructive tests (NDTs). Most important, they are essential for determining the performance of plastic materials to be processed and of the finished fabricated products. Testing refers to the determination of properties and performances by technical means, applying established scientific principles and procedures.

In the familiar form of testing known as destructive testing, the original configuration of a test specimen and/or product is changed, distorted, or usually destroyed. The test provides information such as the amount of force that the material can withstand before exceeding its elastic limit and permanently distorts (yield strength) or the amount of force needed to break it. These data are quantitative and can be used to design structural products to withstand a certain load, heavy traffic usage, etc.

NDT examines material without distorting the specimen or impairing its ultimate usefulness. NDT allows suppositions about the shape, severity, extent, distribution, and location of such internal and subsurface residual stresses, as well as, defects such as voids, shrinkage, cracks, etc. Test methods include acoustic emission, radiography, infrared (IR) spectroscopy, X-ray spectroscopy, magnetic resonance spectroscopy, ultrasonic, liquid penetrant, photoelastic stress analysis, vision system, holography, electrical analysis, magnetic flux field, manual tapping, microwave, and birefringence.

As reviewed throughout this book, the properties of plastics tend to be strongly affected by temperature; therefore, it is important to specify the temperature of the material during a test. Also important is moisture. A number of plastics (nylons, cellulosics, etc.) can absorb small amounts of water (moisture), which affects many of their properties. Plastics have a low thermal conductivity and the movement of water molecules in plastics is also slow. Reaching thermal and moisture equilibrium can therefore be a relatively long process (1 to 4 days), which is referred to as "test specimen conditioning."

7.11 Evaluation of Plastic Processability

The selection of a plastic type for a given BM application is determined primarily by the functional properties of the desired product and cost (Table 7.26). Requirements such as permeation, solvent, or temperature resistance must be met first; the rheological properties are important for defining the process to make the part. Having selected the plastic type, rheological characterization is needed to select (or prepare) the optimum grade of that type. Once this has been done successfully, it is then necessary to test the plastic as produced or received from the supplier to ensure that it is sufficiently uniform from lot to lot to be processible with a minimum of machine adjustment or nonspecification products [47].

The distinction between the types of characterization tests needed for the initial plastic selection and for quality control purposes cannot be overemphasized, and they are therefore considered separately in this section. The initial selection process may well involve a lengthy research program, using sophisticated instruments that can only be operated by skilled technicians. The quality control tests, on the other hand, must be done economically on a routine basis, using simple instruments capable of giving reliable results in the hands of relatively unskilled operators.

Table 7.26 *Examples of Data for Choosing BM Plastics*

Polymer	Appearance in blow molded form	Cost for equivalent stiffness (PE-LD = 100)
High-impact PVC-U	Clear high gloss	104
PE-LD	Transculent, high gloss	100
Low-impact PVC-U	Very clear, high gloss	87
High-impact PP	Rather opaque, low gloss	79
PE-HD	Rather opaque, low gloss	76
Low-impact PP	Transculent, moderate gloss	71

Another point is worth making. It is not always true that the source of a problem encountered during processing, even though it makes its appearance in the melt flow stage, reflects a variation in rheological properties. For example, one might observe a variation with time of the uniformity of the melt produced by an extruder. This might reflect a variation in the melt viscosity. On the other hand, it could also result from a variation in the pellet size distribution if the extruder is being operated near the margin of its plasticating capability, and if the pellet size affects the behavior in the solid conveying zone of that extruder. Diagnosis of such a problem can be a difficult and frustrating exercise.

7.11.1 Plastic Selection Test

Because there is such a diversity of BM processes and of shapes and sizes of blow molded articles it is not feasible to give cookbook recipes for defining the rheology required for a given situation. If there is one single process parameter that is of overwhelming importance, resistance to parison sag, for example, then selection of the appropriate rheological property to measure is straightforward. More commonly, a number of process requirements must be met simultaneously, and the rheology suitable for one requirement may conflict with that for another. Ideally, one would design the machinery and tooling to match the plastic characteristics and optimize all of the process requirements. If it is necessary to use existing equipment, relying only on the range of operating conditions that can be adjusted for that equipment, the problem becomes one of defining the rheology that can accommodate all of the requirements.

An example of one way to handle such a situation is to adjust only the shot pressure and the die gap. It is necessary to select a plastic that would simultaneously meet specifications for parison diameter, bottle weight, melt fracture, and pleating, but some of the requirements are in conflict. Decreasing the die gap, for example, increased parison diameter, but this could lead to unacceptable curtaining. An understanding of the mechanics of each stage of the process is useful for defining the effects of the rheological characteristics; without such an understanding, empirical trial-and-error techniques are necessary.

It is useful to break the BM process down into component steps, each with its own requirements for matching the properties of the plastic to the configuration of the machinery: (1) plastication of melt in an extruder, (2) parison formation, and (3) parison shaping.

Plastication may be by continuous extrusion or by an intermittent process, like injection blow molding as described in Chapters 2 and 3. The extrusion process is described extensively in the literature, and computer programs for screw and die design are well known. This aspect of the BM process is covered in Chapters 1 to 5. With regard to the rheological requirements, it is necessary only to know the shear rate and temperature dependence of the viscosity. In principle, one also needs to know the effect of pressure on viscosity; in practice this is rarely measured but rather is estimated.

Injection molding is also a widely studied process, and the reader is referred to the specialized works [6] in this field. One consideration that is specifically applicable to injection blow molding (IBM) is that the preform is shaped to its final form after the IM. Because orientation and/or crystallization may be induced by the IM, and this may affect the subsequent shaping, it may be important to control the extent of orientation more closely than is normally required for many conventional injection molding operations. Low viscosity at mold filling shear rates and low melt elasticity tends to minimize orientation. The presence of long chain branching generally broadens MWD and thereby tends to cause shear thinning at relatively low shear rates, and this may be helpful. However, long chain branching can also increase significantly the rate of crystallization of an otherwise slowly crystallizing polymer, such as PET, in extensional flow.

Parison formation is, at the same time, the most critical and the most difficult to analyze step in extrusion blow molding. The rheology of parison formation is discussed in detail in a previous section. The specific rheological properties that must be considered for a given application depend on the parison geometry and on the machine used for the process. As a guide for test selection, Table 7.27 summarizes various properties and their influence on parison formation; these are discussed in some detail below.

It is reemphasized that many of the properties affect more than one aspect of parison formation. For example, the viscosity at the shear rate range corresponding to flow through the die obviously determines the pressure required to extrude the parison. If the extrusion rate is limited by the pressure capability of the machine, this limitation will determine the time required to form the parison. In turn, if this time is sufficiently long, excessive sag and curtaining may occur, and these will affect parison diameter, length, and appearance. Similarly, if the high shear rate viscosity is too high, the stress at the desired extrusion rate may exceed the critical stress for melt fracture of the plastic.

The viscosity at low stress, the drop time, and the parison length will determine how much sag occurs. The stress corresponds to the weight of the parison divided by its cross sectional area; that is, it is of the order (qgL), where q is the density of the melt,

***Table** 7.27* *Examples of the Influence of Properties on Parison Formation*

Property	Effect on parison formation
High shear rate viscosity	Extrusion pressure and/or rate (length, diameter and thickness, melt fracture, curtaining)
Low shear rate viscosity	Parison sag (length, diameter and thickness, curtaining)
Die swell	Parison length, diameter, thickness, curtaining
Critical stress	Melt fracture
"Relaxation" time	Sag, stretch orientation

g is the acceleration due to gravity, and L is the parison length. In principle the proper viscosity to consider is the extensional viscosity. Further, one should consider that the gravity deformation consists of a recoverable elastic deformation plus a permanent viscous flow. The relative contributions of these processes are determined by the elastic compliance of the melt and its relaxation time. If the drop time is long compared to the relaxation time, viscous flow is the major contribution; at shorter times, the elastic deformation dominates.

For practical purposes sag is probably most important in the extrusion of very long parisons. The drop time is then likely to be long compared to the melt relaxation time, so that the viscosity contributes more than the elasticity to the total sag. Also, at very low stresses, the extensional viscosity will be approximately equal to three times the shear viscosity. If one restricts the problem to the case of small sag, the case of the most interest, the maximum sag occurs near the middle of the parison, and is given by the quantity $(qgLt/2\acute{\eta})$, where t is the drop time and $\acute{\eta}$ is the shear viscosity at the stress qgL.

Die swell is one of the most important properties to be determined for any parison extrusion process. Die swell is not a property that can be predicted accurately from measurements of molecular structure or from other rheological measurements. Furthermore, it depends strongly on die geometry, extrusion rate, and time after extrusion. It must therefore be measured directly in deciding on a plastic for a given application or for designing tooling for a given plastic. The die swell should be measured under the conditions anticipated in the actual BM process, that is, at shear rates corresponding to die flow, with a capillary having an aspect ratio similar to that of the BM die, and over a range of times corresponding to the parison drop time (Chapter 3). The measurement of drop time may make it necessary to obtain apparatus specially designed for this purpose. A segmented or flat parison pinch-off mold may be attached to a conventional capillary rheometer. The more closely one can simulate the BM conditions, the better.

It should also be noted that die swell as measured with simple capillary may not give all of the information required. It is possible that two plastics with similar capillary swell may have different extents of diameter swell for a given weight swell. In many cases the discrepancy can be accommodated by adjusting the die gap, but if this is not possible or if control of parison diameter is critical, it may be necessary to measure diameter and area swell from an annular die.

Melt fracture is a problem that is likely to be encountered with very high viscosity plastics or in high shear rate processes. One can estimate the likelihood of encountering it from the die geometry, the flow rate, and the viscosity-shear rate dependence of the plastic. As a rule of thumb, the critical shear stress for melt fracture is on the order of 10^6 dynes/cm^2 (14.5 psi). If the stress in the die is estimated to approach this value, direct measurement of the critical stress by capillary rheometry is indicated.

Extensional viscosity has been touched on as possibly important with regard to sag. Much more important is its role in the parison inflation process, as this is a rapid extensional flow (Chapter 3). The extensional viscosity will determine the parison inflation rate and pressure requirement. Also, the dependence of extensional viscosity on time and strain rate will affect the stability of the inflation process. Extension thickening favors stability, because an initially thin spot is subjected to a higher than average stress and tends to stretch faster. But, if the extensional viscosity increases with the strain, this increase tends to counteract the increased stress and to allow the rest of the material to catch up with the thin spot.

Direct measurement of extensional viscosity is not a simple task. Recourse to the less rigorous methods, such as estimation from end correction or orifice flow, may be satisfactory, especially if there is a background of information on materials whose processing behavior is known.

To summarize, the minimum information that is needed for selecting a plastic for a BM application is the dependence of viscosity on shear rate at processing temperatures. For free parison processes, knowledge of the die swell, measured at the appropriate conditions, is essential. For very rapid processes and for those in which orientation is important, a measure of melt elasticity is desirable; combined with viscosity, this also determines the relaxation time scale. Extensional viscosity is useful for comparing inflation behavior of various polymers. Finally, from the viscosity, it is possible to estimate whether measurement of the critical stress for the melt fracture is necessary.

7.11.2 Quality Control Test

Once a suitable combination of plastic and machine have been selected for a given application, it is necessary to establish routine tests to assure continued satisfactory operation with a minimum of machine adjustments. Batch-to-batch variation is a fact of life. It is the function of a QC (quality control) program to make sure that a given lot of plastic does not fall outside the acceptable range of variation.

Among the requirements for a QC test are reliability, speed, and cost, both of the test equipment and its operation. Above all, however, the test must relate to a rheological property for which control is needed. The range of acceptability depends on the particulars of the process and must be defined for each case. Obviously the test must be sufficiently sensitive to this range.

Some of the plastic selection tests discussed measure a well-defined rheological property. This may be a simple, well understood property, whose value is independent of the instrument used for its measurement. Shear rate dependent viscosity is an example. This can be measured by capillary flow or by a rotational test, and if the tests are done properly the values are the same. This is not true for many of the QC tests in common use. For example, in the measurement of the melt index (MI), the die swell or the critical stress for melt fracture, the values may depend on the details of the test method, but if these are fixed, a characteristic material response can be determined.

It is not essential that a QC test be sensitive to only a single rheological property. It may well happen that only one of the properties that affect a given test fluctuates significantly during normal plastic production. In that case, these fluctuations will correlate with the results of the control test, and the control test, even though it is affected by more than one property, will still be useful in monitoring the fluctuations. The catch in this argument is that the resin manufacturer may change the production process, deliberately or inadvertently. If such a process change occurs, it is quite possible that two batches can have identical properties as measured by the QC test but respond very differently in processing.

7.11.2.1 Melt Index

These rather general statements will be made explicit by considering the melt index (MI) or melt flow index (MFI) test, one of the earliest and still one of the most widely used QC tests (Fig. 7.35). The MFI test meets many of the above requirements (cost, speed, reliability, and sensitivity) quite efficiently. As a measure of melt viscosity it suffers one major defect. As an example the aspect ratio (length-to-radius) of the die used for PE is only 7.65. This ratio is comparable in magnitude to the capillary flow end correction of many plastics. Therefore, if a change in the plastic manufacturing process causes a significant change of the end correction, a variation of the MFI may reflect not only a change in viscosity but also the change in the end correction.

Typically the end correction may be changed by alteration of the MWD and/or long chain branching of the polymer. These changes also affect the shear rate or shear stress dependence of viscosity. By measuring the MFI at two different loads, one would expect to see this different shear stress dependence. Clearly, it is advantageous to

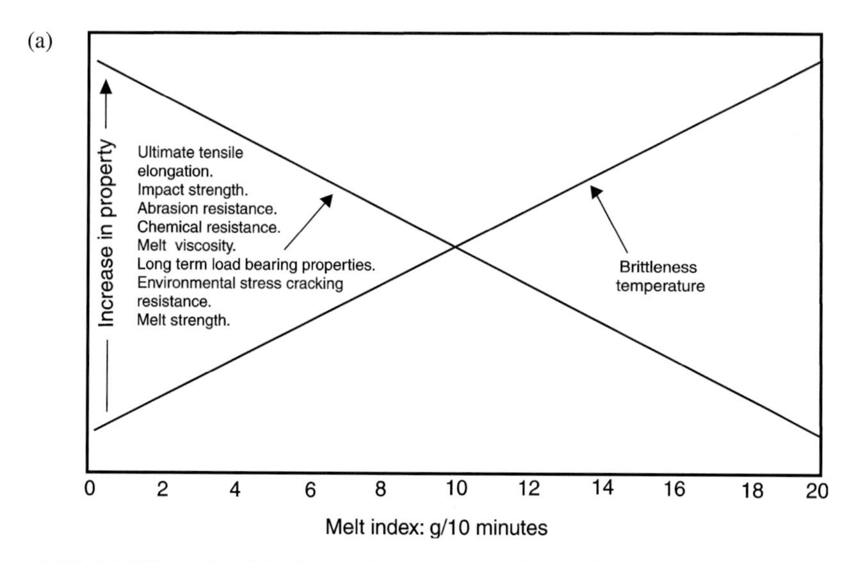

***Figure** 7.35 (a) Effect of melt index on the properties of polyethylene*

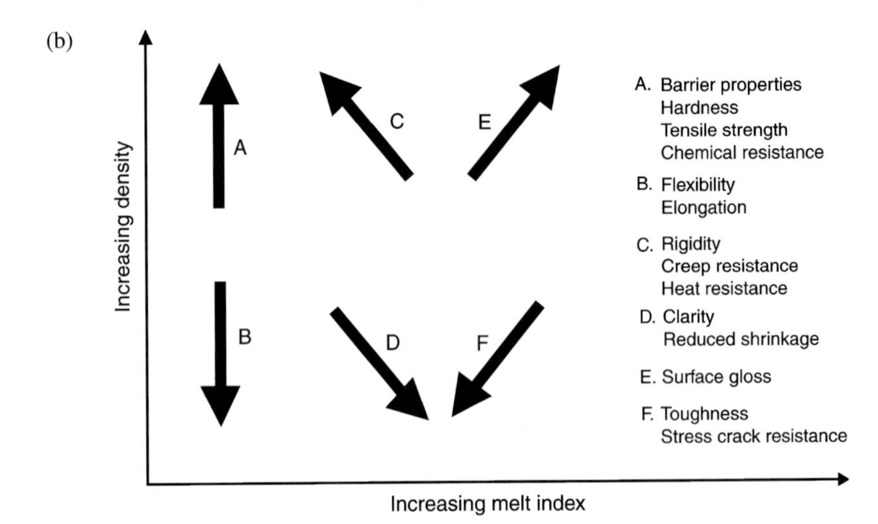

Figure 7.35 *(b) Effect of density and melt index changes on the properties of PE with properties increasing in the direction of the arrows*

measure the MFI at two loads rather than one, with the expectation that two plastics that are identical at the two conditions will in fact process similarly. However, there is no assurance that this will be the case. It is possible that for these two plastics the shear stress dependence of both the viscosity and of the end correction may compensate, in such a way, that the MFIs at both the loads tested are in fact equal but that the viscosities are different. If such a situation occurs, it can be resolved only by making corrected viscosity measurements, supplemented if possible by characterization of molecular structure by gel permeation chromatography, or some other technique.

An additional "free" bit of information one can obtain from the MFI test is a measure of die swell, simply by use of a micrometer or by weighing the extrudate. The test conditions are not as rigorously defined as one might wish if control of the die swell is crucial to the particular process. In particular, because die swell varies with shear rate, it will vary with the MFI of the resin. However, presumably the MFI is fairly closely controlled anyway, so this effect may not be important under QC conditions.

As reviewed, the extensional viscosity is difficult to measure for many plastics even under ideal laboratory conditions. If its control is important for a particular process, it will probably be useful to consider one of the nonrigorous tests that have been shown to correlate with it. A test that indicates the resistance of a melt to extension and is simple enough to be considered for QC purposes is the "melt strength" test. This test may also be useful for determining the ease of stripping a parison from a die. It has been found empirically to relate to an observed difference in the ratio of weight swell to diameter swell for two plastics that could not be distinguished rheologically by other tests. It should be noted, however, that this test is sensitive to many variables, including die swell, cooling rate, and the temperature dependence of viscosity.

Melt elasticity and relaxation time are not easily measured in steady shear, particularly at high shear rates, with instruments suited to QC purposes. Perhaps the best approach to their determination for control purposes is to make use of the analogy between steady and dynamic flows. Instruments are available to measure the dynamic modulus and viscosity over a wide range of frequencies [2]. Although these instruments are more expensive than a MI tester and require more care in their operation, the test can be operated fairly quickly. They are operated by microprocessor control and the results are available both graphically and numerically without need for operator computation. The dynamic tests are likely to be most useful for linear homogeneous polymers. The results on highly branched polymers or on multiphase polymers such as blends or grafts are likely to be affected by the degree of branching or the sizes of the phase domains, and may therefore be more difficult to interpret and use.

To conclude, here are some general principles or recommendations applicable to the choice and operation of all QC tests. First and foremost, the tests should be chosen to control the variables that an analysis of the process shows to be important, and to measure at conditions comparable to those of the process. In other words, for a high shear rate extrusion the viscosity at that shear rate is important, and a measurement at a much lower shear rate is irrelevant. To control parison sag, it is important to measure the viscosity at a stress corresponding to the length of the parison. The more closely the test conditions approach the processing conditions, the more likely the test results will be of value.

Second, it is important to isolate, measure, and control other variables that interact with the rheology. The effect of pellet size distribution on extrusion performance has been cited. In a process in which stress crystallization occurs, not only the melt rheology but also other factors such as the presence of nucleating agents may be relevant. Under this heading, we can also consider chemical effects. The rheological properties may be affected by thermal and/or oxidative reactions, or by hydrolysis. Condensation polymers (polyesters, polycarbonates, nylons, etc.) are susceptible to hydrolytic degradation. Proper drying and control of exposure to atmospheric moisture and oxygen are essential for these plastics. One must be aware of these possibilities, both for the operation of the process and for making the rheological measurements.

Last, maintenance and calibration of instruments is essential for reliable control testing (Chapter 12). Dies must be kept clean and their dimensions verified periodically. Transducers drift and require calibration, and so on. Perhaps the best control for the control test is periodic measurement, preferably daily, of one or more standard plastics that are kept especially for that purpose. Maintenance of a log of the results on such standards will reveal when the test equipment or procedure is going out of control.

7.11.2.2 Melt Index Testing

Different test methods are used to characterize plastics for melt flow that relates to shear melt processing. The major method used is the melt index. The more exact methods to improve quality and process control is the rheometer type. General characterization of flow behavior is offered by the steady shear test such as a capillary viscometer or a rotational rheometer. However, some tests are conducted with a variable force [3].

MFI testing uses basically a ram extrusion plasticator. This so-called rheological device is used for examining and studying TPs in many different fabricating processes. It is not a true viscometer, as a reliable value of the viscosity cannot be calculated from the flow index. However, it does measure isothermal resistance to flow using an apparatus and test method that is standard worldwide. In this instrument, the unmelted-solid plastic is contained in a "barrel" with a temperature indicator. An electrically controlled heater that melts the plastic basically surrounds it. A weight drives a plunger that forces the melt through the die opening; the orifice is usually 0.0825 in. (0.2096 cm) in diameter when subjected to a force of 2,160 g at 190 °C (374 °F). The usual procedure involves the determination of the amount of plastic extruded in 10 min (after initial flow starts). The flow rate is expressed in g/10 min. More than one test is conducted and an average is reported. As the flow rate increases, the viscosity decreases (Chapter 6).

This MI makes a single point test that provides information on the resistance to flow only at a single shear rate. Because variations in branching or molecular weight distribution (MWD) of the plastic can alter the shape of the viscosity curve, the MI may give a false ranking of plastics in terms of their shear rate resistance to melt flow. To overcome this situation, extrusion rates are sometimes measured for different loads and other modifications made to the instrument such as changing the size of the orifice.

7.11.2.3 Other Melt Flow Tests

The Bartender plasticorder rheometer is an instrument which continuously measures the torque exerted in shearing a TP, thermoset (TS), or compounded material over a wide range of shear rates and temperatures, including those conditions anticipated in actual manufacturing. It records torque, time, and temperature on a graph called a plastogram, providing information regarding the processability of the plastic. It shows the effects of additives and fillers including lubricity, plasticity, scorch, cure, shear, heat stability, plastic consistency, and so on.

The Canadian melt flow test, developed by the Canadian Industries, Ltd. (CIL), is a method of determining the rheology or flow properties of TPs. In this test, the amount of melt that is forced through a specified size orifice, per unit of time, when a specified, variable force is applied, gives a relative indication of the flow properties of the plastic.

The melt flow plastometer instrument is used for determining the flow properties of a TP by forcing the melt through a die or orifice of specific size, at a specified temperature and pressure.

The Rossi–Peakkes melt flow instrument is designed to measure the temperature at which a plastic will flow a definite amount through a special orifice when subjected to a prescribed constant pressure.

The spiral melt flow test method is used for finding flow properties of either TPs or TS plastics, formulated by the distance of flow under specified pressure and temperature along a spiral runner in a mold. The test is usually performed using injection molding for TPs. With TS plastics, a compression or transfer molding press may also be used.

The melt rheometer test is a “real” method to characterize the plastic’s melt behavior. Capillary melt rheometers are fully automated and cover a wide range of shear rates. When used on-line with extruders, many operating benefits are gained such as improving/ensuring quality, reducing scrap, and lowering costs. This instrument is far superior for understanding melt behavior than others.

7.12 Recycled Plastic

When plastics are granulated, the probability is its processability and performance when reprocessed into any product is reduced—could be significantly reduced. Thus, it is important to evaluate what the properties of the recycled material provides. The size reduction exerts a substantial influence on the quality of the recycled plastics. Recycled plastics material is usually nonuniform in size so that processing with or without virgin plastics is subject to operating in a larger fabricating process window.

Different approaches are used to improve performances or properties of mixed plastics such as:

- additives, fillers, and/or reinforcements (use specific types such as processing agent, talc, short glass fibers);
- active interlayers (crosslinking, molecular wetting); and
- dispersing and diffusing (fine grinding, enlarging molecular penetration via melt shearing).

Most processing plants have been reclaiming/recycling reprocessable TP materials such as molding flash, rejected products, blown film trim, scrap, and so on. TS plastics (not remeltable) have been granulated and used as filler materials. If possible the goal is to significantly reduce or eliminate any trim, scrap, rejected products, etc.

in an industrial plant because it has already cost money and time to go through a fabricating process; granulating just adds more money and time. Also, it usually requires resetting the process to handle it alone (or even when blending with virgin plastics and/or additives) because of its nonuniform particle sizes, shapes, and melt flow characteristics. Keeping the scrap before/after granulating clean is an important requirement.

Figure 7.36 shows how regrind levels affect the mechanical properties of certain formulations of plastics "once through" the fabricating process and blended with virgin material. The regrind, or scrap, amount is a percentage by weight. Figure 7.37 is an example of the potential effects of the number of times regrind influences the performance of an injection molded TP mixed with virgin plastics (tensile strength-top curves and impact strength-bottom curves). In each set of curves the top contains 25 wt% regrind, middle curve at 40%, and bottom curve at 60%.

When fiber RPs are granulated, the lengths of the fibers are reduced. On reprocessing with virgin materials or alone, their processability and performance definitely change. So it is important to determine if the change will affect final product performances. If it will, a limit for the amount of regrind mix should be determined or no recycled RP is to be used. Consider redesigning the product to meet the recycled performance or use it in some other product.

7.12.1 Handling Regrind

BM, particularly extrusion blow, due to the nature of the system, will generate a certain amount of regrind in the trim areas of the parts being produced. In some parts, regrind can account for as high as 50% of the total part weight. It is an economic necessity that this material be reused. To use the regrind, it is necessary that it be kept clean (free from dirt, oil and grease).

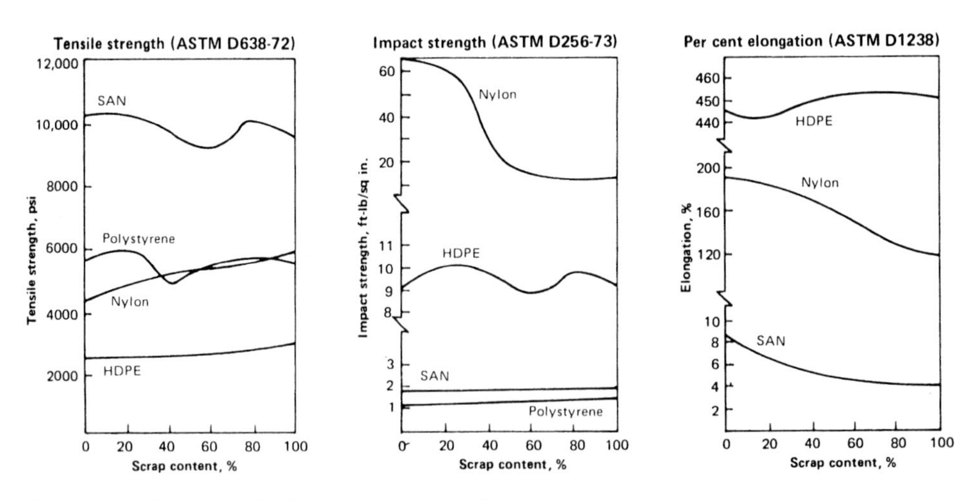

Figure 7.36 *Example of effect of regrind on plastics*

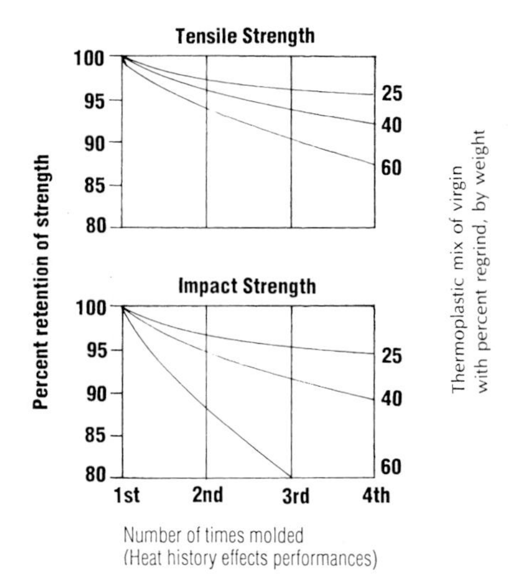

***Figure** 7.37 Example of effects of number of times in recycled injection molded plastics*

Cleanliness of regrind cannot be overemphasized because any foreign material, however small, is a contaminant. It will mar the appearance and degrade the properties of a blow molded part or hang up in restricted areas in the BM machine head and impair the performance of the machine over a period of time.

Average BM systems could produce up to 30 wt% regrind in making parts. This trim material is usually allowed to cool, ground in a granulator, and proportioned into the extruder feed. In this manner, the regrind is used as it is generated. If the plastic is fed into the granulator while still hot, it will stick to the granulator knives and stall the motor. Cleaning a granulator under these conditions is a difficult task.

In some instances, it is more desirable not to use any regrind in making certain critical parts, yet the regrind must be utilized for economic reasons. In these instances, a machine may be designated to run on 100% regrind for making other less critical parts. It has generally been found that when 100% regrind is used, the extruder temperatures can be reduced by 10 to 20 °F relative to virgin resin and still operate satisfactorily. With some grades of HDPE, regrind will have slightly lower die swell characteristics than the virgin resins from which the regrind originated. Therefore, it will be necessary to adjust the die gap when using 100% regrind to provide the same part weight that would result from virgin plastic alone or virgin plastic with lower proportions of regrind.

7.12.2 Size Reduction

The economics of the plastics industry, where raw material costs form a significant part of the price of a finished article, has always provided an incentive for recycling [74]. As with the metal, glass, and paper industries, efforts have been concentrated on the reworking of clean and homogeneous waste collected at the site of manufacture. There is a basic difference between the nature of recycling of plastics and the recycling of metals, paper, glass, and some other materials.

For example, metal production begins with an impure ore that is progressively concentrated, smelted, refined, and has impurities removed from it. The production of plastics, however, begins with high-purity virgin polymer to which various additive colorants and reinforcing material are added. In the metals industry, basic technology concerns itself with purifying and upgrading ores and concentrates. The plastics industry, by contrast, progressively modifies its raw material during its evolution into a finished product. Technology to purify waste plastic has not yet been established. One major obstacle is that different polymers may not be compatible with each other and must be separated as part of the recycling process. The variety of different plastic materials is extensive and unlimited. Through polymer chemistry, new materials are appearing frequently. This increases the markets for the plastics industry.

The current estimate of post-consumer reclaim of plastics is less than 2%, which indicates that there is a vast amount of a valuable resource being wasted. Plastics are quite complex compounds, and size reduction is a very complex art if we are to reclaim a useful product.

Machines used for size reduction in the plastics and rubber industries may be broadly classified as granulators, dicers, pelletizers, die face cutter, hammer mills, attrition mills, pin mills, and pulverizers. These machines are universally used to produce a granulate, pellet, or powder as required, making it possible for the scrap to be reused or reclaimed into a processable material.

7.13 Drying

Plastic materials absorb moisture that may be insignificant or damaging. All plastics, to some degree, are influenced by the amount of moisture or water they contain before processing; moisture may reduce processing performances. With minimal amounts in many plastics, mechanical, physical, electrical, aesthetic, and other properties may be affected.

In the past, probably 80% of fabricating problems was due to inadequate drying of all types of plastics. Now, it could be reduced to 50%.

The hygroscopic plastics [such as nylon, PC, PMMA, polyurethane (PUR), PET, and acrylonitrile–butadiene–styrene (ABS)] are capable of adsorbing and retaining atmospheric moisture. They require proper (special) drying procedures prior to being processed to meet product performance requirements. There are certain plastics that, when compounded with certain additives such as color, could have devastating results. Day-to-night temperature changes is an example of how moisture contamination can be a source of problems if not adequately eliminated when plastic materials are exposed to the air; otherwise, it has an accumulative effect.

Although it is sometimes possible to select a suitable drying method simply by evaluating variables such as humidities and temperatures when removing unbound moisture, many plastic drying processes involve removal of bound moisture retained in capillaries among fine particles or moisture actually dissolved in the plastic. Knowledge of internal liquid and vapor mass transfer mechanisms applies. Measuring drying rate behavior under controlled conditions best identifies these mechanisms. A change in material handling method or any operating variable, such as heating rate, may effect mass transfer (Chapter 10).

During the drying process at ambient temperature and 50% relative humidity, the vapor pressure of water outside a plastic is greater than within. Moisture migrates into the plastic, increasing its moisture content until a state of equilibrium exists inside and outside the plastic. But conditions are very different inside a drying hopper (etc.) with controlled environment. At a temperature of 350 °F (170 °C) and dewpoint of −40 °F (−40 °C), the vapor pressure of the water inside the plastic is much greater than the vapor pressure of the water in the surrounding area. Moisture migrates out of the plastic and into the surrounding air stream, where it is carried away to the desiccant bed of the dryer.

Before drying can begin, a wet material must be heated to such a temperature that the vapor pressure of the liquid content exceeds the partial pressure of the corresponding vapor in the surrounding atmosphere. Different devices, such as a psychometric chart, can conveniently study the effect of the atmospheric vapor content on the rate of the dryer, as well as, the effect of the material temperature. It plots moisture content dry bulb, wet bulb, or saturation temperature, and enthalpy at saturation.

First, one determines from the material supplier and/or experience, the plastic's moisture content limit. Next, determine which procedure will be used in determining water content, such as weighing, drying, and/or reweighing. These procedures have definite limitations. Fast automatic analyzers, suitable for use with a wide variety of plastic systems, are available that provide quick and accurate data for obtaining the in-plant moisture control of plastics.

7.13.1 Drying Hygroscopic Plastic

Drying or keeping moisture content at designated low levels is important, particularly for hygroscopic types where moisture is on the surface and particularly collected internally. They have to be dry prior to processing. Usually, the moisture content is >0.02 wt%. In practice, a drying heat 30 °C below the softening heat has proved successful in preventing caking of the plastic in a dryer. Drying time varies in the range of 2 to 4 h, depending on moisture content. As a rule of thumb, the drying air should have a dew point of −30 °F (−34 °C) and the capability of being heated up to

250 °F (121 °C). It takes about 1 ft^3/min of plastic processed when using a desiccant dryer. The pressure drop through the bed should be less than 1 mm H_2O/mm of bed height. Simple tray dryers or mechanical convection, hot air dryers, while adequate for certain plastics, are incapable of removing enough water for the proper processing of hygroscopic plastics, particularly during periods of high humidity.

Hygroscopic plastics are commonly passed through dehumidifying hopper dryers before entering a screw plasticator. However, except where extremely expensive protective measures are taken, the drying may be inadequate, or the moisture regained may be too rapid to avoid product defects unless barrel venting is provided. To ensure proper drying for "delicate" parts such as lenses and compact disks, the combination of drying the plastics and using vented extruders provides a double check.

With hygroscopic materials the predrying systems are used but a more viable approach can be the use of a vented barrel without predrying (Chapter 2); however, in certain operations to ensure fabricating precision products meeting very tight performance requirements, predrying with the vented barrel is used. With the vented barrel, the plastic is devolatilized after it has been melted, and because the vapor pressure of water at typical melt temperatures is high, devolatilization can be accomplished rapidly. Moreover, at typical melt temperatures, other undesirable nonaqueous volatiles may also be removed by using the vented barrel IMM. Devolatilization from the melt stream is made possible by the use of a two-stage screw and barrel incorporating a vent port. The first stage of the two-stage screw accomplishes the basic plasticating functions of solids feeding and melting. During this first-stage process, significant material pressures are generated.

Plastic usage for a given process should be measured so as to determine how much plastic should be loaded into the hopper. Usually the hopper should hold enough dried plastic for $\frac{1}{2}$ to 1 h of production. This action is taken so as to prevent storage in the hopper for any length of time, eliminating potential moisture contamination from the surrounding atmospheric area. Care should be taken to ensure that hygroscopic plastics are in an unheated hopper for no more than $\frac{1}{2}$ to 1 h, or as specified by the material supplier (and/or experience).

7.13.2 Drying Nonhygroscopic Plastic

The drying operation for nonhygroscopic plastics is different. They collect moisture only on the surface. Drying this surface moisture can be accomplished by simply passing warm air over the material. Moisture leaves the plastic in favor of the warm air resulting in dry plastic.

Many factors have to be considered when sizing a hot air drying system. The first is the plastic material. The material should have a specified residence time (length of time the material should be in a hopper dryer). The temperature of the drying air, as well as a certain temperature, at which the material should be dried, is also critical to prevent melting or plasticizing of the material in the hopper. Chapter 11 further discusses drying systems.

Another consideration when drying nonhygroscopic plastics is the production rate or, in simpler terms, quantity of plastic used in a 1-h time limit. If these two factors, residence time and production rate, are taken into consideration, a proper plenum hopper can be selected. The proper selection permits the plastic to enter the hopper and slowly work its way down to the bottom of the hopper for a 1-h residence time (most hygroscopic plastics have a residence time of 1 h) and keep up with a steady production rate. A hot air dryer can now be chosen based on the cubic feet per minute rating needed to dry the plastic.

7.13.3 PET Drying

PET plastic drying technique has received extensive attention. PET is a hygroscopic material, that is, it absorbs a great deal of moisture from the air around it, and this moisture becomes tightly bound with the molecules of which the material is made. Unless most of this moisture is removed, the PET will not be of the proper viscosity to be fabricated for extruding film, IM parisons for BM beverage bottles, and many more products.

It is not difficult to remove moisture from a hygroscopic plastic material such as PET. The material is simply passed through a drying system, where it is exposed to a stream of hot dry air for as long a time as is necessary to reduce its moisture content to the correct level. It is generally agreed that crystallized PET chips should be dried to a moisture level of 0.002% so that they will have the intrinsic viscosity of about 0.7 needed for injection molding.

Other things being equal, the higher the temperature of the hot air that passes through the material into the drying system, the more rapidly will drying be accomplished. For example, when a 300 °F (149 °C) air temperature is used, a total of $3\frac{1}{2}$ h of drying time is needed to reduce the moisture level of PET material from 0.3% to 0.001%. If the temperature in the air is raised to 350 °F (177 °C), the same job can be completed in less than 2 h, resulting in overall cost savings.

On the basis of this kind of information, it is all too easy to assume that raising drying temperatures as high as possible is desirable, as higher temperatures mean faster drying. However, this is not the case. Excessively high drying temperatures can create serious problems. At temperatures of 400 °F (204 °C), PET materials undergo a physical breakdown and become discolored. At unit temperatures higher than 350 °F (177 °C) acetaldehyde will be developed in excessively high quantities in the fabricated product. In the case of blow molded bottles, it can ultimately affect the flavor of the beverage, and so on.

PET plastic when received may contain about 0.05 wt% moisture. To prevent loss of clarity and some physical properties, the plastic must be dried so that it contains less than 0.005% moisture (target 0.002%) before being fabricated into products.

To achieve this very low moisture content, the plastic must be heated to 350 °F (177 °C) and exposed to air having a moisture content of –40 °F (–40 °C) dewpoint at a flow rate of about 1 ft/s. This low dewpoint and high temperature create what is referred to as a "pressure–dewpoint differential." More simply, in the drying hopper, we create conditions within and around the pellet that virtually boil out the moisture.

At ambient temperature and 50% relative humidity, the vapor pressure of water outside the pellet is greater than that within. Moisture migrates into the pellet, thus increasing its moisture content until a state of equilibrium exists inside and outside the pellet.

However, inside the drying hopper, the controlled environment, conditions are very different. At a temperature of 350 °F (177 °C) and –40 °F (–40 °C) dewpoint, the vapor pressure of water inside the pellet is much greater than that of the surrounding air, so moisture migrates out of the pellet and into the surrounding air stream where it is carried away to the desiccant bed of the dryer.

As a result of prolonged exposure to high temperatures, the plastic releases some small amount of acetaldehyde into the air stream. A special molecular sieve is used that has pore diameters of sufficient size to allow this substance to be trapped and later expelled from the closed-loop system. Acetaldehyde content must be controlled to meet established standards. Most plastics, when received from the supplier, contain less than 2.5 ppm of acetaldehyde.

The strict moisture control required to successfully process PET plastic is achieved by the use of a closed-loop, hopper/dryer system. This system generally includes a self-regenerating desiccant dryer, high-temperature process air heaters, a return air filter, an after cooler (heat exchanger), and a well insulated drying hopper. A good rule of thumb used to determine the cfm [cubic foot per minute or ft^3/min air flow (m^3/min)] requirements for a PET system is 1 cfm lb/h processed. As an example, if a machine processes 250 lb (114 kg) of material per hour, one would select a dryer that produces an air flow of not less than 250 cfm (7.1 m^3/min) at –40 °F (–40 °C) dewpoint.

To determine the capacity of the drying hopper (and thus the exposure, or residence time), simply multiply the residence time by the use rate (expressed in lb/h). Using the 250-lb/h (113-kg/h) use rate from the previous example and a 4-h residence time, there is 250-lb/h × 4 h or 1,000-lb (454-kg) usable capacity. One would select a 1,000-lb capacity hopper to ensure proper exposure time.

7.13.4 Nylon Drying

When processors refer to moisture problems in relation to nylon, usually the material is considered to be too moist. It may come as a surprise, therefore, to learn that serious problems also can arise if the nylon is molded too dry. Over drying has been found to be a factor in product failures. The widespread notion of the drier, the better often results in unsatisfactory flow characteristics, poor fill, and nonuniform crystallization. It is not uncommon in the winter months, when the humidity is low, for the nylon to be dried to 0.04% to 0.08% moisture content.

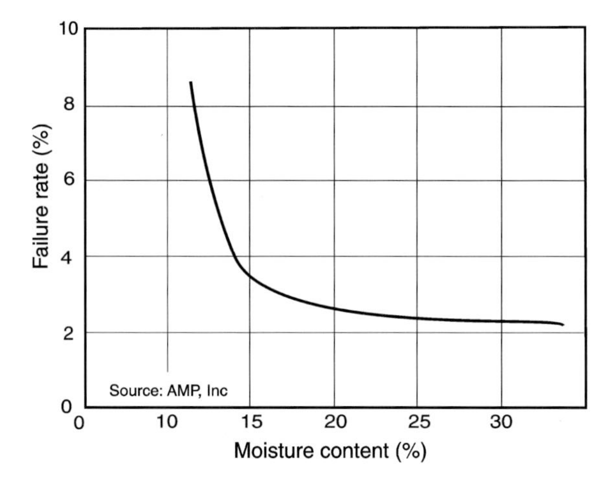

Figure 7.38 *Flexing-failure rate in connector legs vs. dryness of nylon before molding*

As an example for most IM applications, nylon 6/6 should be dried to a range of 0.15% to 0.3%. A producer of nylon connectors found that, by controlling moisture content in that range, they could reduce brittle failure in gripper fingers by 70% (Fig. 7.38). Nylon 6/6 usually contains about 0.1% to 0.3% moisture before the package is opened and the material exposed to the outside air.

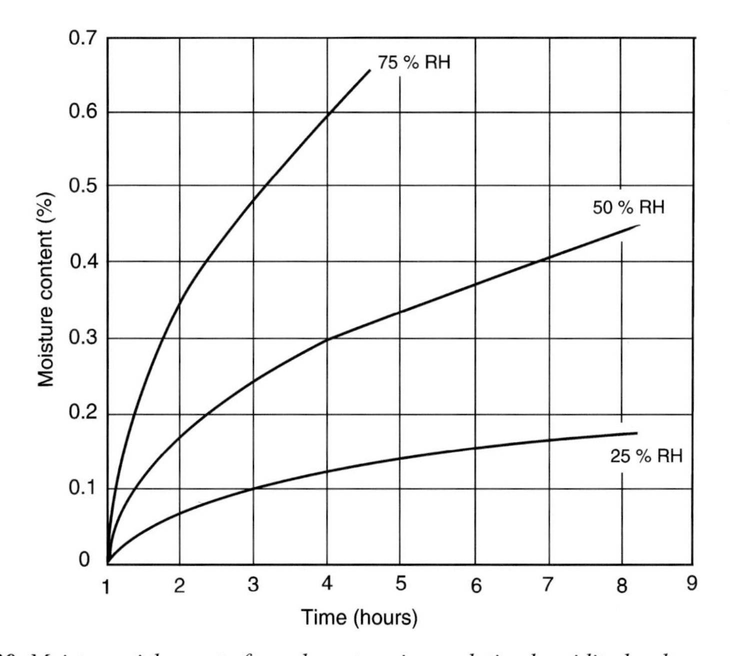

Figure 7.39 *Moisture pickup rate for nylon at various relative humidity levels*

The plastic often is used just as it comes out of a sealed container properly dried. However, after a container has been opened and the nylon exposed to the air, it starts picking up moisture and some type of drying may be required. Figure 7.39 shows the moisture pickup on initially dry nylon 6/6 at room temperature while exposed for up to 4 h at 75% relative humidity, and up to 8 h at 25% and 50% relative humidity. In damp weather (relative humidity of 75%), the moisture level of the nylon 6/6 will rise by 0.35% in 1 h. This additional moisture by itself is enough to cause product brittleness, irrespective of how little moisture was present in the nylon originally. On the other hand, when it is very cold and dry outside and the relative humidity inside the plant is 25% or less, the time required to pick up 0.35% moisture would be about 40 h.

8 Fundamentals of Product Design

In the creation of products to meet performance requirements at the lowest cost, a number of steps must be coordinated and interrelated to ensure success, starting with the design that specifies the plastic and the manufacturing process. Usually, even though more than one material and/or process approach exists, analyzing the type and degree of materials and process variables can direct the final approach with costs an important factor (Fig. 8.1).

The specific process is an important part of the overall scheme and should not be problematic. Unfortunately many fabricating facilities (custom or captive) have only one processing technique; thus, one should ensure that the complete processing requirements for the product can be met with the current system available in the plant.

Following the product design is designing and producing a tool (mold, die, and so forth) around the product. Depending on the design approach, all kinds of variables and problems can develop. Toolmakers usually are experts in specific areas of tool design and manufacture, but not all have the same capabilities. Thus, familiarity with the toolmakers' capabilities is necessary to find the best toolmaker for the project if no in-house capabilities exist.

Many different processing factors could influence the repeatability of meeting performance requirements. Computer programs have been developed to provide the capability of integrating all the applicable factors, thus replacing traditional trial-and-error methods.

The basic design concept involves: (1) the end-use requirements for the part or product; aesthetic, structural, mechanical etc., (2) the number of functional items that can be designed into the part for cost effectiveness, and (3) consideration if multiple parts can be combined into one blow molded part. The third factor provides a very important design concept or characteristic that has already been used. However, the real new growth for BM is in this arena. It is having a significant influence on industrial and commercial blow molded parts. Therefore, it is important for the designer to recognize what can be achieved with BM where all the basic concepts can be applied, particularly replacing assembled parts.

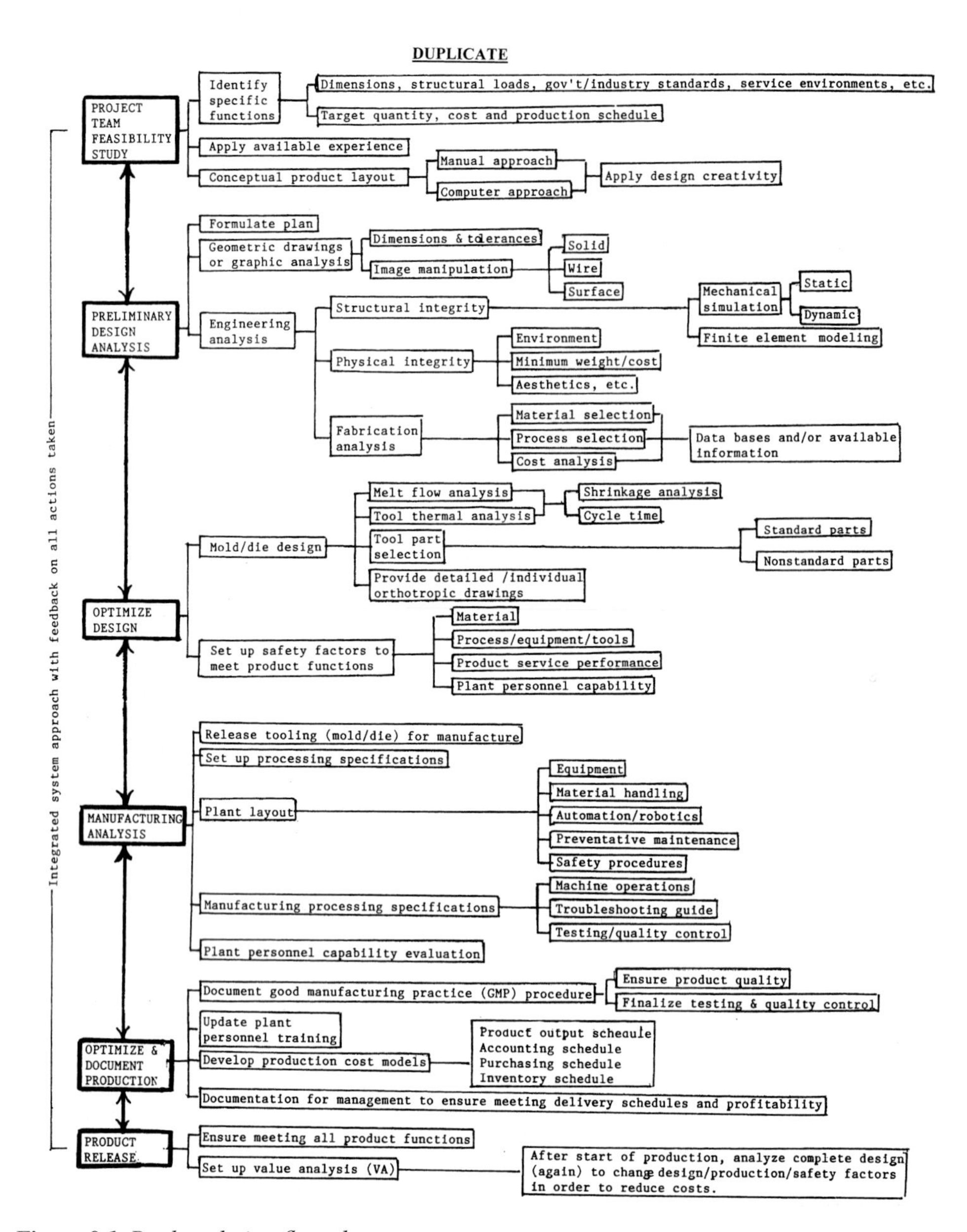

Figure 8.1 *Product design flow chart*

8.1 Overview

Blow molded plastic products have certain inherent characteristics that in most cases are the principal reasons for their application. An important aspect concerns shapes ranging from very simple to extremely complex configurations resulting in manufacturing economies that are now being used more frequently for cost reduction. The conventional relatively "simple" shapes that are produced to meet packaging, medical, auto, electronic, toy, leisure and other markets include mostly complex shapes. The important products now on the increase are those that have extremely complex shapes. They are targeted for all markets, particularly irregular hollow-shaped industrial parts. Often, a single blow molded part can replace an assembled part made of other materials, substantially reducing costs. The process adapts to both low and high volume production runs. The process is flexible as to part size and shape.

Other important characteristics include low weight, thermal and electrical insulation, corrosion resistance, chemical resistance, permeability resistance, transparency, low coefficient of friction, color, aesthetics, capability of monolayer and multilayer, and so on. In regard to flammability, all plastics, like other materials, can be destroyed by extreme heat. Some burn readily, others slowly, others with difficulty, while still others do not support combustion on removal of the flame.

Blow mold design is highly dependent on the type of machine in which the mold will be used. As in all molding operations, good mold design is of great economic importance for obtaining good product quality and fast production rates. The target is to meet performance requirements at the lowest cost.

There is a practical and easy approach in designing with plastics: it is essentially not different from designing with other materials—steel, aluminum, glass, wood, concrete, and so on. Blow molded plastics have been designed into many different shapes and applications for over a century. They have been used successfully and have provided exceptional cost advantages compared to other materials.

8.1.1 Interrelation of Material and Process with Design

With plastics, more than with other materials, the opportunity exists to optimize design by focusing on material composition and orientation, as well as on structural member geometry. With plastics especially, there is an important interrelationship between shape design, material selection (including orientation), and manufacturing selection. This interrelationship is different from that of most other materials, for which the designer basically is limited to obtaining specific prefabricated forms or profiles that are bent, welded, and so on.

Many different products can be designed using plastics—to take low to extremely high loads and to operate in widely differing environments, ranging from highly corrosive to "under-the-auto-hood" conditions. The major consideration for a designer is to analyze properly what is available and develop a logical selection process to meet performance requirements, which generally are related to cost factors.

Part design is a combination of understanding both the practical and theoretical viewpoints. It is usually a compromise between the artist who emphasizes aesthetics, the production person who desires economical and efficient production, and the user who must have the proper end use properties.

As an example, the simplest shape to blow mold is the vertical closed cylinder. The parison is equidistant from the vertical mold walls, giving a uniform blowing radius and resulting in uniform wall thickness. The top and bottom will always be thicker, since the blowing radius is less in these areas. (Actually, this shape should be considered a "simple" complex blow molded product.) The farther the design varies from the cylinder, with varying blowing radii, the greater will be the variation in wall thicknesses. A parison wall thickness sufficient to give strength to the greatest blowing radius will place excessive material at lesser radii. As reviewed throughout this book, there are design and process techniques for retaining strength and rigidity with reduced part weight.

Designers, in an effort to capture more of the market, are constantly designing more blown products with what might be called "exotic" shapes, which are appealing to the customers or functional in industrial applications, but are difficult to make by conventional blow molding techniques. In order to produce many of these containers economically, it has become necessary to modify blow molding equipment in both extrusion and injection.

The continuing developments of higher performance resins and barrier materials is one of the factors that make blow molding equipment so much more effective. Machinery builders have come up with the primary equipment and accessories that enable such materials to be molded at rates that are both cost effective and adequate to the demands of volume markets. They also meet ever increasing performance and quality standards.

Blow molding processing techniques have been devised which will control the parison wall thickness vertically and horizontally. This technique permits equal wall thickness with differing blow ratios. It provides design versatility without the waste material and longer cycles of too much material at short blowing radii but necessary to get enough wall thickness at long blow radii. Design is much less restricted by function, strength and economy.

The range or properties literally encompasses all types of environmental conditions, each with its own individual broad range of properties for the different plastics. Unfortunately, no one plastic can meet all property requirements, but the designer has the option of combining different plastics (keeping them as separate materials by processes such as coextrusion or coinjection). Plastics are also combined with other materials such as steel. Any combination requires that certain aspects of compatibility exist, such as thermal coefficient of expansion or contraction, etc.

Combining or mixing plastics to produce what is generally referred to as "alloys" creates plastic compounds with properties that differ from those of the components (Chapter 7). Compounding certain plastics can produce synergistic results, enhancing properties. The designer must be familiar with the techniques used in processing plastics.

8.1.2 Design, Shape, and Stiffness

One advantage of BM plastics is their formability into almost any conceivable shape. Plastic shapes can be almost infinitely varied in the early design stages, and for a given weight of material, can provide a whole spectrum of strength properties especially in the most desirable areas of stiffness and bending resistance.

In all materials, elementary strength of material theory demonstrates that some shapes resist deformation resulting from external loads or residual processing stresses better than others. Deformation in beam and sheet sections depends upon the product of the modulus and the second moment of inertia, commonly expressed as EI. Physical performance can be changed by varying the moment of inertia or the modulus or both [9].

Using thick plastic walls to meet stiffness requirements is an expensive design method because inefficiently large quantities of material are used, and the long heating and cooling times required affect the economics of production considerably by increasing cycle times.

Therefore, the desired stiffness is better obtained by the use of double curvature, shaping and even ribbing. Although these devices tend to increase tool costs, in long production runs, these investments are more than offset by shortened molding cycles and less raw material use. Also, such techniques as using sandwich wall construction further extend the stiffness of the thinner sheet.

Adding material to a structural component does not always make it stronger. For example, curvature/shaping is often added to stiffen a structure, particularly in large molded plastic parts. However, the addition of a shape to a plate loaded in bending may cause stress to be increased rather than decreased. Although it increases the moment of inertia of the structure, the distance from the structure's center of gravity to its outer edge may become relatively greater, and stress increases accordingly. The key to reducing stress is adding shapes in the correct proportions and position.

Lengthy equations for moments of inertia, deflection, and stress normally are required to determine the effect of shapes on stress. However, nondimensional curves have been developed to allow a quick determination of proper proportions.

With the latitude that exists in designing shapes with plastic materials, designs using large amounts of materials do not necessarily work the best or give the best physical performance per unit weight of material. Sometimes quite minute amounts of material judiciously placed can make an important difference to the cost.

When designing shape and stiffness requirements, plastic parts, in most cases, can take advantage of a basic beam structure in their design. Much of the conventional design with other materials is based on single rectangular shapes or box beams. Blow molded hollow-channel, I and T shapes designed with generous radii (and other basic plastic flow considerations) rather than sharp corners are more efficient on a weight basis in plastics because they use less material, providing a high moment of inertia. The moments of inertia of such simple sections, and hence stresses and deflections, can be fairly easily calculated using simple engineering theories.

Processing any plastics into curved parts is relatively easy and inexpensive. Such parts conform to recognized structural theories that curved shapes can be stiffer in bending than flat shapes of the same weight. Both single and double curvature designs are widely used to make more effective use of plastics materials.

An example of single curvature in a structural element is a corrugated roofing panel, which is inherently much stiffer than the same volume of material would be in a flat sheet. To improve the stiffness further, the corrugated panels can sometimes be slightly curved along the length of the corrugations.

Double curve blow molded corrugated shapes can take the form of spherical domes, be saddle-shaped, or use hyperbolic shapes. Some of these features are found in well known architectural designs using plastics materials. These shapes can also be made in modular form molded of composite/reinforced plastics, providing an efficient structural shape with a higher buckling resistance.

8.1.3 Residual Stress and Stress Relaxation

Given a fixed mold design and material, the influence of processing on container properties and appearance is not as susceptible to definite guidelines as resin parameters on properties or processability. One of the greatest influences of processing is on the level of residual stress in the container, which in turn affects stress crack resistance and heat resistance. The greater the level of residual stress, the poorer these properties tend to be.

Stress is placed in the blown objects by the molecular orientation of extrusion, then by the orientation of inflating the parison, and finally by differential shrinkage between thick and thin blow sections. Hotter stock and warmer mold temperatures increase shrinkage, and the greater the variation in wall thickness, the greater the residual stress due to differential cooling rates and resulting differential shrinkage. This condition also exists with injection blow molding.

The properties of plastics are strongly dependent on temperature and time, compared to those of other materials such as glass and metals. This dependence is due to the viscoelastic nature of plastics. Consequently, the designer must know how the product is to be loaded with respect to time.

In structural design it is important to distinguish between various failure modes in the product. The behavior of any material in tension, for example, is different from its behavior in shear. For viscoelastic materials such as plastics the history of deformation has an effect on the response of the material, since viscoelastic materials have time and temperature dependent material properties.

Whether the part is deformed in tension or shear, continued straining under a sustained load is called creep (Fig. 8.2a). Either type of deformation leads to a material property called the creep compliance. There is a creep compliance that describes the tensile behavior of the material, and there is a creep compliance for shear behavior.

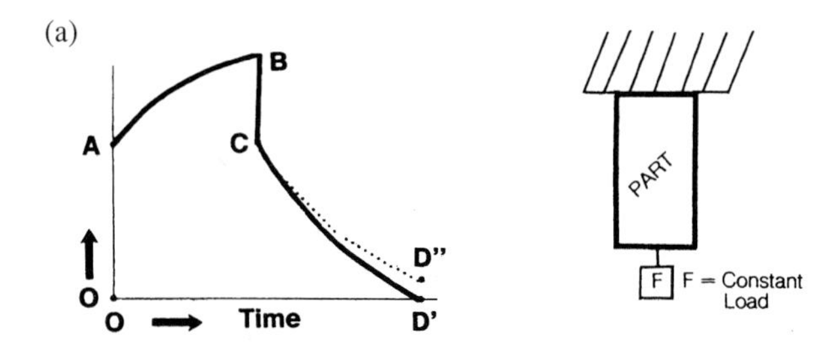

Figure 8.2 (a) Viscoelastic behavior of plastics—relation of stress–strain–time; creep
O–A: Instantaneous loading produces immediate strain
A–B: Viscoelastic deformation (or creep) gradually occurs with sustained load
B–C: Instantaneous elastic recovery occurs when load is removed
C–D: Viscoelastic recovery gradually occurs; without permanent deformation (D′) or with a permanent deformation (D″–D′). Any permanent deformation is related to type plastic, amount and rate of loading, and fabricating procedure

The creep behavior of a plastic can go through certain specific phases. Load can be applied instantaneously, resulting in strain point *A*, a reaction that can be called the elastic response. With the stress maintained, the plastic deforms or creeps with time to point *B*. The viscoelastic deformation that occurs identifies the strain from *A* to *B*. When the load is removed, the plastic strain is reduced immediately to point *C*; this phase can be identified as the elastic recovery. The next phase has the plastic gradually relieving its strain (with time) to point *D* (viscoelastic recovery). Based on the type of plastic used and the amount of load applied, the final recovery could be a return to zero strain.

On the other hand, if the material is subjected to a sustained constant strain rather than constant stress, the force or torque necessary to sustain this constant strain decays with time. This process is called relaxation (Fig. 8.2b). As in creep, there is a

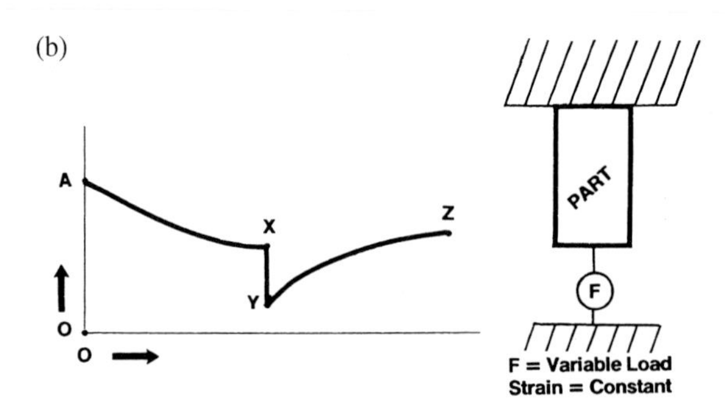

Figure 8.2 (b) *Viscoelastic behavior of plastics—relation of strain–stress–time; stress relaxation*
O–A: Instantaneous loading produces immediate strain
A–X: With strain maintained gradual elastic relaxation occurs
X–Y: Instantaneous deformation occurs when load is removed
Y–Z: Viscoelastic deformation gradually occurs as residual stresses are relieved. Any permanent deformation is related to type plastic, amount and rate of loading, and fabricating procedure

fundamental material property associated with relaxation, the relaxation modulus. There are practical problems in which relaxation is important, just as there are practical problems where creep matters. A designer must be able to recognize which of these fundamental modes is involved in a particular situation.

These stress relaxation modes can occur in all types of design and in different materials, including the low performance commodity plastics. Note that with the very high strength or high modulus of elasticity composite plastics, this stress relaxation condition is easier to analyze than those of the high performance steels.

Stress relaxation behavior of a plastic can be analyzed after load is applied instantaneously to stress point A (elastic response). As the strain is maintained, the plastic relaxes with time to point X. With load removed, the plastic recovers elastically (immediately) to point Y, less than point X. With time, residual stress induces viscoelastic deformation to point Z. The amount of strain due to this type of stress relaxation behavior will depend on the type of plastic used and the amount of load applied.

8.1.4 Design Parameters

Design analysis involves selecting the proper plastic, process equipment, and process conditions based on known and available plastic and process variables. The product objectives will vary from a simple cost, appearance requirement to a complex interrelated combination of cost, appearance, and properties, depending on the end use of the item, and then the specific function of the molding within its end use category. No one single application can be selected to illustrate the complete interplay of variables on product objectives.

As an example, the PE bottle with its extensive use appears to be simple to design. However, like all blow molded products, complex situations exist since many requirements have to be met and compromises usually have to be included. What happens in a "simple" complex shape can be applied to any shape. The polyethylene bottle (resembling the vertical closed cylinder previously reviewed) has a shape that is one of the more simplified categories, but its basic design principles can be applied to other shapes.

Property requirements vary with the specific application of the bottle, whether squeeze, detergent, bleach, carboy, etc. General requirements relate to rigidity and "feel", stress crack resistance, impact resistance, permeation, crush resistance, chemical resistance, etc. Meeting design objectives of the bottle requires a consideration of both the plastic material, and the conditions under which it is processed. Resin parameters and test results are a guide to anticipated bottle properties, but the processing stage influences these properties in bottle form.

Resin parameters are guides to the processability of the resin, but bottle requirements may place limitations on how the resin may be processed, or on process design which then may influence resin parameters. In short, bottle properties, resin, and process are closely interrelated, and one cannot be ignored in considering the relationships between the other two.

The appearance of the bottle not only covers basic design, but such factors as gloss, smoothness, lack of imperfections in the bottle wall, color, and absence of molded distortion. The cost of the bottle is basically a function of the cost of the plastic, weight of the bottle, and cycle at which it is blown, exclusive of posttreating costs.

Design of a blow molded bottle requires consideration of the following factors: material to be blown; size and weight of product and mold; contour; variations of design desired, e.g. neck inserts, bottom plates, or side inserts; surface texture and engraving; sharp corners and straight angles; blow openings available and location; and parting lines.

Most blow molded articles perform better (particularly to improve toughness) with rounded, slanted, and tapered surfaces. Square or flat surfaces with sharp corners are undesirable. Wall thickness can vary considerably from side panels to corners. Corners become thin and weak, while heavy side panels become thick and distorted. Flat wall sections are not uniform, and flat shoulders offer little strength. Highlight accent lines should be "dull" with a radius of 1.5 mm (0.060 in.) or more. If they are sharper, the parison does not penetrate the mold, and trapped air marks result along the edge.

8.1.4.1 Biaxial Orientation

An important design characteristic for the designer to understand and apply that will significantly improve performance and/or cost involves biaxially stretching the plastics during processing. Details on how to process biaxially stretched containers are reviewed throughout this book, particularly in Chapter 4. To date, the primary application for stretching is in injection stretch blow molded 2-liter (0.5 gal) polyethylene terephthalate (PET) beverage bottles. The designer can take advantage of this characteristic in many different products molded by extrusion or injection. Equipment, including large size machines and many different materials, can be used.

Basically, biaxiality of the stretching process permits the two individual degrees of controlled longitudinal and circumferential (or hoop) stretching to overrule the overall degree of stretching. In a pressure beverage bottle, one will always try to achieve the strength in the circumference to be approximately twice as great as in the longitudinal direction, as the load on the material under internal pressure corresponds to just this distribution pattern.

However, with a bottle which relies more on stacking strength, the emphasis will be more on the longitudinal than on the circumferential degree of stretching. As the cooling time required is a square function of the wall thickness to the degree of stretching is in fixed ratio to the wall thickness of the hollow molding, it can be understood that the planned degree of stretching quite significantly determines the production output of a BM machine.

8.1.5 Interrelation of Design and Resin with Processing

Container rigidity is related directly to the crystallinity of the polymer from which it is formed. Generally speaking, the higher the crystallinity of the polymer, as indicated by density, the more rigid the bottle. Corollaries to rigidity are flex life and snap back—the lower the density, the longer the flex life. Snap back requires a density not so low as to be limp but not so high as to be rigid. Two other properties, almost wholly controlled by density, are permeation and abrasion resistance, both improving with increasing density, as does vertical crush resistance.

Melt index (Chapter 5) and density influence heat resistance, with density being the dominant parameter. As an example, the higher the heat resistance requirement, the higher must be the polyethylene density. Bottle clarity, however, will decrease with increasing density. Gloss will increase with increasing melt index, and with increasing density. However, high density polyethylene is not as glossy as low density blow molded items under most conditions.

Another property melt index and density influence is impact strength. Resins with a lower melt index and a lower density have improved impact resistance. Stress crack resistance is also a function of both melt index and density. Higher molecular weight (low melt index) polyethylenes are usually superior to lower molecular weight polymers of equal density.

With PEs, the higher density with the same melt index, when stressed to the same level, will usually show superior stress crack resistance. The higher density, however, being more rigid, develops more stress at equal deformation, and shrinks more to yield higher residual stress. Thus, in most practical applications, an increase in density results in a decrease in stress crack resistance. Resin additives will also influence bottle properties. Examples are antistatics, ultraviolet inhibitors, colorants and the influence of antioxidants on heat embrittlement. The additives may also influence resin processability, but proper formulation should not result in any severely detrimental processing features.

It is important for the designer to interrelate properties of plastics with method of processing—how the machine operates to melt the plastics and temperature/time/pressure settings in the die and/or mold. Resin parameters of melt index and density strongly influence cycle, but within the interrelationships of minimum stock temperature without parison roughness, minimum mold cooling time without distortion, flow, and parison sag. Higher melt index indicates greater flow at a given stock temperature. Conversely, resins having a higher melt index permit lower stock temperatures for equal flow. This will usually yield faster cycles, since less heat will be used to flow the resin, and therefore, less heat must be removed by the molds.

When not considering end use property requirements, an upper limit on melt index is imposed by parison sag which increases with increasing melt index. Parison sag is a thinning of the parison nearest the die before blowing, due to a lack of hot melt strength resisting the weight of the parison. Even with higher melt index resins, a limit may be imposed on faster cycles due to extruded parison roughness, although increasing melt index at constant temperature and extrusion rates usually reduce the tendency of parison roughness.

Cycle time is directly influenced by density for two basic reasons. First, heat distortion increases with increasing density, and, therefore, moldings may be ejected hotter and more quickly without distortion. Second, the greater rigidity of higher density PEs may permit thinner blown walls for reduced mold cooling times. At equal melt indices, density increases tend to reduce flow.

Pinch or weld line quality may also be affected by resin parameters, although proper process and mold design will not greatly limit melt index or density in this respect. Pinch quality usually improves with increasing melt index and decreasing density.

For the designer to meet dimensional requirements, understanding shrinkage of plastics is important. Molded wall thickness differences between the parting line and adjacent areas may increase with increasing density due to greater shrinkage. If the material at the parting line is not in intimate contact with the mold surface, it will take longer to cool. Material may be drawn from the parting line to the more rapidly cooling and shrinking adjacent areas (thermodynamic heat transfer behavior changes). Adjustment of processing and design conditions can remove this limitation on density.

In contrast to conventional materials such as metals and glass, design in plastics cannot, except in rare elementary design cases, be based on one key property (e.g., tensile or shear stress with a steel or metal and compression with glass). The designer rarely can call on standard sections (as in structural steel work) or on standard metal components for mechanical engineering applications. Also, plastics are constrained by a much larger set of production variables, all of which must be examined by the designer. Based on the above review for the PE bottle, the bottle properties, appearance, and cost interrelate with the plastic density, melt index, molecular weight distribution and additives as well as processing temperature, pressure, cycle time, and design of die and mold. Performance can change just based on melt index and density (Table 8.1).

Table 8.1 *Performance Influenced by Plastic Properties*

	With increasing melt index	With increasing density
Rigidity	—	Increases
Heat resistance	Decreases	Increases
Stress crack resistance	Decreases	Decreases
Permeation resistance	—	Increases
Abrasion resistance	—	Increases
Clarity	—	Decreases
Flex life	Decreases	Decreases
Impact strength	Decreases	Decreases
Gloss	Increases	Increases
Vertical crush resistance	—	Increases
Cycle	Decreases	Decreases
Flow	Increases	Decreases
Shrinkage	Decreases	Increases
Parison roughness	Decreases	Increases
Parison sag	Increases	Decreases
Pinch quality	Increases	—
Parting line difference	—	Increases

8.1.6 Design Approach

A general design program includes both practical engineering and advanced engineering approaches. With the practical engineering approach (predominantly used), property requirements may include mechanical strength and stiffness, chemical resistance, permeability, heat resistance, electrical insulation, etc. Short-term, conventional static tests generally suffice.

The advanced engineering approach is used when the plastic product will be exposed to long-term static or dynamic loads based on varying environmental conditions such as time and temperature (large chemical tanks, boat hulls, automobile parts, pressure containers, etc.). When a part is exposed to any load for a long period, the design engineer is provided long-time data, such as creep, fatigue, tensile temperature/time, and other data. These data provide common stress analysis, using creep or other data for determination of permissible stresses or strains.

8.2 Design Guidelines

Part design must incorporate the factors pertaining to functional and appearance requirements, the properties of the material, and the feasibility of processing in arriving at an acceptable solution in terms of performance and cost. As mentioned at the outset of this discussion, the designer has choices that sometimes require compromising requirements with processing. The choice will determine, to a large extent, the type of problems to be solved regarding not only part configuration, but also mold design which must be considered as a corollary in arriving at a useful, economically acceptable part.

Blow molded products should be designed with generous radii at corners and edges. Fillets and rounds should be employed wherever possible in corners, ribs and edges. Such parts will possess more uniform wall thickness, and as a result of more uniform and faster cooling, internal stresses, as well as product distortion, will be reduced.

In the case of bottles (Fig. 8.3), it is important that a generous radius in the chime area of bottles exists. Thinning causes stress cracking between the heavy pinch-off sections and the thinner edges. Rounded bottom edges contain lower levels of stress than do sharp edges and, therefore, have more resistance to stress cracking agents. A concave base in bottles acts as a shock absorber and improves impact strength. The entire bottle design should be as elastic as possible so that impact energy can be absorbed. For this reason, cylindrical bottles are preferred to containers of rectangular cross section, and the joint between the side wall and the base should have as large a radius as possible.

At the shoulder side wall and the shoulder neck joint, the same design approach exists. In waisted bottles, a gently curved waist is better than a sharp, well defined waist. Sharp corners must be avoided at the attachment of any vertical or horizontal ribbing. Large undercuts should also be avoided. These may prevent removal of the product from the mold or cause excessive distortion during removal. When large undercuts are necessary, it may be possible to provide for their removal with moving parts in the mold design. Generally the ratio of the diameter of the finished part to that of the parison should be in the range of 4 : 1, the exact value depending on material, process, and design details.

Varying degrees of wall collapse may occur. This is due to a negative pressure within the container and may arise from diffusion of the contents outward through the wall, from absorption of one or more of the atmospheric gases within the bottle, or from premature sealing of a container which has been filled at an elevated temperature. By proper design, a certain amount of wall collapse can be accommodated without noticeable distortion.

While the primary goal of design must be to produce an article that will be satisfactory in appearance and in performance, other factors cannot be ignored.

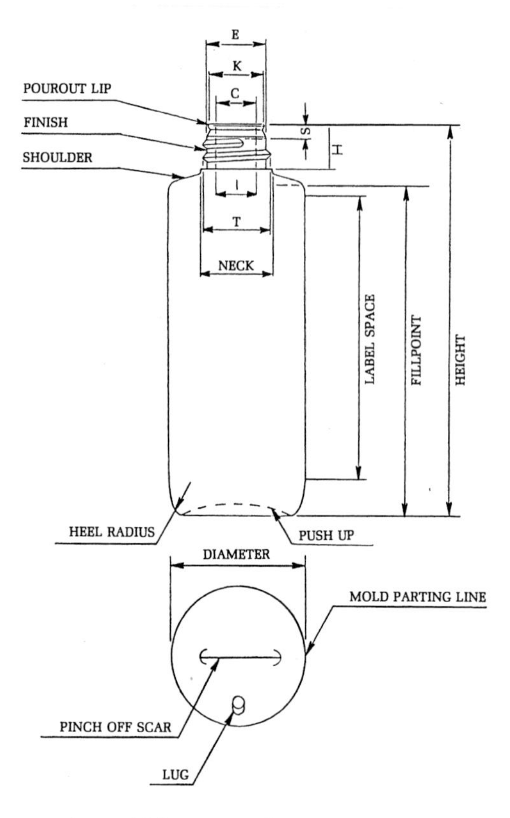

Figure 8.3 *Nomenclature of round blown container*

The designer must constantly keep in mind that the shape of the article must be such that a cavity to produce it can be cut into a mold, that this cavity must be of such configuration that plastic can flow freely into all of its irregularities, and that removal of the molded article from the cavity must be possible. Mold action can include side action, eject snap fits, etc.

Techniques used for strengthening parts include:

- Recessed panels may be concave, convex or flat, but their lower blow ratio also makes for thicker wall sections. The greater the span of the panel, the thicker should be the wall or the greater the tendency toward a convex or concave surface.
- Flutes are concave ribbing and serve the same function. The horizontal is also preferable to the vertical direction.
- Horizontal ribbing adds to container strength without an increase in wall thickness. It is superior to vertical ribbing for drop impact and container collapse resistance.

- Because an item molded from certain plastics such as polyethylene terephthalate (PET) must be cooled below its glass-transition temperature (T_g) before it can be removed from the mold, undercuts of any significant depth will not strip from a mold. Moldings of PET usually can be stripped from undercuts as deep as 0.010 in./in. (0.010 mm/mm) of cavity width, but probably not much greater. The maximum allowable undercut is influenced by the design of the undercut and the usual factors such as the thickness of the sidewall.
- Diameter change applies to round containers, and increases rigidity and strength by increasing wall thickness at the reduced diameter or blowing radius. Increased performance or stiffness is achieved using elementary strength of material theory using shapes such as corrugation, I- and T-shapes, ribs.
- Taper imparts an angular support at its junction with a walled section differing in shape or direction. Wall thickness also increases with the decreasing diameter and blow ratio.

With PE, as an example, blow ratios of 4 : 1 or less should be employed where possible for ease in wall distribution control. (Each plastic has its optimum blow ratio requirement.) All molded radii should be in the range of $\frac{1}{4}$ in. This is more important with HDPE than with LDPE as its greater crystallinity can cause environmental stress crack or impact failures at short radii high residual stress areas. Excessively short radii also result in thin areas. The bottom of a container should be concave rather than flat, to avoid a slight bottom distortion resulting in a bottle that would not stand upright. A radius of curvature about 1.5 times the bottle diameter is suggested. Place solid lugs at the parting line. The maximum thickness will be twice the parison wall thickness. Lugs may be placed elsewhere than at the parting line, but they will be essentially hollow (Fig. 8.3).

The neck and thread design and size may be selected from standards of the industry, unless a snap-on closure is to be employed. A screw type buttress thread is recommended. A buttress type with approximately a 10° base angle will provide maximum cap holding power. The resiliency of polypropylene (PP) or PE is such that the cap may slip up and over the bottle threads of other designs. A minimum 360° closure to thread engagement is also recommended to avoid lifting or cocking of the cap.

Proper mold design for a desired container capacity is difficult due to the influence of PE shrinkage. Past experience and/or shrinkage data from test cavities are the best guide. There can be no exact rule of thumb on blow molded PP or PE shrinkage as it is influenced by the interaction of resin type, mold temperature, stock temperature, mold time, and part thickness. In addition surface treating and the oven cure cycle for decorating increase shrinkage. Tolerances may range ±5% to 10%, depending on the size of the part.

As previously stated, bottles must be compatible with the product they hold. Designers must consider such factors as top load, blowup ratios, fill temperatures, bottom support, labeling, capping, line handling, wall thickness, sharpness or corners, drop strength, embossing and consumer appeal and utility. As an example, resistance to top load begins in the shoulder design immediately below the top section (Fig. 8.4). Choice of method of producing the bottle is dictated to a degree by the blowup ratio, usually identified as the root diameter of neck thread to maximum body diameter. Injection blow molders usually stay under 2.5 : 1; extrusion blow molders generally stay within a 4 : 1 blowup ratio. This ratio affects bottle detail, wall material distribution and quantity of material required to make a good bottle.

If recessed, the bottle area used to include the label should have a gradual change and be shallow, so top load is not affected (Fig. 8.5). Bottom support design covers the requirement for allowing the bottle to be stable and for preventing easy bottle tipover (Fig. 8.6). It also covers sharp corners and pushups.

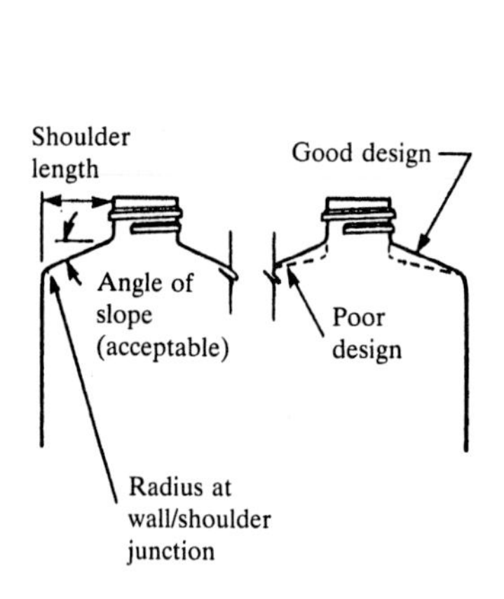

Figure 8.4 Shoulder design for top load and stacked strength

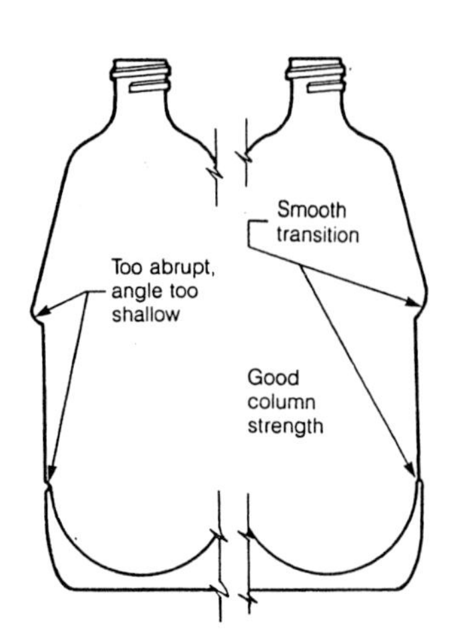

Figure 8.5 Recessed label for column strength

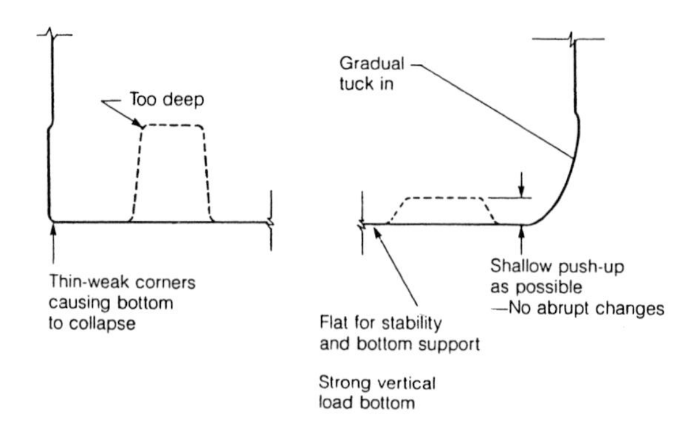

Figure 8.6 Bottom design for support and stability

Designers can take advantage of blow molding machine capabilities to obtain different type neck and shoulders held to tight tolerance with high strength. (It is generally recognized that this capability exists for injection blow, but not for extrusion blow.) These machines have calibrating devices which take the hot melt parison that is gripped by the mold and simultaneously shape a neck that is flash free. Examples of the shapes molded are shown in Fig. 8.7.

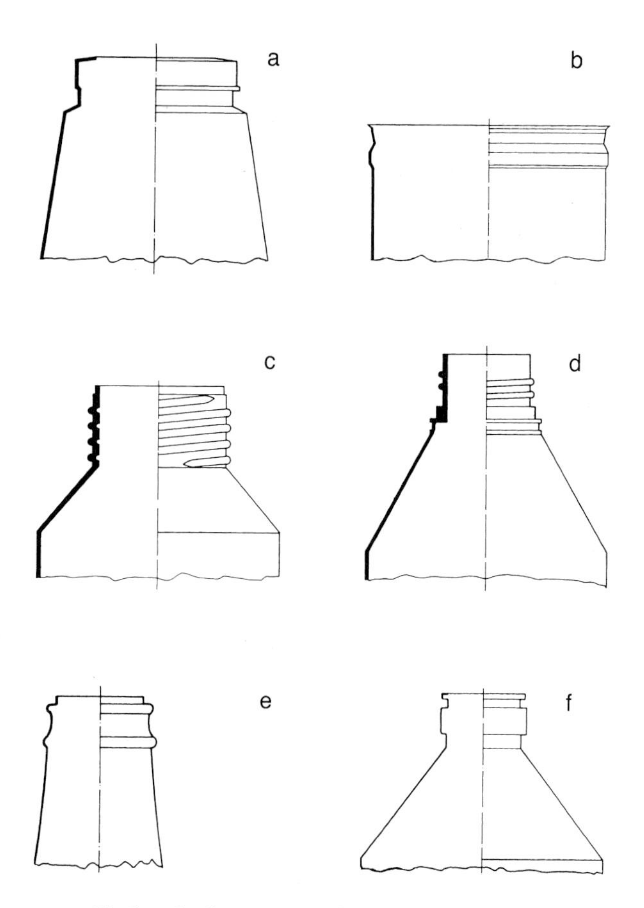

Figure 8.7 *Precision molded necks that are scrapfree. (a) Wide mouth with internal lip; (b) wide mouth with external lip; (c) calibrated mouth, thread area blown, for standard screw closures; (d) calibrated thread mouth for standard screw closures; (e) calibrated mouth for cap closures; and (f) blown and cut mouth for snap-on closures*

It is important for the designer to recognize what can be done in the mold in order to take advantage of advancing technology. The calibrating devices with the extrusion blowing mandrels are operated hydraulically or pneumatically, dependent on machine size and customer's choice.

The individual blowing mandrel receptacles, designed as construction modules, are adjustably mounted on a common base plate. In this way the center-to-center distance, the so-called caliper measurement, is alterable. For the first stage of disengagement of the molding from the blowing mandrel, $\frac{2}{3}$ before mold opening, the blowing mandrel is lifted by a first, steplessly adjustable stroke in accordance with the neck design.

Apart from the mold care aspect, this is also important for the quality of the blown article, as otherwise, the total force required for releasing the blowing mandrel would have to be absorbed by the mask of the transfer station. This, in turn, would result in distortion of the molding. Following the initial stroke of the blowing mandrel, the mold opens and swings back under the head to take over the new parison.

Simultaneously with the closing of the mold, the attached transfer device closes around the preloosened molding still suspended from the blowing mandrel, which then returns to its starting position with the second stroke.

With the smooth movement, the mold gripping the new preform swings down into the calibrating position, and the molding from the previous cycle, retained by the transfer device, is deposited on the guide rails precisely in the working position of the punch unit and is deflashed. Scrap removal is carried out by punching or cutting, operated pneumatically or hydraulically depending on machine size. Logically arranged guide baffles lead the scrap to a centralized recycling unit for re-use.

The design can include aesthetic features that have structural benefits. Embossing and debossing, for example, can stiffen side walls. Decorative ribs, vertical or circumferential, can rigidize a container (Fig. 8.8).

Depending on bottle design, mold shrinkage calculations are normally figured from 0.018 to 0.025 in./in. In bottle design, it may be necessary to consider a bulge factor. The primary effect of this is on the bottle's fill point. It can be controlled by

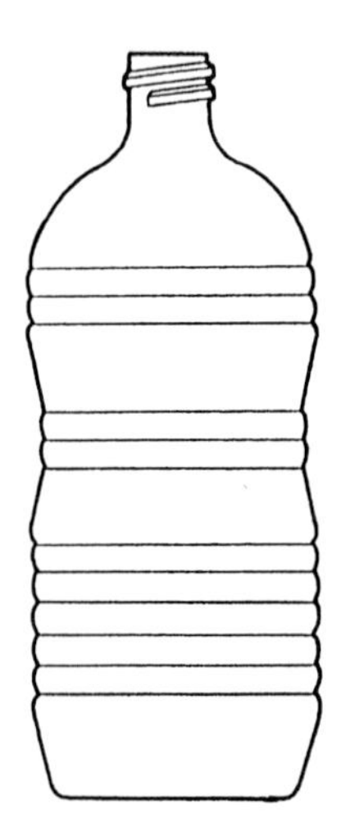

Figure 8.8 *Rib design for hoop stability*

controlling the pressure being exerted on the plastic bottle while it is being filled with product. This also applies to capping or sealing of the bottle. Capping pressure must be controlled.

8.2.1 Design Basics

Vertical strength or top load strength of the plastic bottle functions to: fill the bottle, cap the bottle, to transport the bottle, and to store (warehouse). In the filling cycle, the vertical strength of the bottle will need to actuate the filling nozzle. The nozzle may be equipped with a spring mechanism requiring 20 to 30 lb (9 to 14 kg) force to start product flow. Bottles with flat shoulders, corrugations and similar design efforts may be strained during the filling cycle. Vertical strength plus resistance to torsional force is required in the fast capping operation. Vertical strength of a container is surely tested during the application of a snap-on closure.

Due to high cost of warehousing floor space, cases of filled bottles may be stacked 10 to 20 high. The bottom case is subjected to the weight of the balance of cases above this case. The bottle may have to support some of this load. Cases stacked similarly in a box car or a truck trailer are subjected to an even greater vertical load when the truck or train are in motion.

Seasonal high humidity in some geographical locations will dilute the strength of the paper board of which the cases are made. Here, a danger exists that excessive vertical load on the bottle will cause the bottle to stress crack. The resultant leakers will damage the cases affected by stress cracked containers. Where such conditions are possible, bottles should be checked for environmental stress cracks subjected to a specific top load. Test procedures will vary according to requirements. Over designing or over weighting the bottle costs money.

HDPE, PP, and polyvinyl chloride (PVC) are the most popular, with polyesters as the material of choice for soda pop bottles. HDPE is probably the easiest resin to handle on a production line. The properties of HDPE are acceptable to package bleach, gasoline and radiator antifreeze, food products, various detergents, fabric softeners, etc.

PP is blown as biaxially oriented and as nonoriented resin. One of many applications of the blown nonoriented PP containers is for pancake syrups. The high fill temperature and the favorable contact clarity makes this resin well suited for this application.

The biaxial orientation process is basically reheating a cooled parison to temperature below PP extrusion temperature and stretching it longitudinally prior to blowing. Next is blowing, which is a lateral stretch. Consequently, orientation in the longitudinal and lateral axis. This procedure has a substantial effect on clarity and strength. The weight of a biaxially oriented bottle is lower. This process has been employed for surgical irrigation containers, intravenous liquid containers and in detergent bottles.

The outstanding characteristic of PVC resin is its water clarity. It is an obvious alternative to glass. It does not shatter like glass and the weight of a PVC bottle is much lower than glass. The potential "drawbacks" to PVC over PE or PP are that for one, the resin is corrosive in the plastic (molten) state. Molds and die head tooling are attacked by the fumes from the hot material. The corrosive activity is greater in a humid atmosphere. (Proper manufacturing procedures are used to eliminate these "drawbacks.")

Secondly, the molds for PVC are more expensive due to the molds being made of more expensive beryllium–copper or stainless steel. Aluminum may be utilized for lesser quantities. The molds need high polish and have to be equipped with a cavity venting system.

The usual PVC resin does not possess the heat stability of PE and tends to "burn" if allowed to stay hot for more than 2 or 3 min. Since PVC is a "rigid" material, thinner walls are practical.

Blow ratio can have reference to the ratio of the "E" dimension of the neck (minor thread diameter) to the largest lateral dimension on the body of the bottle. The blow ratio can also be the comparison of the width (widest point) of the bottle to the thickness (front to back) of the bottle.

The effect of one or the other ratio indicated has a very wide range from no problem to outright inability to blow the bottle without flashing, depending on the various manufacturing systems, materials, bottle weight, and so on.

On bottles with necks that exceed a 4 : 1, it is a good probability that the parison will not be contained within the neck of the bottle. The excess material will need to be trimmed from the neck and the shoulder of the bottle. There, the ratio of the "E" dimension to the lateral body dimension is at or near one to one (as in a jar). The neck will tend to have a thin neck wall, threads that could accordion and a thin wall on the top sealing surface, even with programming.

The ratio of "front to back" to the "side to side" dimension does not present the same processing problems to the various bottle blowers. However, an extremely flat bottle is subject to increasing difficulty in maintaining wall uniformity. Base corners tend to be weak unless they are brought in and slower cycles are probable.

The flat oval bottle requires much more material to "cover" a specific volume of product as compared to a cylinder or a "mild oval." The ratio above is three to five in favor of the cylinder in regard to material consumption. However, from a marketing point of view, the flat bottle gives an illusion of being much more product. It provides a panel for a large label area. Both are valuable tools for marketing. The extra cost for the bottle may be well justified for sales promotion.

Another category of bottle shape that requires flashing, regardless of blow ratios or manufacturing processes, are bottles with handles or bottles with the neck close to the side of the bottle. These bottles also are subject to the principle: the flatter the bottle, the greater the amount of plastic required to cover a specific volume of product.

However, the image is favorable for market consideration, size impression and a large billboard. A round bleach gallon bottle will weigh around 90 g. Meanwhile, a narrow rectangular shaped bottle, packaging a national brand of antifreeze (gallon) weighs 145 g. The 55 g differential allows the antifreeze container to display a large prominent label extolling the qualities as well as instructions.

Pancake syrup is one of the better known hot fill products. Within the past 5 years, chocolate syrup made its debut on the market in a plastic bottle. With temperatures ranging 165 °F to 185 °F (74 °C to 85 °C), problems had to be faced that are not common to room temperature fill, namely: material design selection, bottle bulge, volumetric calculations, vertical strength loss, effect on capping and label wrinkle.

PP is widely used for pancake syrup due to good performance in the elevated temperatures and the favorable contact clarity.

Bottle bulge is a problem with hot fill bottles. A flat panel will bulge more than one with curvature. Hot fill will soften the plastic bottle. Front and rear panel bulge will affect the volumetrics and possibly the label quality. Thermal expansion of the product can be established. As an example, product fill is at 165 °F to 185 °F. The product is sold at room temperature. If the label reads 12 fl oz, the bottle must contain 12 fl oz at room temperature.

The expansion of the bottle due to hot fill calls for immediate capping. The capped bottle will be showing fill level dropping. However, as the bottle cools, the cooling product starts to contract. The internal pressure drops with a partial vacuum in the head space. This will draw in the bulge and the fill level will come up almost to the point as it was when capped hot.

Movement within the label area caused by bulging and the draw-in can cause label wrinkle and voids. Bottle design must account for a label panel of no more than one axis of curvature. Bulge compensation in the mold can be utilized. Predecoration (prior to filling) is a better route for label appearance on the store shelf.

Several chocolate syrups are packaged in oblong and in cylindrical PE bottles. The following refers to oblong containers. PE's hot strength is less than PP's, but, through proper design considerations, the PE bottles are very functional. Most PE syrup bottles are colored and opaque. This masks the internal walls of the bottle partially coated with chocolate.

Both PE and PP are subject to loss of rigidity; therefore, vertical strength and thread strength lessen when hot filled. PP has the edge over PE in this area. Thus, PP is a bit more forgiving where the desired aesthetics may cut into a design intended for good mechanical qualities.

The capping of the hot filled bottle requires consideration for the softened neck and threads. Excessive "on" torque may strip, or at least damage the thread. A minimum of 360° of thread engagement is a necessity. The wall of the neck should be thick enough to effect a substantial wall in the thread to prevent accordioning of the thread.

This type of failure will not have an appreciable, if any, “off” torque. This thread is also highly stressed.

Note that rolled metal caps are not a good closure for a hot fill bottle. The thread contour and thread projection are subject to tool wear. Metal caps do not render 360° of full thread projection. If possible, the “M” style thread should be used. The closure should be molded PP with at least 360° thread engagement.

The plastic bottle industry, after many years of investigation, has recommended two contours for threads for plastic bottles. The designations are “M” style and “L” style. The “M” style is a modified buttress thread. This contour is intended for molded plastic caps with a complementing thread contour on the engaging surfaces. This thread form will not reach its full potential with metal caps due to contour differences in the thread shape.

The “L” style thread is sometimes referred to as the all purpose thread contour. It is best to use this thread form with a cap having a 30° engaging angle for best “on” and “off” torque values. The 30° engaging angle is also better for centering with the cap as opposed to the “M” style thread form. The “L” shape is also a compromise thread form for rolled metal caps. The metal thread form was basically intended for the half round (bead) thread form normally found on glass containers.

Some bottle/cap combinations (carried over from glass) use the roll-on thread. A blank cap is a cap without the thread included and is placed onto a bottle. Three contoured rollers engage the assembly with side pressure pushing the soft metal into the voids between the threads. Thus, the rolled threads. In this application the neck wall must be rigid enough to be the anvil for the rolling process.

To have the true form, the “M” style and “L” style must be made in the more expensive unscrewing molds. That is: the core is turned to be threaded out of the cap. The stripper mold method requires a less expensive tool. However, a stripper mold cannot produce a true “M” or “L” contour. The thread form is a compromise of moldability and function [1].

The thread projection is generally less than from unscrewing molds. The thread form has a steep contour to facilitate stripping the cap off from the core. Extra fast cycles in molding may cause “thread drags” or other distortion. This thread form has been used successfully on many applications, especially on home fill such as mustard and ketchup dispensers. While this cap has wide usage, it is not for high torquing.

8.2.2 Closures

Closures are important to ensure that contents in blow molded bottles meet certain requirements, such as a positive seal, to prevent product contents from escaping and to allow no outside substances to enter the container, yet are easy to open or close (Fig. 8.9). The type of plastic and how it is processed in both the blow molded container and closure are of utmost importance to ensure positive seals.

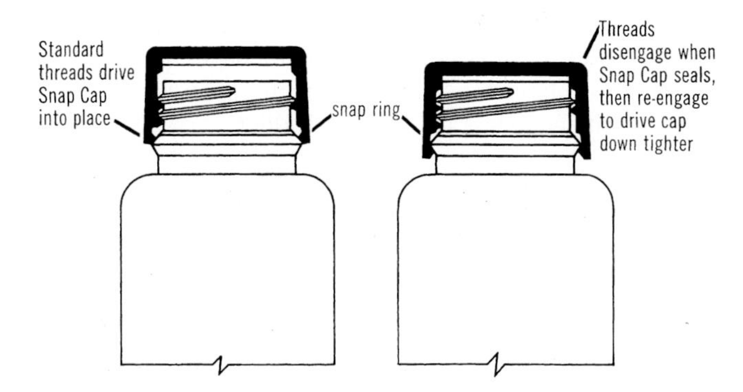

Figure 8.9 *The NBO, or non-backoff, cap from Sunbeam goes on like a screw cap and seals like a snap fit cap to provide a liquid-tight closure. The NBO caps are injection molded from Marlex PP supplied by Phillips Chemical Company. Although conventional screw caps are easy to use, they can work loose, or backoff, during shipping and handling. Snap caps, although sometimes difficult to operate, provide a positive seal: the cap is either open or shut. The NBO cap from Sunbeam is an example of combining the advantages of both*

With use, environmental conditions (heat, moisture, etc.), or contents (corrosive liquids, etc.), improperly processed plastics will no longer hold their preform shape correctly. Cap or bottle neck section could become elliptical in shape, flexible, etc. Proper quality control procedures can eliminate these problems.

Plastic closures are principally molded from TPs (injection molded) and TSs (compression molded). TP closures are primarily PE, PS, and PP. These three resins account for about 90% or more of all TP closures. Other materials are styrene acrylonitriles (SAN) and ABS copolymers; they are basically part of the PS family and are used for special requirements, such as clarity, high stress crack resistance, product compatibility, ability to be electroplated or metallized.

The non-rigid plastic closure has many advantages over the thermoset closure, such as closure design, closure specifications and other advantages. The closure has made considerable inroads not only in toiletry and cosmetics, medicinal and household product packaging, but has also figured prominently in the packaging of liquors and other distilled products. In the last several years, it has recently started to move into and take over a dominant position in the instant coffee field among others.

TS materials produce extremely hard, rigid parts with excellent dimensional stability and outstanding chemical resistance. TS closures—ureas and phenolics—were first used in the early 1900s and are still used because of their outstanding chemical resistance. TS closures also accept vacuum-metallizing decoration in silver and gold with superior adhesion qualities.

For economical production, most thermoset closures are plainly shaped, straight sided or slightly reverse tapered.

Principal closure types are classified by method of application: (1) screw-on threaded or lug, (2) crimp-on (crowns), (3) press-on, (4) roll-on, and (5) special snap-on.

Variations include vacuum, tamper-evident, child-resistant, linerless and easy opening closures and dispenser applicators. Primarily of metal or plastic, closures are also made of paper, cork, and rubber for special applications on dairy products, wines and liquors, drugs and pharmaceuticals and other products.

While certain types of closures are traditional for various containers or products, packagers examining closure options should weigh ease of handling on the filling line, consumer convenience in opening and reclosing, econony, decorative appearance of dispensing qualities, as well as product protection requirements.

Increased public awareness of packaged product safety has encouraged suppliers to develop new security-type closures and refine existing closures which offer evidence of breached packages.

Breakable caps of metal or plastic separate when opened so that the top part of the cap is removed and the bottom rim remains on the neck of the bottle retained by a special finish with a molded-in retaining ring. Variations on the breakable cap include aluminum roll-on closures and a cap which employs a tear-strip that must be pulled and torn away from the closure body, separating it into two parts, so the upper part can be removed.

Thread or lug type closures in plastic, metal or composites, are widely used on vacuum packed bottles and must be broken loose in order to pull up and snap off the lid.

Valve type closures, now in common use for plastic milk containers, employ a lower skirt on the PE closure that overrides ratchets on the container finish. The cap can be removed only by tearing away the band.

Protection, decoration and added barrier properties are achieved by special supplementary closures. The paper disc or diaphragm, sealed over a bottle neck opening, protects food and pharmaceutical products. Outer foil and shrink-on band are used for both protection and decoration.

PVC heat shrinkable seals, which protect against tampering and provide a moisture barrier, are supplied dry and shrink up to 50% of the original diameter within 3 s when exposed to heat from a heat gun or tunnel.

Closure trends are toward larger sizes and increased use of common stock caps to avoid higher costs of creating private mold closures. Special emphasis is being placed on improved designs for tamper-evident, child-resistant, and convenience closures.

8.2.3 Handles for EBM and IBM Containers

Many designers are familiar with the fact that the extrusion blown process can include a "blown" handle; typical product is the HDPE milk bottle. What many do not recognize is the fact that injection blown process also can readily include a handle that is not blown as shown in Fig. 8.10a.

Figure 8.10a presents figures which show the incorporation of a traditional jug handle above the blown portion of an injection blow molded container. The handle is molded

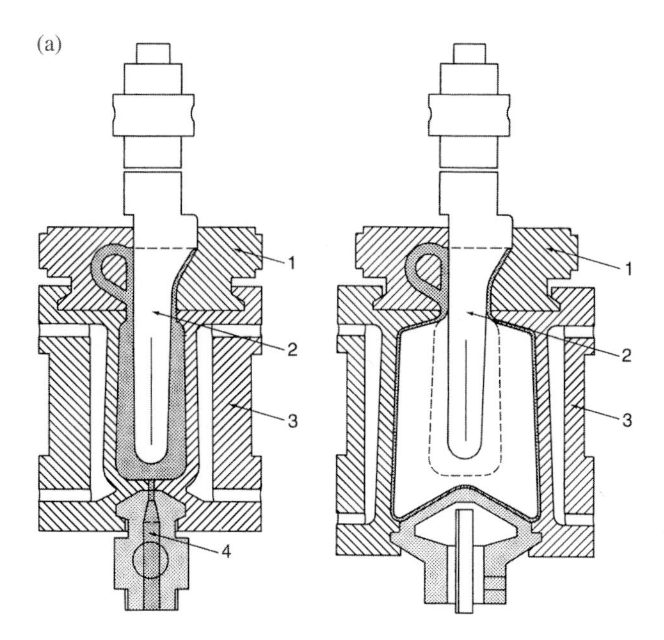

Figure 8.10 (a) Integral carrying handle for an injection blow molded product: (1) Precision "neck" mold that includes solid handle; (2) preform core and blow pin; (3) basic water-cooled bottle female mold; (4) injection nozzle of the injection molding machine

as part of the preform and is undisturbed when the container is blown. A direct extrapolation to stretch BM technology would incorporate the jug handle immediately below the neck finish of an injection molded preform. It would be necessary to mold such a preform in a split mold which suits itself to production on certain rotary type injection stretch blow equipment.

Another approach to a carrying handle provides for gripping features molded to a ring or strap placed just below the neck finish. Various carrying handles are shown in Fig. 8.10b.

Such a preform does not require a split injection mold and adds only a nominal amount to the clamp force and platen area of an injection molded preform. Attention

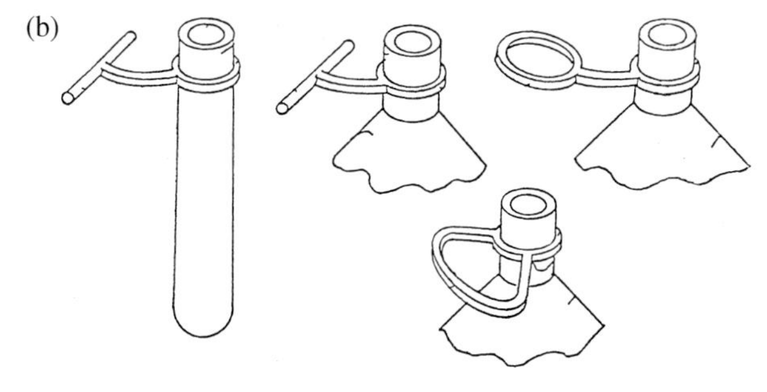

Figure 8.10 (b) Ring or strap type carrying handle

would have to be given to the arrangement of the cavities in the injection mold with the possibility that the number of cavities could be restricted in some situations. Such a handle would not affect bottle shape or height, which provides design freedom for either aesthetic or functional purposes.

An approach to producing an integral pouring handle or strap is shown in Fig. 8.11. There is an appendage in this design to the preform below the neck finish which can be attached at its lower end to the blown portion of the container. The appendage may be a portion of an injection molded preform or it may be attached to the preform prior to stretch BM.

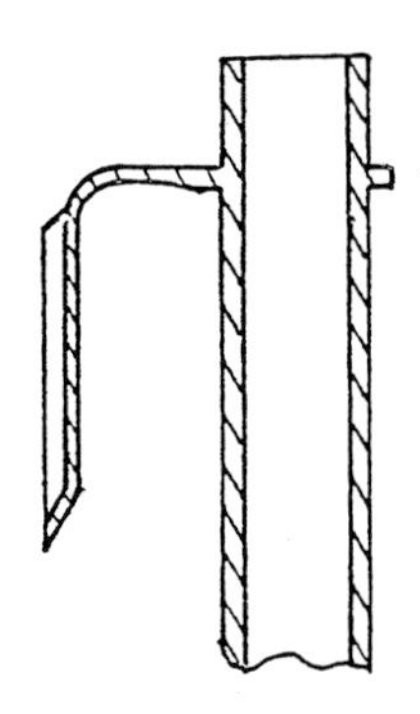

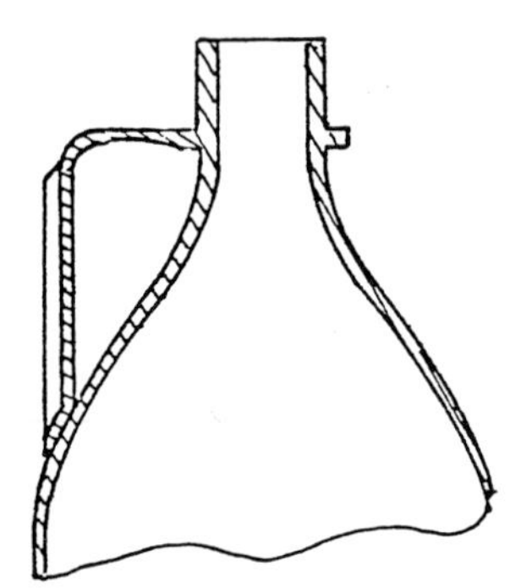

Figure 8.11 *Integral pouring handle*

Attachment of the bottom end of the handle can be made in many different ways including welding, gluing or use of a restraining means such as a wraparound label. The lower end of the handle may be used as a part of the blow mold surface and, if suitably configured, may produce a mechanical interlocking to the blown portion of the bottle. The handle may be semielliptical in cross section to conform comfortably to the hand.

8.2.4 Hinges

Hinges and straps are used in different blow molded products, such as containers and medical devices. A hinge is a device that is compliant in rotation about a fixed axis, but strong in tension, bending, and torsion about any other axis. Plastic hinges made from different polymers may endure from a few to thousands of flexes. The most popular material for hinges is PP. PP hinges that are properly made can last for over one million flexes.

Integral plastic hinges are also called living hinges. PP hinges can be manufactured by processes such as hot-stamping, extrusion, injection molding, and blow molding. Other plastics such as polyvinyl chloride, nylon, acetal, and even high-impact polystyrene (PS) can be molded or stamped into hinges. Some hinges are needed only in the assembly process, not in the finished product. There is just one cycle of flexing required of such hinges.

Basically, to produce a hinge during extrusion BM, the hinge is formed perpendicular to the parison flow from the die (Fig. 8.12). With injection blow molding, the hinge is perpendicular to the melt flow in the preform mold [1]. Guidelines to mold the living hinge with PP (other plastics are similar but may require their own specific dimensions that are dependent on their viscoelastic characteristics) are as follows:

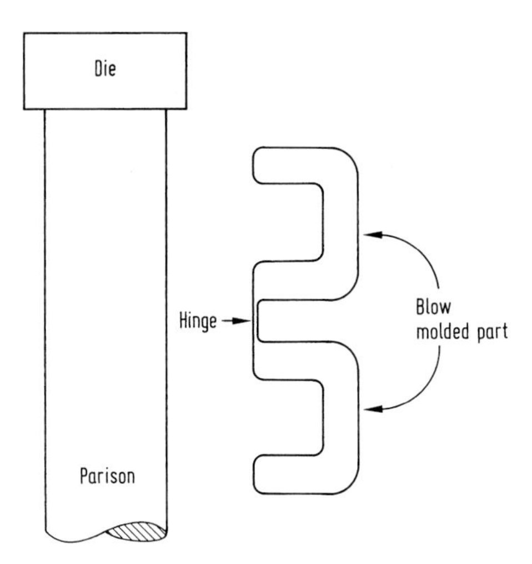

Figure 8.12 *Blow molded container with a living hinge in the as-molded position*

- The land of the hinge (Fig. 8.13) should be at least 1.5 mm (0.06 in.) wide for proper flow pattern and at least wide enough so that when the part is bent in service it will not develop strains: too short a land length will cause the hinge to have limited flex life.
- The minimum plastic thickness or pinch-off gap at the center of the hinge should be 0.25 to 0.38 mm thick (0.010 to 0.015 in.) and 0.5 mm wide (0.020 in.).
- When the plastic melt flows across the small hinge gap, frictional heat will be generated. There should be sufficient cooling of the mold around the hinge area.
- With injection blow, the hinge gap is a difficult area to flow across. Therefore, the gate should be placed so that the molten plastic can flow perpendicular to the hinge to ensure a good fill. If the melt flows along the length of the hinge, there is bound to be short-shot or cold-weld at the hinge.
- Shoulders and lips should be included in the two mating parts to help alignment.
- The finished piece should be flexed immediately upon ejection from the mold, while the heat from the mold is still present. A flex angle between 90° and 180° is recommended. The flexing action can stretch the hinge area by 200% or more;

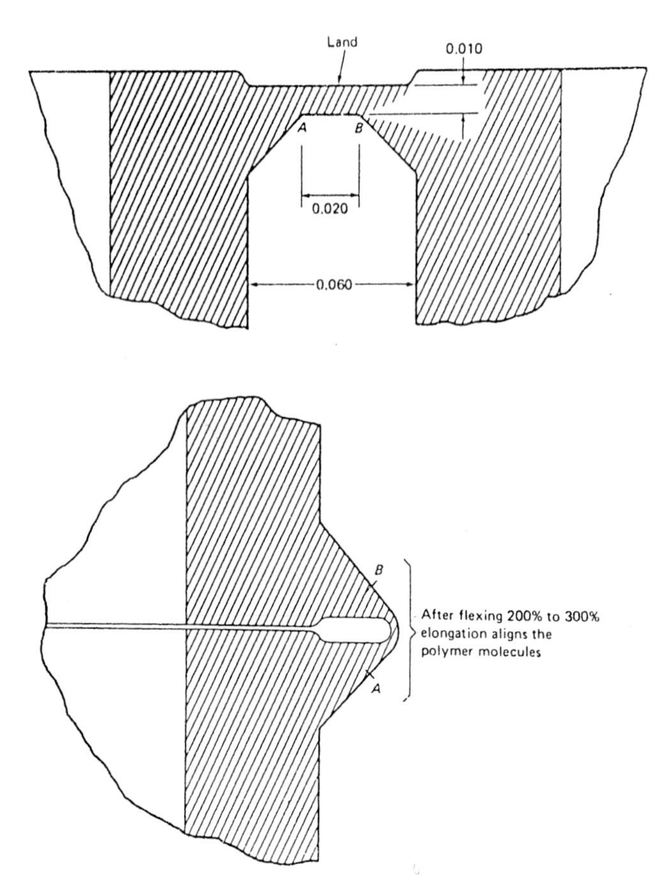

Figure 8.13 *Molded plastic hinge in as-molded position (top) and in a flexed position (bottom). Dimensions are in inches*

thus the initial 0.25 to 0.38 mm thickness (0.010 to 0.015 in.) will be thinned down to less than 0.13 mm (0.005 in.). This elongation aligns the plastic molecules and increases the tensile strength from 34×10^6 to 552×10^6 Pa (5000 to 80,000 psi).

The extrusion blown hinge can be compared to stamping. Hot and cold stamped hinges are made by compressing a sheet of material down to the desired thickness (about 0.25 mm or 0.010 in.). Stamped hinges are less durable in flexing, but have good tear strength.

Even though blow molding provides the capability of providing multiple parts combined into one, by including hinges the designer has an added feature. The ability to produce many articulated parts in one shot opens new design possibilities.

8.2.5 Snap Fits

Snap fits are widely used today for temporary and permanent assemblies principally in injection molded parts, but they are finding more use in blow molded parts [1]. Besides being simple and inexpensive, snap fits have many superior qualities. Snap-fitting can be applied to any combination of materials, such as metal and plastics, glass and plastics, and others. All types of plastics can be used.

The strength of a snap fit comes from mechanical interlocking, as well as from friction. The pull-out strength in a snap fit can be made hundreds of times larger than the snap-in force. In the assembly process, a snap fit undergoes an energy exchange with a click sound. Once assembled, the components are not under load. Unlike the snap fit, a press-fit component is constantly under the stress resulting from the assembly process. Therefore, over a long period of time, stress relaxation and creep may cause a press fit to fail. The strength of a snap fit will not decrease with time.

When used as demountable assemblies, snap fits can compete very well with screw joints, which under vibration can loosen bolts and screws. A snap fit is vibration proof because the assembled parts are in a low state of potential energy. There are fewer parts in a snap fit, which means savings in component costs and inventory costs.

The successful design of snap fits depends on observing a set of rules governing the shapes, dimension, material, and interactions of the mating parts. The interference in a snap fit is the total deflection in the two mating members in the assembly process. Too much interference creates difficulty in assembly, and too little interference causes low pull-out strength. A snap fit can also fail from permanent deformation or breakage of its spring components. A drastic change of friction due to abrasion or oil contamination may ruin the snap.

A snap fit can be characterized by the geometry of its spring component. The most common snaps are cantilever type, the hollow cylinder type as in the lid of a pill bottle (Fig. 8.9), and the distortion type. Such classification is rather nominal because cantilever is used loosely to include any leaf spring components, and cylinder is used loosely to include noncircular section tubes. Distortion-type snaps include any shape that is deformed or deflected to pass over an interference. While the shapes of the mating parts in a hollow cylinder snap are the same, the shapes of the mating parts in a distortion snap are different by definition. As an approach in designing snap fits the design engineer applies engineering equations for spring action, such as spring force, spring rate of snap, lead angle, and so on.

There is always some part of a snap fit that must flex like a spring, usually past a designed-in interference, and quickly return, or at least nearly return to its unflexed position, to create the assembly of two or more parts. The key to successful design is to provide sufficient holding power, without exceeding the elastic limits of the plastic.

The following guidelines are recommended regarding the position of the snap joint: (1) there should be no binding seams at critical points; (2) avoid binding seams created by stagnation of the melt during filling; (3) the plastic molecules and the filler should be oriented in the direction of stress; and (4) any uneven distribution of the filler should not occur at high-stress points [9].

8.2.6 In-Mold Labeling

In-mold labeling involves preprinted and stacked hot melt adhesive coated labels being individually placed into the mold halves. The hot surface of the melt activates the adhesive and the label is bonded into position. This process may add one second or more to the molding cycle, but can provide important savings in other areas.

In-mold labeling eliminates the need for pad printing, hot leaf stamping, and other labeling methods. Further, due to the flush inlaid position of the label, it allows existing filling lines to be run at higher speeds. An advantage of direct interest to the blow molder is that in-mold labeling allows the bottle weight to be reduced. The combination of the stiffness of the label and the prestressing that occurs in the bottle wall as it cools and contracts differentially relative to the label, allows thinner blown wall sections to be specified.

Molds that have been modified or designed for "in-mold labeling" are vacuum fitted to both vacuum position and retain the "in mold" labels. Due to the negative air pressure effect (vacuum), such molds will usually have increased capacities proportional to the shrinkage reduction obtained resulting from the vacuum. Compensating to recalculated filling levels is mandatory to balance the resultant volumetric increase of the container.

Under most circumstances, molds which have been normally vented at the parting line, then modified to accept in mold labels via vacuum clamp of the label will run without further modification. However, the vacuum draw may create cosmetic parting line problems, necessitating the elimination of face vents on the mold parting line.

Where volumetric piston filling of a blow molded container is used, filling speeds are often restricted due to lack of available head-space allowance in the container. Three options are open: (1) restrict line speeds; (2) rework or remake molds to increase capacity; (3) adapt the molds for vacuum assist pressure molding.

The latter (vacuum assist) often offers the benefits of capacity (head space) increase without the time and cost of cavity reworking or full mold replacement.

The vacuum assisted mold ensures closer contact of the parison to the mold wall. Effectively, the rate of heat transfer is accelerated, which usually results in cycle reductions of up to 10%. Table 8.2 shows the effective cycle gains made using vacuum in conjunction with various wall thicknesses in a HDPE container. The advantage of vacuum assist is further enhanced by utilizing maximum blow pressures. Note in Table 8.2 that the "vacuum only" cycle was only on a par with the 60 psig blow pressure. The maximum benefit occurred at full vacuum *plus* 125 psig blow

Table 8.2 *Cycle Time Decrease [Cooling Input Kept Constant at 28°F (−2°C); G.P.M. Coolant Flow Kept at Constant G.P.M.; Vacuum Set at 27 Hg Gauge; Surge Tank Cap. 10 ft³ at 27 Hg]*

Wall thickness (in.) (mm)	Blow pressure (psi)[a]	Normal cycle time (s)	Cycle time with vacuum assistance (s)	Cycle time vacuum only (s)
0.040 (1.0)	40	14	12	12
	60	12	11	
	80	11	10	
	100	10	9.2	
	125	10	9	
0.050 (1.21)	40	16	14	15
	60	15	13	
	80	14.5	11	
	100	14.5	10	
	125	14	10	
0.060 (1.52)	40	20	16	18
	60	18	15	
	80	18	15	
	100	17	14	
	125	16.5	14	

[a] psi $\times$ 6.8966 $\times 10^{-3}$ = MPa
psi $\times$ 68.966 $\times 10^{-3}$ = Bar

pressure. In the case of containers which are trimmed "in mold," the problem of deflashing using reduced blow cycles and vacuum assist is substantially improved. This is due to the more rapid rate and degree of container cooling.

In the case of stretch-blow containers, whether generated from injection or extruded parison, the requirement for thermally controlled parisons at the lowest feasible temperature places a heavier demand on internal pressure/volume to ensure adequate movement or expansion of the parison in the transverse direction.

In recent studies, there is evidence that vacuum assisted stretch blowing provides faster expansion rates of the parisons and improved performance at higher stretch-blow ratios.

The early work in vacuum assist blow molds has now reached a point in time where the technology may resolve many problems, particularly those associated with new practices such as direct "in-mold" labeling.

8.2.7 Large Industrial Containers

Large containers from at least 5 gallon (18.9 liter) to 55 gallon (208.2 liter) are fabricated replacing metal. The US Department of Transportation (DOT) continues to approve more of these plastic containers for hazardous material. The light weight of the plastic as well as the number of trips that it can make continues to provide an attractive 55 gallon drum for the industry.

No reconditioning is necessary, such as in steel drums. Plastic containers up to the 55 gallon size can be made on-site as opposed to metal containers that are made and shipped to the consumer. This type of BM is done with an accumulator type BM machine, as the large weight of the parison precludes the use of continuous or intermittent type machines. The accumulator machines extrude the parison at a high speed so that the parison does not sag.

The automotive industry is one of the largest markets for plastic products. Replacement of metal products by plastics reduces the weight of the car and eliminates the rusting problem. In addition to custom BM plants, BM machines can be installed at the car manufacturing site and plastic products can be made as required. This eliminates the need to transport metal products from a supplier company. The projected growth for plastics in the automotive industry are tremendous and it is projected that there will be few metal parts in the future automobile.

8.2.8 Double Wall Molding

The discovery of industrial BM by the automotive industry has led everyone who works or is involved in plastic development to exploit and expand its market opportunities. There are three basic design approaches in industrial BM: hollow, double-wall and structural shapes, which today are used to market a variety of consumer and industrial parts and products. For many applications, all these design methods offer protection, versatility, and economy to such an extent that other processes have difficulty competing with them [18].

Hollow and structural industrial shapes are blow molded into such forms as automotive sun visors and chair seats. Double-wall structure concepts are, however, directly attributable to the work of Peter T. Schurman. In December 1963, Schurman developed double-wall cases that required a substantial male mold member or “core,” instead of just two flat cavities with incidental contours therein. Schurman explained that Illinois Tool Works molded, on a development basis, flat step treads for use on Hamilton Cosco kitchen stools. They actually were double-wall parts, about 13 × 8 × 1 in., with a level, textured top surface and a ribbed underside. Joining two treads with an integral hinge created the first double-walled product, with the smooth side on the outside cavity, the ribbed side on the inside core. From this exciting beginning, the concept of double-wall cases has been used wherever there was a need for special products such as products to store and protect high-value instruments and major tools.

In addition, the sales appeal of a permanent storage case developed the market opportunities that corrugated packaging could not offer. Double-wall cases are being used for chain saws, small power tools, monitors for heart pacers, computers, security chests, musical instruments, carrying cases (Fig. 8.14), and many other applications.

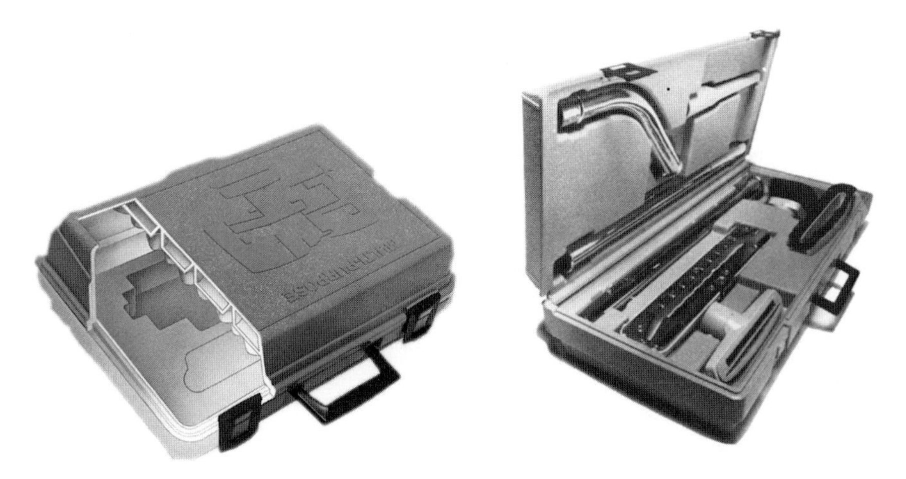

Figure 8.14 *Examples of BM double wall products: HDPE carrying case protects and simplifies part storage*

HDPE and propylene copolymers, with or without fillers, are the resins of choice, concurrent with new resin developments such as advanced engineered thermoplastics alloys.

In 1982 and 1983 research carried out by Phillips Chemical resulted in the use of large seat back panels for the Chevrolet Z28, the Pontiac Firebird and the Chevrolet Camaro which were blow molded. This exciting step to blow mold double-wall parts with high structural strength, using filled HDPE with mica, opened the door to new ventures, first in the automotive industry (Fig. 8.15), later with spillover to the construction and furniture segments of the industrial market.

Continuous-extrusion machines with large accumulator-head systems are utilized for polyolefinic as well as the new engineered resins. Specifications for the machines vary according to the needs of the processor. Following are the maximum requirements for a state-of-the-art machine which can handle the most sophisticated applications being performed today:

- Accumulator head machine, 15 to 20 lb (6.8 to 9 kg) plus
- Oversize head tooling, 18 to 20 in. (46 to 51 cm) wide
- Wide platens, 54 in. wide by 60 in. (137 to 152 cm) long
- 120 ton clamp pressure
- 1200 in./min (3,048 cm/min) clamp close speed
- $4\frac{1}{2}$ in. (114 mm) extruder
- Multipurpose barrier screw
- Additional features: Parison programming, fast response hydraulics, adequate torque

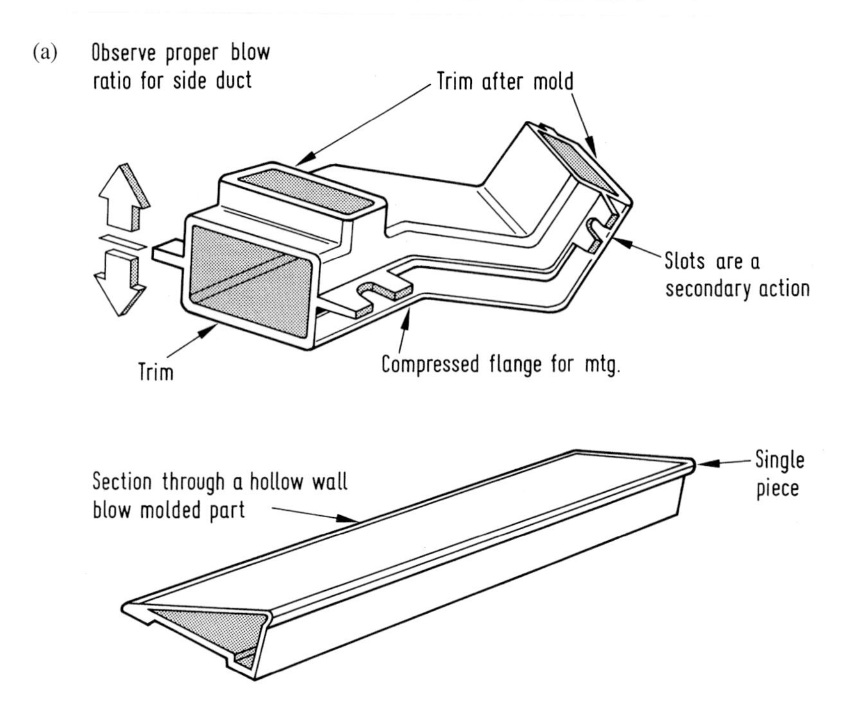

Figure 8.15 *(a) Spoiler duct*

- 1,000 lb (454 kg) dryer
- Mold heater

The manufacturers of these machines are the same companies who ventured into the building of large machines for polyolefinic materials. They must concentrate not only on larger extruders, heads and platen sizes, but more important, on specifics to process the new resins. Faster closing, high tonnage clamps and programmed blowing, in conjunction with specialized barrier screws using grooved feed sections are a few of the challenges which face the manufacturers. Automatic ovalization with limit switches or servo-controls is essential. Hydraulic or pneumatic part-takeoff systems are being installed dependent on the needs of the particular processor. Following is one set of detailed guidelines:

- Rugged, efficient first-in–first-out (FiFo) accumulator heads. By using accumulator heads designed for first-in, first-out material flow, change of color or material is done with ease. Optimum structural weight and close control with conventional or engineering-grade materials is accomplished with the same screw and accumulator head. For maximum control, each accumulator head has its own servocontroller and parison weight and program control. Advanced design eliminates material degradation in flow areas, helps to avoid sagging, stretching, and weld lines.

(b)

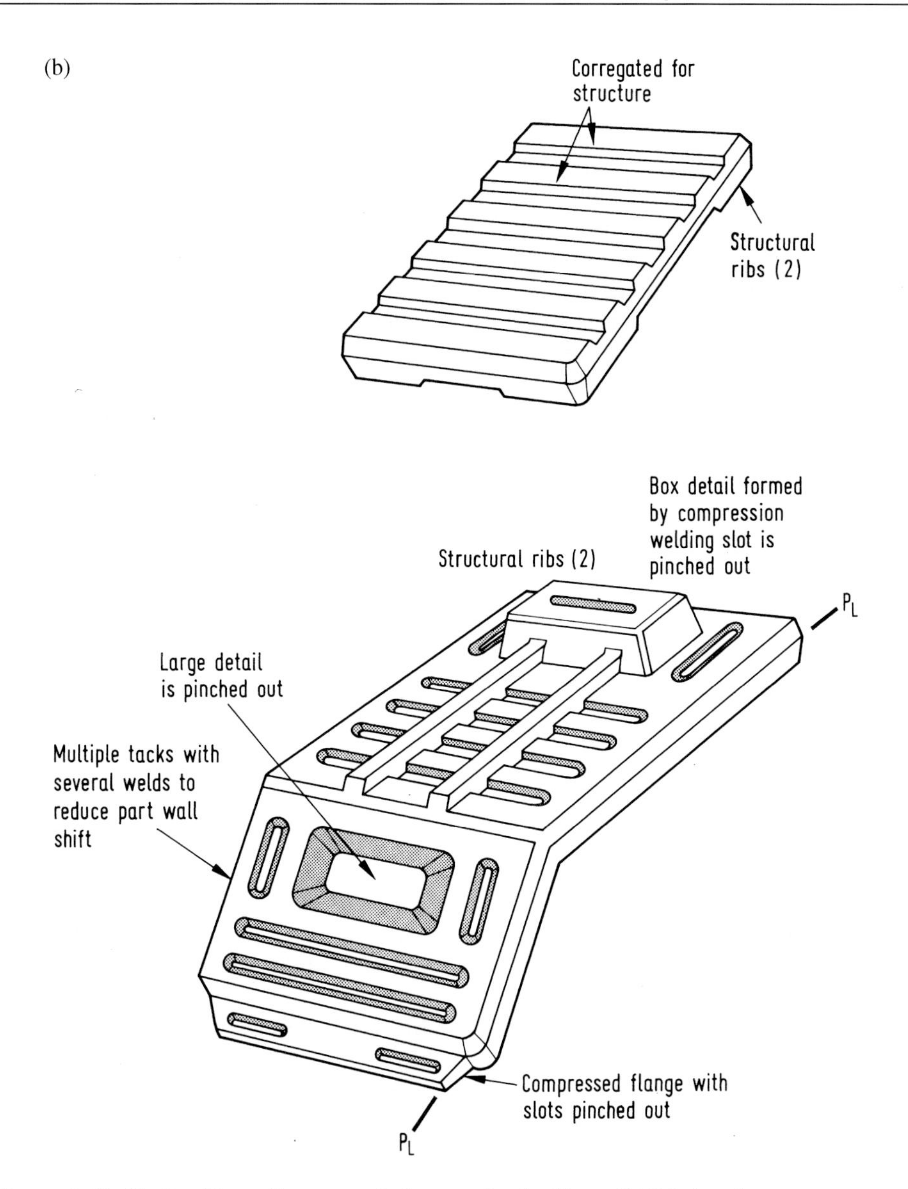

Figure 8.15 *(b) Double-wall structural designs. In double-wall BM the ultimate structure is achieved by ribbing and compression molding with two walls of the parison together. However, depending on part size, the compression area is limited to a total clamping capacity of the blow mold equipment. Tacking the two surfaces without compression provides extra strength without creating clamping pressure problems*

- Precise, time-proven microprocessor control. Electronic production control assures that every part will conform to the most rigid specifications for sharp production definition. It is also possible to reduce product weight and cut cycle time up to 25%. One can electronically control temperature, pressure, all sequential operations, linear positions and count. There is a built-in diagnostics system.

  The easy-to-program microprocessor lets one set a job once and retrieve key data any time to rerun. Set up time is minimized with optimal molding cycle stability and repeatability. Superb parison control is made possible by state-of-the-art parison programmer with parison display. One has full control with: (a) precision adjustments to run a wide range of engineering plastics; (b) uniform temperature control for energy savings, minimum waste and scrap, and quick part cooling time; (c) meticulous regulation of wall cross thickness and material distribution by control of weight, length of parison, die gap and shot size during parison extrusion. Part weight can often be reduced by 5% to 30%; (d) optimum strength-to-weight adjustments for smooth profiles, uniformity, stability, high-impact tolerance, and better looking, stronger parts; (e) production flexibility for running unusual resins, shapes and features such as core slides, knockouts, needle blow, inserts, and so on.
- Long-life high performance feed screw. Feed screw features unique barrier flight that melts polymer in the first two thirds of the screw. With durable SAE 4140 chrome molybdenum steel construction, it greatly outperforms conventional or even fluted mixing screws. Energy savings up to 20% can be achieved with efficient homogeneous melt and can be used with any feedstock: pellets, powder beads, flake, regrind or a combination. Power, versatility and higher output rates at lower screw speeds and stock temperatures are available with screw.
- Continuous extrusion for optimum control.
- Large accessible clamp area for a variety of product applications. Customized platen sizes meet large mold production needs. Each arrangement offers 1,200 in./min (3,048 cm/min) and superior accessibility for fast mold changes. Microprocessor controls allow one to set up new mold positions in seconds. Guide bar design assures precise platen parallelism. Platen equalization is controlled by a mechanical rack and pinion assembly. Motorized jacks allow incredibly precise vertical height adjustments of ±4 in. (10 cm). Horizontal adjustments of ±3 in. (8 cm) front-to-back and ±3 in. side-to-side are controlled manually. Automatic gates are air cylinder controlled.
- Quick setup for optimum productivity. Clamp systems are specially engineered for precision large part blow molding, such as double wall parts, where high tonnage flow is critical to success. To assure true equal clamp tonnage, hydraulic cylinders are all push type and the main hydraulic system has a large reservoir capacity. To minimize oil leaks, all valves are manifold mounted. The manifolded air blow system gives unheard-of flexibility to apply any kind of blow molding technology to challenging parts or products. One can select and fully control any kind of blow required for precision results: top blow, preblow, needle blow, bottom blow, gates, prepinch circuit, air-core pull circuit, and stripper.

- Special considerations. Equipment should include control with digital readout of either pressure or flow (speed) of the accumulator. The same control is used on clamp speeds with pressure or flow regulation. Air blow pressure capability of at least 150 psi (1 MPa) with clamp tonnage of at least 150 ton on 54 in. × 64 in. (137 × 163 cm) press. Use of proportional air valves to allow for controlled formation of parison. Rectangular tooling to produce parisons with flat rectangular finished parts. Special coated surfaces on the internal parts of the head to allow for smooth and hard finishes of the steel. Quick disassembly/assembly of head subcomponents to allow for rapid inspection and cleanout.

Most other manufacturers have their own designs and innovative approaches to mold double wall products on their machinery, but essentially the above-mentioned guidelines give a glimpse of what has been accomplished or what is to come.

8.3 Mold Preparation and Other Considerations

The practice of using aluminum molds (forged aluminum 6061, 7075) is still viable. However, considerable venting is usually required and the surface appearance is much more critical compared with molds for polyolefinic parts. Also, the higher tonnage required to mold engineering resins, resulting in larger lands of pinch-off, coupled with the toughness of the materials, have forced the mold builders to use steel pinch-offs. Complete steel molds (P-20) have been designed for large volumes and/or high-finish applications. Casting is sometimes considered.

The molds must be heated to a degree depending on the choice of resin being used. In designing the molds, the double-wall aspect must be reviewed in detail because the rigidity in the final part is of utmost importance. Ribbing and compression moldings are two approaches that mold the two walls of the parison together. Draft angles and undercuts must be modified from standard practice for olefinic parts.

In selecting a texture, patterns should be etched to a depth of 0.015 in. to 0.020 in. (0.38 to 0.5 mm) to obtain an even 0.010 in. to 0.012 in. (0.25 to 0.3 mm) when using polyolefinic (PE, PP) materials and aluminum molds. As soon as some of the engineered TPs are combined with the structure of a P-20 steel mold and the state-of-the-art BM machinery, the rules for injection-molded parts are being approached. Then one can employ again the rules of $1\frac{1}{2}^{\circ}$ draft for every 0.001 in. (0.025 mm) of texture depth and average depth of 0.0025 in. to 0.003 in. (0.06 to 0.07 mm). New 3D processes are providing double-wall blow molded parts with new alternatives today. The controlled disintegration of metal methods is continuously expanding to reach new markets and applications, from high-tech finishes with matte and spatter-paint type finishes to graphics, logos and other designed styles.

8.3.1 Complex Irregular Shape

Very complex, irregularly shaped products have been extending BM technology because conventional mold designs fail to produce acceptable wall thicknesses. Using mold halves with moving sections, some complex products are routinely and successfully molded. When the product geometry involves reverse folds or inwardly protruding channels and narrow projections, conventional molds cannot guarantee acceptable wall thicknesses. In plastic-starved areas, particularly corners, product walls are thinner than in other sections and are thus subject to rupture. The solution is in using an alternative to the conventional moving section [18].

8.3.1.1 Moving Mold Section

Technology has been developed and applied to the use of moving mold sections to blow mold highly irregular product shapes. The significance of the technology is that it provides a means to BM complex products in one piece. Although it is fairly common to move mold sections to release undercuts which permits product removal, molding techniques that affect or control wall thickness are used by relatively few blow molders. The ability of these new molders to blow mold uniquely shaped products has paid dividends, by eliminating or reducing costly secondary operations.

8.3.1.2 Integral HDPE Handle Lid

HDPE water cooler lids with an integral handle have been blow molded commercially since 1965. Several million lids of different sizes have been produced using movable mold sections of blow molds to form handles. The mold consists of two mold halves, one half for the top of the lid and the other half for the bottom. The top half contains the movable "slides" to form the handles. The bottom half of the handle cavity is located in the movable slides and the top half in the standard part of the mold. The horizontal slides are located under a face plate as shown schematically in the "open" position (Fig. 8.16).

The molding sequence starts with a pre-pinched and slightly preblown parison between the two open mold halves. The handle slides are open at this time. The mold halves are then closed on the parison. The closing action of the mold compresses and forces part of the parison into the handle cavity. A horizontal cut through the mold showing the parison "bubble" during mold closing is illustrated in Fig. 8.17.

The blow air is introduced through a needle into a flash "pocket" under the handle. The blow air forms the parison to the flash pocket shape, and then flows through two open pinch blade spots under the handle, resulting in the lid body being blown through the handle passage as illustrated in Fig. 8.18. The handle slides open with the mold opening to release the undercut and allow the part to eject.

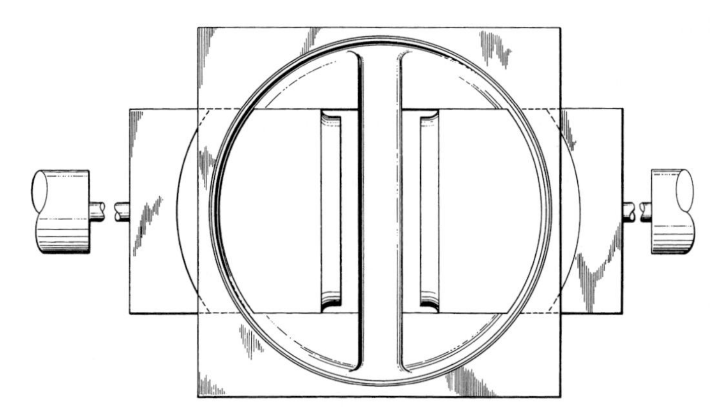

Figure 8.16 *Integral handle mold half—slides open*

The processing efficiencies are excellent, but require precise control of the following variables: timing, distance of slides opening, and speed of slides movement. The amount of preblow air and blow air timing also must be regulated and synchronized accurately with respect to both mold closing and slides closing.

The resultant lid has flash which must be trimmed around the periphery and under the handle; otherwise the handle is completely finished as the slides and stationary mold cavity form the handle without flash.

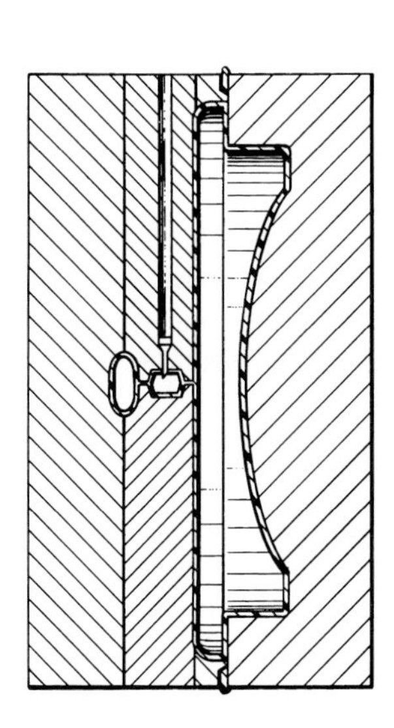

Figure 8.17 *Integral handle mold cross-section—slides closed*

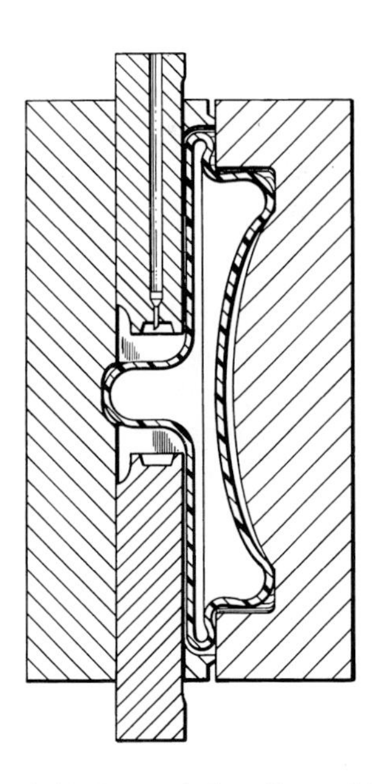

Figure 8.18 *Integral handle mold cross-section—slides open*

8.3.2 Integral Handle, Double Wall, Internally Threaded HDPE Lid

Further development work resulted in a second HDPE one-piece water cooler lid also having an integral handle plus an internally threaded double wall skirt.

Again there are two basic mold halves—one half to form the top and outside skirt and the other to form the internally threaded inside wall of the lid. The top half of this mold differs considerably from the first lid mold. This top mold half is split into two horizontally movable quarter mold sections to form the handle as shown in Fig. 8.19. This is in contrast to the first lid mold in which the slide sections only are movable. The bottom half is also quite different in that double wall blow molding technique is utilized to form the double wall skirt (Fig. 8.19). A third distinction is the molding of the internal threads on the inside of the double wall. This necessitates an unscrewing core which releases the threaded lid off the core on mold opening.

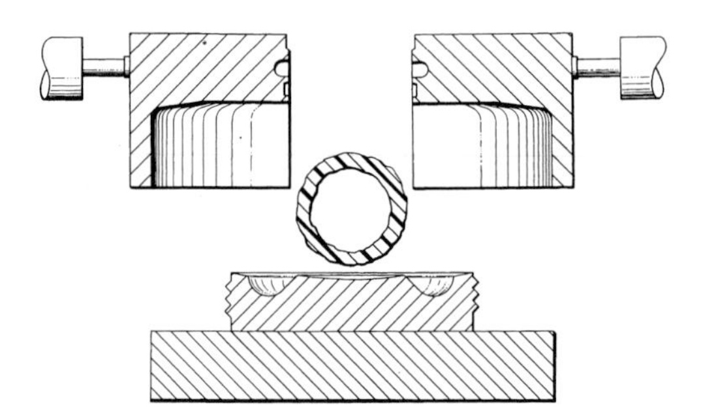

Figure 8.19 *Open quarter mold sections and thread forming core*

The molding sequence requires that a prepinched and preblown parison be dropped between the mold halves. The quarter molds of the upper half are open at the time the parison is extruded. The press closes and almost simultaneously the quarter molds are closed by air or hydraulic cylinders to pinch the parison sector which forms the handle. The closing of the mold halves compresses and "balloons" the prepinched parison over the threaded core, "flashing" the cavity periphery. Blow air is introduced, forming the part as shown in the horizontally sectioned illustration, Fig. 8.20

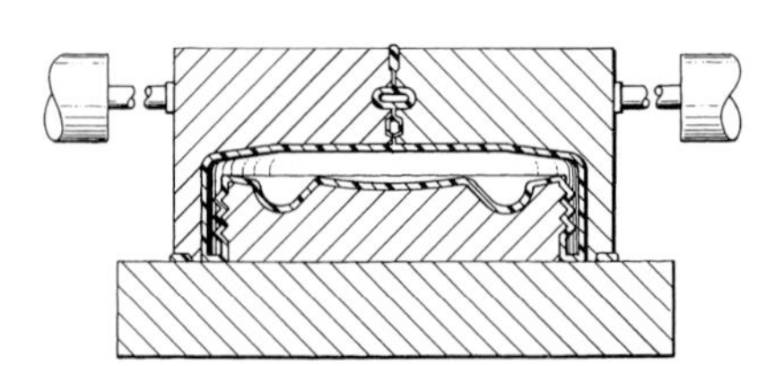

Figure 8.20 *Open mold sections closed on thread forming core half*

At about the same time the press opens, the threaded core rotates to unscrew the part off the core. After a short delay, the quarter mold sections open. The delay on the quarter mold opening "holds" the lid so that the threaded core can unscrew rather than rotate the entire lid. Splitting the handle section of the mold into quarter sections results in the advantage of being able to mold a complex, irregularly shaped part with excellent wall distribution. The disadvantage is "flash" in two planes which requires more trimming and regrinding.

The internally threaded lid can also be molded with internal threads in a compression molded solid wall. This type of lid has excellent thread strength but it does not have the rigidity of a lid with a double-wall skirt; it tends to warp, and the outside appearance is inferior to the blow molded double-wall part because of "sink" and "drag" marks.

Over forty million of water cooler lids produced in the past two decades have proven the practicality and economics of using moving mold sections to blow mold irregularly shaped one piece parts otherwise not blow moldable. The technology should be applicable to other parts now made in two or more pieces, and to improve wall distribution in some blow molded parts produced conventionally.

8.3.3 Drum Integral Handling HDPE Ring

A 30 gallon (114 liter) HDPE drum with a handling ring integrally molded in the bottom chime area has been developed and repeatedly molded to meet performance requirements. The handling ring allows metal drum lifting equipment to be used. A "double wall" relatively sharp radius chime was molded in the top (bung end) of the drum. The handling ring and sharp radius corner was located in their respective ends for molding building convenience and would be reversed in a production mold.

Although similar to the European "L Ring" drum in appearance and function, the Plastics Technical Center of Phillips 66 Co. drum mold was designed and built independently. This drum mold has been used to develop molding techniques, to mold drums for testing (including handling), and for the evaluation of resins.

Each end of each mold half has a movable section referred to as a "plug." The movable plug in the end of each mold half is semicircular. The mating plugs of two closed mold halves form the head and bottom of the drum mold. The plugs are moved by means of hydraulic cylinders on each end and both sides of each mold half. The $3\frac{1}{2}$ in. diameter hydraulic cylinders are securely mounted to the mold body and plug assemblies. Large cylinder rods ($2\frac{1}{2}$ in. diameter) are used to assure smooth plug movement with no "cocking" caused by possible variations of hydraulic flow or pressure in the cylinders. Good wear plates are required between the plugs and mold body to allow plug movement under machine clamp and blow air pressures.

As is normal in large drum BM, the mold is placed in the press with the bung opening down. The molding sequence starts by extruding a parison between the mold halves. The parison is dropped over the two bung opening blow pins which serve as a parison spreader device. The blow pins are in the "together" position when the parison is extruded and then move outward to a position corresponding to the bung openings in the mold halves.

The mold halves are closed on the blow pins, either (1) compression molding bung threads on the outside diameter of the stub neck, or as in this case, (2) the blow pins hold injection molded high density polyethylene bung inserts which fusion weld to the parison forming the drum head. During mold closing, the plugs are open (extended). With the plug ends *open*, low pressure blow air is introduced forming the parison to the cavity shape. A short time *after* the blow air is introduced the plugs are closed (retracted) about 2.9 in. The extra length of parison is compression molded into the handling ring for the drum bottom. On the bung end, the extra parison length is moved into the drum top corners.

The next step is to apply high pressure blow air into the drum and the part is cooled. The plugs retract on mold opening to release the undercuts created by the handling ring and top corner reverse radii. Separate timing was required for the plug movement on the bung and relative to the timing for the handling ring plug movement.

A combination of die shaping, parison programming rate and distance of plug travel results in a solid handling ring with uniform circumferential thickness. The distance of plug travel is extremely important—increasing the plug travel from 1.9 in. to 2.9 in. increased the wall thickness and improved wall thickness uniformity in both the handling ring and blown top corners.

Reproducing wall distribution from part to part requires high precision timing of the blow air with respect to the plug movement. The handling ring can be made solid, semisolid or hollow simply by changing the timing of the plug movement. Early plug movement results in hollow handling rings and later plug movement in solid rings. Although the plug timing is an easily adjustable process control, it is also a process variable as normal operating "drifts" can cause the ring to vary in wall thickness from solid to semi-hollow, etc.

The work at the Plastics Technical Center indicated that a solid ring can probably be molded consistently with good equipment, careful monitoring of the molding conditions and good quality control; however, molding a hollow ring with acceptable reproducibility is more difficult and possibly would require more precise controls and/or more consistent melt viscosity in resins than presently available.

The same plug movement on the bung end of the drum resulted in good wall thickness and in a sharp radius chime (no compression molding). This is significant in that drums with straight walls and sharper radius corners can be designed for good top load strength.

Drums molded with solid handling rings can be lifted with conventional drum lifting equipment; however, the hollow ring does not have enough wall thickness to withstand the frequently used “parrots beak” lift.

HDPE resins with about a 10 HLMI (High Load Melt Index as measured by ASTM D1238 test method) have very good drop impact performance in conventional drums and are the standard material used in the United States. These same resins fail dismally in the solid handling ring drum. The poor impact results are a combination of three factors:

- There usually is a slight “vee” notch between either the drum head and solid ring, or the drum side wall and the solid ring.
- The wall thickness in the “vee” often is slightly less than the nominal wall thickness of the drum wall.
- The slower cooling of the thick wall handling ring adjacent to the faster cooling walls probably results in molded-in stresses.

Poor drop impact properties of drums with solid rings molded with nominal 10 HLMI can be overcome by using higher molecular weight resins. HLMI vs. drop impact of 30 gallon (114 liter) drums filled with water shows the effect of molecular weight:

High load melt index	*Drop impact (Ambient temp.)*
10	2.5 ft (0.76 m)
2.2	7.2 ft (2.2 m)
1.9	9.3 ft (2.8 m)

HDPE resins with a HLMI of about 2.0 are considered optimum for drums with solid handling rings. These resins are available only in fluff (powder) form.

Development work has shown that drums with an integral handling ring which can be handled with conventional lifting equipment can be blow molded in one piece using moving plugs. One of the most important keys to achieving desired wall thickness is the distance of plug travel. The molding process requires good equipment, careful monitoring of production, and high molecular weight high density polyethylene “fluff” resins with great inherent toughness. Plastics Technical Center work on blow molded drums with solid handling rings generally agrees with the known parameters used in molding the European “L-Ring” drums.

The ability to mold relatively sharp radii corners with good wall thickness using moving plug technique is significant. The resulting greater design freedom should improve part geometries and spawn new applications, particularly in the medium to large size parts. The higher precision of microprocessor controls on blow molding equipment should result in wider usage of moving sections blow molding technology.

8.4 Multiple Parting Lines

A multiple parting line technology has been developed which allows the blow molding of part shapes not possible with previous moving sections methodology. Articles with a reverse "fold" or inwardly protruding channel at the parting line of the mold can now be blow molded. Similarly, parts with narrow, double or single wall projections on either side of the parting line can be made with good wall distribution.

8.4.1 Moving Section

At first glance, it seems possible to make the part with the mold shown in Figs. 8.21 and 8.22, where the moving section, which forms both the channel and the recess, is attached to one mold half. Less obvious is the difficulty of molding the long, inwardly protruding channel, which forms legs along both edges of one surface of the part, and the deep recess transverse to the direction of mold travel. A look at the molding sequence reveals the flawed design of the mold. In the first stage, the preblown

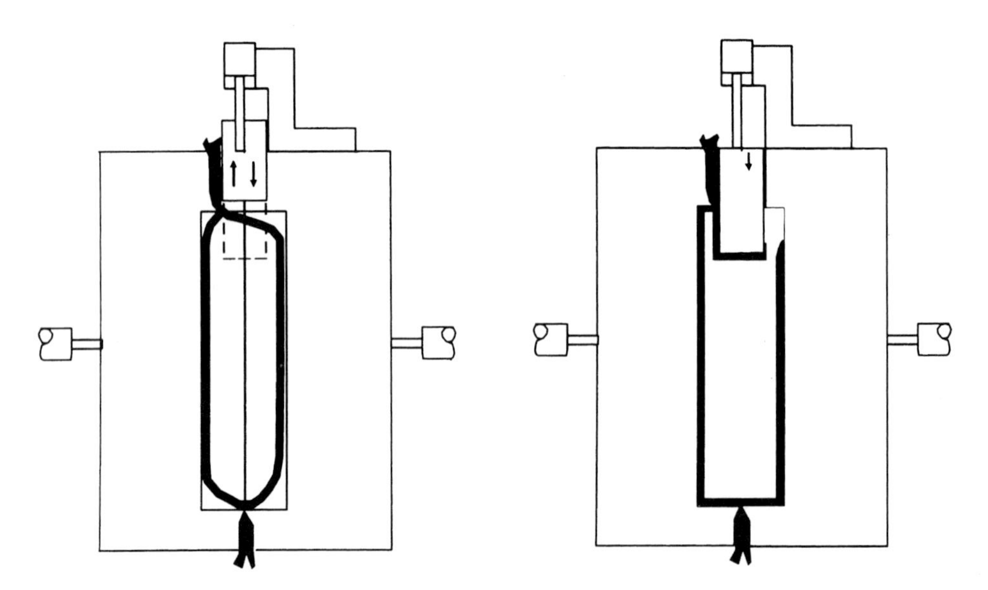

Figure 8.21 *A mold incorporating a moving section (shown in the retracted position). The leg formed between the moving section and the mold half to which it is attached will be thinwalled and prone to rupture, the leg formed on the other side will be strong, with good wall-thickness distribution*

Figure 8.22 *Basically, the mold has a movable member attached to a mold half*

parison flows (or flashes) unimpeded into the space between one mold half the retracted moving section, as the mold closes. When blow air is introduced, a strong leg with good wall thickness distribution forms on the side where the flashing (good flow) has occurred. The other leg—formed by blowing the resin into the narrow space between the moving section and the mold half to which it is attached—is paper thin and usually ruptures. Even with much greater wall thicknesses (and proportionally greater part weights), the results are the same. Eighteen months of trials, modifications, and further trials proved that the only way to blow mold the separator in one piece and with good wall thickness distribution throughout was to design a mold that permits the preblown parison to flow unimpeded around both sides of the third mold section.

The key to unrestricted flow around the third mold section is to detach it from the mold half and mount it to a beam affixed to the press frame. A stationary third mold section is thus established between the mold halves, creating a second parting line to balance the flow.

8.4.2 Fixing the Moving Section

A company was contracted to blow mold a one piece high density polyethylene refrigerator compartment separator. The part was to replace an existing eighteen piece combination metal/ABS separator. The flat, planar part required "legs" to protrude from each edge of both sides of one parting line side. The general configuration of the item is shown in Fig. 8.23. Overall dimensions were about $29 \times 23 \times 2\frac{1}{2}$ in. ($74 \times 58 \times 6$ cm) thick.

A mold as built to blow mold this shape using a conventional moving section to try to form the reverse fold. The moving member was attached to one mold half as shown in the cross section (Figs. 8.21 and 8.22). The molding sequence was to close the mold halves on a preblown parison with the moving member (bar) in the retracted position.

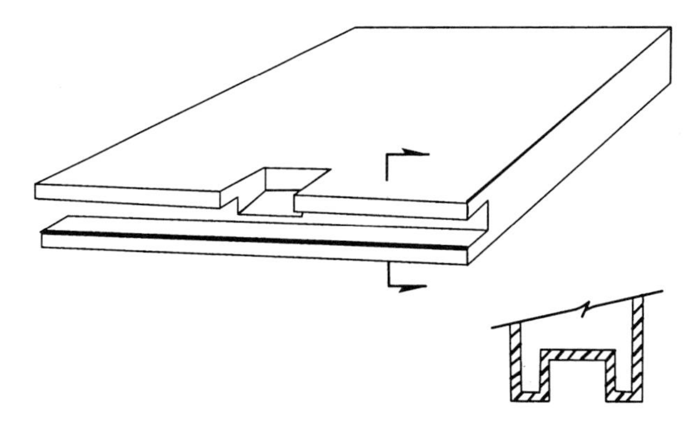

Figure 8.23 *Flat planar part with legs*

The preblow pressure causes the parison to "flash" between one mold half and the movable member, resulting in good wall thickness in that leg. Blow air is introduced and the bar moved inwardly in an attempt to form the second leg. Because the movable bar is fastened to this mold half, plastic must blow into the narrow width "leg" between the bar and mold face.

A complete refrigerator compartment separator was not made by a custom molder in $1\frac{1}{2}$ years of trials, modifications, and further trials. Blowouts or ruptures always occurred in corners of the leg formed between the moving member and that mold half (Fig. 8.24). Wall thicknesses of up to $\frac{1}{2}$ in. (1.3 cm) with corresponding part weights of 17 lb (7.7 kg) still resulted in ruptured and/or paper thin corners. At this time the mold was evaluated at the Plastics Technical Center. Although much better parts were made at half the weight, ruptured or extremely thin corners still occurred.

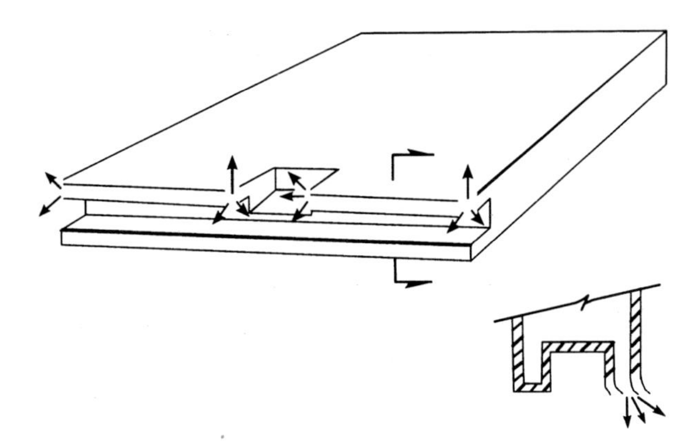

Figure 8.24 Ruptured corners encountered with moving member attached to one mold half

8.4.3 The Newer Technology

The new technology was simply to detach the moving member from its location on one mold half and locate it in a stationary position midway between mold halves. The member (hereafter called a third member or a "bar") was mounted to a beam which was affixed to the press frame. This arrangement results in a second parting line—one on each side of the third member.

The molding sequence consists of extruding a high density polyethylene parison between the mold halves and adjacent to the vertical side of the third member (Fig. 8.25). The parison is then prepinched and preblown while the mold halves are open. This results in the parison starting to fold or wrap around both sides of the third member (Fig. 8.26). Because there is open space between both sides of the third member and the mold halves, the flow of the preblown parison around the member is unimpeded. This is the principle design feature of this technology.

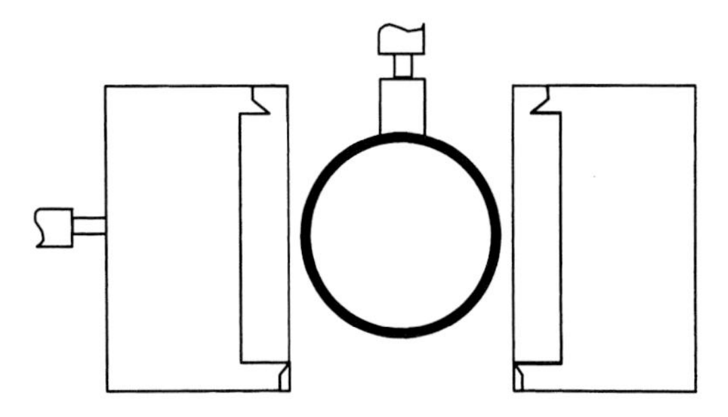

Figure 8.25 *Parison extruded adjacent to third member*

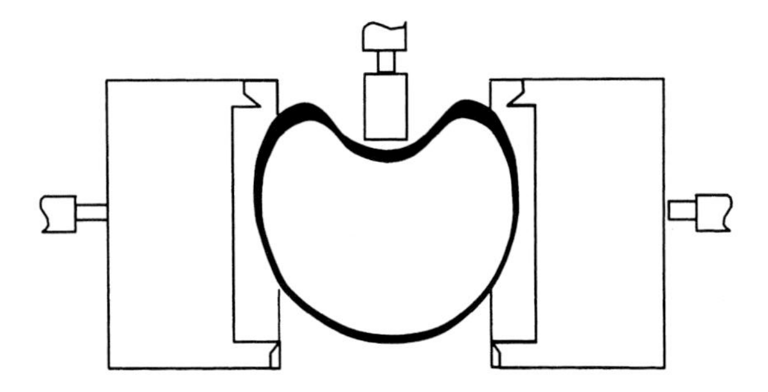

Figure 8.26 *Prepinched and preblown parison starts to fold around third member*

The closing action of the mold halves further compresses the trapped air inside the parison forcing the molten plastic tube around both sides, top and bottom of the bar (Fig. 8.27). Adequate material is trapped around the third member to form legs having relatively uniform thickness (Fig. 8.28). The mold halves have pinch blades on the sides, top and bottom of the parting line cavity edges and index at the appropriate area of the third member. The mold halves then close on the third member,

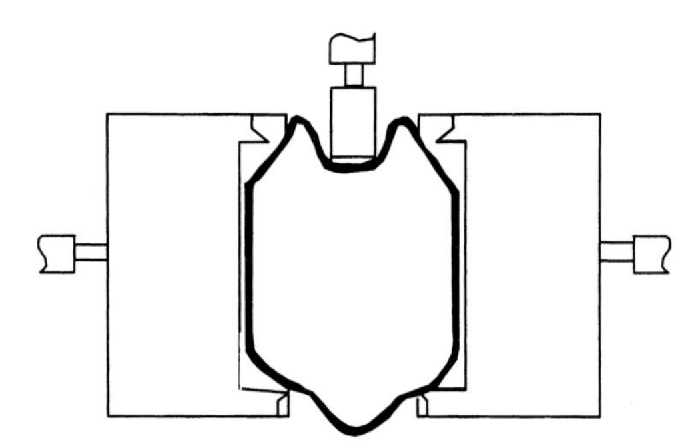

Figure 8.27 *Closing action compresses trapped air inside the parison, forcing molten plastic tube around both sides, top and bottom of bar*

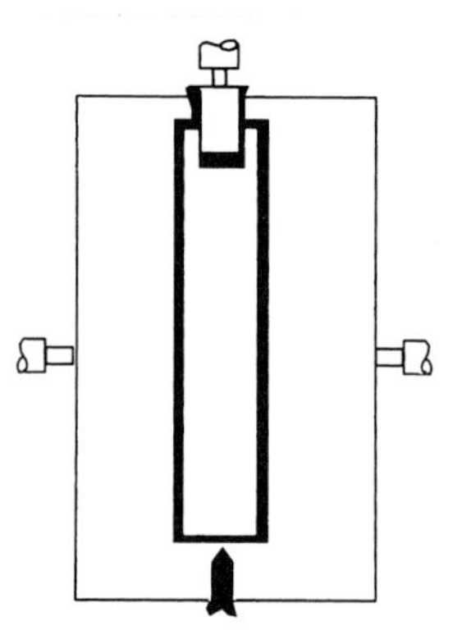

Figure 8.28 *Formed legs are relatively uniform*

conventionally pinch welding the “flashed” plastic on all sides of the bar. The part is blow molded. Part ejection is achieved by pulling the separator off the third member or by retracting the third member out of the part undercut prior to mold opening.

Reciprocating the third member is an option which would be desirable to form deep undercut parts. This was evaluated and found unnecessary for this particular part.

Production rates are the same as with any other part that is prepinched and preblown. Rates depend on wall thickness, part geometry, surface appearance, flatness, etc.—all the same factors that control cycle times for any large blow molded part.

Because of the greater amount of pinch welding in this technology, greater press clamping force than usual would be helpful. The amount of flash trimming is proportionally increased by the amount of pinch blade length. Good pinch blade quality is important to this methodology.

Preblowing parisons between split mold sections results in parts that inherently have more trimmed offal to be granulated. There is an added cost for the extra regrinding; however, this is not a high cost if normal blow molding temperatures are used, the trimmed offal is kept clean and handled efficiently.

8.4.4 Orienting 3D Parison

An overview of this EBM technique is presented. It is also called 3D (three-dimensional) BM and nonaxisymmetric BM. In conventional EBM, the parison enters the mold vertically rather in a straight tube. In 3D BM, the parison is oriented in the open or closed mold. It is manipulated in the tool cavity, providing complex geometric products that can have uniform or nonuniform wall thicknesses, corrugated and noncorrugated sections, and so on. It provides a means to significantly reduce scrap/flash (etc.) waste and quality.

Different techniques are used for placing the monolayer or coextruded plastic parison into 3D positions such as:

- articulate the extruder nozzle,
- articulate the mold platen,
- robotically orient the parison (Fig. 8.29a), and suction BM (Figs. 8.29b and c).

Sequential BM can be used to integrate hard and soft regions of different plastics on a single tubular structure (parison) as shown in Fig. 8.29d. This diagram shows two extruders controlling the sequence coextrusion (SeCo) of operation. It can use rigid–soft–rigid, soft–rigid–soft, and so on plastic combinations via accumulators "1" and "2."

As reviewed by Dr. Michael Thielen and Frank Schuller, 3D was introduced some years ago [83]. In the meantime a certain number of different systems became established in the market: suction BM, 3D BM with parison manipulation, and a split mold, horizontal machine with vertically opening mold and a six-axis-robot laying the parison into the cavity or a machine without using a closing unit. All these systems can be combined with six- or seven-layer coextrusion or with sequential coextrusion running hard–soft–hard plastics one after the other.

If strongly bent, 3D curved products, such as filler pipes for automobiles, are produced with traditional EBM technology. Flash areas near the mold parting line cannot be avoided and result in high flash percentages and surrounding pinch lines. In extreme cases the amount of flash can reach several times the actual article weight. The very long pinch lines lead to extremely high clamping forces.

With 3D, substantial savings are possible that may be achieved by using BM machines specifically designed for the production without (or at least with significantly reduced) pinch line. In these processes the extruded parison (with a diameter smaller than the article diameter) is deformed and manipulated and then moved directly into the mold cavity so that the remaining pinch line length is reduced to a minimum. The 3D manipulation of the parison with programmable manipulators or six-axis-robots and special devices allows the pinchless production of complex articles. These advantages require a higher effort as to the machine technology, depending on the article quality required and therefore also to type of process applied.

8.4.4.1 Advantages

In comparison to conventional BM technology with surrounding flash, the 3D technology presents a number of advantages. They include lower clamping force required, less effort for deflashing necessary, no refinishing work on outer article diameter, and improved quality of the article owing to wall thickness distribution and no reduction of strength due to pinch lines.

The significantly reduced flash weight brings further advantages such as smaller extruders can be used, less effort is necessary to grind flash and for recycling of regrind, and less degradation of sensitive materials.

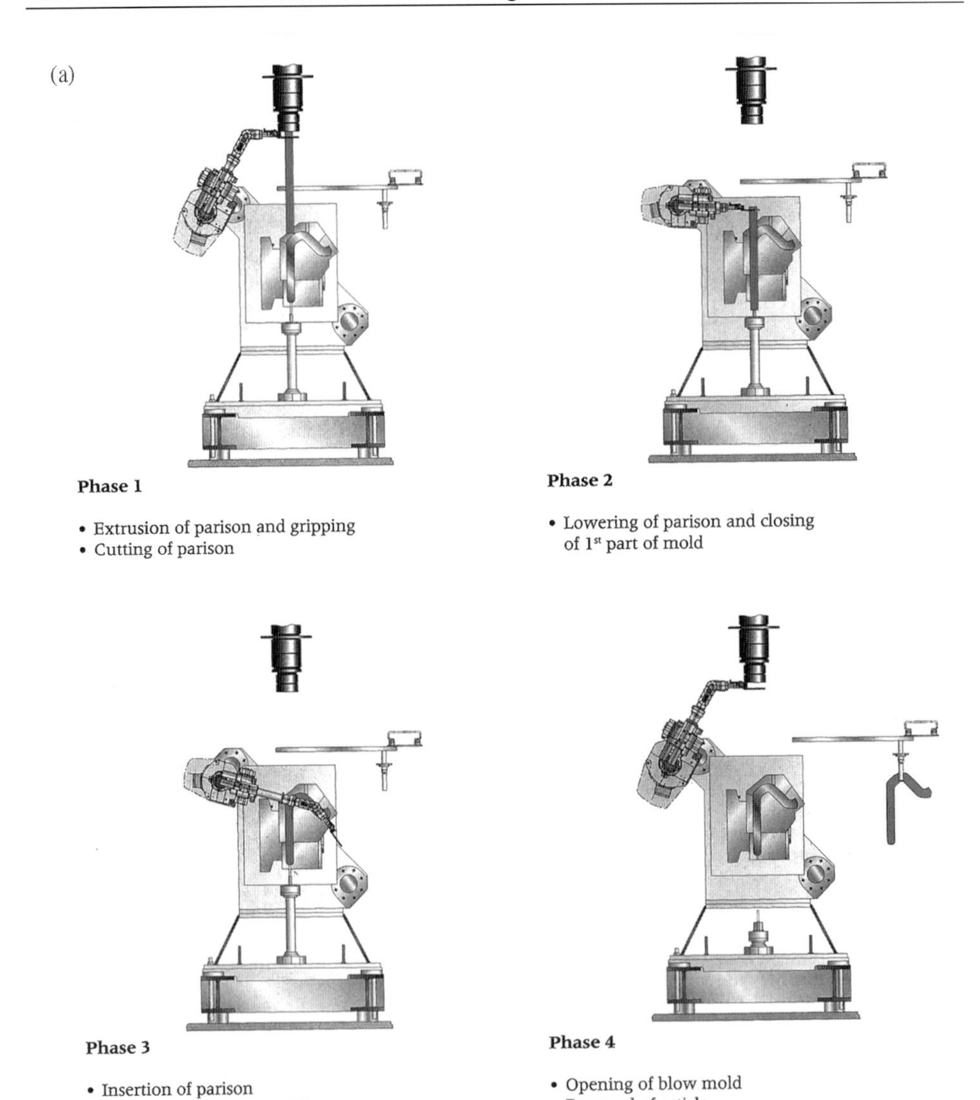

Figure 8.29 *(a) SIG Plastics' schematic example of a six-axis robot control that manipulates a parison in a 3D-mold cavity to BM a 3D product*

8.4.4.2 Available System

Due to the problems of conventional BM of certain 3D products and the advantages of 3D BM as described in the preceding, quite a number of different machine technologies were developed in recent years. Not all of them have reached significant market maturity. The most important developments are the Placo system, where an inclined clamp is moved in coordinates while the parison is placed into the mold

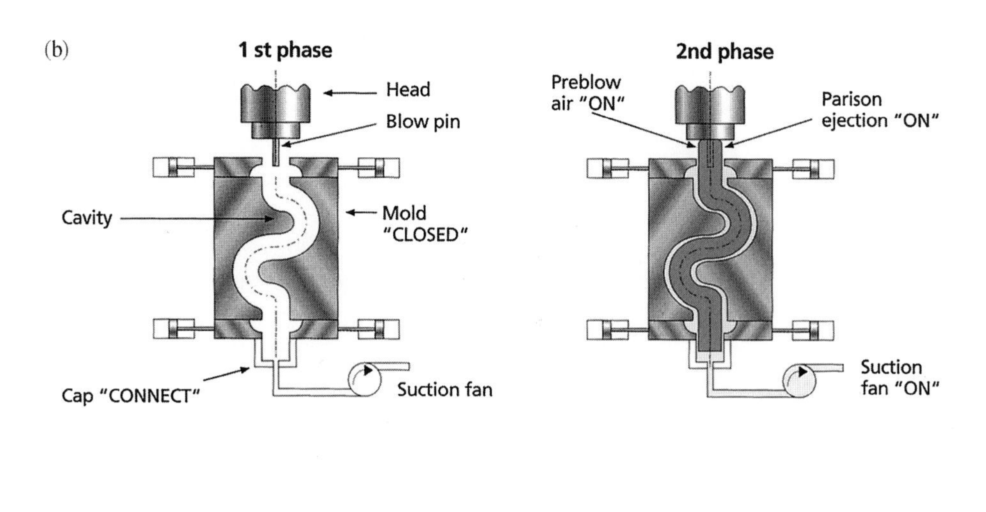

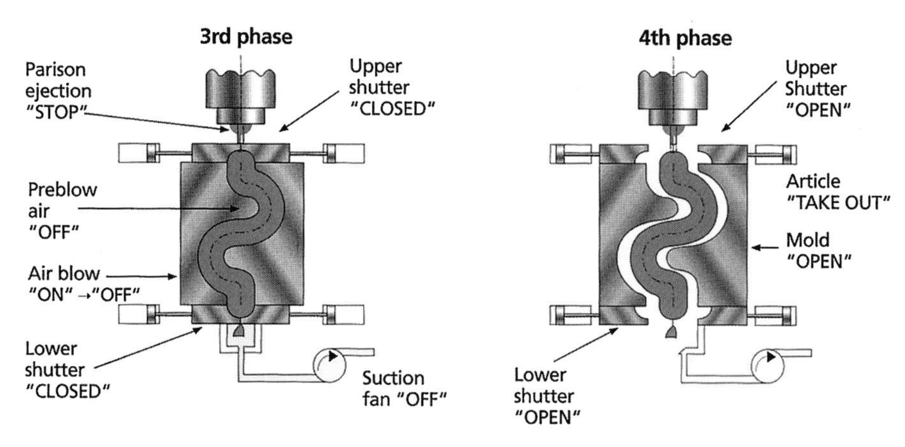

Figure 8.29 *(b) A basic suction blow molding schematic from SIG Plastics*

cavity. This Excel I system lays down the parison moving the extruder or the clamp in coordinates. The Yamakawa system uses a moving die to lay down the parison in coordinates, whereas, the 3D system by Etimex grabs the parison over its length to deform it and lay it down. The ABC system developed a special version of the suction blow system that was also developed by Fischer W. Muller (now SIG Blowtec of SIG Plastics International, Germany) in joint cooperation with Sumitomo of Japan.

Various different systems were developed by SIG Blowtec (formerly Krupp Kautex and Fischer W. Muller) using a parison manipulation with split mold. The systems consist of parison manipulation with a gripper or six-axis-robot combined with mold and machine technology (Fig. 8.29a). Grippers and parison bending devices adapted to the requirement of the article in addition to mold movements (segment-wise and to different times) allow manipulation of the parison in a wide range. This variant is limited because a complicated blow mold for parison manipulation is needed.

(c)

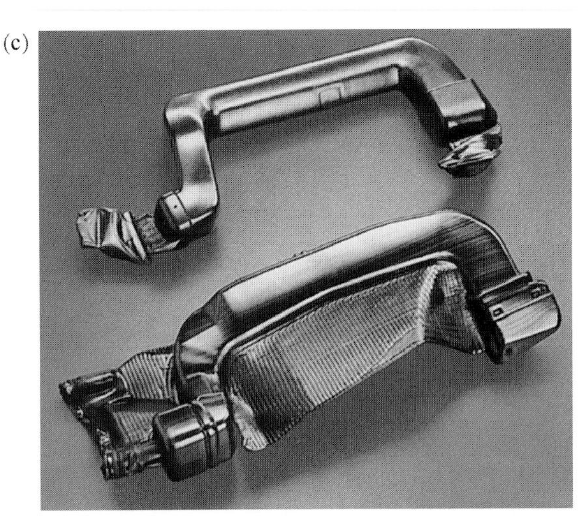

Figure 8.29 (c) An impressive difference occurs when comparing suction (top) with conventional (bottom) BM in respect to flash that is eliminated and surface quality. (Courtesy of SIG Plastics International)

Depending on the application a multiaxial operating handling device and/or parison bending device is needed.

While the molds in the above process are comparably expensive, the suction BM process requires simple and inexpensive molds only (Fig. 8.29b). Here the parison is ejected into the closed mold and "sucked" through the mold via an air stream provided by a blower system. After exiting at the lower end of the mold, the parison is squeezed off by closing devices and the inflation and cooling process can follow. With very complex geometries or with materials with very high melt viscosity, difficulties may occur in bending the parison.

In addition, a simple and inexpensive mold can be installed on a horizontal 3D machine, in a vertically opening clamp. The lower mold half slides out under the head where a manipulating robot places the parison into the cavity. After laydown the mold half slides back under the top mold half. The clamp closes and the inflation and cooling process follows.

The latest development is the Flat Desk (FD). It is a new kind of 3D technology that is based on the experiences with the 3D parison manipulation described previously. FD is a 3D BM machine without any clamping unit at all. The mold, which consists of a one-piece lower mold half and movable slides as the upper half of the cavity, is mounted onto a table of the FD unit.

By means of a multiaxis gripper system (six-axis robot), the parison is inserted in the lower half of the cavity. The slides are closed according to the motion of the parison manipulation. As the construction of the blow mold is rather simple and there is no clamping unit, the investment cost is reduced. Furthermore, the FD technology can be

(d)

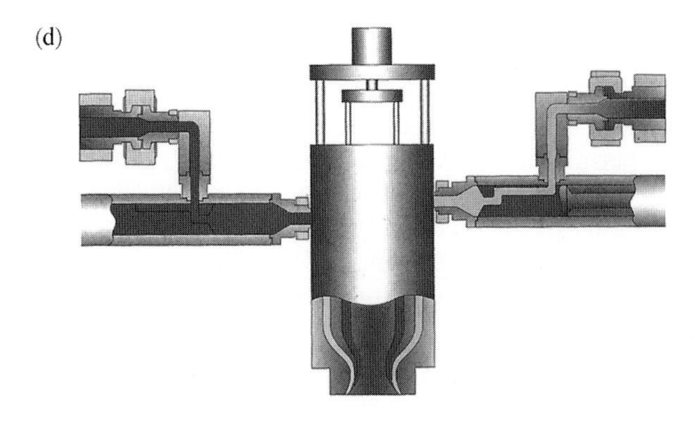

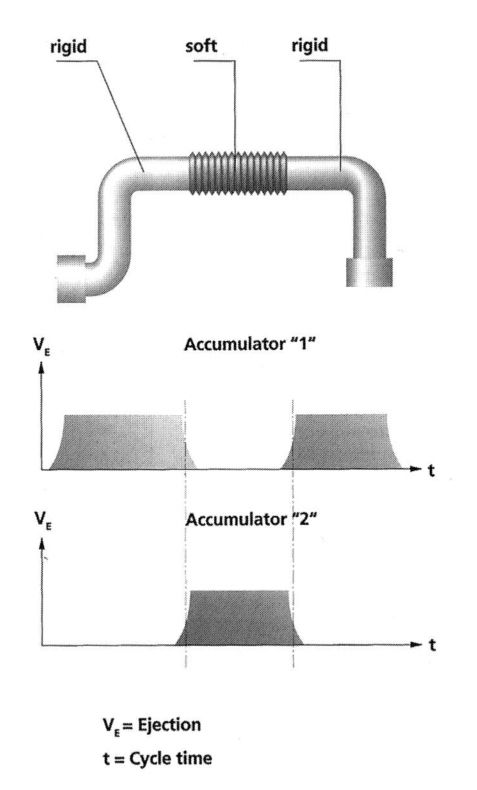

Figure 8.29 *(d) Sequential coextrusion (SeCo) from SIG Plastics allows 3D BM products*

extended to a double or triple station production unit that offers a high level of flexibility.

The spaces around and above the table (FD unit) offer an optimal area for manipulation and permit an easier, but also, stronger blow mold construction. Therefore, a clamping force up to 160 kN for each pair of slides is possible to achieve a desired quality of the welding line if there are article areas that have to be squeezed (such as tank filler pipes). Applying the FD process, it is even possible to use a calibration blow mandrel (including integrated spreading device, if necessary) within the area of the first pair of slides.

The table area of a FD unit is equipped with positioning holes/positioning pins that, in case of a mold change, guarantees a precise and repeatable positioning of the new blow mold without any trouble. The mold change can be moved either by crane or forklift truck. At any rate, the easy accessibility allows shortest changing times. The producer has equipped the FD unit with quick couplings for all connections between table and blow mold, which also helps to reduce the setup times.

With both of the aforementioned horizontal technologies, materials with low crystallization speed are to be used so as to minimize marks on the part surface due to mold contact.

8.4.4.3 Coextrusion

All 3D systems as described in the preceding can be combined with sequential coextrusion (Fig. 8.29d). Here two different materials can be extruded alternating one after the other so that a parison with different sections of material in the extrusion direction can be produced. A specially developed first-in-first-out (FiFo) accumulator head was developed for this process.

Also, very important is the transition zone of the sequentially coextruded material in the parison. Special design peculiarities enable production of reproducible material transition zones. The right choice of the material combination is also very important for a good change of materials. The closer the melting point and the stretch ratio (strain viscosity) of the combined materials are to each other the more favorable are the conditions for short and reproducible transition zones.

In practice, materials with different hardnesses are combined in sequential coextrusion. This leads to articles with soft ends and a hard middle section. Application examples are air ducts in the automotive industry where the soft ends can perform the scaling functions at the connection points. The hard middle section offers sufficient stiffness against deformation due to the vacuum and overpressure.

Other examples for applications are connecting hoses or bellows for machine industry household appliances or the automotive industry. Another aspect that can lead to the use of sequential coextrusion is the cost reduction for articles that are today completely made out of thermoplastic elastomer (TPE) materials. If the soft sections of the article are not necessary some areas of the article can be replaced by hard material that often costs 25% that of the soft material types.

The combination of the 3D BM technology, with the six-layer coextrusion, with embedded barrier-layer of ethylene–vinyl alcohol copolymer (EVOH) is now well established in the market.

8.4.4.4 Radial Wall Thickness Control

One of the major problems involved in high strength/strong curved blow moldings is irregular wall thickness at inner and outer radii, a result of the different stretch ratios. This effect is known as the use of a partial wall thickness control (PWDS) with flexible die rings. It did not offer a solution because of the small parison diameters.

For this reason so-called radial wall thickness distribution systems (RWDS or DDD^r) were developed that permit a balancing out of the wall thickness between the inner and outer radii of the curved section, even with small parison diameter. To guarantee a regular wall thickness profile of the final product, the wall thickness must be adjustable not only with regard to the length of the parison but also its circumference.

The die will be displaced with the help of two hydraulic cylinders in a rectangular position to each other. If both cylinders are in central/neutral/position, the die is in the center position and the wall thickness profile constant over the whole circumference. With the help of both cylinders, the mouthpiece can be displaced in a way that results in an eccentric die gap. Thus, the wall thickness of every circumferential point can be influenced effectively so that the wall thickness ratio between the outer and inner radius can even be > 1 for certain special applications.

8.4.4.5 Literature

Available literature on this subject includes:

Balzer, M., Daubenbüchel, W., Kulik, M., Trends in Extrusion Blow Molding during the Nineties, Polimeri 6/1993, p. 241 ff.

Daubenbüchel, W., Blasformen von thermoplastischen Elastomeren, Kunststoffe 85 (1995) 5, p. 640 ff.

N. N., Neue Blasformtechnologien erweitern den Anwendungsbereich, Plastverarbeiter, 12/94.

Thielen, M., Blow Molding, Sequential Coextrusion: Advanced Technology Opens Door into a New World of Tailored Parts, Modern Plastics Encyclopaedia '99, P. D5.

Schüller, F., Developments in 3D Blow Molding Technology, Plastics News International (Australia) July 2000, pp. 18–19.

Thielen, M., The Blow Molding of Multilayer Plastic Fuel Tanks, SPE Automotive Blow Molding Design Technologies Conference, Dearborn, Michigan, USA, June 1999.

Thielen, M., Parison Wall Control Systems for Extrusion Blow Molding Equipment, SPE-ANTEC, Orlando, Florida, USA, May 2000, Paper 222.

Thielen, M., 3D-Blow Molding—The Way Ahead, Paper: SPE Processors Conference ProCon 2000, October 2000, Cincinnati, Ohio, USA.

Thielen, M., et al., 3D Blow Molding Today, An Overview about Different Systems which are established in the Market Today, SPE-ANTEC, May 2001.

8.4.5 Other Approaches

In addition to the products, processes, and designs already reviewed, BM continues to offer definite advantages through the elimination of secondary operations including consolidation of parts and modular approaches. Figure 8.30 provides examples of BM two (or more) products or parts during a single BM operation followed with manual or automatic cutting action to separate the parts.

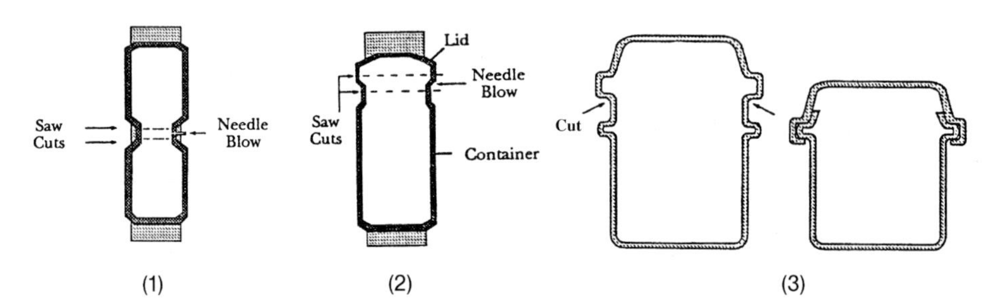

Figure 8.30 *Views show: (1) two containers, (2) container with lid, and (3) container with snap fit*

8.4.5.1 Collapsible Container

An interesting and practical design involves a bellows-collapsible bottle produced in conventional EBM equipment (Patent No. 4,492,313). These "foldable", in contrast to "passive" bottles, provide advantages and conveniences such as (1) reducing storage, transportation, and disposal space; (2) prolonging product freshness by reducing oxidation and loss of carbon dioxide (CO_2) because as the contents are removed its level is also reduced; and (3) providing continuous surface access to foods such as mayonnaise and jams. The marked advantages of the collapsible bottle over conventional packages justify the product's trade name: "The Smart Bottle."

The bellows overlap and fold to retain the folded condition without external assistance, thus providing a self-latching feature. The latching is the result of bringing together, under pressure, two adjacent, conical sections of unequal proportions and of different angulations to the bottle axis.

On a more technical analysis, the latching is due to the swing action of one conical section around a fixed pivot point, from an outer to an inner rest position.

The two symmetrically opposite pivot points and rotating segments keep a near-consistent diameter as they travel along the bottle axis. This explains the bowing action of the smaller conical section as it approaches the overcentering point (Fig. 8.31).

Initial collapsing of the bottle should occur no sooner than three to ten hours after manufacture. Additional pressure is needed for this first time collapse in order to create permanent fold rings. Subsequent collapsings and expansions of the same bottle are substantially facilitated by the fold rings which serve as hinges. Collapsing and expansion of the bottles before filling can be performed at a recommended ambient temperature of 20 °C or higher.

In most disposable applications, the bottles would undergo three changes of volume:

1. Initial collapsing of the container before shipping or storage
2. Expansion of the container at destination, before or during filling
3. Finally, collapsing the bottle for disposal

The fold rings designed on the folding bellows have proven to be very durable and sturdy. Prototype bottles, made from PETG and 75 durometer PVC, were able to withstand dozens of collapsings and still pass stress tests.

Reusable latching bottles made from polypropylene are presently being sold around the world for backpacking or camping purposes.

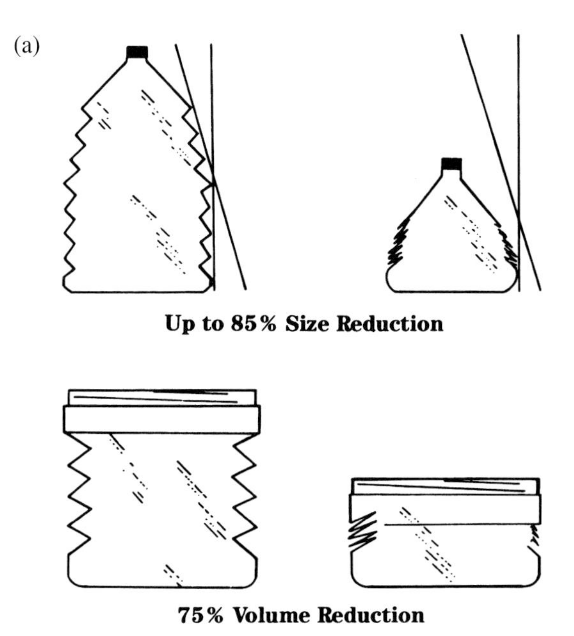

Figure 8.31 *Design ideas for the collapsible bottle: (a) Continuous latch bottle; (b) skip latch bottle; (c) section through bellows showing collapsing and latching mechanism of collapsible bottle*

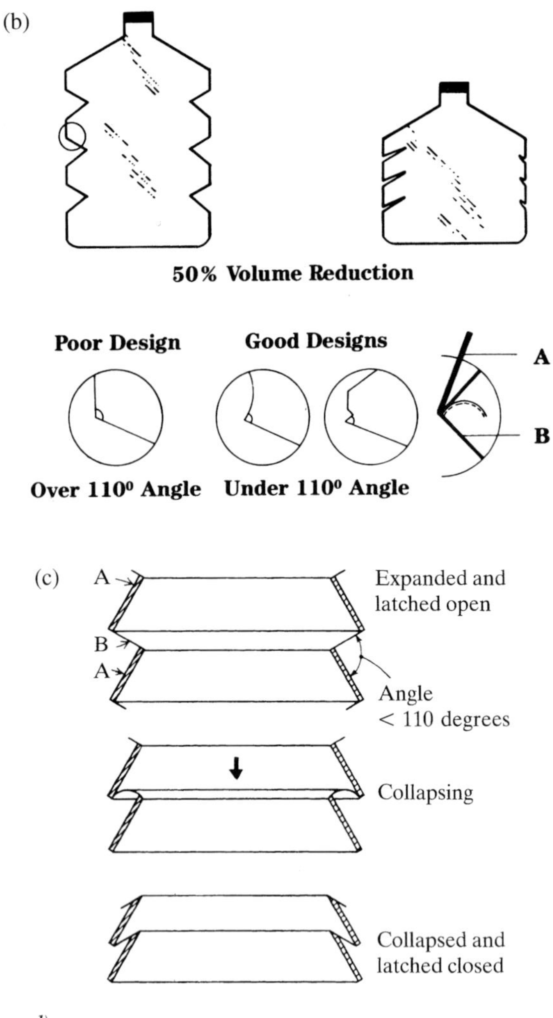

Figure 8.31 *(continued)*

The two adjacent conical sectors providing the latching should not exceed 110° angle in order to make a sharp fold ring (see Fig. 8.31b). The size of the rotating conical section B should not exceed 80% of conical section A in order to prevent confusion and wobbling as the bottle is being collapsed.

Product labeling can be accomplished, utilizing a floating sleeve attached to the neck or shoulder section of the bottle. The cylindrical sleeve accommodates the bellows as they fold from bottom up. The bellows are contained within a cylindrical sleeve as the jar is collapsed. The maximum length of the sleeve is limited by the collapsed dimension of the jar. An extended cap also can be used to hold the label to the side of the bottle.

8.4.6 Hot Fill PET Bottle

Injection blow molding (IBM) of PET bottles is an important and major market worldwide. This review concerns the use of Nissei ASB Machine Co., Ltd. development and equipment used to IBM hot fill PET bottles [18]. The shape and thickness of the neck, panel, and bottom shown in Fig. 8.32 are important sections of the hot fill PET bottle. Let us consider the neck section first. The neck should not (1) deform under rolling pressure of a capping machine and (2) suffer heat distortion when the bottle is hot filled to the brim or when subjected to sterilization by turning the bottles upside down.

The neck can be made to withstand higher fill temperatures by insert molding with a heat resistant resin like polycarbonate. Such a process is described in Tech. INF ASB-8301.

Another way to improve neck properties is to use a three-layer neck such as PET/polyarylate/PET as described in Tech. INF ASB-8401. The following example shows the better heat resistance (i.e. less neck shrinkage) of bottles produced from PET/polyarylate/PET as compared to those made from PET. The testing method is illustrated in Fig. 8.33. In this case, each bottle had a neck size of 28 mm. The relative amounts of shrinkage of neck area is shown in Table 8.3.

The panel section illustrated in Fig. 8.32 should not shrink due to hot filling, and the volume shrinkage of the contents should not cause deformation. These objectives can be achieved by (1) heat setting the panel area using high temperature molds (ref.

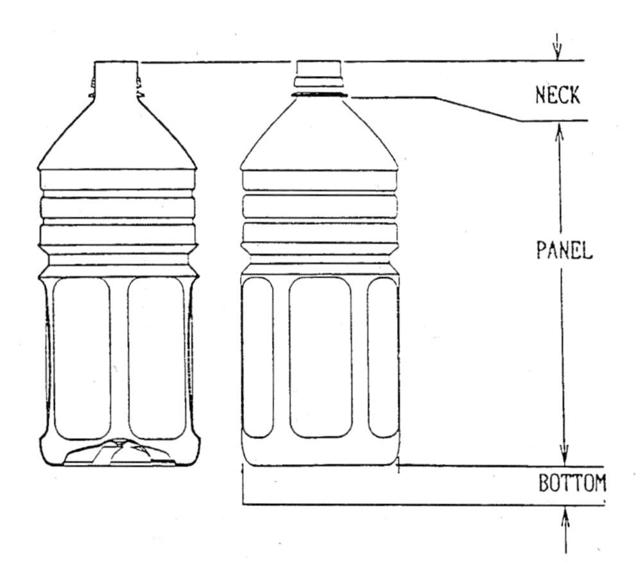

Figure 8.32 *Designing a hot fill PET bottle. The shapes and thickness of the three sections of the hot fill bottle are important*

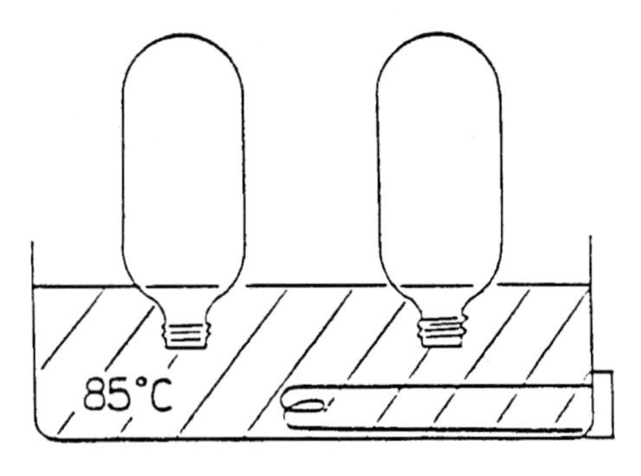

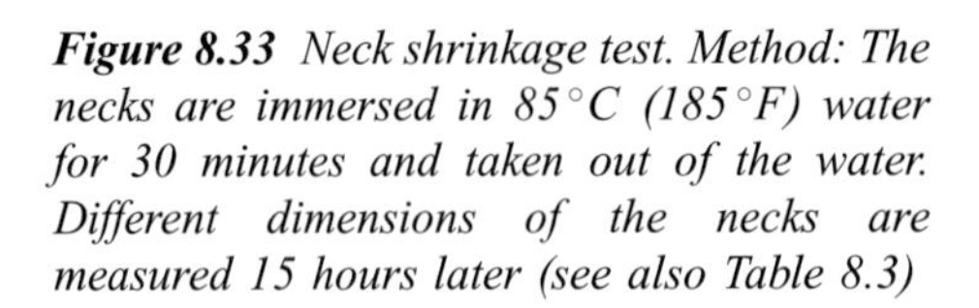

Figure 8.33 Neck shrinkage test. Method: The necks are immersed in 85 °C (185 °F) water for 30 minutes and taken out of the water. Different dimensions of the necks are measured 15 hours later (see also Table 8.3)

Table 8.3 Neck Shrinkage Test Results

	PET	PET/Polyarylate/PET
T	0.622%	0.308%
E	1.174	0.408
I	1.256	0.586
B	0.748	0.210
N	1.914	0.094
F	2.450	1.244

Tech. Data ASB-8301) or (2) by using a three layer blow molding technique (e.g. PET/polyarylate/PET).

A comparison of the latter technique is illustrated in Table 8.4. In this comparison, shrinkage data was obtained from a single layer PET bottle and from two three-layer PET bottles which had either SAN or polyarylate as the center layer. All the bottles were fabricated using a blow mold temperature of 90 °C (194 °F).

The bottles are first filled to the brim with 20 °C (68 °F) water of 0.99717 specific gravity to measure the content volumes. The bottles are emptied. Then, the bottles are filled with 85 °C (185 °F) water; capped with an Alcoa capping machine; left in the

Table 8.4 Shrinkage in Percentage Based on Volume Change after Filling with Hot Water (%)

	Filling temperature		
Resin	85 °C (185 °F)	90 °C (194 °F)	95 °C (203 °F)
PET (Single layer)	1.60	3.80	—
SAN (8.4 W%)	0.25	2.40	—
Polyarylate (7.0 W%)	0.21	1.43	3.43

ambient temperature until the bottle contents cool down to room temperature; and emptied. The bottles are refilled with 20 °C water to measure the content volumes. The content volumes measured before and after hot filling are compared to determine the shrinkage.

The paneling effect after filling with hot water is shown in Table 8.5. Note that bottles of PET deform at 80 °C (176 °F) while three-layer bottles having a center layer of SAN or polyarylate withstand higher temperatures.

Table 8.5 *Paneling Effect after Filling with Hot Water*

	Filling temperature			
Resin	80 °C (176 °F)	85 °C (185 °F)	90 °C (194 °F)	95 °C (203 °F)
PET	X	X	X	X
SAN (8.4%)	○	△	X	X
Polyarylate	○	○	△	X

○: keep original shape; △: slightly paneling; X: deformed

Note: Tables 8.3, 8.4 and 8.5 show the results of tests with 1 liter bottles with 28 mm necks weighing 48 g

Finally, the bottom section shown in Fig. 8.32 should be designed to allow the maximum stretch ratio, or to increase the biaxial stretch ratio of the preform surface area and be as thick as possible. To satisfy these requirements, Nissei ASB Machine Co., Ltd. is developing its own unique bottom design (Fig. 8.34).

8.4.7 Wide Mouth PET Containers

Some applications for wide mouth PET containers, having neck sizes larger than 40 mm (1.6 in.), include containers for cosmetics (skin cream), agricultural chemicals, nuts, candies, salt, liquor, etc. One unusual application is the production of electric light diffusers using polycarbonate and biaxial draw blowing. For cosmetics, both PET resin and polypropylene resin are used with biaxial draw blowing. Table 8.6 gives a rough classification of the diameter of the neck and examples of applications.

Wide mouth PET containers can be molded for cold filling at temperatures below 70 °C (158 °F), as for example liquor containers. If the temperature exceeds 70 °C (158 °F), the neck may be deformed. The deformation depends on the shape of the neck. A screw-type neck is deformed only slightly as the tightening torque is applied to the entire thread. With twist-off necks, the tightening torque of the cap is applied locally to four to six points and this is likely to cause deformation. For instance, for a four-point contact type that is a four-start twist-off thread, the neck is deformed to a rectangular shape by the filling heat and the tightening torque of the cap.

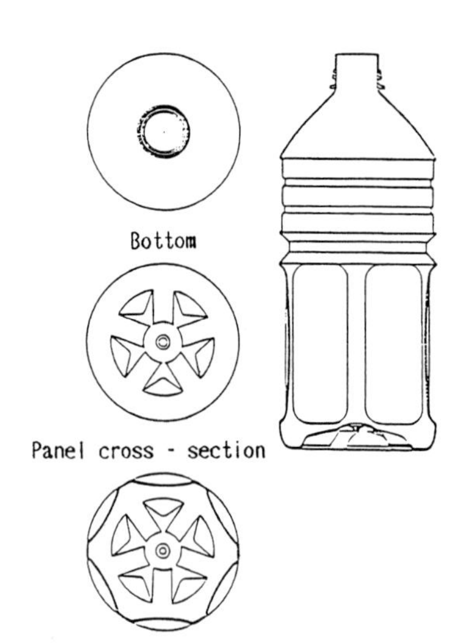

Figure 8.34 Nissei ASB unique bottle design

Table 8.6 Examples of Applications by Neck Size

Neck	Examples of applications
40–53 mm	Liquor, medicine, cosmetics, agricultural, table salt etc.
53–75 mm	Pickles, honey, nuts, candy, liquor, electric light globe etc.
75–120 mm	Pickles, salted greens, peanut butter etc.

Other than deformation, another problem to be taken into account is the shrinkage caused by pressure reduction of the contents. When 50% orange juice is cooled from 90 °C (194 °F) to 5 °C (41 °F), a volumetric shrinkage of about 16 cc occurs for a 16-oz (423-cc) container. Therefore, the shape of the container must be designed so that volumetric shrinkage is absorbed by the shape and the entire container, including the paneling, is not deformed. To achieve this, the design of wide mouth containers requires:

- Adding vertical rib-shaped panels to the container body.
- Making the base of the container flexible between concave and convex. It should be convex during filling and concave after shrinkage of the contents.

The employment of PET and biaxial draw molding method will enable the production of wide-mouth containers with increased similarity to present glass containers. Heat resisting properties will be added to wide mouth containers in the near future and this will further promote the substitution of PET containers for glass containers. Wide-mouth PET containers will be required to fulfill the following requirements:

- Withstand filling temperature of 85 to 95 °C (185 to 203 °F).

- Provide an oxygen barrier (at least two to four times greater than that of the current PET containers).

8.5 Shrinkage

Generally shrinkage is the difference between the dimensions of the mold cavity at room temperature (72 °F) and the dimensions of the cold BM product checked 24 h after production. The elapsed time is necessary to allow the thermoplastic product to shrink since these materials have a rather large shrinkage factor due to their large coefficient of expansion and the different shrinkage behavior depending on whether they are crystalline or amorphous plastic materials (Chapter 7).

The shrinkage behavior of different thermoplastics and the part geometry must be considered. Without experience, trial and error determines what shrinkage will occur immediately at the time of fabrication and what time period is required after molding (usually up to 24 h) to ensure complete shrinkage. Lengthwise shrinkage tends to be slightly greater than transverse shrinkage.

Most of the lengthwise shrinkage occurs in the blow molded wall thickness rather than a body dimension. As an example, in PE, higher shrinkage occurs with the higher density plastics and thicker walls. Lengthwise shrinkage is due to a greater crystallinity of the more linear type plastics. Transverse shrinkage is due to slower cooling rates that results in more orderly crystalline growth. Part shrinkage depends on many factors such as plastic density, melt heat, mold heat, cooling rate and uniformity, part thickness, pressure of blown air, and control or capability of the BM production line.

Once operating conditions are established, tolerances of ±5 % may be expected. Typical PE shrinkage is as follows:

LDPE:
Thickness up to 0.075 in.: 0.010 to 0.015 in./in.
Thickness over 0.075 in.: 0.015 to 0.030 in./in.

HDPE:
Thickness up to 0.075 in.: 0.020 to 0.035 in./in.
Thickness over 0.075 in.: 0.035 to 0.055 in./in.

The cavity defines the shape of the article. As with other molding processes, cavity dimensions are enlarged slightly to compensate for resin shrinkage. For polyolefins, particularly in the neck finish, slightly higher shrinkage rates are used for EBM than for IBM. The amount for the body is 2% and as high as 3.5% for the neck finish. Rigid resins, however, remain about the same.

The most common special feature of the mold is the "quick-change" volume control insert. Rigid volume control is a must for certain products such as dairy containers. Unfortunately, an HDPE container slowly shrinks and changes in size for many hours after molding. Because of production "volume control" requirements, some dairies fill containers molded half an hour before and then switch to fill containers molded several days before. Volume control inserts that displace the difference in size are added to the mold (usually as a disk in the sidewall) to ensure that volume and fill levels are the same in both containers at the time of filling. The device works because HDPE shrinkage is reduced to virtually zero for the life of the container, when filled with milk or juice and stored at cold temperatures. Several sizes are available to suit particular molding needs.

Raising the blowing pressure and lowering the temperature of the mold may reduce shrinkage. Rapid cooling is desired to reduce processing cycle time, but cooling too rapidly can cause surface imperfections, distortion, and frozen in stresses. Raising the stock temperature, while it may not appreciably affect outside dimensions, causes more of the shrinkage to occur in wall thickness, because the higher melt temperatures lessen strain recovery and reduce blowing stresses.

Using the calculated shrinkage theory can dictate how much oversize to cut the tool (mold or die) if a product has a relatively simple shape. For other shapes, some critical key dimensions of the product will not be as predicted from the shrink allowance, particularly if the item is long, complex, or require tight tolerance. The important factor that influences the shrinkage of a specific plastic in using a specific machine (such as extrusion and injection) is causing it to vary and not follow the values of flow direction. Other influences include wall thickness, flow distance, and the presence of reinforcing fibers when the material contains short reinforcing fibers (glass, etc. [8]).

Determining shrinkage involves more than just applying the appropriate correction factor from a material's data sheet. Shrinkage is not a single event but occurs over a period of time. Most of it happens in the mold or die, but it can continue for up to 24 to 48 h after being molded or extruded. This so-called postmold shrinkage may require a constraining cooling fixture. Additional shrinkage can occur when annealing relieves frozen-in stresses or by exposure to high service temperatures.

The main considerations in mold or die design affecting shrinkage are to provide adequate cooling, and structural rigidity. Cooling conditions is the most critical especially for crystalline plastics. The cooling system must be adequate for the heat load. Slow cooling increases shrinkage by giving plastic molecules more time to reach a relaxed state. In crystalline types, having longer cooling time leads to a higher level of crystallinity, which in turn accentuates shrinkage. Proper cooling, along with having an overall melt-flow analysis of how the material will react in the mold, by the mold designer, will eliminate or at least be capable of controlling the potential problems of shrinkage and warpage. This analysis can also include the best gate locations in molds.

A number of the computer-aided flow-simulation programs now offer modules designed to forecast shrinkage and, to a limited degree, warpage from the interplay of plastic and melt temperatures, cavity pressures, stress, and other variables in mold- or die-fill analysis. The predicted shrinkage values in various areas of the product should be used as the basis for sizing the mold or die cavity, either by manual input or feedthrough to the mold dimensioning program. Computer-aided flow-simulation programs are also available for dies. All the programs can successfully predict a certain amount of shrinkage under specific conditions that can be applied to experience.

8.5.1 Tolerance

Inspection variations are often the most critical and most overlooked aspect of the tolerance of a fabricated product. Designers and processors base their development decisions on inspection readings, but they rarely determine the tolerances associated with these readings. The inspection variations may themselves be greater than the tolerances for the characteristics being measured, but without having a study of the inspection method capability, this can go unnoticed.

Inspection tolerance can be divided into two major components: the accuracy variability of the instrument and the repeatability of the measuring method. The calibration and accuracy of the instrument are documented and certified by its manufacturer, and it is periodically checked. Understanding the overall inspection process is extremely useful in selecting the proper method for measuring a specific dimension [2]. When all the inspection methods available provide an acceptable level of accuracy, the most economical method should be used.

As the overall fabricating tolerance is analyzed into the sources of its variation components, the potential advantage of analytical programs comes into play with their ability to efficiently process all these factors. All the empirical tolerance ranges for each tooling method and inspection method are stored in data files for easy retrieval. For each critical dimension, the program sums all the component tolerances and computes a ± overall tolerance for each critical dimension. The program then provides a tabulated estimate of the achievable processing tolerances and pinpoints the areas that contribute most of the overall required tolerance. This information is useful in identifying the needed tolerances that, in practice, can be expected to exceed the initial design tolerance.

Establishing initial design tolerances is often done on an arbitrary, uninformed basis. If the initial estimated tolerance proves too great, a lower shrink plastic could be used to reduce the shrinkage range. However, if the key dimension was across a main parting line, the tooling could be redesigned to eliminate the condition and consequently reduce variation from tool construction. Even with all these data processed by a computer, estimating tolerances is difficult if they are not properly interrelated with the highly dependent factors of the product and tooling designs.

8.6 Behavior of Plastic

The purpose of this section is to acquaint the designer with the structural behavior of plastics. It provides concepts as background for estimating and anticipating behavior in design situations. Obviously, there are no universal methods to describe the behavior of all plastics, just as there is no single reference that covers completely the behavior of all metals and their alloys. Nonetheless, all plastics show many similarities in behavior. The differences are frequently in terms of magnitude, not of kind.

The stress–strain and strength behavior of plastics varies widely, depending on the generic type, or family of plastic, and on the specific composition of compounds within that family. Many factors interact to alter behavior of a given compound over a very wide range. Key parameters that must be considered in structural design of plastics are:

- *Magnitude and duration of stress, strain and temperature:* At a given temperature, both the magnitude and duration of stress or strain affect structural response and strength behavior. Conversely, at a given magnitude and duration of stress or strain, a shift in temperature can produce marked changes in structural response and strength behavior.
- *Environment:* The environment interacts with the magnitude and duration of stress, strain, and temperature to further alter material response and strength of plastics. Chemical environment permeability, ultraviolet (UV) radiation, sustained elevated temperature, and even water, for example, can have profound influence on performance and hence may dominate the design problem. Environmental effects, or the failure to properly design for them as they interact with sustained stress or strain, have been a chief cause of failure of plastics products.
- *Additives and modifiers:* Fillers and plasticizers alter the basic response and strength of plastic materials. Particulate fillers (wood flour, flake, clay, limestone, etc.) are introduced to reduce resin shrinkage, increase stiffness, improve processing characteristics, or to lower cost. Most fillers also lower impact resistance. Plasticizers increase flexibility and toughness; impact modifiers, such as rubber blends, are more permanent and improve toughness without significant sacrifice in stiffness. Thus, the properties of a basic generic plastic, such as PVC, can be varied to provide either rigid sewer pipe or flexible rubberlike water stops and liners.
- *Reinforcement:* Stiff strong fibers and flakes incorporated into plastics improve stiffness, strength, and dimensional stability. An increase in their proportion relative to the resin matrix results in a corresponding improvement in these properties [10].
- *Process:* The blow mold process used to convert plastic materials into structures may dictate the structural performance of the finished product, as illustrated by the following:

- Orientation of the molecular structure of TPs may strengthen the product in the direction of orientation. Such preferential orientation can be used to advantage. The designer, however, must be on guard to ensure that orientation does not prove to be the cause of failure. For example, flow patterns and knit and weld lines developed during molding can cause orientation that creates zones of weakness.
- Oxidation and crystallization induced during processing can embrittle an otherwise ductile material. PE, for example, held too long at high temperatures may either oxidize in the air or develop excessive crystallization.

• *Similarities with conventional materials:* The plastics behavior introduced here is not necessarily any more complicated than the behavior of conventional structural materials. Overall, the engineer must design both plastics and conventional materials to meet the criteria of time, temperature and environment [9]. The similarities are as follows:
  - *Time:* In designing with conventional materials (steel, aluminum, glass, wood, and concrete), effects of load duration are recognized in terms of creep and ultimate strength. Time effects are frequently more prominent in plastics, particularly in terms of strength.
  - *Temperature:* In designing with steel and glass, brittleness at low temperatures and loss of yield strength at high temperatures must be considered. Similarly, in designing with plastics, both stress–strain behavior and strength can vary with temperature, and the low and high temperature limits of certain plastics are frequently in the range of the normal outdoor environmental limits.
  - *Environment:* The strength, stiffness, dimensional stability, and useful life of wood are a strong function of moisture content. Steel rusts, glass cracks, aluminum corrodes, and so on. The problem is similar for plastics, but because composition can vary widely, each plastic compound must be considered for its strengths and weaknesses on exposure to various environments.

Design criteria for plastics are generally not established in standard specifications. A myriad of different materials and variations is available. Only a relatively few compounds drawn from several generic classes of plastics have been characterized structurally for sustained loads.

8.6.1 Plastic with a Memory

TPs can be bent, pulled, or squeezed into various useful shapes. But eventually, especially with added heat, they return to their original form. When most materials are bent, stretched, or compressed, they alter their molecular structure or grain orientation to accommodate the deformation–permanently. Not so with plastics. Plastics temporarily assume the deformed shape but always maintain internal stresses that tend to force the material back to its original shape. This behavior, known as plastic memory, can be annoying. But, when properly applied, plastic memory offers some interesting design possibilities for blow molded parts.

The time/temperature-dependent change in mechanical properties results from stress relaxation and other viscoelastic phenomena typical of plastics. When the change is an unwanted limitation, it is called creep. When the change is skillfully adapted to the overall design, it is called plastic memory.

Most plastic parts can be produced with a built-in memory. That is, the tendency to move into a new shape is included as an integral part of the design. So then, after the parts are assembled in place, a small amount of heat can coax them to change shape. Plastic parts can be deformed during assembly, then allowed to return to their original shape. In this case, parts can be stretched around obstacles or made to conform to unavoidable irregularities without permanent damage.

Potential memory exists in all thermoplastics. Polyolefins, neoprene, silicone, and other crosslinkable polymers can be given a memory either by radiation or chemical curing. Fluorocarbons, however, need no curing. When the phenomenon is applied to fluorocarbons such as TFE (polytetrafluoroethylene), FEP (fluoronated ethylene propylene), ETFE (ethylene tetrafluoroethylene), CTFE (chlorotrifluoroethylene), and PVDF (polyvinylidene fluoride), interesting high temperature or wear resistant applications are possible.

8.6.2 Viscoelastic Behavior

The relationship between stress and strain or structural response of plastics varies from viscous to elastic (see Fig. 8.2). Most plastics display a structural response that is intermediate, between the viscous and elastic states; they are viscoelastic materials. The type of plastic compound, stress, strain, time, temperature, and environment all play a significant role in determining whether the response is mostly viscous, elastic, or viscoelastic.

Viscoelasticity tends to be a complex subject. Rigorous approaches are not usually needed in practical structural design; hence, simple models and analogies are used below to demonstrate viscoelastic behavior. The terminology used herein to characterize viscoelastic behavior of plastics, including the idealized components of viscoelastic models, is given below:

- *Elastic response* is represented by a Hookean solid, as modeled by a linear spring [9]. Stress is proportional to strain and independent of time; response to stress is instantaneous; there is no permanent or irrecoverable deformation; all energy used to deform the spring is stored and is fully recoverable.
- *Viscous response* is represented by a Newtonian fluid, as modeled by a dashpot. Stress is proportional to strain rate, making behavior time-dependent. No recovery when stress is removed. Energy to deform the dashpot is dissipated completely during the deformation process.

- *Creep* is the time-dependent increase in strain of a viscous or viscoelastic material under sustained stress. Some of the time-dependent deformation is recoverable with time after release of stress. Creep experiments are usually performed under constant load conditions, and when stresses are high, the sample may "neck" and the cross section supporting the load may be reduced significantly at some point during the test. Unless otherwise indicated, the creep stress based on original cross sectional area (engineering stress), will be used rather than the true creep stress which is based on the reduced cross sectional area, that occurs on necking [42, 66]. This reflects typical practice, which is, to report engineering rather than true creep stress in creep experiments, but this practice is not universal, particularly in scientific behavioral investigations.
- *Relaxation* is the time-dependent decay in stress of a viscoelastic material under sustained strain. Some of the deformation is recoverable with time after release of the sustained strain. Unless the imposed initial strain is above the yield point, the cross section remains fairly close to the original, throughout the test. This differs from creep behavior as noted above.
- *Recovery* is the extent to which an element returns to its original configuration after release of stress or strain.
- *Linear viscoelastic response* refers to viscoelastic response in which stress and strain are related by a single modulus which depends only on duration of the applied stress and strain for a given temperature. This differs from nonlinear viscoelastic response in which the modulus depends on the magnitude and duration of stress or strain.

Viscoelastic stiffness response, traditionally, has been expressed in terms which relate to the manner of loading. That is, the time dependent apparent modulus, or ratio of the (decaying) *stress* to the (constant) *strain* imposed during relaxation experiments, is termed the "relaxation modulus," $E(t)$. If the test is performed in creep, the "creep compliance," or the ratio of (increasing) *strain* to applied *stress* $D(t)$, is used to define response. This notation is extremely useful in the study of nonlinear viscoelastic behavior which occurs at the stresses and temperatures above the range of interest in structural engineering.

It can be shown that for small strains at temperatures in the useful range, $E(t) \approx 1/D(t)$. Thus, providing the range over which this assumption can be made is known, a single modulus can be used to describe time-dependent stiffness response under both creep and relaxation loading conditions.

A term, "viscoelastic modulus," E_v, defines the ratio between stress and strain after any duration of stress or strain. This ratio is frequently referred to as the "apparent modulus." The "viscoelastic modulus" terminology, however, appears to be a more descriptive and appropriate companion to the term "elastic modulus," which is well known to the structural designer. The approach taken by some designers is to establish conditions over which $E_v \approx E(t) \approx 1/D(t)$, for practical purposes; in essence this establishes the linear viscoelastic range.

Whether or not behavior is in the linear or nonlinear viscoelastic range decides whether or not behavior can be modeled by simple adaptations of conventional, readily available analysis methods based on elastic theory.

Linear viscoelasticity means that deformation is a function of time, and time only, under a given magnitude of sustained stress (creep). Alternately, stress is uniquely a function of time under sustained strain (relaxation). In both instances, temperature is assumed constant. In essence, the assumptions of linear viscoelasticity are similar to those of Hookes Law (stress is proportional to strain, or vice versa) except that time dependence is also involved.

By contrast, in nonlinear viscoelastic behavior, the relation between stress and strain (modulus) at a given time is dependent on the magnitude of the applied stress or strain. The complications introduced by such nonlinear stress–strain behavior should be obvious.

Creep curves given in Fig. 8.35 for three levels of sustained stress illustrate linear and nonlinear viscoelastic behavior. For the lower stress, σ, the strain response is proportional to stress; the strain at a given time at 2σ is twice that of σ. For the highest stress, 3σ, the strain response is greater than 3 times that at σ. Thus, at 3σ, linear proportionality between stress and strain is violated, and viscoelastic behavior is nonlinear.

Viscoelastic behavior for real plastics becomes increasingly nonlinear under the following circumstances:

- *High stress of strain levels:* Stresses or strains approaching yield or failure levels produce nonlinear viscoelastic response. Most plastics display nonlinear visco-elastic behavior even at low stress levels, but viscoelastic behavior may be assumed linear with certain limits.
- *Elevated temperatures:* Increasing temperature promotes nonlinear viscoelastic behavior and, hence, lowers the limits of stress over which linear viscoelastic response is displayed. Associated strain may either remain the same or increase.
- *Environment:* Certain solutions, gases, solvents or plasticizing agents, for example, promote nonlinear viscoelastic behavior for certain plastics.

8.6.3 Isochronous Stress–Strain Relationship

An isochronous stress–strain curve is a plot of stress vs. strain at a given time, under sustained stress or strain. For example, in a creep test, several levels of sustained stress are applied. The isochronous stress–strain curve is obtained by plotting strains measured at a given time under these stresses against applied stress.

Figure 8.35c shows isochronous curves cross-plotted from Fig. 8.35a which describes behavior in creep tests. The isochronous plots show the nonlinearity associated with the high stress level, 3σ, discussed in the preceding. They also show the linear proportionality between stress and strain at stresses up to 2σ, for all three loading times.

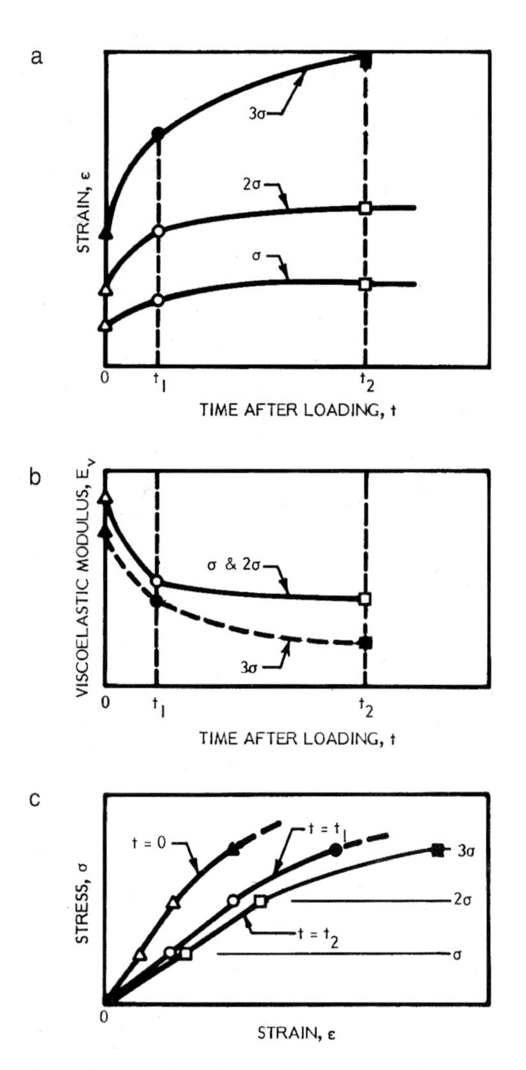

Figure 8.35 *Linear and nonlinear viscoelastic behavior during creep. (a) Creep behavior (linear time scale); (b) variation of viscoelastic modulus with time (linear time scale); and (c) isochronous stress–strain curves*

Isochronous relationships can also be obtained in relaxation experiments. In this case, several levels of constant strain are imposed on identical samples, and the stresses measured after a given time under these strains are plotted against the respective constant strains.

There is an important distinction between the isochronous stress–strain curve and the familiar stress–strain curve obtained from a "typical" stress–strain test. The isochronous curve captures a time frame, in a creep or relaxation experiment, involving many samples, held under several levels of sustained stress or strain. In contrast, the "typical" stress–strain curve depicts the response of a single specimen to a load or strain, which is continuously increasing with time. These curves are not interchangeable.

Isochronous stress–strain curves can be used to determine E_v. The initial slope (up to 2σ) of the isochronous plot shown in Fig. 8.35c is equal to E_v in the linear range. The secant modulus which connects the origin and the strain at 3σ defines E_v for that unique point only.

Overall, isochronous stress–strain relationships provide the most direct indication of nonlinearity of the three plots shown in Fig. 8.35. Thus, they are valuable in determining the limits of linear viscoelasticity. Such limits can be detected in creep tests lasting only a few minutes. Therefore, the nonlinear viscoelastic range can be defined rather quickly, and the designer may use this information as a guide in initial design estimates and materials selection and in the design of any longer term experiments which may be required to verify behavior.

A powerful tool for the designer of structural plastics, who must deal with variations in stress and strain input with time, is the Boltzmann Superposition Principle [8]. Simply stated, this principle holds that the effect of a change of stress or strain imposed on a linear viscoelastic material is independent of stress or strain history, and can be added, algebraically to the response of the system had the change not been imposed.

The Boltzmann Principle can be used to estimate recovery behavior or the effects of superimposing a transient live load upon a sustained dead load. This principle may also be used to handle such effects as a temperature (modulus) change on the deformation of a linear viscoelastic system under constant load.

8.7 Decorating

Because most fabricated products are attractive as well as inherently corrosion and rust resistant, they usually do not require any finishing or decoration. For others, there are treatments mainly used to enhance eye appeal. The finishing of certain plastic products requires different methods of adding decoration including printing or functional service effects. Plastics are unique in that decoration and color can be added prior, during, or after fabrication.

The different decorating methods provide various capabilities and benefits (Fig. 8.36 and Table 8.7), including better quality and second surface graphics, decoration on subtle curves and embossing, design versatility, dry processing eliminating volatile organic compound issues, durability and tamper proof, economic advantages for multiple colors, graphic flaws eliminated prior to molding, offer no-label look, recessed label area eliminating collection of dirt, scrap reduced or eliminated, and secondary step to printer eliminated.

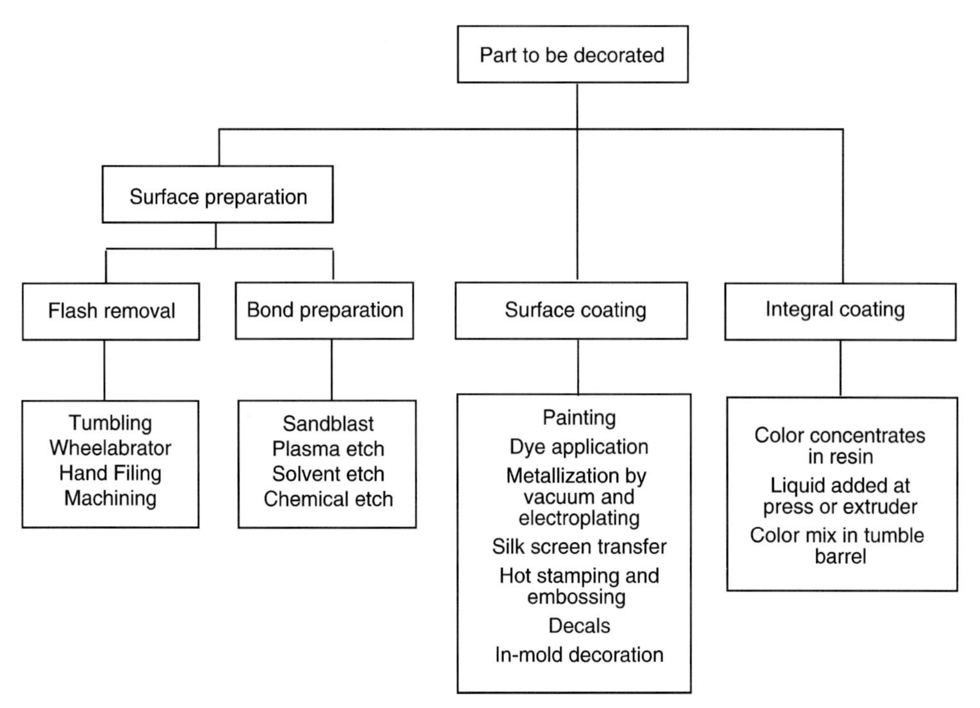

Figure 8.36 *Guide to decorating selection*

With surface decoration, an important requirement is that the surface be cleaned and prepared in the correct manner for the decorating method used. Some of the common causes of failure are contamination from processing lubricant, dust, natural skin oil, excess plasticizer on the surface, moisture, frozen-in strain. Decorating any material should be carried out in clean, controlled room conditions.

Each plastic (material type with their different additives, fillers, etc.) is to be considered separately when selecting coatings, thinners, decals, and foils. Some surfaces are prone to solvent attack, particularly thermoplastics. Frequently, it is necessary to preheat, flame heat, or chemically treat a plastic surface before applying the decoration. Static electricity on surfaces tends to cause major problems; airborne impurities become attracted and settle on the surfaces. Equipment is available to eliminate static charge.

Factors including molecular weight, method of manufacture, processing conditions, processing flash, and the form and shape of the part all affect decorating such as paint adhesion. This may involve nothing more than water cleaning the dirt or other residue. For some plastics, such as the polyolefins with their wax-like surfaces, more involved treatments are required. However, many plastics require special types of surface treatment prior to decorating. Methods used include solvent treatment (hot or cold), etching using strong oxidizing agents, flame or heat treatment, mild

Table 8.7 *Printing and Decorating Systems*

The process	What it's about	Equipment	Applications	Effect
Painting				
Conventional spray	Paints sprayed by air or airless gun(s) for functional or decorative coatings. Especially good for large areas, uneven surfaces or relief designs. Masking used to achieve special effects.	Spray guns, spray booths, mask washers often required; conveying and drying apparatus needed for high production.	Can be used on all materials (some require surface treatment).	Solids, multi-color, overall or partial decoration, special effects such as woodgraining possible.
Electrostatic spray	Charged particles are sprayed on electrically conductive parts; process gives high paint utilization; more expensive than conventional spray.	Spray gun, high-voltage power supply; pumps; dryers, pretreating station for parts (coated or preheated to make conductive).	All plastics can be decorated. Some work, not much, being done on powder coating of plastics.	Generally for one-color, overall coating.
Wiping	Paint is applied conventionally, then paint is wiped off. Paint is either totally removed, remaining only in recessed areas, or is partially removed for special effects such as woodgraining.	Standard spray-paint setup with a wipe station following. For low production, wipe can be manual. Very high-speed, automated equipment available.	Can be used for most materials. Products range from medical containers to furniture.	One color per pass; multicolor achieved in multistation units.
Roller coating	Raised surfaces can be painted without masking. Special effects like stripes.	Roller applicator, either manual or automatic. Special paint feed system required for automatic work. Dryers.	Can be used for most materials.	Generally one-color painting, though multicolor possible with side-by-side rollers.

Screen printing	Ink is applied to part through a finely woven screen. Screen is masked in those areas which won't be painted. Economical means for decorating flat or curved surfaces, especially in relatively short runs.	Screens, fixture, squeegee, conveyorized press setup (for any kind of volume). Dryers. Manual screen printing possible for very low-volume items.	Most materials. Widely used for bottles; also finds big applications in areas like TV and computer dials.	Single or multiple colors (one station per color).
Hot stamping	Involves transferring coating from a flexible foil to the part by pressure and heat. Impression is made by metal or silicone die. Process is dry.	Rotary or reciprocating hot stamp press. Dies. High-speed equipment handles up to 6,000 parts/h.	Most thermoplastics can be printed; some thermosets. Handles flat, concave or convex surfaces, including round or tubular shapes.	Metallics, wood grains or multicolor, depending on foil. Foil can be specially formulated (e.g., chemical resistance).
Heat transfers	Similar to hot stamp but preprinted coating (with a release paper backing) is applied to part by heat and pressure.	Ranges from relatively simple to highly automated with multiple stations for, say, front and back decoration.	Can handle most thermoplastics. A big application area is bottles. Flat, concave or cylindrical surfaces.	Multi-color or single color; metallics (not as good as hot stamp).
Electroplating	Gives a functional metallic finish (matte or shiny) via electrodeposition process.	Preplate etch and rinse tanks; Koroseal-lined tanks for plating steps; preplating and plating chemicals; automated systems available.	Can handle special plating grades of ABS, PP, polysulfone, filled Noryl, filled polyesters, some nylons.	Very durable metallic finishes.
				(continued)
Metallizing Vacuum	Depositing, in a vacuum, a thin layer of vaporized metal (generally aluminum) on a surface prepared by a base coat.	Metallizer, base- and topcoating equipment (spray, dip or flow), metallizing racks.	Most plastics, especially PS, acrylic, phenolics, PC, unplasticized PVC. Decorative finishes (e.g., on toys), or functional (e.g., as a conductive coating).	Metallic finish, generally silver but can be others (e.g., gold, copper).

(continued)

Table 8.7 *Printing and Decorating Systems*

The process	What it's about	Equipment	Applications	Effect
Cathode sputtering	Uniform metallic coatings by using electrodes.	Discharge systems—to provide close control of metal buildup.	High-temperature materials. Uniform and precise coatings for applications like microminiature circuits.	Metallic finish. Silver and copper generally used. Also gold, platinum, palladium.
Spray	Deposition of a metallic finish by chemical reaction of water-based solutions.	Activator, water-clean and applicator guns; spray booths, top- and base-coating equipment if required.	Most plastics. For decorative items.	Metallic (silver and bronze).
Tamp printing	Special process using a soft transfer pad to pick up image from etched plate and tamping it onto a part.	Metal plate, squeegee to remove excess ink, conical shaped transfer pad, indexing device to move parts into printing area, dryers, depending on type of operation.	All plastics. Specially recommended for odd-shaped or delicate parts (e.g., drinking cups, dolls' eyes).	Single- or multi-color—one printing station per color.
In-the-mold decorating or in-mold labeling	Film or foil inserted in mold is transferred to molten plastics as it enter the mold. Decoration becomes integral part of product.	Automatic or manual feed system for the transfers. Static charge may be required to hold foil in mold.	Most plastics, especially polyolefins and melamines. For parts where decoration must withstand extremely high wear.	Single- or multi-color decoration.
Flexography	Printing of a surface directly from a rubber or other synthetic plate.	Manual, semi- or automatic press, dryers.	Most plastics. Used on such areas as coding pipe and extruded profiles.	Single- or multi-color.

Offset printing	Roll-transfer method of decorating. In most cases less expensive than other multicolor printing methods.	Ranges from low-cost hand presses to very expensive automated units. Drying, destaticizers, feeding devices.	Most plastics. Used in applications like coding pipe.	Multi-color print or decoration.
Valley printing	Uses embossing rollers to print in depressed areas of a product.	Embosser with inking attachment or special package system.	Used largely with PVC, PE for such areas as floor tiles, upholstery.	Generally two-color maximum.
Labeling or post-mold labeling	From simple paper labels to multi-color decals and new pre-printed plastic sleeve labels.	Equipment runs the gamut from hand dispensers to relatively high-speed machines.	Can be used on all plastics. Used mostly for containers and for price marking.	All sorts of colors and types.

sandblasting, corona or arc discharge, mechanical abrasion, primer coat, gas plasma, and machining.

The type of release agent used during the fabrication of products can cause poor bonding of decorations or finishing of the product's surface. Zinc stearates are the least harmful while silicones could be very damaging. Treatments are used to remove the undesirable agents. If possible, do not use an undesirable release agent and regardless of type used, apply sparingly.

Some common decorative finishes applied to plastic are spray painting, vacuum metallizing, hot stamping, silk screening, metal plating, sputter plating, flame spray/arc spray, electroplating, printing and the application of self-adhesive label, sublimation printing, decal, and border stripping. In some cases, the finish will give the product added protection from heat, ultraviolet radiation, chemicals, scratching, or abrasion.

In-mold decoration is popular. This term is used both to describe designs that are etched or engraved in the mold surface and the process of inserting a printed film into the mold, to be produced as an integral component of the finished product. Etched surfaces can be drawn both parallel and perpendicular to a parting line of molds or postforming in an extrusion line. However, in the case of molds, be alert to the fact that parallel to the parting line, additional draft is required. A wide selection of patterns is available and new ones can be readily created.

Engraved designs and lettering normally have greater depth and fine detail. Parallel to the parting line, a side action (to clear the engraving) will be required in most cases. Therefore, they should be used only when absolutely necessary. Recessed letters and designs are to be avoided whenever possible. They collect dirt and are costly to put into the tooling. Raised letters are less expensive to make and to maintain. In either case, sharp points, such as those found in the letters N, M, and W, are prone to breaking out when subjected to molding pressures over a period of time.

Hence, it is wise to place designs and lettering in an insert in the mold. This will create an outline around the lettering (which can hopefully be incorporated into the design) and will also make repairs and revisions far less costly. Generally, one should avoid the use of serif typefaces unless the letters are very large indeed. Artwork is prepared for engraving in the same manner as for printing.

Inserting printed film in a mold provides some advantages in appearance, such as hiding defects. In addition, it can provide increased strength and other properties in the plastic, thus reducing the amount of plastics required. Required is compatibility with the fabricated plastic material. The inserts can range from flat to rather complex shapes.

Regardless of which finishing method is used, it is important to consider its design requirements from the beginning. Many designers, preoccupied with the mechanical requirements of the product, postpone consideration of this aspect until the very end of the project, only to discover that major design revisions are necessary to meet appearance requirements.

9 Process Control

9.1 Overview

As blow molding (BM) became more complex, molders required greater accuracy and increased variations in the types of cycles that could be adapted to their machines. Different types of machine process controls can be used that provide the means to meet the requirements based on the molders' operating needs. Process control systems can monitor (provide alarm buzzes or lights flash on deviation), provide feedback (deviation sets up corrective action), and act as program controller (computers interrelate all machine functions and all melt process variables) [92]. Knowledge of the machine and its operating needs is a prerequisite before an intelligent process control program can be developed. The control unit is composed of input, signal processing, and power stages.

Some controls are of an open-loop type and merely set a mechanical or electrical device to some operating temperature, pressure, time, or travel. They will continue to operate at their setpoints even though the settings are no longer suitable for making quality products. The problem is that during molding the total process is subject to a variety of hard-to-observe disturbances that are not compensated for by open-loop controls. Process control targets to close the control loop between some process parameter and an appropriate machine control device to eliminate the effect of process disturbances due to different variables.

With closed-loop controls properly installed and applied, the performance of the plastics in the machine can be controlled within limits to produce zero-defect products meeting performance requirements at the lowest cost. In developing the "best setting" of machine controls, the target is to operate within controllable/ repeatable parameters interrelating the controls to meet performance requirements and operate the machine at the lowest cost. In its simplest form, one starts with a two-dimensional molding area diagram (MAD) approach either for extrusion or injection molding. Limits have to be set based on test and evaluation of blow molded parts, followed by an analysis of the effect of interfacing the different variables.

Fabricating controls involve many facets of the machine operation and the behavior of the plastic. Most important is the interaction between the machine operation and the behavior of the plastic. Basically, the processing pressure and temperature vs. time provide the direction in determining the quality of the product. The design of the control system has to take into consideration the logical sequence of all these basic functions and their ramifications.

9.2 Control Flow Diagram

Developing a process control flow diagram requires a combination of experience (at least familiarity) with the process and a logical approach to meet an objective that has specific target requirements. Process controls range from very simple/standard types to advanced complex types.

Continual new developments in control of machines dramatically improve ease of machine setup, allow more uninterrupted operation, simplify remote handling, reduce fabricating times, cut energy costs, boast part quality, and so on. The process of making a product has many dynamic fragments that must come together properly for successful results. Lack of sufficient control over each of these fragments will result in a less than desirable product. The three key ingredients for success are sufficient dynamic performance, sufficient repeatability, and, very important, the selection of proper control parameters. A lack of these can result in unacceptable products, higher scrap rate, longer cycles, higher part cost, and so on.

All process controls monitor the process variables, compare them to values known to be acceptable, and make appropriate corrections without operator intervention. An acceptable range of values can be determined by using melt flow analysis software and/or trial and error when the machine first starts production. Using the software approach, the acceptable process values are known before the mold or die is ever built. It should be noted that most of the process control systems available today are rather complex and require skilled people to set them properly and/or use them efficiently.

Adequate process control and its associated instrumentation are essential for product quality control. The goal in some cases is precise adherence to a control point. In other cases, maintaining the temperature within a comparatively small range is all that is necessary. For effortless controller tuning and the lowest initial cost, the processor should select the simplest controller (of temperature, time, pressure, melt flow, rate, etc.) that will produce the desired results.

One example of controls used with injection molding is seen in Fig. 9.1. Based on the process control settings, different behaviors of the plastics will occur.

9.3 Control Problem/Solution

Purchasing a sophisticated process control system is not a foolproof solution that will guarantee perfect products. Solving problems requires a full understanding of their

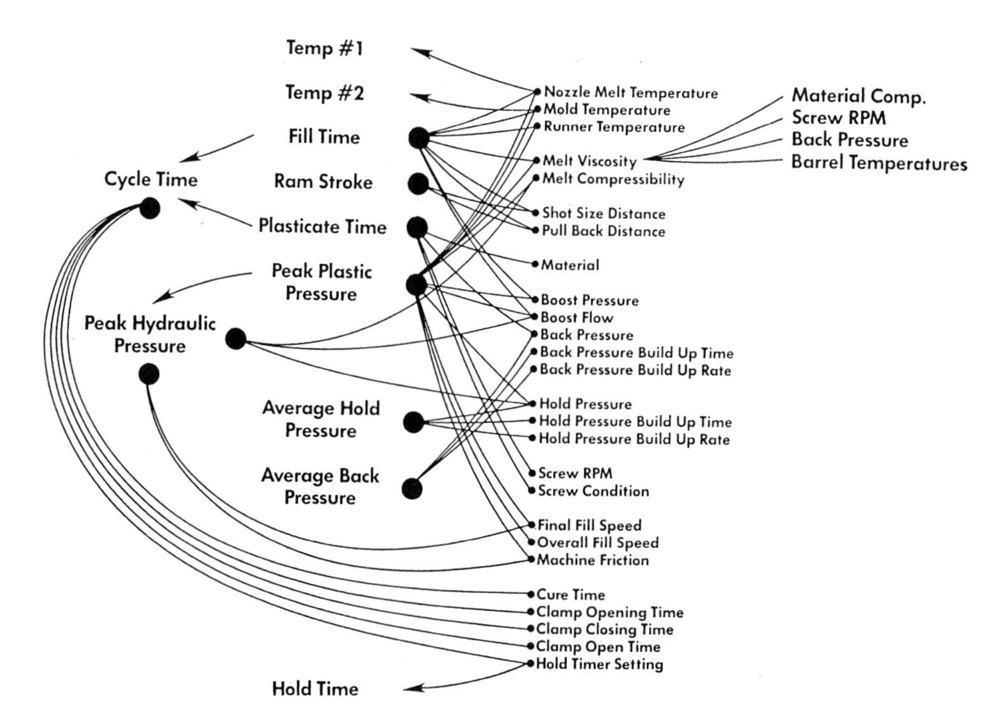

Fig. 9.1 *Injection molding machine controls*

causes, which may not be as obvious as they first appear. Failure to identify contributing factors when problems arise can easily result in the microprocessor not doing its job. The conventional place to start troubleshooting a problem is with the basics of temperature, time, and pressure requirement limits (Chapter 12). Often a problem may be very subtle, such as a faulty control device or an operator making random control adjustments. Process controls cannot usually compensate for such extraneous conditions; however, they may be included in a program that provides the capability to add functions as needed.

There are two basic approaches to problem solving—either to find and correct the problem, applying only the control needed or to overcome the problem with an appropriate process control strategy. The approach one takes depends on the nature of the processing problem and whether enough time and money are available to correct it. One must systematically measure the magnitude of the disturbances, relate them to product quality, and identify their cause so that proper action can be taken. Process controls may, in most cases, provide the most economical solution.

Before investing in a more expensive system, the processor should methodically determine the exact nature of the problem to decide whether or not a better control system is available and will solve the problem. For example, the temperature differential across a mold can cause uneven thermal mold growth. The mold growth can also be affected by uneven heat in the clamping system. One side can be hotter, causing platens to bend and the change could be reflected on the mold operation. Perhaps all that is needed to correct the mold heat variation is to close a nearby large garage door eliminating the flow of air upon the mold. In the case of air conditioning, all that may be required is to change the direction of air flow.

9.4 Basic Control Element

9.4.1 Sensor and Gauge

The ability to measure parameters is a key element of effective process control. Accurate determinations of processing variables such as temperature, pressure, time cycle, and so on are important. Sensors have traditionally played an important role in measuring and monitoring a broad range of parameters. All sensors perform the same basic function of the conversion of one type of measurable quantity, such as temperature, into a different but equally quantifiable value, usually an electrical signal. Although the basic function remains the same, the technologies used to perform that function vary widely.

Sensitivity and complexity increased as advances were made in electronics technology, including innovative circuit designs and more efficient power sources that followed the invention of vacuum tubes and transistors. Microfabrication and nanotechnology, improved materials, and new design capabilities today have a significant impact on sensors, which are evolving more rapidly than ever. Microfabrication technology can be used to produce geometrically well-defined, highly reproducible structures and surface areas. Consequently, this may simplify or minimize the need for individual calibration.

Although a wide range of sensors can be used, they all can be categorized generally as either physical or chemical in nature. Physical sensors are used to measure a range of physical responses such as temperature and pressure. In addition to the most common types of physical sensors, optical and electrical, the category includes geometric, mechanical, thermal, and hydraulic types. Sensors that detect electrical activity include electrodes. Optical sensors are being used in a number of applications in which light is used to collect physical data. They are a key element of certain new technologies.

Chemical sensors, which include gas and electrochemical devices, are designed to detect or measure the presence of specific chemical compounds. Photometric sensors, which are optical sensors used to measure chemical presence, are also included in this category.

There are sensors designed to respond to a physical stimulus (temperature, pressure, motion, product gauging, product weight, etc.) and transmit a resulting signal for interpretation, measurement, and/or operating a control. There is a very broad selection of sensors with extremely different sensitivities, capabilities, and repeatabilities.

To select the correct sensor, it is important to know something about how the different sensors work, and which is used for what application, as not all sensors measure the same way. The three most common sensors used downstream are nuclear, infrared, and caliper. There are also specialized types such as microwave, laser, X-ray, and ultrasonic. They sense different conditions for operating equipment (temperature, time, pressure, dimensions, output rate, etc.) and also sense color, smoothness, haze, gloss, moisture, dimensions, and many other characteristics.

9.4.2 Transducer

The term *transducer* is frequently interchangeable with sensor. It is a device that converts something measurable into another form and is often a physical property such as motion, pressure, temperature, and/or melt flow.

The use of linear displacement transducers (LDTs) has evolved over the past decades. They are used extensively in injection molding machines. They are used to perform different functions such as control mold closing, determine the location of feed screws, and track the position of part ejectors. In high-performance systems, they have largely displaced mechanical limit switches and potentiometers. Although these contact methods of switching and linear measurement worked satisfactorily, they did have their drawbacks, such as mechanical wear and the limitation of determining only the end position of the components being monitored.

As the LDT provides continuous position feedback and is a noncontact method of measurement, it is ideally suited for either open- or closed-loop fabricating operations. Various types of outputs are available that can enhance both the positioning accuracy and processing speed of machines.

The linear velocity displacement transducer (LVDT) is used for measuring relatively small amounts of movement in the vertical or horizontal plane. The amount of movement is detected by means of a change in an electrical signal caused by the movement of an iron core within a coil; this change is then amplified and converted into a linear measurement. Very accurate readings of tie bar extensions and downstream roll positions can be obtained when using these devices.

The specifications on pressure transducers from different manufacturers can vary significantly, so it is important to understand their accuracy. An ideal device would have an exactly linear relationship between pressure and output voltage. In reality, there will always be some deviations; this is referred to as nonlinearity. The best straight line is fitted to the nonlinear curve. The deviation is quoted in the specifications and expressed as a percent of full scale. The nonlinear calibration curve is determined in ascending direction from zero to full rating. This pressure will be slightly different from the pressure measured in the descending mode. This difference is termed hysteresis; it can be reduced via electrical circuits.

9.4.3 Pressure Sensor

Pressure sensors are extensively used such as in plasticators to improve output and melt quality, enhance production safety, safeguard, and so on. They aid in obtaining optimum processing pressure to ensure product quality such as output dimensions and surface finish, and minimize material waste. Sensors play a critical role at every stage of the fabricating process, and because the pressure of the process affects the physical properties of a plastic material, controlling pressure is critical. Pressure translates into heat and shear, which can significantly change physical and even chemical properties in the finished product. Pressure sensors can help solve these control problems.

In many processes, the critical pressure points to measure depend on technical knowledge and ability to justify the expense of time, personnel, and equipment. If one runs a highly technical and close-tolerance product, one can ill afford to live with the reject rate of a noninstrumented system.

If, however, one runs a product that does not require very close tolerances, but does require high product output and high value-added characteristics, one can use a less expensive and less technical approach to quality monitoring. Still, measurement of the pressure is important. Small changes in pressure such as in the extruder die or injection molding nozzle can cause warpage and poor physical properties of the finished product.

For each fabricating machine and die/mold combination, there is an optimum number and location for pressure sensors. They provide the means to ensure that plastics and machine time are put to optimum use, and that production of quality parts is maintained.

Two types of pressure sensors are used: strain gauge and piezoelectric pressure sensing devices. Both technologies have strengths and weaknesses. The strain gauge characteristics include the following:

- they are typically cheaper than piezoelectric sensors and should be used when cost is an issue;
- they can be used without special wiring;
- they are best with long cycle times; and
- their ruggedness, the ability to splice wiring and low replacement costs, make them ideal for rapid die/mold changes and harsh environments.

Characteristics of piezoelectric sensors include the following:

- they have quicker response times, typically greater than 20 kHz, vs. 0.2 kHz for strain gauges;
- they require balanced wiring, charge amplifiers, and careful attention to ground loops;
- they are best for short cycle times, high temperatures, and critical part control;
- they are smaller than strain gauges so they can be used in tight spaces; and
- they are immune to high electrical discharge and radiofrequency noise.

9.4.4 Temperature Sensor

Fabricating plastic products is a thermal process with the major objective of controlling temperature. Too much or too little heat at the wrong place can cause many problems. One cannot see this thermal energy, only its effects. Thermal energy radiates in the infrared (IR) spectrum, outside the spectrum of visible light. IR video cameras can detect energy color patterns in all locations around the primary and auxiliary equipment. With this IR thermography, every plastic has its own wavelength and temperature readings that are related to the IR color patterns [2, 18].

Common sensors used for temperature include the thermocouple (TC), resistant temperature detector (RTD), and thermister (TM). Each has technical and cost advantages and limitations. TCs tend to have shorter response time, while RTDs have less drift and are easier to calibrate. The TC is the more common in use. RTD provides for stability; its variation in temperature is both repeatable and predictable.

A TC is a thermoelectric heat sensing instrument used for measuring temperature in or on equipment such as the plasticator, mold, die, preheater, melt, and so on. TC depends on the fact that every type of metallic electrical conductor has a characteristic barrier potential. Whenever two different metals are joined together, there will be a net electrical potential at the junction. This potential changes with temperature.

TMs are semiconductor devices with a high resistance dependence on temperature. They may be calibrated as a thermometer. The semiconductor sensor exhibits a large change in resistance that is proportional to a small change in temperature. Normally TMs have negative thermal coefficients. Like RTDs, they operate on the principle that the electrical resistance of a conductive metal is driven by changes in temperatures. Variations in the conductor's electrical resistance are thus interpreted and quantified as changes in temperature occur.

Temperature control units give operators more oversight during processing with the result of being able to increase product precision. The demands on temperature controllers to obtain quality molded products require accuracy on their dynamic behavior such as response time and transient performance. They differ widely depending on requirements such as short response time, minimal overshooting, high circuit stability even with system (equipment and plastic) variations, transient suppression, and sluggishness to keep temperature variations small. Most controllers provide a derivative term that is called the rate term. This is an anticipatory characteristic that shortens the response time to changing conditions [92]. There is also a circuit that limits overshooting.

As not all demands can be met in an optimal manner at the same time, tradeoff has to be accepted in the adjustment of controllers and in the selection of suitable control elements. Controllers operate in a digital mode and employ microprocessors. The input signal is converted into a numerical value and mathematically manipulated. The results are summarized to obtain an output signal. This signal regulates the power output in such a way that the temperature is maintained at the set value.

Previous proportional controllers did not allow the actual temperature to be at the set point. For compensation, the offset was squared, resulting in a larger deviation that carried more weight than a smaller one. Regardless, the temperature always differed by some fraction of the proportional band and the offset was always constantly changing. Because the input signal is continuous, the microprocessor performs an integration and adds an averaging term that permits the temperature to be kept at a preset point from ambient temperature to full heating even if larger time lags are present. This function is called automatic reset.

Most processes operate more efficiently when functions must occur in a desired time sequence or at prescribed intervals of time. In the past, mechanical timers and logic relays were used. Now electronic logic and timing devices predominate. These programmable logic control devices provide sophisticated operations with controllable functions.

9.5 Computer Controller

9.5.1 Overview

The solid state controller consists of a set of base functions [92]. They all require a programming method, logic engine, operator interface, communication interfaces to I/O with a programming system, an operating system, and a hardware platform/package. To make an informed decision on which controller is best for an application, it is important to understand that for each controller there is a choice regarding their different levels of risk or responsibility. With low user risk, there is higher vendor responsibility; with higher risk there is lower vendor responsibility.

This range of vendor-to-user responsibility is an essential consideration for making an informed hardware choice. There are choices available for both controllers and the functions within them. The fundamental set of requirements involves speed, scale, packaging, reliability, and peripherals. These requirements must be satisfied before successfully applying any control system technology.

9.5.2 Control Choice

Controls are any instrumentation such as pressures, temperatures, timers, etc. used to control and regulate the fabricating cycle.

As control choices continue to expand, the user has to recognize what to specify such as soft, programmable, hybrid, or entirely new architectures. Amidst this potential confusion, control venders are waging their own debate over which technology is least expensive, most popular, or will outlast others.

One should define control not by the "box" performing it, or which solution is typically used, but rather by virtue of the problem it solves. To do this, users must consider the choices available at each level within a controller. The choice requires personal knowledge or help from a reliable source to determine which type of controller is appropriate for a specific application.

Today's programmable microprocessor controller operating system, like the hardware platform, is the result of over a quarter of a century of evolution to provide the available repeatability and reliability required on the plant floor. In the past, achieving these objectives meant choosing a vendor-specific operating system and choosing that vendor's entire control system. This can be a benefit if risk is to be avoided, but can be detrimental if a high degree of in-house customization and integration is desired.

Intel Corp. basically started logical circuits of a microprocessor using a single silicon chip in 1969 in controlling injection molding machines (IMMs). Up to that time in conventional logic control all sequential functions were programmed by means of an electrical wiring matrix [1]. In a microprocessor control system, this matrix is replaced by software that contains the instructions. The multiprocessor (transputer version) is told how to process the signals from the operator panel. This signalling action is linked with other process parameters that are stored in electronic memory modules. This sequence logic can be divided into individual independent blocks with separate microprocessors running programs. With this multiprocessor technique, which is a modular design, the microprocessor takes over the entire logic of the IMM whereby it controls the operation of the machine.

The controllers are fairly simple devices, but problems can develop. A checklist for eliminating problems includes: heater element burnout, location and depth of sensor as related to response time, type of on/off control action such as proportional controller, set point control, and proper selection of basic electrical components. It is important to have the proper depth of the sensor in a barrel to obtain the best reading for the melt; the deeper the better. Where water is involved, as in mold cooling controllers, with improper construction the most common problem (due to expansion/contraction not properly incorporated) are water leaks.

Computer coordinator controllers are connected together so that they may all be changed at the same time from a single point. Also used are multizone microprocessors that monitor temperature, pressure, output rate, and such signals from several sensors to achieve more reliable and efficient performance, either independently or coordinated.

A microprocessor or multiprocessor system has to carry out control and monitoring functions such as

- standard functions (sequence control, timer, malfunction indication, etc.),
- monitoring functions (self-diagnosis of malfunctions, control of setup procedures, calculation of operating data, etc.), and
- control functions (different temperatures in the IMM, process control of speed and holding pressure, etc.).

9.5.2.1 Programmable Controller Safety

Guarding of safety circuits is used in programmable controllers to protect people and equipment. The purpose is to establish a procedure for the guarding of OEM (original/equipment manufacturers) supplied safety circuits on machines equipped with programmable logic controllers. It is imperative that OEM supplied circuitry not be subject to modification or deletion by the end user.

9.5.3 Control System via Website

From hundreds and thousands of miles away, using a Web camera feed, a computer screen diagnoses faults in a plant operation. Using a standard Web browser, the engineers jog the axis, run profiles, and watch the action occurring in the plant. The entire process is accomplished using proven, readily available commercial technology that allows fabricating equipment and components from different manufacturers to be mixed and matched freely, easily, and inexpensively.

The system features programmable logic controllers (PLCs) with embedded Web servers that make it easy to display information in graphic form, as well as, serve as drives communicating with controllers over a fiber optic link, and so on. Anyone with proper security clearance and network access can pull information from the PLC, display it on custom Web pages, and change the parameters from anywhere in the world. This system is a marked departure from proprietary plant networks (Interbus, Profibus, ControlNet, etc.), that provide a heavily customized, one-way architecture that collects production and diagnostic data, interrelates them, and displays it on a Web server. Such systems provide no means to react to the data and send back commands that are then implemented automatically.

Fieldbus developments involve process controls. In the ideal world, PC integration would be simple. One would buy all the computers, instrumentation, and electrical equipment from a single vendor and connect it all using a single network standard.

But rarely are the PLCs, distributed control systems (DCSs), drives, motor controls, field instrumentation, and computers all purchased from the same vendor. Even when they are, the vendor usually has a number of different vintages or models of equipment that do not interconnect cleanly.

9.5.4 Fieldbus Communication Control

The major buzz in the industrial automation business is networking communications on the factory floor and in the process environment. Both big and small businesses are seeking to connect sensors to actuators and everything else in between, including programmable logic controllers, I/O, computers, data acquisition, alarms, displays, recorders, and other controllers. Companies of all sizes and disciplines seem to be waiting for fieldbus, the bidirectional communications protocol used to connect field-level instruments and controls.

End users expect interoperability as the primary benefit of standardization. But that remains elusive as vendors push to promote their proprietary differentiation. Even if or when a fieldbus standard is complete and products are available, this will not eliminate the need for other industrial networks for a variety of reasons that include cost, speed, complexity, compatibility with already installed devices, and future expansion requirements.

A wide variety of networks are available in the industrial automation environment; however, confusion arises because their capabilities overlap. At the lowest level are the sensor networks that were originally designed primarily for digital (on/off) interface. These are fast and effective, but with only limited applications beyond simple machine control. Actuator sensor interface is still popular in Europe, while Seriplex is a US development. At the next level in the hierarchy are the device buses, which provide analog and digital support for more complex instruments and products. DeviceNet interfaces very well with programmable controllers; however, it is still only a local, device-level network, operating typically up to 64 points within about 1,000 ft. Profibus-DP and Interbus are somewhat comparable to DeviceNet and are excellent general-purpose, device-level networks. LonWorks operates over greater distances and is currently the only practical peer-to-peer network extendable to many thousands of points, although it is still comparably slow and more complex.

Profibus-PA provides improved performance at the fieldbus level for instruments and controls, replacing the features and functions provided by HART (originally developed for transmitter calibration and diagnostics). The primary differences are in speed, complexity, and distance. LonWorks also extends into this realm. Fieldbus SPSO overlaps somewhat with Profibus-PA and Profibus-FM, and this is the primary area of contention in the ongoing IEC/ISA SPSO committee discussions.

At the next level are the control networks, which include ControlNet (developed by Allen-Bradley and utilized by Honeywell), clearly overlapping with some of the functionality and performance intended to be provided by Profibus-FMS and Fieldbus SPSO. The highest level of fieldbus concerns information and Internet connectivity, for which Ethernet TCP/IP is clearly the standard. During the year 2000, Ethernet has gained wide acceptance, even at the lower levels, because its performance is well established at the higher end, while commercial proliferation and plummeting prices have extended usage at all levels.

9.6 Preform/Parison Thickness Control

Thickness control of the nonstretched or stretched injection blow molding (IBM) preform or extrusion blow molding (EBM) parison directly relates to the thickness control of the blow molded products [51, 55]. Control of the preform requires that the cavity in the preform mold relate to the product requirements, which will include stretching action if used.

Parison control or programming is a method of controlling weight and wall thickness of the blow molded product. As an example, when the parison is blown into a complex shape, the wall thickness of the product will vary according to the area and contour that are formed. The widest part of a container, for example, will have the thinnest wall section as the same amount of plastic has to cover a greater area (parison blowup ratio). Also, when extruding a large parison, the wall thickness will vary as the weight of the material increases and begins to sag.

To maintain accurate wall sections and control weight, the parison wall thickness is varied as it extrudes by an electrohydraulic mechanism. Many changes, such as up to hundreds of points, can be made in the parison wall thickness as it extrudes. Programmers, both linear and nonlinear point distribution, are available that provide a range of point settings. Redistributing the material in the parison makes it possible to selectively strengthen areas of the product. Conversely, excessively thick walls can be reduced. Adjustments to the programming can be made without stopping the machine. Programming is used for high-specification type containers, most industrial parts, and to obtain cost savings. Fig. 9.2 shows parison programming to fit specific mold configurations.

Two methods of parison programming are in use. One varies the orifice size of the extrusion die and the second varies the extrusion rate of the parison. A variable orifice programming system can be used on either continuous or intermittent extrusion machines while the variable extrusion rate system can be used only with intermittent EBM machines.

The extruder and die head can have a continuously variable central height adjustment mechanism, with which the distance between die and mold can be set very accurately. This ensures perfect matching in every case, even with very long molds. In addition

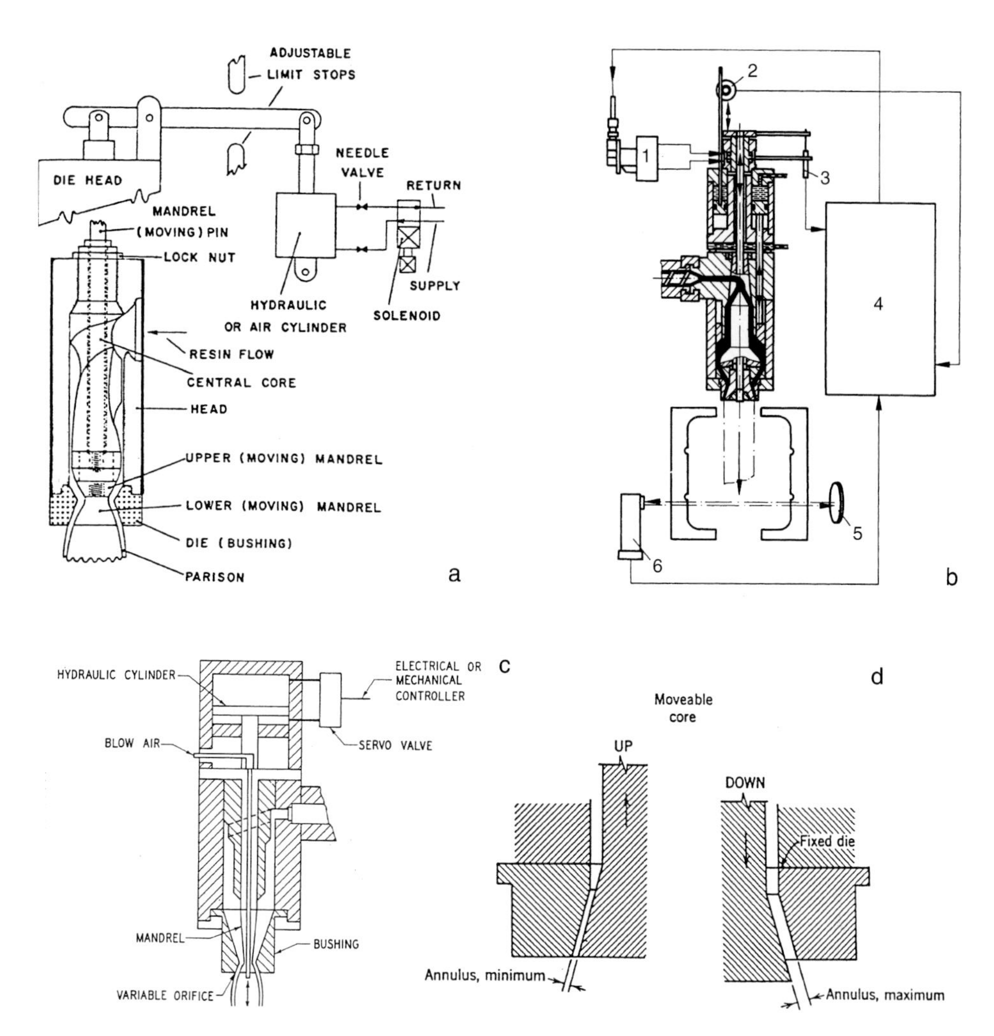

Fig. 9.2 Automatic variable orifice programmed die heads where in (b): (1) servo valve, (2) accumulator position, (3) accumulator position transducer, (4) wall thickness programmer, (5) reflector, and (6) photocell

to programming wall thickness, systems are also available to control parison length, a factor of special importance in blow molding large industrial pieces.

The method most commonly used varies the annular opening between the die and mandrel, causing changes in the parison thickness as it extrudes. Moving the mandrel up and down while the die bushing remains stationary varies the annular opening.

Variable parison extrusion rate is achieved by controlling the pressure of the flow of the melt flow (Fig. 9.3). While the variable orifice programming system is used on all

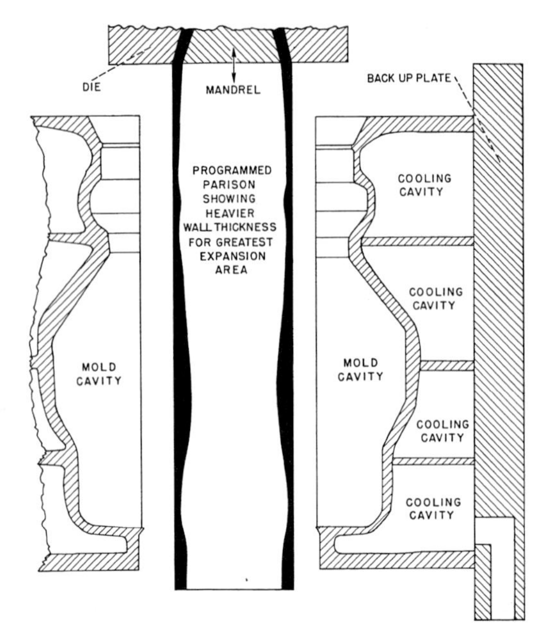

Fig. 9.3 *Programming parison to fit specific mold configurations: wall thickness distribution (mm) on BM articles with parison programming*

EBM machines, the variable extrusion rate system is used usually only on the large ram accumulator BM machines.

The method of parison programming widely used involves varying the position of the internal mandrel, as shown in Fig. 9.4. The actual method of varying the mandrel

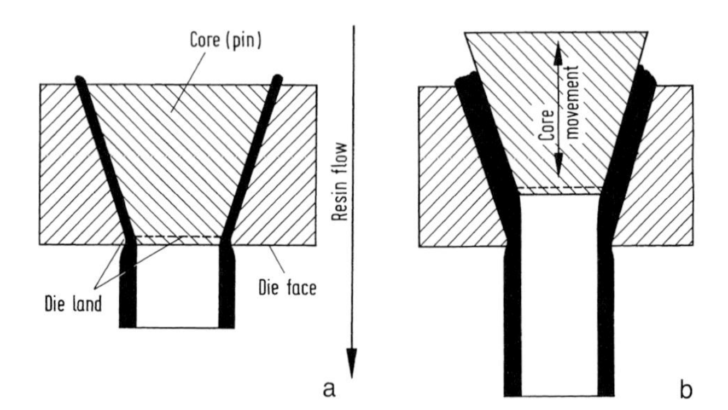

Fig. 9.4 *Basics in varying position of parison programmed internal mandrel to overcome or at least limit sagging is by automatically raising the core inside a variable die so that it is flush with the die face (a). Such gradual increase of the die clearance (b) takes place during the parison-forming stage. Wall thickness increases gradually compensating for sagging at the upper end of the parison*

position is the use of an electrical hydraulic assembly that automatically programs the parison. Thus, the parison may be extruded with an even wall thickness to compensate for sag, or it may be programmed to have several variations in thickness along its length for irregularly shaped objects (Chapter 2).

Finished necks can be produced on bottles during the molding cycle by the use of mandrel calibration. In this process, a tube larger in diameter than the desired bottle neck is extruded down over a specially designed blowing mandrel. After the mold is closed and the bottle is initially blown against the mold walls, the mandrel is forced further into the neck of the bottle forming both internal and external dimensions of the neck in conformity to the machined shape of the driven pin. To speed the molding cycle the calibrating pin is usually cooled to give cooling on both sides of the plastic in the neck finish.

More difficult and also more expensive is the influence of parison wall thickness on the circumference, so called partial wall thickness control. To achieve this, the die is elastically deformed into an elliptical shape so that two strips, lying opposite each other, are formed, which are either thicker or thinner. The length and the thickness of these strips are controlled by normal wall thickness control instruments. (Another method makes use of partial coextrusion of the same material. In this way, the parison can be made thicker for any required section of the circumference. The unit can be controlled independent of parison length.)

An even wall thickness distribution on the parison circumference and wall thickness control in longitudinal direction are in many cases, no longer sufficient to guarantee processing quality. The increased requirements and ever more unusual geometric shapes require a partial wall thickness adjustment, which is realized by shaping the flow channels in the die region.

Each shaping is based on a varying flow channel design on the circumference of the ring die. Shaping is carried out on the die gap. This is possible at the core as well as at the mouthpiece and can be realized with diverging as well as converging dies. This shaping has more or less effect, depending on the position of the core.

Diverging dies allow a shaping in cylindrical part whose effect increases with increasing die gap. This kind of shaping is suitable, for example, for the strengthening of corners which extend over the whole length of the article.

Converging dies have the advantage of blowing regions of one and the same parison with shaping and others without shaping, with the help of wall thickness control.

The manufacture of such dies proves to be most difficult and generally leads to long optimization times. Therefore, trials nowadays turn toward devices termed Static Flexible Deformable Ring (SFDR) (Fig. 9.5) and Programmable Wall-Thickness Distribution System (PWDS) (Fig. 9.6). Similar concepts are used in extruded film and sheet dies using programmable controlled lip adjustments [10].

The conventional process of shaping tooling, in which the tooling is brought up to running temperature, tested, cooled, removed from the machine, and remachined can

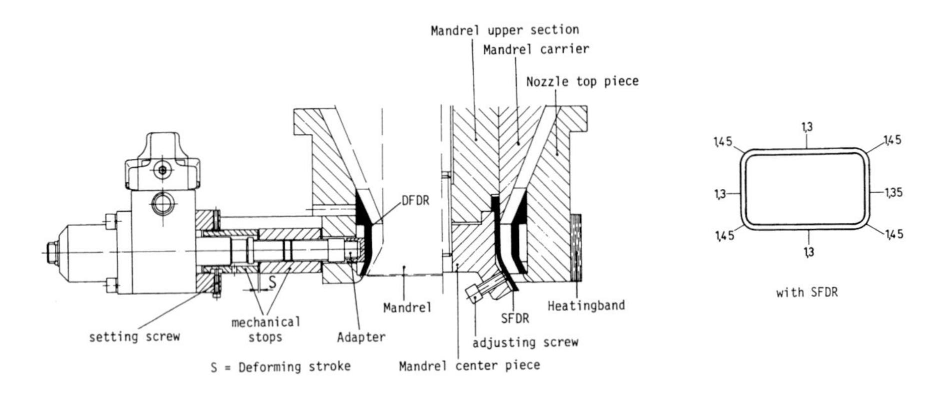

Fig. 9.5 *Details on static flexible deformable ring (SFDR) mandrel unit (Kautex/Moog)*

represent a significant expense and loss of production. There is the additional risk that overcutting may require that the tool be scrapped and the process be started over again.

With the SFDR it is only necessary to interrupt the machine cycle and adjust the deformation of the SFDR using nothing more than hand tools, without removal from the head (Fig. 9.5). Should an overadjustment occur, it can be quickly and easily corrected. In several cases where a number of similar machines are producing the same product, the profiled SFDR on one machine has been used as a model for the conventionally profiled mandrel on the other machines. The construction of the SFDR is such that it is operated in the same manner as a conventional mandrel.

In the PWDS system, a Dynamic Flexible Die Ring (DFDR) is deformed according to the profile set by a programmer in a manner similar to that of an axial wall thickness control system. Two servoactuators with integral position transducers are attached directly to the DFDR to affect ovalization. The basic configuration for an accumulator head machine is shown in Fig. 9.6.

A significant feature of the system is that the stroke of the PWDS actuators can be set separately via the die gap and program range potentiometers while they follow the same preset profile. Should asymmetric ovalization or shifting be desired, then each PWDS actuator can be controlled by its own parison programming unit. This has proven most useful in various applications such as automotive gasoline tanks, canisters (Fig. 9.7), etc.

It has been demonstrated that PWDS and, if present, SFDR enable the molder to implement the corrections that are frequently required rapidly and easily, specific to the parison, by entering clear, reproducible information directly to the parison wall thickness controllers.

As an example, consider the case in which fluctuations in material batch or type have caused the circumferential distribution at the center of a simple part (i.e. a closed

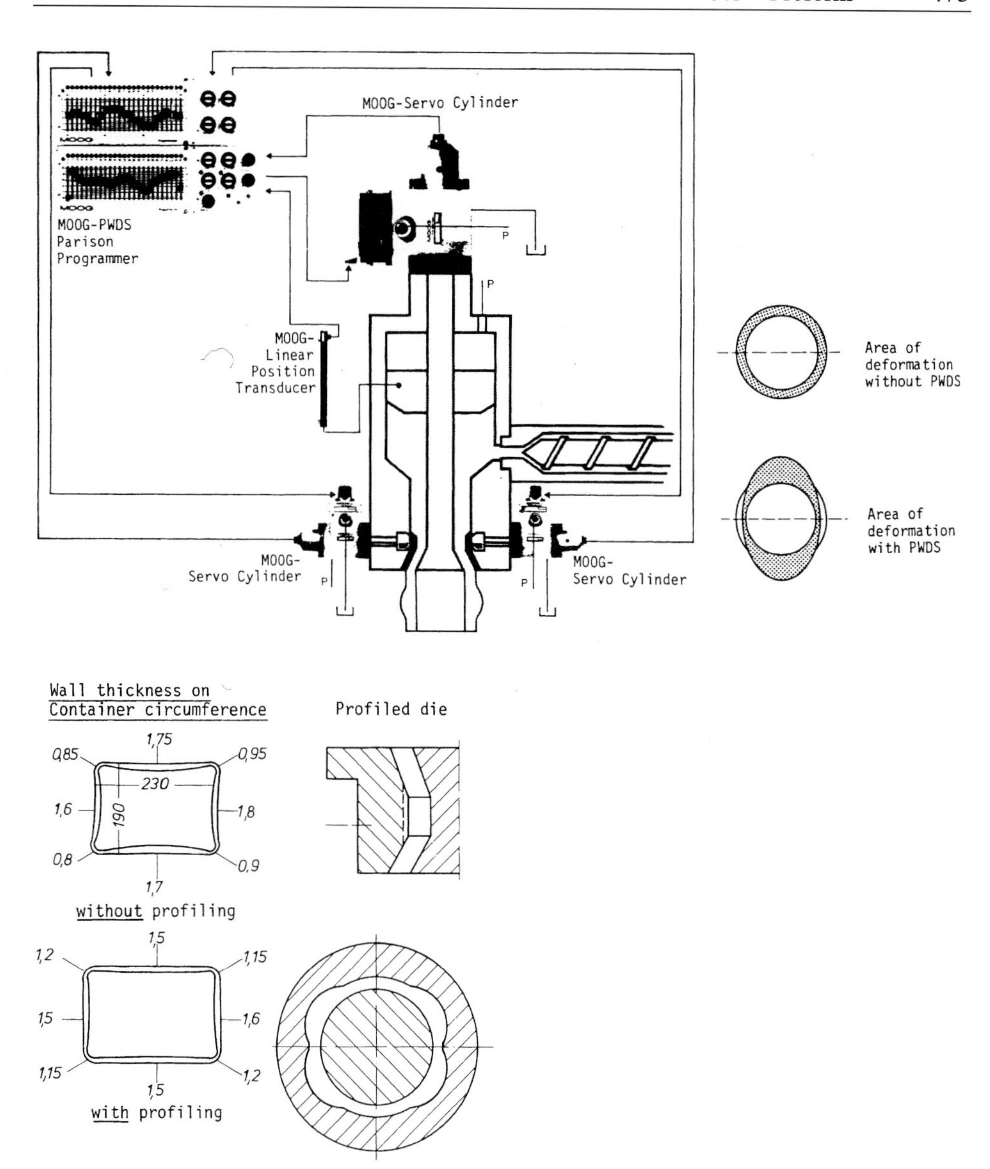

Fig. 9.6 Basic configuration of a programmable wall thickness ring (PWDS) system on an accumulator head; units in mm (Kautex/Moog)

drum) to go out of tolerance. This is certainly caused by the change in flow behavior of the material, which in turn results in a change in stretchability. The consequence is a nonuniform wall thickness.

Previously, the operator varied the stock temperature, die temperature, head temperature or pushout rate, etc. to re-establish a uniform wall. If, for example, he has raised

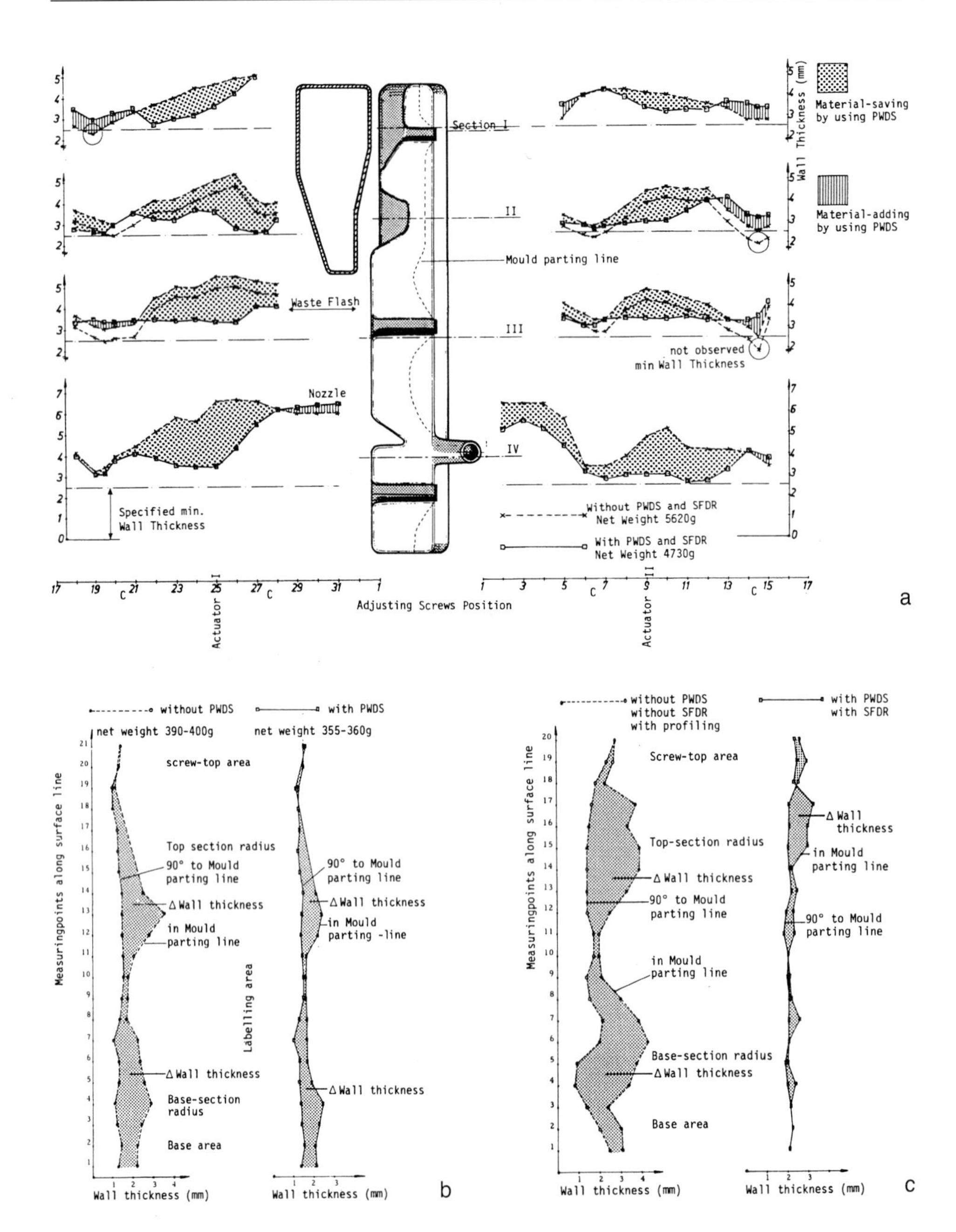

Fig. 9.7 Parison profiles using the SFDR system: (a) four critical sectional planes of a BM fuel tank that correspond exactly to four program points of the wall thickness control programming system. Sectioning at the die circumference by the 32 adjusting screws present on the SFDR is shown on the abscissa. (b) Comparison of the wall thickness configuration along the length of a 10-liter canister. With and without radial wall thickness programming and with various net weights. (c) Comparison of the wall thickness configuration along the length of a 25 liter canister. With and without radial wall thickness programming and with the same net weight (1,040 g)

the stock temperature, he will also experience a higher mold release temperature risking his production.

Now, the stock temperature can be set toward the lower limit recommended by the material supplier. It should be adjusted upward only enough to ensure good welding characteristics in the head and an acceptable surface. The low mold release temperature ensures maximum possible production. Shot rate, head temperature, die temperature, and if applicable, cooling can now be freely selected to control surface quality.

With PWDS and/or SFDR, the machine setter has three additional measures and can thus influence wall thickness directly without disturbing the basic process.

Wall thickness distribution along the length of the parison is controlled by the conventional wall thickness programming unit. Wall thickness circumferential distribution at the center of the part is controlled by the predeformation of the die ring.

The predeformation is set individually by the "Base Gap" potentiometers of the PWDS programmer for each of the PWDS actuators. These allow the molder to directly influence the wall thickness in the center of the part. Where the part exhibits differing stretch in various areas, these can be accommodated by the "Program Range" potentiometer on the PWDS programmer. On products with noncircular cross section or poorly running heads, the basic circumferential distribution is set via the SFDR adjusting screws. Circular products do not generally benefit from SFDR.

In regard to parison control, the successful production of blow molded parts is a compromise between the desire to reduce net weight and the need to maintain a sufficient safety margin over a set of minimum specifications. These include: minimum wall thickness, compressive strength, drop strength, dimensional stability, stress cracking, fluctuations in net weight, leakage, surface characteristics, and contamination.

Most of these can be directly affected by the molder's ability to control the parison wall thickness. The most common method to do this has been to adjust the gap between the extrusion die and mandrel. Table 9.1 shows the chronological development of this technology.

9.7 Processing Window

Regardless of the type of controls available, the processor setting up a machine uses a systematic approach based on experience or outlined in the machine and/or control manuals. Once the machine is operating, the processor methodically makes one change at a time, to determine the result for each change.

Table 9.1 Technology Development in Adjusting the Gap Distance between Die and Mandrel

Type die	Feature	Advantage/disadvantage
Simple die	Fixed die gap	Simple, inexpensive: no adjustment facility
Die profiling	Permanently profiled: preferred in die land area	Fixed circumferential wall thickness change: time consuming, complex
Die centering	Can be permanently shifted laterally to correct parison drop path	Compromise between required drop path and equal wall thickness
Open-loop axial die gap control	Can be axially shifted during extrusion	Equal circumferential wall thickness change possible: no feedback
Servohydraulic closed-loop axial gap control	As above, with greater speed, accuracy, and flexibility	Equal circumferential wall thickness, change possible, with feedback
Stroke-dependent die profiling	Permanently ovalized die gap	Fixed, unequal circumferential wall thickness change possible: affects entire parison length
Die/mandrel adjustable profiling SFDR (Static Flexible Deformable Ring)	Settable adjustment of die gap profile	Settable, unequal circumferential wall thickness change possible: rapid optimization
Servohydraulic closed-loop radial die gap control PWDS (Programmable Wall-thickness Distribution system)	Programmable ovalization and shifting of die gap	Programmable circumferential wall thickness change possible, independent of parison length

Two basic IM examples are presented in Figs. 9.8 and 9.9. They show a logical approach to evaluating the changes made with any processing machine (injection or extruder) targeted to meet the product performance requirements when operating only within that window. This approach is applicable to the other BM processes. Within the area (Fig. 9.8) or volume (Fig. 9.9), basically, all products meet the performance requirements. However, because of machine and plastic variations, rejects can develop at their edges. As the processing machines are rather complex, with all the controls required to set it up as well as the variables that exist with plastic materials and machines, these examples refer to the overall fabricating process. Note that a major cause of problems is not poor product design but rather processes operating outside of their required operating window.

The size of the diagram (Fig. 9.8) shows the molders' latitude in producing good products. To mold products at the lowest cycle time, the molding machine would be

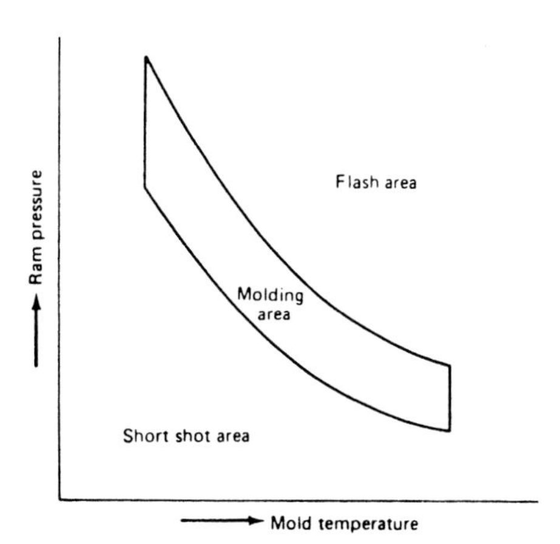

Fig. 9.8 *A 2D molding area diagram (MAD) relates injection pressure (ram pressure) to mold temperature*

set at the lowest temperature and highest pressure location on this two-dimensional molding area diagram (MAD). If due to machine and plastic variables, rejects develop, move the machine controls so that higher heat and/or lower pressure are set to produce only quality parts. This is a simplified approach to producing quality products, as only two variables are being controlled.

The next step in the molding area technique is to use a three-dimensional (3D) diagram (Fig. 9.9). By plotting melt temperature vs. injection pressure vs. mold temperature, one obtains a molding volume diagram (MVD), providing more precision control in setting the machine.

Developing the actual data involves slowly increasing the ram (injection) pressure until a value is obtained where the mold is just filled out. This is referred to as the minimum fill pressure for that combination of material, mold temperature, and melt temperature. Ram pressure is then increased until the mold flashes. This is logged as the maximum flash pressure. These two pressure values then represent a set of data points for one combination of melt and mold temperature.

Next, melt temperature is changed (leaving mold temperature constant), and a new set of minimum and maximum pressures is determined. This is continued until maximum and minimum melt temperatures are found (to meet specific requirements). Then, mold temperature is changed, and all the above repeated until the maximum and minimum mold temperatures are found. Once the data are obtained, MVDs are constructed.

The significance of the MVD approach lies in the fact that one ends up with a dramatic and easily comprehended visual aid to analyzing three of the most important variables for injection molding, namely, injection pressure, mold (or barrel) temperature, and melt temperature.

Using this 2D and 3D approach for making molding diagrams one can analyze injection rate, cavity pressure, and so on. Also consider whether to use manual or automatic process controls.

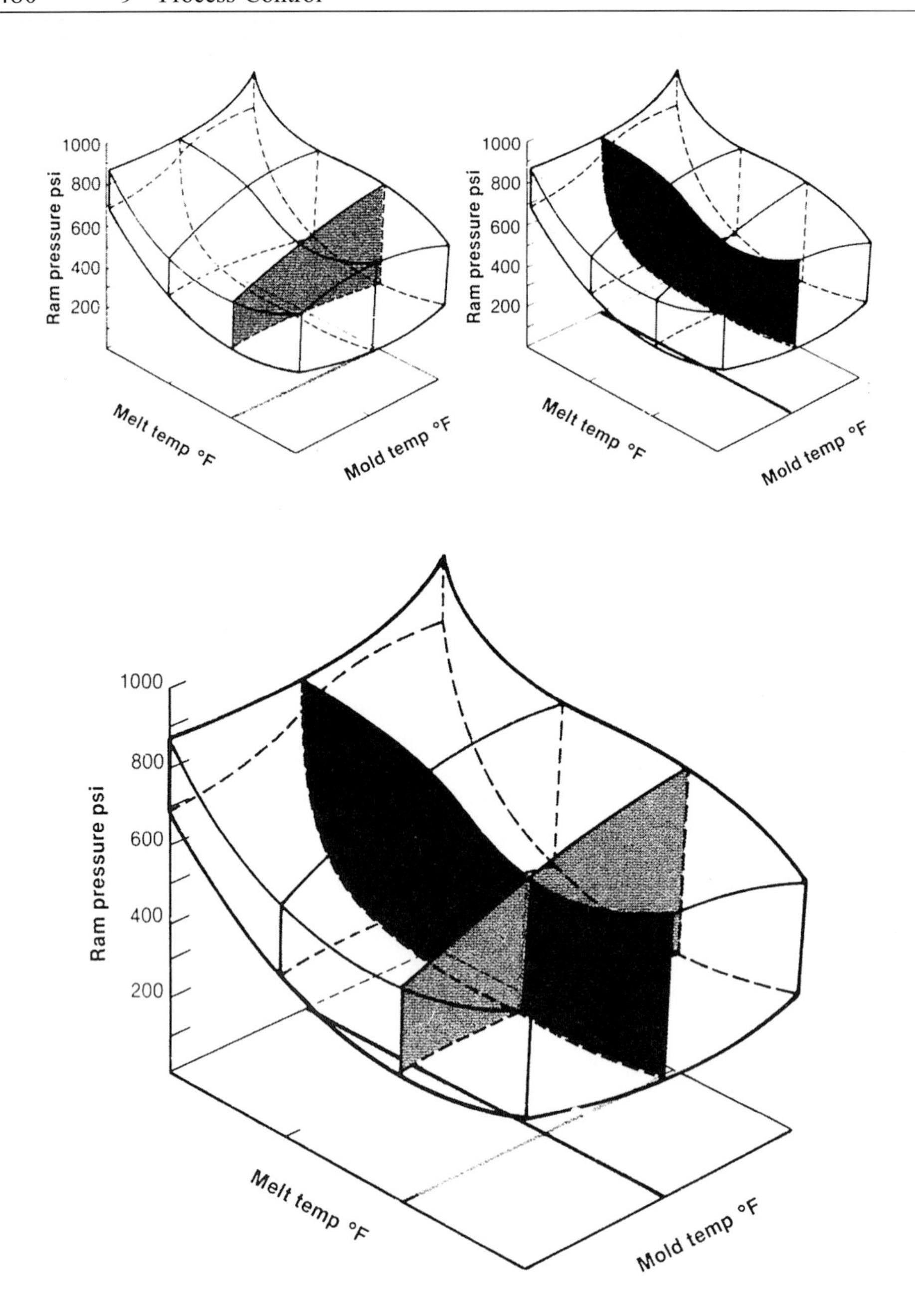

Fig. 9.9 *A 3D molding volume diagram (MVD) can provide the optimum combination of melt temperature, mold temperature, and injection pressure*

This diagram approach can also be applied to other control factors (injection and extrusion) such as analyzing output rate, screw/barrel temperature profile, melt swell, melt sag, blow mold temperature, blow pressure, and so on.

9.8 Logic Control System

The term process control is often used when machine control is actually performed. As the knowledge base of the fundamentals of the molding process continues to grow, the control approach moves away from press control and closer to real process control where material response is monitored and then moderated or even managed. The designer should note that changes in process parameters, such as melt flow rate, can have dramatic effects on moldings, especially mechanical properties, meeting tolerances, and surface properties.

9.8.1 Fuzzy Logic Control

Different logic control systems are used. An example is the fuzzy logic control (FLC) that provides a way of expressing nonprobabilistic uncertainties. Fuzzy theory has developed and found application in database management, operations analysis, decision support systems, signal processing, data classifications, computer vision, and so on. However, the application that has attracted most attention is control. FLC is being applied industrially in an increasing number of processing plants. The early work in FLC was motivated by a desire to directly express (mimic) the control actions of an experienced operator in the controller and to obtain smooth interpolation between discrete controller outputs.

Classical control methods have shown their use in many practical control problems in the industry. It has been shown that the FLC system approach can be used to solve their problems. Many applications of FLC are related to simple control algorithms such as the PID (proportional-integral-differential) controller [59, 84]. In a natural way, nonlinearities and exceptions are included that are difficult to realize when using conventional controllers. In conventional control, many additional measures have to be included for the proper functioning of the controller: anti-resist windup, proportional action, retarded integral action, etc. These enhancements of the simple PID controller are based on long-lasting experience and the interface of continuous control and discrete control. The fuzzy PID-like controller provides a natural way to apply controls. The fuzzy controller is described as a nonlinear mapping [59, 84].

FLC introduction has been a controversial subject in the control theory community resulting in misunderstandings between the FLC and the conventional control proponents. These are partly due to exaggerated claims made by the FLC community and partly due to the presumptuous attitude of those in the traditional control camp (So what's new?). However, in practice, FLC is becoming increasingly popular, partly because of the commercially available programming tools. Nonlinear models play an important role in many advanced controllers because of commercially available programming tools. As reported, the need for control methods for nonlinear systems is becoming very important because of modern production methods and innovative operations.

9.9 Intelligent Processing

What is needed is to cut inefficiency, such as the variables, and cut the costs associated with them. One approach that can overcome these difficulties is called intelligent processing (IP) of materials. This technology utilizes new sensors, expert systems, and process models that control processing conditions as materials are produced and processed without the need for human control or monitoring. Sensors and expert systems are not new in themselves.

What is novel is the manner in which they are tied together. In IP, new nondestructive evaluation sensors are used to monitor the development of a material's microstructure as it evolves during production in real time. These sensors can indicate whether the microstructure is developing properly. Poor microstructure will lead to defects in materials. In essence, the sensors are inspecting the material on-line before the product is produced.

The information these sensors gather is communicated, along with data from conventional sensors that monitor temperature, pressure, and other variables, to a computerized decision making system. This decision maker includes an expert system and a mathematical model of the process. The system then makes any changes necessary in the production process to ensure the material's structure is forming properly. These might include changing temperature or pressure, or altering other variables that will lead to a defect-free end product.

A number of benefits can be derived from IP. There is, for instance, a marked improvement in overall product quality and a reduction in the number of rejected parts. The automation concept behind IP is consistent with the broad, systematic approaches to planning and implementation being undertaken by industries to improve quality. It is important to note that IP involves building in quality rather than attempting to obtain it by inspecting a product after it is manufactured. Thus, industry can expect to reduce post-manufacturing inspection costs and time. Being able to change manufacturing processes or the types of material being produced is another potential benefit of the technique.

9.9.1 The TMConcept

This section on injection molding can be related to extrusion and other processes. When compared to other processes, IM requires rather more sophisticated controls to produce products meeting performance requirements at the lowest cost. The need to change the way injection molding machines (IMMs) are initially set for a new job was recognized by the industry in the mid-1980s. As reviewed by Anne Bernhardt, initially, this was accomplished through direct integration of computer-aided engineering (CAE) analysis with machine setting. While this process has produced very exciting results, its application has proven to be inaccessible to many industrial applications. For this reason, an expert system was developed to assist the settings of IMMs by an operator interface that requires a description of the mold and the part features. The following describes the evolution of this system together with examples of its application in production [32, 33].

The IM industry has always been pushed by demands for higher levels of product quality and productivity, both obtained and maintained without the costly errors and corrections experimented with by many users and processors. So it is not surprising that the need to rationalize the molding machine settings and process controls, considered core to the above target, has long been a priority for all molders. Unfortunately, its implementation is very challenging due to the complexity of the molding process. The governing input variables (material properties plus local pressure, temperature, and their evolution with time) cannot generally be measured directly, and the output (product properties in terms of dimensions, appearance, strength, etc.) are difficult to measure directly, especially in real time.

The solution to the problem requires:

- determining, for each specific application, all parameters of the process physics and their correlation with part quality;
- assessing the relationship of these parameters to the ones used by the machine controls;
- informing the machine controls on all possible quality problems in the shortest possible time; and
- programming the IMM controls to react to each situation.

To complete the list, we should add to the system a strategy for a continuous productivity optimization during production. These ambitious targets can come only from an evolutionary process requiring the contribution of many technologies.

Brief comments on the technical evolution of the past years, and its limits in confronting this challenging task, can help focus on the reasons behind the continuing design of the new developments. The evolution has taken place on two fronts:

- *on the machine capability*; more repeatability, more precise (multipoint) settings, self-tuning (closed loop) controls, statistical process controls (SPCs), and the possibility to fully automate a molding cell with all the necessary auxiliaries, and

- *on the determination of the molding conditions*; more precise modeling of the relationship between part/mold design, processing parameters, and quality/productivity values.

The benefits derived from these improvements were generally evident at the time of their commercial introduction. It also became clear that the self-regulation of machines could be effective only when the part/mold design were optimized, and when at least the major parameters required for the correct processing conditions have been predetermined. It opened the way to computerized process simulation to replace the traditional trial and error method for molding optimization.

The interest developed was identified by the authors as a post-processor program, capable of translating the results of a complete TMConcept (Total Molding Concept) analysis of filling, packing, and cooling into the major parameters in a microprocessor controlled IMM. It is important to understand that such activity is not confined to a simple reorganization of data, but also requires an additional set of calculations.

For example, whereas the CAE calculations for the filling phase are based on the melt temperature entering the nozzle (in the best case where the FEA algorithm allows it), the machine settings to reach this temperature are affected by barrel temperature profile, screw speed, and back pressure. It calls for the use of empirical models to set parameters not considered by the "typical" CAE programs. A statistical method based on experimental data, previously developed for the TMConcept-MCO (Molding Cost Optimization) package for molding optimization, was taken for these calculations. The machine characteristic database, already part of the package to estimate the moldability, was enlarged to define more precisely the process capability essential to the settings, for example, plasticating rate at different rpm.

While the technical results of the programs were well within expectations, at the end of the 1980s, the market was considered insufficiently mature for its introduction because of the following:

- the need for an interface between the FEA workstation and the molding machine;
- the need to replace the IMM with new generation machines equipped with advanced microprocessor controls;
- the lack of the complete set of CAE analyses for many commercial parts; or
- the recurrent problem of part design modifications during mold construction not included in the CAE analysis.

What made the authors invest in the development of a stand-alone system for the computation of the initial IMM settings was the inherent advantage in limiting errors when using a computerized system. Errors in machine settings are difficult to identify, and may not only cause production inefficiencies, but may even damage the mold and the machine. Two types of errors can be recognized: mistakes in data entry and mistakes due to insufficient knowledge.

The basic concept behind the development of the CAMMS (computer aided molding maching setting) program was the realization that the initial setting of IMM parameters should be based on a description of the mold and part specifications instead of the process. This has the potential to eliminate two steps from the current traditional three-step process that requires the operator:

- to understand the "features" of the mold, material, and part requirements;
- to look into one's experience for the parameters to be set for the above; and
- to enter the settings in the IMM controller.

With this approach, it is sufficient for the operator to comprehend how to describe the mold and part topology. It will be superfluous for the operator to know the optimal value for each material and situation. The two major implementation difficulties are the need for an extensive database and the user interface strategy. The previously mentioned TMConcept-MCO calculations, used for some variables of the first automatic setting based on CAE analyses, were extended to model the aspects of filling, packing, and cooling. The machine capability database evolved accordingly.

The resulting CAMMS program created a link between IMM capabilities, machine settings, and mold/part characteristics. The need to detect if a particular machine was certainly not, or may not be, suitable for a specific job, was clear from the beginning. This situation is not typical, as the causes of machine inadequacy are not always directly evident. While the size of the tool is an obvious factor in the machine suitability, the clamping force requirements can remain questionable even with a complete set of the most advanced FEA analysis. The latter must be integrated with part quality, cost considerations, and skilled judgment of the complete project.

Those requirements called for a system capable of anticipating the greatest number of potential problems. It also demanded that a strategy be defined to determine:

- the number of inputs and the level of precision necessary to start the calculation of settings, and
- the area of the identified moldability window, a function of the previous input definition, to be used for the settings.

The CAMMS program was designed to work with a simple mold/part input description. The output strategy was to provide setting values that would increase the probability of obtaining acceptable molding, without further manual modification.

Warning messages were an important part of the program from the beginning of its development. These messages indicate, with various degrees of probability, the potential for the occurrence of a particular problem, to help the operator make a decision on how to handle a specific situation.

These settings were integrated with problem analysis and strategy for continuous improvements into KAMMS (Knowledge-base Molding Machine Setting). Along with the CAMMS development, a quick setup was added to the machine microprocessor so that the operator did not have to adjust all the parameters not strictly relevant to the molding process, but key to machine operation. In this way, "reasonable" values to avoid the risks due to unnecessary machine overcharging are always set. Examples are the ejection system and the motion of the plasticating group. Whenever necessary, the operator can fine-tune these parameters.

This procedure highlights that no system can claim any initial settings as best IMM settings, regardless of the quality of previous calculations and operator experience; the complexity of the molding process makes this claim inappropriate to any experienced engineer. Fine-tuning is always possible. For critical production, one also need to consider the possibilities for continuous productivity improvement and molding process robustness.

These two aspects are part of the new KAMMS. Greater sophistication requiring some forms of reasoned numeric and qualitative knowledge was necessary to perform these tasks. For this reason, and to distinguish it from the previous program, -K (Knowledge) was used to replace the C (Computer) used by the "old" program that was "simply" based on a "quantitative" knowledge derived from experiments.

To accomplish this, action was taken on two major developments made in the CAE programs. One was the application of fuzzy logic in "expert" software to examine mold-filling simulations for potential processing problems and advise on ways to rectify them. The fuzzy logic was required because the correct judgment of a specific part quality parameter (shear stress) cannot be made without weighing other parameters such as local melt temperature and its variability.

The second development related to troubleshooting of defects providing guidance to test and debug startup and also solve problems in operations that had been running well. These defects are considered "qualitative" in nature because they cannot be detected by the machine, but require operator judgment.

Experience has shown that, once the problem is sufficiently well described, the ability to find the right solution instead of providing a set of possible remedies is strictly connected to the knowledge of existing molding conditions. Therefore, if we have a way to keep track of the evolution of these conditions, we have the tool for continuous improvement of the robustness and, probably, also of the cycle time.

KAMMS starts the settings like its predecessor by providing a direct comparison of evolving parameters any time required by the operator, and asking to enter the quality conditions.

At current development, the solution must be implemented by the operator even in cases in which it could be made directly by the machine controls: For example, "too low shrinkage" can be corrected by a "decrease in holding pressure."

Concern exists with the risks of automatic implementation, and it is believed that for a long time to come there will always be problems requiring manual operations. An example is material clogging the throat of the hopper, a problem that can be minimized by a proper equipment design but never eliminated.

Regardless, one should think that an intelligent machine control should always explain the reasoning behind its suggestion. This is the essence of all intelligent actions, which by definition, require judgment to solve situations that do not have a simple exact answer.

It is believed that those IMM settings, through the description of product/mold, are certainly in the near future of IM. If this approach is linked with an extensive knowledge base system, its advantages may become so important that many molders may justify the acquisition of the new generation of machines prior to traditional obsolescence.

9.10 Startup and Shutdown Procedure

Steps can be taken to ensure proper process control and a clean and efficient BM operation. There is a need for startup and shutdown procedures. Optimum procedures are important in minimizing material degradation that causes black speck and/or gel contamination upon startup. As an example, high-density polyethylene (HDPE) is much more stable than polyvinyl chloride (PVC), which requires flow passages with no dead spots, although minute amounts of carbon build-up will occur with HDPE.

HDPE is stabilized with antioxidants to help prevent its oxidation, but even the best additive package will not protect the material if it is abused by being subjected to high temperatures and/or long residence times. The lowest usable temperature profile and optimum shutdown and startup procedures will minimize material degradation. Use of improper procedures will become apparent when large quantities of degraded material are generated. Dismantling and cleaning up the equipment is the best choice for fixing the problem.

The optimum procedures are normally found through experimentation. Each piece of equipment is unique and may require slightly different steps, which is not uncommon. The following are examples of suggested procedures for startup and shutdown (also applicable to injection processes), but may need to be altered to provide optimum results.

9.10.1 Startup (HDPE)

- Turn on power to heaters and set temperatures to the desired setting.

- After the temperatures have reached set point, start continuous extrusion at minimum rpm. The extruder amperage and head pressure should not exceed recommended maximums. Gradually increase the rpm to the desired speed.
- If contamination is experienced at an undesirable level or time frame, do the following:
  - Reduce barrel heats in the transition and metering zones to 325 °F.
  - Continue until contamination is eliminated or reduced to an acceptable level. During this step, be sure not to introduce contamination back into the system via the regrind. All containers should be inspected. Remove all contamination from flash and side walls before regrinding.
  - Reset barrel temperatures to their normal setting.

9.10.2 Shutdown

- If the equipment will be down for 1 h or less, stop the plasticator and leave the heats on at set point.
- If the equipment will be down for 1 h or more, stop the extruder and reduce all heats to 275°F.
- If the equipment will be down for 8 h or more, use one of the procedures listed below. The following are in order of the safest methods. However, the optimum procedures, from the point of view of minimizing degradation, are reversed from (d) to (a):
  - *Method a.* Stop extruder and turn off all power to the heaters. This procedure allows the material to cool and shrink. This shrinkage causes carbon buildup to be pulled or loosened from internal surfaces and also allows space for the presence of oxygen, which will promote oxidation.
  - *Method b.* Bottle production should be stopped. Barrel heats in the transition and metering zones should be reduced to 325 °F, and the screw rpm lowered to a drool.

    Material should be extruded until the melt temperature reaches 325 °F, and the extruder stopped. Turn all heats off. This procedure allows the material to cool down the extruder and act as a purge, due to the material's higher viscosity. *Caution:* There is a possibility of damaging the equipment at lower than recommended extrusion temperatures. The maximum head pressure and/or extruder amperage should never be exceeded.
  - *Method c.* Stop extruder and turn all heats down to 275 °F. This temperature minimizes the chance of oxidation and still allows the material to remain in the molten stage. This molten stage reduces the chance of air spaces, and carbon being pulled or loosened from the internal surfaces of the equipment. *Caution:* The equipment should not be left totally unattended. An electrical malfunction could cause equipment and/or building damage. If this procedure is adopted,

the equipment should be checked periodically.

- *Method d.* A combination of (b) and (c) [i.e. instead of turning off heats in (b), go to step (c)].

9.11 Quality Control and Statistical Process Control

To reduce BM cycle time and produce quality controlled products, there is a need for more precise control in the fabrication operation. At higher prodcution rates, excessive scrap and rejects during extrusion become less desirable than ever before. With quality control and Statistical Process Control (SPC) properly applied, the performance of the plastics in the machine can be controlled within limits to produce zero-defect products meeting performance requirements.

9.11.1 Quality Control

Quality Control (QC) usually involves the inspection of components and products as different phases of processing are completed. Products that are within specifications proceed, while those that are out-of-spec are either repaired or scrapped. The approach just outlined is an after-the-fact approach to QC; all defects caught in this manner are already present in the part being processed. This type of QC does little to correct the basic problem(s) in production.

One of the problems with add-on QC is that it constitutes one of the least cost effective ways of obtaining a high-quality part. Quality must be built into a product from the beginning such as following the FALLO approach (see Fig. 1.2); it cannot be inspected into the process. The target is to control quality before a product becomes defective.

QC is the regulatory process that measures a product's performance, compare that performance with established standards/specifications, and pursue corrective action regardless of where those activities occur.

9.11.2 Statistical Process Control and Quality Control

Statistical process control is an important on-line method in real time by which a production process can be monitored and control plans can be initiated to keep quality standards within acceptable limits. Statistical quality control (SQC) provides off-line analysis of the big picture such as what was the impact of previous improvements.

There are basically two possible approaches for real-time SPC. The first, done on-line, involves the rapid dimensional measurement of a part or a nondimensional bulk parameter such as weight and is the more practical method. In the second approach, in contrast to weight, other dimensional measurements of the precision needed for SPC are generally done off-line. Obtaining the final dimensional stability needed to measure a part may take time. As an example, amorphous injection molded plastic parts usually require at least a half-hour to stabilize.

The SPC system starts with the premise that the specifications for a product can be defined in terms of the product's (customer's) requirements, or that a product is or has been produced that will satisfy those needs.

Generally, a computer communicates with a series of process sensors and/or controllers that operate in individual data loops. The computer sends set points (based on performance characteristics of the product) to the process controller with constant feedback as to whether or not the set of points are in fact maintained. The systems are programmed to act when key variables affecting product quality deviate beyond set limits.

On-line SPC software excels at monitoring production processes in real time giving a closeup view of what is currently occurring. This important capability uses parameters that have a direct, understandable effect on the process. Even though SPC is very important, what SPC does not do is show the overall operations such as detecting differences over time, look at a complete process, or compare multiple production lines. SPC's classical use is to keep plant-floor operations from over-adjusting their machines. It also helps determine when a significant shift in the process happens and whether it requires corrective action.

Off-line SQC is essential to detecting differences over time such as shift changes, day-to-week problems, differences between suppliers of materials, differences between injection molding machines, and so on. SQC more easily permits combining plant-floor data with results from test stations and the laboratory. Once the process is in control, off-line SQC provides the means to make long-time process improvements. It can uncover relationships, monitor the results of process changes, and provide various other decision-support functions such as whether the process can continually deliver products within specifications.

To target for better yields, higher quality, and increased profits, fabricators should consider the SPC and SQC techniques as standard tools for understanding, validating, and improving processes in all areas of manufacturing including product distribution, transportation, and accounting. Using on-line software, SPC provides the closeup view; using off-line software, SQC detects differences over time. These two techniques provide two different essential functions.

10 Computer Operation

10.1 Overview

Computers permeate all areas of the plastics industry from the concept of product design, to raw material and processing, to marketing and sales, to recycling, and so on. Computers have their place, but most important is the person with proper knowledge in using and understanding hardware and software.

The computer supports rather routine tasks of embodiment and detailed operation rather than the human creative activities of conceptual human operation. If the computer can do the job of designer, fabricator, and others there is no need for these people, but the computer is simply another tool for these individuals. It makes it easier if one is knowledgeable regarding the computer's software capability in specific areas of interest such as design of simple to complex shapes, product design of combining parts, material data evaluation, mold design, die design, finite element analysis, and so on. By using the computer tools properly, a much higher level of product design and processing will result.

The industrial production process, as practiced in today's business, is based on a smooth interaction between regulation technology, industrial handling applications, and computer science. Particularly important is computer science because of the integrating functions it performs including tool manufacture, primary processing equipment, auxiliary equipment, material handling, and so forth up to business management itself. This means that computer-integrated manufacturing (CIM) is very realistic to maximize reproducibility that results in successful products.

The use of computers in design and related fields is widespread and will continue to expand. It is increasingly important for designers to keep up to date continually with the nature and prospects of new computer hardware and software technologies. For example, plastic databases, accessible through computers, provide product designers with updated property data and information on materials and processes. To keep material selection accessible via the computer, there are design databases that maintain graphic data on thermal expansion, specific heat, tensile stress and strain, creep, fatigue, programs for doing fast approximations of the stiffening effects of rib geometry, educational information and design assistance, and more.

Today's software developers face a serious challenge concerning how to produce a safe and reliable product in the shortest possible time frame. This is not a new problem; it has simply been exaggerated in recent years by pressures from the marketplace, and the manufacturing industry certainly is not immune to those pressures. Manufacturers have sought solutions by huge spending on software development tools and manpower. In some cases these approaches have worked, but often they have not.

10.2 Computer and Product

Computer use, with appropriate software, results in a better understanding of the operating requirements for a tool (mold or die), and therefore, results in better process control. This increased process control results in additional benefits:

- there is a faster startup of the tool with less rework required resulting from the simulation performed with mold analysis programs and testing the operating conditions before the mold/die is built;
- there is an overall increase in part quality (Table 10.1).

Other benefits resulting from computer-aided design/manufacturing/engineering (CAD/CAM/CAE) technology for mold/die design include

- fewer errors in drawings, which improves mold quality and speeds up delivery time;
- improved machining accuracy;
- standardization of parts and components that reduces the amount of supervision required in a manufacturing facility;
- improved speed and accuracy in the preparation of the quotation; and
- a faster response to market demand. These benefits, which are more difficult to quantify, must be evaluated by each user individually [3].

CAD, CAM, and CAE are the directions all types of plastics product design, mold or die making, and the fabricating industry continue to take. The number and complexity of plastic products being produced are greater every year, but the number of experienced product designers, mold/die designers, and fabricators generally has not kept pace. The answer to this "people power" shortage has been to increase productivity through the use of CAD/CAM/CAE.

The technology of plastic product and mold/die design has evolved over a period of years to become a multifaceted problem. Many of today's mold/die designs are the results of many years of trial-and-error techniques practiced by a relatively small number of artisans, craftsmen, and specialists involved in an ever-growing plastics

Table 10.1 Examples of Computer Information Generated and Typical Problems It Can Solve

Flow analysis	
Fill pattern	Weld line position
	Air trap position
	Position of vents
	Overpacking: excessive costly part weight
	Overpacking: warpage due to differential shrinkage
	Underflow: structural weaknesses
Pressure distribution	Pressure required to fill
	Clamp force required
	Overpacking: ribs, etc. sticking in mold
Temperature	Poor surface finish
	Weak weld lines
	Distortion due to differential cooling
Shear stress distribution	Quality of part: tendency to distort
	Quality of part: tendency to crack
	Cycle time: low stresses permit hotter demold temperature
Shear rate cooling time	Avoids degradation of material
	Shows tendency to distort due to uneven cooling
Flow-angle packing pressure volumetric shrinkage	Quality of part: molecular orientation
	Under/overpack to poor packing
	Dimensional variations due to poor packing
Cooling analysis	
Mold surface	Part quality: even cooling prevents distortion
Temperature	
Freeze time	Minimizes cycle time
Coolant temperature and flow rate	Optimizes coolant: can eliminate need to chill coolant
Metal temperature	Cooling efficiency: ensures optimum circuit design
Warpage analysis	
Warped shape	Tendency to warp
Single variant	Indicates fundamental causes of warpage
Warpage shape	

industry. This period of sustained growth, of the plastics industry, has been accompanied by an equally sustained growth period in the level of understanding of plastic materials and how they react to various changes in processing parameters. The net result has been the development of a technology base that can explain many of the phenomena that, thus far, were considered the "art" of successful plastic design. The "rules of thumb" that governed designs of the past have given way to sophisticated analysis tools used on high-speed computers enabling major advances in speed, productivity, accuracy, and quality of plastic designs.

High-speed digital computers, once used for the analysis and data processing activities of the past, are now generating graphical engineering information and helping to automate a domain once thought of as totally a creative discipline, the design process itself. The use of computers in manufacturing operations dates back to early work in the 1950s in which the dream was to control metal cutting machine tools by computer. It was hoped that this would eliminate the requirement for many tooling aids, such as tracer templates that helped the accuracy and repeatability of machining operations on the shop floor.

The 1960s and 1970s engineered the continued development of the hardware and software products for numerical control of machine tools. The business climate began to favor companies that could supply a total systems approach to computer graphics problems in three dimensions (3D). This spawned the "turn-key" CAD/CAM/CAE suppliers of today—companies that could supply both the computer hardware and user-friendly software, ready to run, in the customer's place of business.

CAD/CAM/CAE is revolutionizing the speed and efficiency of the plastic design functions. The more the entire design function is studied, the more repetitive tasks are uncovered. The computer's ability to perform these tasks untiringly and with blazing speed is the basis for these productivity gains. At this point, many readers may disagree with the concept that something as complicated as mold design is repetitive in nature. To illustrate the point, some of the major mold elements of a plastic design will be reviewed:

- **Product Model:** The first step of the process is creating a model of the product to be fabricated. Plastic product design is similar to design in other materials. The product designer seeks to provide a solution to a product need, at minimum cost, and with the greatest quality and speed. A design is prepared and detailed, a model or prototype built, and the design revised until acceptable product function and cost are achieved. This process involves many repetitions of creation of drawings and models. Many times, the products being designed include features that exist in other products such as bosses, ribs, snap fits, and so on, incorporating standard design practices.
- **Shrink-Corrected Product Model:** A type of repetition of the original product model, this model is the "pattern" used to develop, as an example, the cavity and core allowing for material shrinkage in the processing operation.
- **Generation of Drawing:** This is the repetition of all the information that may be contained in the die/mold layout.
- **Machining of Mold Geometry:** This is the repetition by a machine tool of all geometry previously described by the shrink-corrected product model and die/mold layout.

The basic point of the preceding description is that there is a great amount of repetitive information flow in the plastic design process. As a result, what is perceived to be an extremely creative process is actually very repetitive in nature.

The types of analytical problems that are encountered in the mold/die design process, generally, fall into the sciences of fluid mechanics and heat transfer theory. These fields encompass many complicated mathematical functions and relationships that were too time consuming to evaluate in manual or conventional mold designs. The ability of the computer to "remember" and execute these computations quickly now adds a new dimension to the mold/die design process, allowing prospective design alternatives to be evaluated and simulated by computer, rather than in the molding press. CAD/CAM/CAE is enabling the creative energies of plastic product and mold designers to be spent in producing better designs, in shorter time periods, rather than performing repetitive mold design tasks.

10.2.1 Communication Benefit

CAD/CAM/CAE has bridged the gap between designer and tool (die/mold) maker, benefiting the plastics industry as a whole. Modeling and machining systems have been within the grasp of even the smallest company. At one time, there was virtually no communication between the design staff of major product manufacturers and their subcontractors, so tool makers would often receive a directive to make a mold/die that could not be manufactured, because the requested design was incompatible with machining operations. This action resulted in the unenviable task of having to modify the design so that the tool could be produced.

Because of the continuing development of CAD/CAM/CAE systems, particularly 3D modeling and machining software for complex shapes, it is helpful to outline some of the latest enhancements now available to the product designer and toolmaker. CAD systems have conventionally split a complex tooling into single patch surfaces that are matched at their edges. Now, it is possible to split such a tool into multiple patch surfaces that remain smooth even though the surface may be highly twisted or contain flat portions, as well as, highly curved ones.

Accordingly, adjacent surface patches remain well matched or have a preset discontinuity of surface tangent, such as at a crease or feature line in the surface. Moreover, if one part of the surface is modified, changes remain localized and continuity is maintained with adjacent surface areas. Multiple patch surfaces can be assembled to make a complete product and a smooth blend or rib can be generated between intersecting surfaces. The introduction of arbitrary trimming curves to form boundaries on the surface, at a curve of intersection or a blend edge has, significantly, reduced the time required to model the complex geometric features of a molding.

10.3 Finite Element Analysis

Product design is usually performed based on experience, as most products require only a practical approach. Experience is also used in producing new and complex shaped products, usually, with the required analytical evaluation that involves stress–

strain characteristics of the plastic materials. Testing of prototypes and/or preliminary production products to meet performance requirements is a very viable approach used by many.

Classical equations and formulas from engineering handbooks for stress–strain static and dynamic loads are utilized. CAD analysis such as the use of finite element analysis (FEA) can be used to perform these functions. With FEA, designers can test design by running simulations of a model. This allows for immediate design modifications until the desired results are achieved. This approach is based less on guessing and more on engineering, thereby, preventing premature product failure, improving reliability, increasing accuracy, eliminating excess material, reducing product weight, and shortening real time to less than half. FEA helps to reduce material cost while lessening the expense of building prototypes and remachining tools. FEA can help a designer take full advantage of the unique properties of plastics by making products lighter, yet stronger while also saving time and money.

The use of FEA has expanded rapidly over the past decades. Unlike metals, plastics are nonlinear, so they require different software for analysis. Early software programs were difficult and complex, but have gradually become easier to use. Graphic displays are better, consume less time, and are easier to understand.

10.3.1 Solving Equations

The computer can solve hundreds of simultaneous equations that would take literally years to solve. FEA can be defined as a numerical technique involving breaking a complex problem down into small subproblems via computer models to be solved. The key to effective FEA modeling is to concentrate element details at areas of highest stress. This approach produces maximum accuracy at the lowest cost.

The first step in applying FEA is the construction of a model that breaks a component into simple, standardized shapes or (usual term) elements, located in space by a common coordinate grid system. The coordinate points of the element corners, or nodes, are the locations in the model where output data are provided. In some cases, special elements can also be used that provide additional nodes along their length or sides. Nodal stiffness properties are identified, arranged into matrices, and loaded into a computer where they are processed, with certain applied loads and boundary conditions, to calculate displacements and strains imposed by the loads [2]. In this method, the modeling technique is critical because it establishes the structural locations where stresses will be evaluated. If a component is modeled inadequately for a given problem, the resulting computer analysis could be quite misleading in its prediction of areas of maximum strain and maximum deflection values. An inadequate model could be quite expensive in terms of computer time.

A cost-effective model concentrates on the smallest elements at areas of highest stress. This configuration provides greater detail in areas of major stress and distortion, and minimizes computer time in analyzing regions of the component where stresses and local distortions are smaller.

Unfortunately, modeling can be a stumbling block because the process of separating a component into elements is not essentially straightforward. Some degree of insight, along with an understanding of how materials behave under strain, are required to determine the best way to model a component for FEA. The procedure can be made easier by setting up a few ground rules before attempting to construct the model.

For example, in the case of plane stress analysis, quadrilateral elements should be used wherever possible. These elements provide better accuracy than corresponding triangular elements without adding significantly to calculation time. Also, 2D and 3D elements should have corners that are approximately right angled, and should resemble squares and cubes, as much as possible, in regions of high strain gradient. Generally, element size should be in inverse proportion to the anticipated strain gradient with the smallest elements in regions of highest stress.

The name, finite element analysis, refers to an object or structure to be modeled with a finite number of elements. Depending on the type of element, each node will have a given number of degrees of freedom. As an example, a 3D element can have 6 degrees of freedom at each node, 3 degrees of translation, and 3 degrees of rotation. Each degree of freedom is described by an equation. Thus, the solution of a finite element evaluation requires the simultaneous solution of a set of equations equal to the number of degrees of freedom. The product element model with 10,000 elements will have more than 50,000 degrees of freedom. Solving such a huge set of equations in a reasonable amount of time, even with the most refined of matrix techniques, requires a great deal of computational power.

10.3.2 Fundamental

In its most fundamental form, FEA is limited to static, linear elastic analyses. However, there are advanced finite element computer software programs that can treat highly nonlinear, dynamic problems efficiently. Important features of these programs include their ability to handle sliding interfaces between contacting bodies and the ability to model elastic-plastic material properties. These program features have made possible the analysis of impact problems that, only a few years ago, were handled with very approximate techniques. FEA have made these analyses much more precise, resulting in better and more optimum designs.

It can be said that FEA is one of the major advancements in engineering analysis. When first introduced, the cost of computer equipment and FEA software limited its use to high budget projects such as military hardware, spacecraft, and aircraft design. With the cost of both computer time and the software significantly reduced, FEA began to be used for high volume product designs in such markets as automobiles, large buildings, and civil engineering structures.

During this period of about two decades, a dramatic increase in the power of desktop personal computers occurred. Simultaneously, advances in relatively low priced software has made it possible for the designer to use advanced engineering techniques that, at one time, were restricted to these high budget projects and designs. All this action has permitted more use of desktop computers running FEA software that can be run with relative ease. At present, helping this expansion is an increase in users and much more competition, both in hardware/software, and design projects to meet quickly manufactured plastic products.

10.3.3 Operational Approach

The opportunity for creative design by viewing many imaginative variations would be blunted if each variation introduced a new set of doubts as to its ability to withstand whatever stress might be applied. From this point of view, the development of computer graphics has to be accompanied by an analysis technique capable of determining stress levels, regardless of the shape of the part; a need that is met by FEA.

The FEA computer-based technique determines the stresses and deflections in a structure. Essentially, this method divides a structure into small elements with defined stress and deflection characteristics. The method is based on manipulating arrays of large matrix equations that can be realistically solved only by computer. Most often, FEA is performed with commercial programs. In many cases these programs require that the user know only how to properly prepare the program input.

FEA is applicable in several types of analyses. The most common one is static analysis to solve for deflections, strains, and stresses in a structure that is under a constant set of applied loads. In FEA, material is generally assumed to be linear elastic, but nonlinear behaviors such as plastic deformation, creep, and large deflections, also, are capable of being analyzed. The designer must be aware that as the degree of anisotropy increases, the number of constants or moduli required to describe the material increases.

Uncertainty about a material's properties, along with a questionable applicability of the simple analysis techniques generally used, provide justification for extensive end use testing of plastic products before approving them in a particular application. As the use of more FEA methods becomes common in plastic design, the ability of FEAs will be simplified in understanding the behavior and the nature of plastics.

FEA does not replace prototype testing; rather, the two are complementary in nature. Testing supplies only one basic answer about a design that either passed or failed. It does not quantify results, because it is not possible to know from testing alone how close to the point of passing or failing a design actually exists. FEA does, however, provide information with which to quantify performance.

10.4 Prototyping

10.4.1 Computer Tool Analysis

Tool analysis programs allow the designer to simulate tool performance prior to cutting steel for the tool. These programs allow one to optimize the tool design without the traditional prototyping trial-and-error methods. Interactive graphics allow the designer to perform such analysis in much less time than if it were to be performed manually.

Tool analysis techniques include plastic flow analysis (such as cavity fill flow front, melt flow orifice, determining venting locations, etc.) and tool cooling and/or heating analysis. By anticipating operating conditions to determine the relative effects of different portions of the heating/cooling system, graphs are automatically generated to illustrate the effects of various ranges of operating conditions on the cooling time, and so on. These types of programs will analyze a series of operating conditions so that the designer can evaluate how the tool will operate once it is in production.

Finalized tool design generally consists of the computer summation resulting from the preliminary tool design and analysis. Dimensions are finalized, section views produced, and the necessary drawings created so that enough information is obtained to proceed with the manufacturing of the tool.

With the tool model constructed, the system will contain all the geometric information required for producing numerical control (NC) input. CAD/CAM/CAE systems can furnish the NC data to postprocessors and generate the NC tapes or electronic systems to drive individual machine tools. CAD/CAM/CAE, with the ability to dynamically simulate tool cutting action, results in a significant reduction in NC programming errors. Additional benefits resulting from NC capabilities include reduced lead time, standardization of machining practices, and reduced cost.

As an example, the process of developing injection or extruded blow molded plastic products has been streamlined by using isolated computer techniques, notably CAD/CAM automated drafting systems for producing engineering drawings faster.

CAE software provides a single tool that ties together all the steps in the plastic products development process. The CAE packages are integrated so that information is passed from stage to stage through a computer database, eliminating redundant effort and misinterpretations. Moreover, with the design analysis capabilities of CAE, the proposed design can be refined in the computer through simulation software. The ability to detect potential problems and correct them at the design stage reduces the heavy reliance on costly, time consuming physical testing.

The most cost effective applications for CAE technology are those with complex product and tool designs, stringent end user requirements, close tolerances, extremely short mold cycle times or fast extrusion rates, and other applications with demanding constraints. In such cases, the return on investment comes quickly because the first prototype is a computer model, resulting in lower tooling costs.

Recent years have seen tremendous growth in the number of software products available to serve the needs of manufacturing functions. The products that have evolved, over this period, are tools that serve to replace the "rules of thumb" of the past with analytical analyses based on sound theoretical principles. These products combine the benefits of relative ease-of-use with the speed of the computer, resulting in tools that can be cost effectively applied to a large number of problems. Over the last few decades, specific software products have arisen to serve the needs of the injection molding industry. These tools are most effectively applied prior to construction of the tool, but they can be applied after the fact to solve process-related problems.

Three types of analysis tools have emerged and are providing major benefits to fabricators:

- flow analyses,
- cooling analyses,
- economic and plant operating analyses.

In general, these analysis tools fall under the domain of CAE. The key word in CAE is aided. Analysis tools in no way replace skill or education in the basics of plastic material properties, tool design, or processing. What analyses do is supplement the knowledge of a trained individual, making him or her more productive and more accurate in predictions.

The basic methodology behind the CAE technique is that a design or process is proposed as the first step. The engineer then constructs a model, or representation, of the specific design using a prescribed method. The computer is then used to rapidly evaluate the results of both the input conditions and model that the engineer has described. The computer lists the output conditions, and the engineer evaluates the consistency of results with his or her experience and then determines how the design must be modified to achieve acceptable results. The process is repeated until a successful design is achieved. In this manner, the computer aids the engineer by calculating results much more rapidly and with greater precision than is humanly possible. The skill and experience of the designer are still reflected in the final results. The CAE technique can be effectively applied in the fabricating field to

- maximize the probability of first-time plastic product or tool functionally;
- solve process problems, such as warping, dimensional inconsistency, and reduction in output; and
- reduce tooling costs, such as tool startup costs, product fabricating costs, material costs, tool rework costs, and scrap and regrind costs.

10.4.2 Prototype Tooling

The basic approach in designing a product made from any material (plastic, steel, aluminum, wood, etc.) involves knowing the behaviors and characteristics of the materials and the effects of manufacturing on the materials. This knowledge can then be correctly applied, such as using the processed material's static and/or dynamic properties. Should a need arise for data under conditions different from those at which test data are available, with few exceptions, it would not be too difficult or costly to obtain via prototyping.

A prototype is a 3D model suitable for use in the preliminary testing and evaluation of a product (also used for modeling a mold and other tools). It provides a means to evaluate the product's performance before it goes into production. The ideal situation is for the prototype to be the actual product, but techniques such as machining stock material to using rapid prototype (RP) methods can be utilized [25].

Conventional machining operations preferably use the same plastic as used in the product. Different casting techniques are used that provide low cost even though they are usually labor intensive. The casting of unfilled or filled/reinforced plastic used include thermoplastic (TP) or thermoset (TS) polyurethane, epoxy, structural foam, and room temperature vulcanization (RTV) silicone. Also used are die cast metals. These materials are reviewed elsewhere in this book with additional information to be reviewed in this section.

10.4.2.1 Rapid Prototyping and Rapid Tooling

As a result of ever-increasing advances in product design and the shortening of this process due to technology and market pressure, prototyping houses, moldmakers, and tool rooms have experienced a mounting urgency to shorten lead times. Various rapid program methods have been successful in offering fast product prototypes and tooling. RP and rapid tooling (RT) is any method or technology that enables one to produce a product or tool quickly. The term "rapid tooling" is derived from RP technology and its application.

They are 3D models suitable for use in the preliminary evaluation of form, design, performance, and material processing of products, molds, etc. When properly used, an automatic/fast RP system can accelerate product development, improve product quality, and time-to-market. 3D products (to date principally for injection molding) are formed from the design concept to production using computer controlled laser beams to produce the final simple to complex shaped products. Models can be made of plastics (cast epoxy, copper–nylon, etc.), metals [steel (including sprayed steel), hard alloys, copper-based alloys, powdered metals, etc.], or other materials (MIT's starch and sugar, etc.). With powder metal molds, they can be used as inserts in a mold, ready to produce prototype products. RT processes work from models to quickly generate tooling suitable for production of up to millions of products. The models include those produced by RP [6].

The RP technology used for building physical models and prototype products from 3D CAD data is popular. Even though these systems are more expensive than previous methods, they provide the desirable end result to the industry that is a much quicker way to obtain prototypes (hours instead of days/weeks). These systems are continuously being updated to expand their capabilities.

Methods used to produce the more costly RPs include those that produce models within a few hours. They include photopolymerization, laser tooling, and their modifications. The laser sintering process uses powdered TP rather than chemically reactive liquid photopolymer used in stereolithography. These systems enable having precise control over the process and constructing products with complex geometries.

10.5 Melt Flow Analysis

Over the last few decades, the field of flow analysis has gained increasing importance in blow molding (BM), injection blow molding (IBM), extrusion blow molding (EBM), and so forth. Flow analysis has provided rational solutions to many of the hard-to-understand effects that cause problems in the fabricating process, including warping, internal stress, excessive fill pressures, product flashing, fisheye, and others. The interrelationships between product design and fabricating process parameters that cause problems of this nature were not well understood in the industry. Practical experience often was insufficient to identify potential problems and too limited to have encountered the full range of fabricating problems that can be addressed by techniques such as flow analysis. Hence, much prototyping and tool "fine-tuning" were necessary before successful fabricated product results could be achieved.

Computerized flow analysis has emerged as a powerful tool to aid in the implementation of fabrication as the production process of choice to a widening spectrum of products. The ability of modern digital computers to perform complex calculations in short periods of time has been the breakthrough that makes flow analysis a tool applicable to increasing numbers of new parts.

Computer simulation of the injection molding process is not new. In fact, virtually since the introduction of the computer, various attempts have been made to develop simulations. Almost all modern computer simulations are based on the work of these early pioneers. The method of analysis is based on a few simple fundamental laws of physics. Unfortunately, the inherent simplicity of the approach is easily lost in the mathematics.

In addition, technological advances in computer hardware have increased the sophistication and accuracy of these flow models, while at the same time bringing the cost of the analyses into a range at which they can be applied to a large number of new designs. The flow analysis tools can be successfully applied and utilized by three different groups in the product development process: the product designer, the tool designer, and finally the fabricator.

10.5.1 Basic Melt Flow Analysis

There are two physical considerations in the flow of hot plastic into an injection mold: (1) the flow equations and (2) the heat transfer equations. The method consists of the solution of the simultaneous equations of heat transfer and fluid flow. Classically, of course, the general equations are written as though they were to be solved by a Newtonian integration technique, and then an approximate numerical solution developed. This approach is used in developing flow analysis for extrusion and other processes.

The mold flow (software) user's major task is to accurately describe the geometry of the product to be molded, so that the program can analyze melt flow through the mold cavity. Simple products present no problem, and sometimes flow length and product weight are all that are needed to balance the runner and gate system. For complex products, such as a headlight, a layflat graphic approach is used. The product is "flattened" to create a 2D graphic presentation. A series of circular flow fronts are drawn, sectioning the mold. These reflect the radial flow of the melt from the gate and could be thought of as the fronts of successively larger short shots.

However, if one skips the stage of formulating actual equations, a practical approach for historical reasons, and instead go straight into a numerical approach, the inherent simplicity becomes obvious.

As an example, the injection molding period is divided into three stages: filling, compression, and compensating flow. The approach is similar in each stage. In the filling stage, either the pressure is set and the flow rate calculated, or the flow rate is set and the pressure calculated. In the compression and compensating stages, the holding pressure is set and the resultant flow calculated.

Consider first the flow equations and take the simple case, for demonstration only, of a thin rectangular section. Assume that the section is symmetric about the centerline. The forces on a small block within the element can be balanced, and the pressure pushing the block along gives a force of P to width to thickness, which is resisted by the shear stresses acting on both faces; that is, shear stress to width to length. This gives the formula that stress = pressure drop to thickness/(2 length).

This gives a relationship between pressure and shear stress across the section. It is a fundamental relation based solely on resolving forces and is quite independent of material or flow characteristics.

Although in practice, the viscosity is a complex relationship of temperature, shear rate, pressure, etc., in a computer program, it is usually read from some subroutine, depending on the required degree of precision. It could be a simple formula tying viscosity to shear rate and temperature, or it could be a complex matrix based on complex experimental data. All that matters for this demonstration is that if temperature and shear stress are known, the viscosity (or shear rate) can be obtained.

If viscosity and shear stress are known, the shear rate can then be calculated, as the definition of viscosity is shear stress/shear rate. Any errors from predicting shear rate arise from the viscosity subroutine and not any mathematical simplification. (Note that the temperature of the element is taken as known at this point.)

Calculations using the marching approach start from the outer edge, where the velocity of the plastic in contact with the wall is assumed to be zero. If we ignore abnormal effects such as jetting, this is true because the plastic is frozen at this point. The block is broken up into a number of thin scales. The velocity of the outer face of the outer block is zero. The increase in velocity over that slice is the shear rate to thickness of slice. (The basic definition of shear rate is an increase in velocity/thickness.)

The velocity on the inner face is now known. Moving into the next slice, one can calculate the increase in velocity in the same way and add to the velocity of the first slice to arrive at the velocity of the second slice. If this is done for every block, the complete velocity distribution is known across the section. Multiplying the velocity of every slice by its cross sectional area gives the volumetric flow in every slice, which can then be added together to yield the total flow rate.

To summarize, if we know the temperature across the section and some relationship between shear rate and shear stress, the flow rate can be calculated for a set pressure.

Now consider the heat flow equations by considering the heat flow into and out of each slice. There are four basic heat flows:

- Heat in by conduction, which can be calculated as heat change (per time increment) equals thermal conductivity to area to temperature difference/slice thickness.
- Heat out by conduction, which can be calculated in exactly the same way.
- Heat in by flow. The plastic entering the section is at a different temperature from that leaving. The change in heat content again can be calculated, as heat change equals velocity to slice area (flow rate) specific heat to temperature difference.
- Heat generated by friction. This is equal to work done, which is simply force times distance moved. The force is shear stress slice area, and the distance moved is velocity to time increment.

One problem still remains. To complete heat transfer analysis, the flow calculation must have been completed, yet to do flow calculations, the temperature distribution must be known. This can be solved by first calculating the velocity of the plastic as it starts to enter the section, when the temperature will be equal to the melt temperature, before any heat transfer has had a chance to occur. This gives a velocity distribution from which a new temperature distribution can be calculated, for a small time difference later. Based on this new temperature profile, a revised velocity distribution can be developed for that small time later and so on until the mold is filled.

Running this type of program has shown that the temperature profile across the section reaches a semiequilibrium state quite early in typical molding cycles. An alternative approach has been to calculate the equilibrium condition. This is mathematically simple and uses less computer time.

If the solution for the equilibrium condition is reached by an iterative procedure, it is possible to stop the iteration before full equilibrium has been reached. This is mathematically equivalent to the real world situation, in which the stable temperature profile is not quite reached. With this approach, quite good results can be obtained. The main errors are in the last section to fill, where the frozen layer has not had a chance to form, so pressures are overpredicted. This seems to be an intrinsic, if minor, weakness of the equilibrium approach that is yet to be solved. There are definite practical advantages to this type of approach, the key ones being that significantly less computer power is required and the solution is more stable. Other programs are available that provide more accuracy in evaluating the melt to solidification process [3].

10.6 Computer Based Design and Manufacturing Aids

10.6.1 Computer-Aided Design

CAD is the process of solving design problems with the aid of computers. This function includes the computer generation and modification of graphic images on a video display, printing these images as hard copy using a printer or plotter, analyzing the design data, and electronic storage and retrieval of design information. Many CAD systems perform these functions in an integrated fashion that can increase the designer's productivity.

It is important to recognize that the computer does not change the nature of the design process; it is simply a tool to improve efficiency and productivity. It is appropriate to view the designer and the CAD system together as a design team. The designer provides knowledge, creativity, and control, while the computer provides accurate, easily modifiable graphics and the capacity to perform complex design analysis at great speeds and store/recall design information. Occasionally, the computer can augment or replace many of the designer's other tools, but it is important to remember that this ability does not change the fundamental role of the designer's innovation capability.

10.6.2 Computer-Aided Design Drafting

CADD, a part of CAD, is the computer-assisted generation of working drawings and other documents. The CADD user generates graphics by interactive communication with the computer. The graphics are displayed on a video terminal and can be converted into hard copy by a printer or plotter.

10.6.3 Computer-Aided Engineering

CAE includes the engineering design analysis, system modeling, simulated structure analysis, finite element analysis, and so forth to improve product quality and lower product development time/cost [61].

10.6.4 Computer-Aided Process Planning

CAPP supports steps of manufacturing planning such as choosing the best and/or available machine for the job, programming delivery of raw materials, etc.

10.6.5 Computer-Aided Manufacturing

CAM describes a system that can take a CAD product, devise its essential production steps, and electronically communicate this information to manufacturing equipment such as robots. A CAD/CAM system offers many potential advantages over past traditional manufacturing systems, including the need for less design effort through the use of CAD and CAD databases, more efficient material use, reduced lead time, greater accuracy, and improved inventory functions.

10.6.6 Computer-Integrated Manufacturing

CIM is the coordination of all stages of manufacturing, which enables the manufacturers to custom design products efficiently and economically, by a computer or a system of computers [63].

This review discusses how CAD/CAM/CAE has been changing in relation to CIM. When addressing the field of CAM, many tend to review solely of numerical control (NC) as the means of utilization of the product database. Obviously, the field of manufacturing is composed of many processes that do not rely on numerical control as the control means. CIM is the natural evolution of CAM into serving an ever-widening scope of manufacturing processes. The unique element of CIM, however, is that it tends to integrate or tie together all the manufacturing processes into one coherent unit, all of which share a common database. CIM involves the appropriate combination of hardware and software that allows the manufacturing processes and functions to draw information from, and contribute information to, the database. This information is shared by all the respective functions, including business functions, as a primary business information system.

A CAD/CAM/CAE system can be envisioned as a central hub of primary information such as the product model and its attributes, and secondarily derived information such as tool paths, bills of materials, and so on. Just as in the manual method of doing business, the engineering information stored in a database begins the product delivery cycle. This basic information about the geometry and attributes of the product is transferred and reused by many subsequent business functions, such as the materials planning function and manufacturing engineering information.

We are moving toward the realization of CIM when we are able to pass information efficiently to and from each of the manufacturing functions to the database. We have discussed in this chapter how product model information for both the tool and fabricated product is passed back and forth. We have also discussed how this model information is traditionally used to drive NC machine tools. There are a number of other manufacturing processes that can serve and be served by the database.

Many of the process parameters we have established via analysis programs need to be expediently transferred to the shop floor. This can be accomplished in a number of ways. For those organizations still using paper as the primary communications medium, process planning software is available to build process plan documents, automatically, from the database and then distribute them to the shop floor. For those firms ready to take a further step in automation, it is now possible to send both graphic and nongraphic information to the shop floor at low cost. The same hardware and software systems that allow this type of distribution may also permit the direct control of manufacturing equipment via high-speed, high-reliability communications links.

In addition, the capability to transfer other types of information, such as bills of materials, to other business functions will make organizations less and less dependent on the expedient handling of paper. By communicating information such as actual molding cycle times or rate of extrusion and computer-assisted inspection results back to the database, it is now possible to close the loop on the entire manufacturing process. Utilization of database management techniques such as group technology allows the quick retrieval of this information such that the results of analysis programs may be tempered with a dose of real world operating environment, with a sound statistical base. The net result of the entire effort will be a more productive manufacturing and product delivery environment, which allows products to be designed and delivered more effectively with less resources.

10.6.7 Computer-Aided Testing

In addition to computer-aided activities, CAT involves the testing that takes place in all stages of product development from design through production to product final evaluation. The advantage of CAT is that the output of sensors measuring the characteristics of the prototype or finished product can manipulate the product model to improve its accuracy or identify design modifications needed. In this way testing integrates design and fabrication into an ongoing, self-correcting development process.

10.6.8 Computer-Aided Quality Control

CAQC performs tests at fast speeds with an unusual high degree of accuracy. They provide information such as fast data collection, data analysis and reporting, statistical analysis, process control, testing, and inspection.

10.7 Computer Optical Data Storage

Optical data storage (ODS) is the technology for storage, processing, and retrieval of vast quantities of data. In various formats, it is suitable for applications in which mass replication of predetermined data are required, long term (over 10 years) archiving, recording of legally nonalterable records, and where finite erasing and recording are required. Because ODS provides very high data density (such as up to 1 gigabyte/s on a 130-mm disk), removal from the drive, random access, and low cost per bit exist when compared to magnetic and micrographic storage devices.

10.8 Computer-Based Training

Affordable computer-based training (CBT) video technology makes possible training a large number of people easily and efficiently. If provides the opportunity to create a plants' own in-house training programs with ease of updating. A major advantage of CBT is that learning can take place at the convenience of the consumer. However, it is a one-way communication tool with the user maintaining a relatively passive role in the learning process.

Multimedia-based training provides a more active learning experience. MMBA uses audio, video, text, and graphics to take full advantage of a PC's ability to capture, reconfigure, and display data. They are efficient and effective, particularly in the area of technical training such as running complex or dangerous equipment in a safe environment. Real life situations are depicted and the learner asked to respond.

10.9 Computer Software Program

Available are relatively thousands of specific software programs that simulate the different design and processing operations such as parison swell [61]. Examples are given in Table 10.2. These guides provide a logical approach in training and conducting research. There are programs that allow fabricating of different designs using different types of plastics. There are also simulated process controls that permit processing operators to make changes, such as thickness or tolerance, and see the effects that occur on a fabricated product.

Basically, a computer software is a set of instruction guidelines that "tells and documents" what is to be done and how to do it. There are many off-the-shelf software instruction programs, with many more always on the horizon, in addition to some operations developing in-house programs. They include product design, mold design, die design, processing techniques, management, storage, testing, quality control, cost analysis, and so on. The software tasks vary so that if a particular program is required, one should be available or close to it. Examples of software follow [2]:

CAD^{Plus} SOLID EDGE Advanced mechanical simulation via finite element analysis by Algor, Inc. Pittsburgh, PA. www.algor.com

COSMIC NASA's software catalog, via the University of Georgia, Computer Software Management and Information Center has 1,300 programs. They include programs on training, management procedures, thermodynamics, structural mechanics, heat transfer/fluid flow, etc.

DART A diagnostic software expert system, developed by IBM, that is used to diagnose equipment failure problems. It is unique in that it does not hold information about why equipment fails. Instead, it contrasts the expected behavior with the actual behavior of the equipment to diagnose the problem.

DFMA Design for Manufacture and Assembly provides determinants of costs associated with processes by Boothroyd Dewhurst Inc., Wakefield, RI. www.dfma.com

EnPlot This is ASM's analytical engineering graphics software used to transform raw data into meaningful, presentation-ready plots and curves. It offers users a wide array of mathematical functions used to fit data to known curves; includes quadratic Bezier spline, straight-line polynomial, Legendre polynomial, Nth order, and exponential splines.

Table 10.2 *Examples of Software for Use in Simulating Injection Molding of Thermoplastics*

Name of program system	Cadmould	C-Mold	Fermould	I-Deas Plastics	Moldflow	PC-Mold	Simuflow	Strim 100- injection molding simulation	TMConcept
Supplier	Simcon	CCMP	Hahn & Kolb	SDRC	Moldflow	KIMW	C-Tech	Cisigraph	Plastics & Computer
A) Simulation of rheological behavior									
A1: Filling phase	C2D C3D-Mefisto C3D-Mestro	C-Flow	C2D C3D-Mefisto C3D-Mestro	Moldfilling[1] (A1 and A2)	MF-2D MFLA-3D	PC-Flow	Simuflow-2D	2D-rheology[1] 3D-rheology[1]	TMC-FA
A2: Pressure holding phase	C2D C3D-Mehold	C-Pack	C2D C3D-Mehold		MFLP-3D		Simuflow-3D		TMC-FA Best
B) Simulation of thermal behavior	C2D C3D-Mecool	C-Cool	C2D	Moldcooling[2]	MF-Cool	PC-Therm	Simucool	Cooling Analysis	TMC-MTA
C) Simulation of mechanical behavior									
C1: of molding			MEF	FE modelling[3] model solution optimization	Fenas			polymer structural analysis	
C2: of mold	C2D C3D-Meclamp (clamp force calculation)		C2D C3D-Meclamp (clamp force calculation)		Moldmech	PC-Mech			
D) Other program modules				warp (shrinkage, distortion)	MF-warp (shrinkage, distortion)		Optimold (design macros from DME and Hasco standards)	3D modelling (for rheological simulation)	TMC-MCO (mold cost optimization) TMC-CSE (shrinkage calculation) PP (postprocessor) SAS (mould fault analysis)
E) Comments	C = Cadmould		C = Cadmould	[1] antecedent: Polyfill [2] antecedent: Polycarb [3] antecedent Superlab				[1] antecedent: Procop	TMC = Total Molding Concept

Injection Molding Operator IBM's molder training programs.

Maintenance Professional-Main Spirex (Youngstown, OH) provides maintenance and inventory control programs for fabricators.

MOLDEST Provides product design, mold design, and process control by Fujitsu Ltd., Tokyo, Japan.

Moldflow This is a series of software modules to analyze melt flow, cooling, shrinkage, and warpage.

MPI LiTE A maintenance scheduling program from Spirex for fabricators.

Nypro Online Nypro (Clinton, MA) fabricator training programs that provide basics to technological advances.

PDM The objective is for product development management and training as opposed to product data or document management. It extends CAD data to a manufacturing organization's nondesign departments such as analysis, tooling development, manufacturing/assembly, quality control, maintenance, and sales/marketing.

PennStateCool Program involves corner cooling to warpage analysis.

PICAT Molders training programs from the British Polymer Training Association (BPTA), Telford, UK

PLA-Ace Software package from Daido Steel Co., Tokyo, Japan. It provides the basic information that encompasses selections that include a mold base, cavity, and core pin(s).

PMP The McGill University, Montreal, Canada PMP Software packages (initially known as CBT) addresses a wide variety of topics associated with plastic materials. They include their introduction, classes/types, processing, technical photographs, and properties (mechanical, physical, electrical, etc.)

Polymer Search on the Internet (PSI) See details following for RAPRA Free Internet Search Engine.

Prospector Examines and provides tabular, single-point (for preliminary material evaluation) and multipoint data (predict structural performance of a material under actual load conditions) for its 35,000 plastics by IDES Inc., Laramie, WY.

SpirexLink An inventory control software package from Spirex for your plant's plasticating components.

SpirexMoldFill A comprehensive, time-saving assistance tool for molders from Spirex with the added advantage of a mold filling analysis program built in.

TMConcept This molding and cost optimization (MCO) software from Plastics and Computer, Inc. is designed as a practical working tool for application by any engineer who bears responsibility for a molding project. It provides a rather complete molding simulation with over 300 variables.

Troubleshooting IM Problems Molders training programs from SME.

SimTech This molding simulator from Paulson Training Programs (Chester, CT) links injection molding with production floor experience. It is designed to provide realistic setup and problem solving training for setup personnel, technicians, and process engineers.

Software programs are useful tools and can perform certain functions. The key to success is the ability to use what is available that includes understanding and putting to practical use applicable software programs.

As an example, since the development of the first injection molding (IM) simulation program modules in the 1970s, computers have provided analyses such as rheology, thermal, mechanical, process control, quality control, statistical analysis, cost modeling, and so on. With time, they have become more and more powerful with the predictive value of their computations continually becoming much greater. The software program systems also include areas of product applications in addition to their performance characteristics. Simulation programs can be clearly differentiated, partly by their user-friendliness and computational precision, and partly on theoretical grounds. Thus a design can be evolved as a planar, 2D model, or 3D model.

10.9.1 Software and Database

Many thousands of software and database programs expand design capabilities, and simplify design analysis, material processing properties, processing capability start-ups, training, and so on. As an example, a worldwide software program that has been extremely useful, based on how it was organized, is the CAMPUS Database software. This Computer-Aided Material Selection by Uniform Standards of Testing Methods (CAMPUS) compares different plastics available from different material suppliers. Special CAMPUS pages on their websites are updated each time they finish further testing of present and new materials. The data can be directly merged into CAE programs. CAMPUS provides a comparable property database on a uniform set of testing standards for materials along with processing information. The database contains single-point data for mechanical, thermal, rheological, electrical, flammability, and other properties. Multipoint data is also provided such as secant modulus vs. strain, tensile stress–strain over a wide range of temperatures, and viscosity vs. shear rate at multiple temperatures.

10.9.2 Rapra Free Internet Search Engine

The number of plastic-related websites is increasing exponentially, yet, searching for relevant information is often laborious and costly. During 1999, Rapra Technology Ltd., the UK-based plastics and rubber consultancy, launched what is believed to be the first free Internet search engine focused exclusively in the plastics industry. It is called Polymer Search on the Internet (PSI) and accessible at www.polymersearch.com. Companies involved in any plastic-related activity are invited to submit their website address for free inclusion on PSI.

11 Process Selection and Auxiliary Equipment and Secondary Operations

11.1 Overview

Primary equipment refers to the basic blow molding machines that start fabricating a product. Auxiliary equipment, also called secondary equipment, supports the primary equipment and production in-line systems.

The machinery and equipment sector is an important part of the plastics industry (Fig. 11.1). The continuing development of more sophisticated processing equipment, in turn, allows the development of more integrated processing equipment. This action results in many improvements such as:

- reduced cost but still meeting product performance requirements;
- higher operating efficiency by reducing tolerances, scrap, and/or rejects;
- higher quality through uniform, repeatable manufacturing procedures;
- improved decision-making and record keeping by converting data to information;
- better access to manufacturing information by supervisors and management; and/or
- improved process control and process management.

To provide the millions of plastic products used worldwide, many different fabricating lines are used. There are different modes of operating these lines.

The automatic line is a complete automatic operation in which after the initial startup is completed, the line is programmed to repeat all functions, stopping only in the event of a malfunction or if manually interrupted.

A process operates automatically taking into account the ability of the system to diagnose problems in operation, the ability of the system to recover from error or fault, the ability of the system to start up then shut down without human intervention, and the like.

The semiautomatic line will perform a complete cycle of programmed functions automatically and then stop. It will then require an operator to manually start another

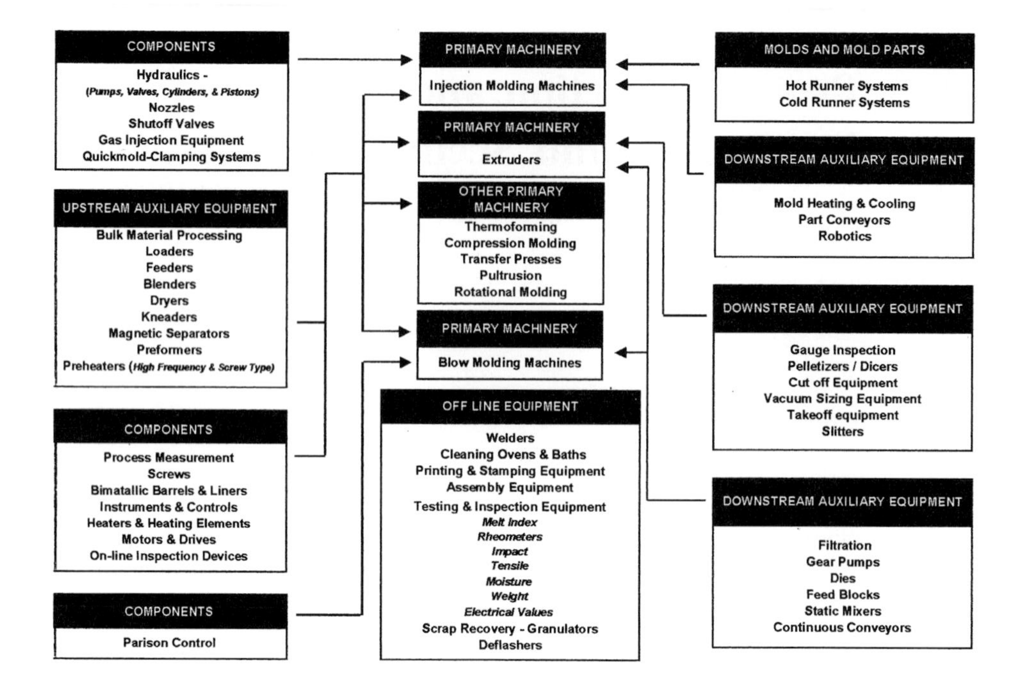

Fig. 11.1 *The plastic machinery and equipment sector*

cycle. The manual mode is an operation in which each function and the timing of each function is controlled by an operator.

There are basically three types of processors. Captive processors are in-house operations of companies who have acquired plastics processing equipment to make components needed for the product they manufacture. Generally speaking, these manufacturers will install a captive operation when their component requirements are large enough to make it economical or a secret product/process exists. The problem with some captive operations is that sometimes “they” do not keep up with new developments critical to their operation.

The custom processor’s operations, like those in the metal working field, might be known as job shops. Custom processors typically have a close relationship with the companies for whom they work. They may be involved (to varying degrees) in the design of the product and the die/mold, they may have a voice in material selection, and, in general, they assume a responsibility for their work.

There is a subgroup in custom processing known as “contract” fabricators. They have little involvement in the business of their customers. In effect, they just sell machine/line time.

The proprietary operation is where the processor makes a product for sale directly to the public or to different companies. They usually have their own trade name.

11.2 Process Selection

When setting up a fabricating line for any given product, the most important processing requirements should be determined based on the plastic to be processed, the quantity, and the dimensions of size and tolerances. Process selection is a critical step in product design. Failure to select a viable process (and material) during the initial design stages can dramatically increase development costs and timing. It is important to recognize that the process can have a significant effect on the performance of the finished product.

In some cases, one will not have the ability to choose freely from all the design, material, and process alternatives. The material and process may be determined largely by the need to use existing materials and facilities. However, to optimize results, one should establish the extent of any design freedom early in the design process and explore the design, material, and process alternatives within these bounds. Before final selection, the entire production process should be considered, including such secondary operations as painting and decorating.

The production flexibility of the fabrication process is often the single most important economic factor in a plastic product. The component's size, shape, complexity, and required production rate can be primary determinants. Many products favor injection molding or long runs in extruders, with their automation capabilities to minimize labor costs. Often the shape dictates the process, such as centrifugal casting, extrusion profile, or filament winding being used for cylindrical parts, rotational molding for extremely large, complex hollow-shaped products, and pultrusion for constant cross sections requiring extremely high strength and stiffness. These general examples should be considered broadly, as individual processes can all be designed for a specific product to meet performance requirements at the lowest cost.

Another important point is that major costs can be incurred in the operations required after fabricating such as trimming, finishing, joining, attaching hardware, and so on. When selecting a fabrication method, observing the following practices will help reduce costs and improve processing and performance:

- Strive for the simplest shape and form.
- Use the shape of the product to provide stiffness, reducing the required number of stiffening ribs.
- Combine parts into single moldings or extrusions to minimize assembly time and eliminate fasteners.
- Use a uniform wall thickness wherever possible, and vary thickness gradually, to reduce stress concentrations.
- Use shape to satisfy functional needs such as slots for hoisting, gripping, and pouring.
- Provide the maximum radii that are consistent with the functional requirements.

- Keep tolerances as liberal as possible; however, in production, aim for tighter tolerances to save plastic material and production cost.

Minimizing cost is generally the overriding goal in any application, whether a process is being selected for a new product application or opportunities are being evaluated for replacing existing materials. The major elements of cost include capital equipment, tooling, labor, and inefficiencies such as scrap, repairs, waste, and machine downtime. Each element must be evaluated before determining the most cost effective process from among the available alternatives. A limited scope of processes will be available for some plastics, but others will be more flexible. Products for limited production or small volumes may require processes with low capital and tooling costs to make them economical.

Preliminary consideration of candidate materials, processes and tooling factors, configuration, thicknesses in section, ribs, bosses, holes, surface characteristics, color, graphics, decoration, assembly, and so on, will begin to impose some discipline on the product design as it evolves. In the middle and latter phases of the design cycle, two or three concepts should make their validity apparent to all involved.

In narrowing the options, if one's experience has not been developed, it is best to seek advice from the appropriate people. When choosing equipment, seek out those with experience close to the materials, end use, categories and sizes of the project.

The manufacture of a product is subject to a range of requirements, the fulfillment of which is decisive to success in the market:

- efficiency of the production process,
- manufactured quality/performance, and
- meeting delivery dates.

In most plants, economic efficiency is always the decisive factor with regards to startup of a product; however, other factors have to be considered such as ensuring just-in-time (JIT) delivery occurs, the product is saleable, etc. A production planning and control (PPC) system has to cope with these requirements. General PPC systems are offered within the scope of commercial electronic data processing (EDP) programs for the operational sequences.

11.3 Auxiliary Equipment

The cost of the upstream and downstream auxiliary equipment can sometimes be more than that of the primary machine. Different performance requirements for this equipment exist, so it is important to use the specific type required in the production line that provides reliability. Its proper selection, use, and maintenance are as important as the selection of the primary fabricating equipment. The material, starting upstream from material storage areas, can progress through the downstream equipment to where the finished product leaves the plant. A set of rules has been developed that help govern the communication protocol and transfer of data between primary and auxiliary equipment. It is important to determine the best location for support equipment and services.

Even though modern fabricating lines are in principle suited to perform flexible tasks, it nevertheless, takes a whole series of peripheral auxiliary equipment to guarantee the necessary degree of flexibility. Examples of the peripheries include:

- upstream material supply systems (dryers, blenders, etc.);
- mold or die transport facilities;
- mold or die preheating banks;
- mold or die changing devices including rapid clamping and coupling equipment;
- plasticizer cylinder changing devices as screws, barriers, etc.;
- molded or extruded product handling equipment, particularly robots with interchangeable arms allowing adaptation to various types of production; and
- transport systems for finished products and handling equipment for passing products on to subsequent production stages.

11.3.1 Drying Systems

Plastic materials absorb moisture that may be insignificant or damaging. All plastics, to some degree, are influenced by the amount of moisture or water they contain before processing; moisture may reduce processing performance. Many plastics contain minimal amounts, and mechanical, physical, electrical, aesthetic, and other properties may be affected or may be of no consequence (Chapter 6).

Nonhygroscopic plastic collects moisture only on the surface. Drying this surface moisture can be accomplished by simply passing warm air over the material. Moisture leaves the plastic in favor of the warm air, resulting in dry products. Hygroscopic plastics refers to a material that tends to absorb moisture from the air. These plastics require extra processing steps of drying to produce high-quality parts. Conventional dryers (not used) waste energy due to heat loss and the use of cooling water. Innovative dryer designs provide improved efficiency and energy savings by using heat exchangers, energy contents of the warm air returning from the hopper, and so on.

11.3.1.1 Hot-Air Dryer

Common types of equipment used with both hot-air dryers and dehumidifying dryers are the air diffuser cones and drying hoppers. An air diffuser assembly is designed to be used in existing loading hoppers. It consists of a football-shaped satellite with four legs that permit it to be placed in the hopper. The drying air is brought into the hopper by a flexible hose to a hood on top of the hopper that connects the flexible tube to the satellite diffuser.

A hot-air dryer and plenum hopper combination or an air diffuser assembly is used to dry nonhygroscopic plastics. A hot-air dryer is a relatively simple machine consisting of heaters and an air blower. Hot-air dryers can deliver hot air thermostatically controlled up to 300 °F (149 °C) at a capacity range from 60 to 1,000 ft^3/min (1.7 to 28 m^3/min).

A hot-air dryer works by pulling ambient air into the air drying filter, through the blower, then across the heating elements. The hot air is then blown to the hopper through a flexible tube. Once the hot air reaches the hopper, it is dispersed through the plastics. The hot air performs two functions. The hot air that passes through the plastic material evaporates the moisture, turning it into steam. The steam then moves out of the hopper back to the ambient air. The hot air also serves to preheat the plastic material, which brings the material closer to the fabricating temperature. When this available heat is used, less heat is required for the fabricating process, and there is a reduction in energy consumption.

In a continuous-flow automated hopper system, an additional amount of 90 lb (41 kg) of PE material will be needed in the plenum drying hopper. As a result, the material that enters the top of the hopper will spend $1\frac{1}{2}$ h in the hopper before entering the molding machine. The hopper with the capacity to handle 270 lb (123 kg) of material would be a 400 lb (182 kg) capacity unit. A hopper must be filled to its capacity for proper operation of the air-trap cone, which prevents contaminated air from entering the hopper. This hopper would have to be filled with 400 lb of PE material.

The hot-air dryer that would be able to handle this 400-lb load has a rating of 150 ft^3/min (4.3 m^3/min). A thermometer would have to be installed in the hopper to obtain a true temperature reading, as a certain amount of heat is lost through the hose leading from the hot-air dryer to the plenum drying hopper. There are two alternatives to this drying system. The first would be to substitute a high-efficiency plenum hopper in place of the standard plenum hopper. Heaters in the base of the high-efficiency plenum hopper make up for the heat lost in the flexible tubing. The other alternative is a high-efficiency dryer that combines a high-efficiency plenum hopper that is an insulated unit with the heaters in the base of the unit. It also has an air blower attached to the plenum hopper. This high-efficiency hot air dryer eliminates the air hoses used in conventional drying systems.

Eventually, the beads become saturated with moisture and have to be regenerated. This is done by blowing air heated to a temperature of 550 °F (288 °C) through the desiccant beds. The elevated temperature drives the moisture out of the beds and into the ambient air. This process varies with the different types of dehumidifying dryers. A multiple desiccant bed absorption system is the most efficient method for drying. A common absorption bed setup is the double-bed system in which one bed is on line drying material, while the other bed occurs in the regeneration cycle.

11.3.1.2 Dehumidifying Dryer

Hygroscopic plastics must be dried by using a dehumidifying dryer. Dehumidifying dryers absorb the moisture within the plastic material by using dry heated air brought down to a dewpoint of −40 °F (−40 °C). This can be done by the use of desiccant beads, which are molecular sieves of synthetically produced crystalline metal aluminosilicates. All moisture is removed from the crystals during their manufacture. The main advantage of these crystals is that there is very little structural change when the water is added or removed. Molecular sieves can dry materials to moisture contents as low as 35 parts per billion (ppb).

There are two classifications for these drying systems: single-bed absorption systems, which use one desiccant bed, and multibed absorption systems, which use two or more desiccant beds. Dehumidifying dryers operate in a closed-loop system. Air is brought in through a filter on the initial startup and sent to the desiccant bed to absorb the water out of the air. The water molecules are absorbed by the desiccant beads. Approximately 1,800 Btu/lb (4.2×10^6 J/kg) of moisture are released, causing the air temperature to rise approximately 19 °F (−7 °C). The air then travels to the heating unit where the air temperature is brought up to the drying temperature specifications. The dehydrated air is then circulated through the plastic in the drying hopper. Then, the air is brought out of the hopper and recycled back through the unit, and the process is repeated.

Dehumidifying dryers are sized similar to a hot-air-drying system. The hopper is sized by the production rate multiplied by the residence time. The dryer is then sized by the corresponding figures from the dryer sizing chart. The dryers are sized on a flow rate of 50 ft/min (0.25 m/s). If the flow rate is more than 50 ft/min, the material will be blown around in the hopper. Any flow rates considerably less than 50 ft/min (0.25 m/s) may not have enough velocity to dry the plastic material. For example, on sizing a dehumidifying dryer system, assume a production rate of 60 lb/h (27.2 kg/h) of ABS material, which has a residence time of 4 h. The 60 lb/h multiplied by 4 h is 240 lb (109 kg). The amount of material that must remain in the hopper to achieve the correct residence time is 240 lb. The correct hopper choice is a 400-lb capacity hopper. The dryer that would dry the plastics adequately is a 50 ft^3/min (1.5 m^3/min) dryer.

Other factors have to be taken into consideration when setting up a total drying system. One factor is the total number of machines that are processing the same type of material. If more than one machine is running the same kind of material, it may be more advantageous to set up a central drying system with one large dryer and a central plenum drying hopper (Fig. 11.2). The central system processes material for several processing machines. A central dehumidifying dryer can also be used with individual high efficiency plenum hoppers. When different types of plastics are used in several individual machines, an individual dryer and plenum hopper for each separate machine is preferred. This configuration offers the maximum flexibility for drying a variety of hygroscopic materials.

Neither conventional dehumidifying dryers nor hot air dryers can automatically limit the degree of drying and prevent overdrying. However, dryers can be modified to

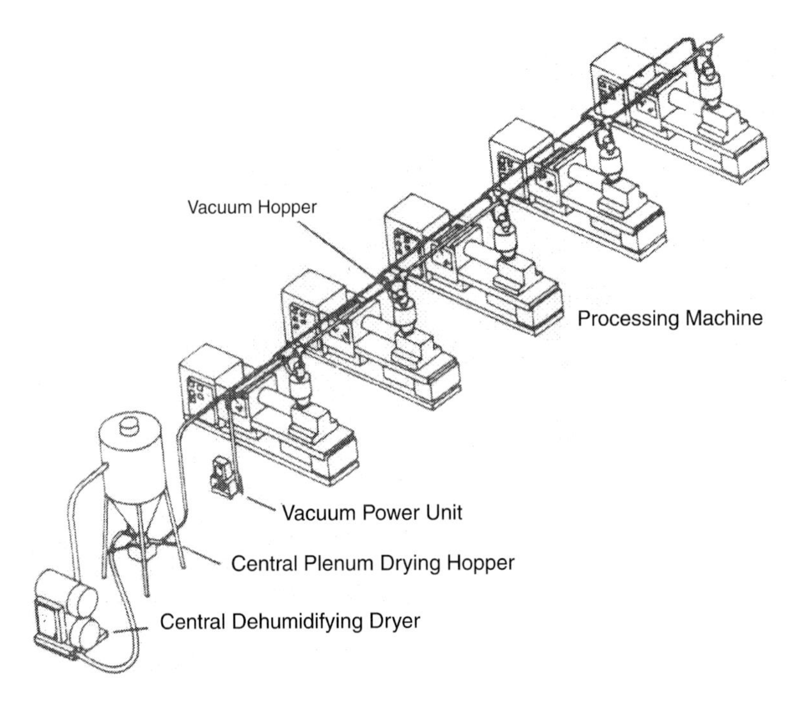

Fig. 11.2 Central dehumidifying dryer with central plenum hopper system

achieve positive control over the dewpoint of the air being recycled to the drying hopper.

Adjusting the proportion of dried and moist return air that is fed to the drying hopper can control the dewpoint. The mixture is controlled by varying the amount of moisture-laden return air that is allowed to bypass the dryer beds, as directed by a moisture sensor at the outlet of the process air heaters. This system prevents overdrying as well as maintains an upper limit on moisture content. Adding moist air as needed conditions the nylon during the drying process. Even when regrind and/or other additives are combined with the virgin nylon, the extra drying time will not result in overdrying.

11.3.2 Water Chilling and Recovery

Highly sophisticated water chilling and cooling techniques are used in controlling equipment temperature such as molds, plasticator barrels, and post-cooling operations. Processors no longer strive to minimize their investment in water chilling and recovery equipment. Instead, they now accept the fact that some type of mechanically refrigerated chilling system is needed to eliminate variables in cooling water temperature, and that evaporative cooling devices, or their equivalent, are vital in eliminating or minimizing unnecessary water waste.

The determination of heat transfer requirements prior to the selection of the equipment or system is contingent on the application (the type of equipment and/or process to be cooled and/or type of material that is being processed).

11.3.2.1 Heat Transfer

Probably the simplest way to regard the heat content of a plastic material in a process is to think in terms of balance. On the one hand, we have the heat supplied and/or generated, and on the other we have the heat removed. A proper heat balance is achieved, of course, when both sides are equal.

The heat content (or enthalpy) of virtually all materials passing through the process consists of latent heat and sensible heat. Latent heat (relevant for all materials except polystyrene) is the heat gained by the material as a result of a change in state in the mold/die without any change in its temperature. Sensible heat develops from a change in the temperature of the materials and is relevant for all materials. Table 11.1 shows the values of a number of commonly processed materials and processes in terms of heat removal requirements. The cooling of thermoplastic materials essentially is the removal of all the heat that was previously put in. The amount of heat put in or removed is expressed in Btu's. As a rule, a nearly continuous process is involved, so we speak of the removal of Btu/h (J/h).

The Btu/h that must be removed from process constitutes the load, which the chilling equipment must have the capacity to remove. Refrigeration capacity is measured in tons. One ton of refrigeration capacity equals a heat load of 12,000 Btu/h (127.7 J/h)

Table 11.1 *Heat Transfer Values for Plastics Processing*

Process	Temperature of material (T_1, into Mold, °F)	Guideline/ton of Chilling (lb/h)	Enthalpy Btu/lb (Final T_1, 125 °F)
Injection molding			
Polyethylene	425	30	300
Polypropylene	400	35	275
PVC	375	45	200
ABS	425	50	175
Polystyrene	425	50	175
Sheet extrusion, calendering			
ABS		60	
Polystyrene		60	
Pipe, profile extrusion			
Polyethylene		50–65	
PVC		60–75	
ABS		60–75	
Compounding			
PVC (cool from 230 to 100 °F		170	

lb/h × 0.454 = kg/h

and is commonly used for refrigeration equipment rating. Table 11.1 shows the heat content of some of the more common plastic materials.

Because of the many mathematical calculations involved, it has proved advantageous to simulate alternate heat transfer situations on a computer and thereby determine the optimum combination of mold cooling channel configuration, diameter, and coolant characteristics (flow rate temperature and pressure). The target of such an analysis is the reduction of cycle time and, therefore, greater productivity of injection molding machines (IMMs). This is, of course, the critical factor to bear in mind.

11.3.2.2 Water Recovery

Although chillers make money, water recovery systems save money so it is important they not be overlooked. The calculations applying to water recovery are identical to those described earlier. Generally, we are dealing with two basic heat inputs: the chiller condensing load assuming that the chiller sized earlier is to be water cooled, and the hydraulics or machine-cooling loads. Also, included on this type of system are air compressors, general extruder cooling, and, occasionally, temperature control units.

In all cases, of course, we are concerned with heat balance, and the input can be obtained from the amount of energy introduced to the system expressed in watts or horsepower. Remember that just because a motor is applied to a process (such as a hydraulic system), the total motor horsepower is not converted to heat. Work is performed by that motor, and in fact not all the motor energy goes into heat to be removed by the water recovery system. The load for the chiller condenser should be calculated on a one-for-one basis, that is, one tower ton for one refrigeration ton.

We will not attempt here to show the mathematics involved in this process because of the many variables involved, which in reality change for each machine considered. Age, duty cycle, design, and components all have a bearing. The point to remember is the need for a safe and practical allowance for all plastic processes normally encountered.

The importance of getting all the capacity available is particularly evident when the nature of the plastics process being cooled limits heat transfer efficiency. Cooling is more efficient in processes such as injection molding—which utilizes high pressures and cooling on two sides under controlled conditions—than processes that involve cooling of only one side of a thick object, such as a pipe that is immersed (under no pressure) in an uncontrolled cooling bath.

Table 11.2 shows the relative characteristics of various processes as they apply to cooling. Note that shell and tube heat exchangers are relatively high-efficiency devices when compared to molds; however, they are limited in their application.

Chillers fall into two basic physical categories: portable and stationary. Portable, self-contained chillers, which can be moved from one point in a plastics processing plant to another, are generally used only when capacity requirements are 15 tons or less.

Table 11.2 Cooling Characteristics

Process or system	Heat transfer coefficient (Btu/°F/h/ft^2)	Limiting factors
Injection molding	30	Thickness and area available
Blow molding	25	One-side cooling and pinch-off cooling
Foam injection molding	20	Thickness and cellular structure
Jacketed vessels	50	Agitation and area available
Bath or trough	100	One-side cooling, thickness, and area available
Shell and tube exchanger		
Oil to water	100	Area and temperature difference
Water to water	250	Area and temperature difference

Portable chillers offer a number of advantages directly stemming from three small capacity increments. They minimize the effects of downtime, can be purchased as needed, and provide very flexible temperature control. In addition, they are advantageous for low-temperature cooling applications, which involve a higher cost per delivered ton whenever a chiller is used to circulate coolant at lower than the design temperature. Portable chillers are available with both air- and water-cooled condensers.

For larger capacity requirements, some type of centrally located chilling system is used. One of the more practical central system designs is the energy-conserving air-cooled type, which supplements plant heating during winter months by reusing the heat recovered from the process. From an economical and ecological standpoint, this type of chilling system provides dual benefits: conservation of water and recycling of energy to conserve increasingly scarce heat-producing fuels. Through an adjustment in the duct work of this air-cooled chiller, it can be used to ventilate processing plants during the summer. This type of equipment is obviously not compatible with air-conditioned spaces.

Other types of systems separate the air-cooled condenser from the chiller with remote mounting of the condenser outside the plant, perhaps on the roof. Such an arrangement is thought by some to be advantageous because of the cooling properties of outside air. On the other hand, fluctuations in ambient temperature may interfere with condensing efficiency, resulting in adverse effects on the system's capacity and performance. Also, the installation of components requires more than the usual plant personnel.

In general, very large process chilling requirements are most economically handled by a central water-cooled chilling system. Central chilling systems are generally limited to one temperature and must be designed and sized with production requirements in mind. In a central system selection, the processor usually has a choice of open or semihermetic type compressors.

Multiple-circuit chilling units can be selected to provide some of the downtime protection mentioned above as an advantage of portable chillers. Many configurations of both air- and water-cooled central systems are available.

In comparing air- and water-cooled chillers, it is important to remember that water-cooled units are more efficient on a horsepower-to-ton basis and less expensive on a dollar per ton basis. In some cases, the factor may be counterbalanced, particularly in small- and medium-sized installations, by the cost of the evaporative cooler or other water source needed to cool the chiller.

A cooling tower uses the evaporative-cooling principle to cool water that has picked up heat in plastics processing equipment. After passing through the tower, the water is recirculated through the system. Theoretically, the water can be cooled to a temperature equal to the prevailing wet-bulb temperature if enough surface area is available. But, actually, an infinite area would be required. The wet-bulb temperature is normally established at 78 °F (26 °C) for most areas, and practically speaking, the water will be cooled to a temperature approximately 7 °F (−14 °C) higher than the wet bulb [85 °F (29 °C)]. Dry-bulb temperature is not a valid rating basis for cooling tower efficiency or its sizing. Because of evaporative cooling, the system will require makeup water. Also, evaporation of the water concentrates its mineral content.

Realistically, it is important not to overlook alternative sources of evaporative cooling, such as wells, ponds, and city water. However, in general, city water is prohibitively expensive for this purpose. Wells, particularly deep wells, when there is only a 5 °F (− 15 °C) spread in water temperature from winter to summer, constitute a good source of cooling if available. Open ponds are another possible source, but only if the ponds are big enough to provide water in sufficient quantities and sustain a relatively uniform temperature. Finally, it should be pointed out that maintenance requirements are usually higher with these alternate sources.

11.3.2.3 Water Treatment

Water treatment must be considered a valuable adjunct to water cooling systems. The justification for water treatment is oriented to both performance and economics. Untreated water, which results in scale deposits and corrosion, can be quite costly. In terms of production, a mere 0.006 in. (0.0152 cm) of scale will reduce refrigeration compressor capacity by 30%. The thin deposit from scale, corrosion, slime, or algae in the cooling tower system makes it necessary for a pump to work from 15% to 50% harder.

Maintenance to counteract fouling resulting from untreated water can be quite expensive and dangerous, as the acid used in cleaning heat exchange equipment removes galvanizing and metal. It is also a hazardous process for personnel. Also, the additional load imposed on pumps, motors, and compressors by deposits from untreated water tends to make them work overtime to compensate for the insulating effects of the deposits. The results are breakdowns, need for costly replacement parts, and labor costs, as well as lost revenues on production equipment because of downtime for service and repair.

As a rule, plastics processing requirements can be met through the use of a water softener or ion-exchange system. In the latter process, calcium and magnesium ions in the water are exchanged in a plastic bed for more soluble sodium ions, resulting in water with minimum hardness, and thereby, preventing carbonate scaling.

Silicate scaling in such a system can be controlled by water system bleed-off. Essentially, this is a matter of diluting the mineral concentration to a safe level of less than 125 ppm of silica. The same results will be achieved through hot-lime softening and demineralizing. It should be pointed out that even soft water is corrosive. Corrosion can be controlled by adding a corrosion inhibitor, such as certain commercially available polyphosphate chemicals fed into the system. Slime and other organic growths can be controlled by the addition of microbiocides and algaecides to the water.

Suspended solids collected in the evaporative-cooling device from the cooling air can be removed from the cooling system water by filtration. Water filtering is of particular importance when in-plant contamination of cooling systems is likely, as in the case of plants using polyvinyl chloride (PVC) powders, doing pipe extrusion, or producing cast film. Plastics material contamination of cooling system piping is worse than scale; it is far more difficult to remove because of its affinity for the piping. No matter how minor this type of contamination, it will build up to a point where it seriously interferes with heat transfer in the chilling system.

Secondary circuits should be considered to protect the refrigeration equipment against contamination, which can reduce its efficiency by as much as 50%. These secondary circuits can be cleaned similarly to the way hydraulic heat exchangers are cleaned. Many other process control considerations strongly favor the secondary circuit approach, and processors should definitely investigate it.

11.3.2.4 Water Cooling

Over the years, the economic justification for the use of refrigerated chillers has been discussed extensively in the literature. It is worth noting, however, that chillers can increase production revenue sufficiently to produce the total recovery of their purchase cost in less than a year. Over and above increased revenue, there are the savings that result from the conservation of water and maintenance of molds/dies.

A close approximation of the cooling load for the modern plastics processing plant is a necessity before any attempt is made to select the mechanical equipment that is to handle it.

The proper approach in making a cooling survey is to first determine the various water temperatures needed for each group of equipment. The cooling load can then be calculated for each temperature level, and a determination made as to the type and size of the mechanical cooling equipment needed.

We can generalize the cooling water temperatures required sufficiently closely to enable us to select the equipment type required to do the job. Once this has been done, the individual needs of the process can be evaluated and the specific equipment

Table 11.3 *Cooling Temperature Requirements*

Eddy current drive	80–85 °F
Extruder barrels	80–85 °F
Hydraulic oil coolers	80–85 °F
Air compressors	80–85 °F
Vacuum pumps	80–85 °F
Refrigeration condensers	80–85 °F
Temperature-control units	80–85 °F
Molds on injection molding machines	20–55 °F
Molds on blow molding machines	20–55 °F
Molds on bottle molding machines	50–55 °F
Extruder troughs and cooling baths	50–55 °F

20 °F (−7 °C; 50 °F (10 °C); 55 °F (13 °C; 80 °F (27 °C; 85 °F (29 °C; 250 °F (121 °C).
Note: High-temperature cooling requirements for special components may require temperatures in excess of 250 °F.

selected. Table 11.3 gives temperature brackets for various equipment types. A study of this table will indicate that there is a temperature range that, in most cases, can be met with evaporative type equipment such as a cooling tower.

Mechanical refrigeration equipment is required to produce chilled water in the lower temperature ranges. Chillers can furnish water temperatures in the leaving water temperature range needed for each group of equipment. The cooling load can then be calculated for each temperature level and a determination made as to the type and size of the mechanical cooling equipment needed from 60 to 20 °F (16 to −7 °C).

As a general rule, supply water temperatures down to about 45 °F (7 °C) can be furnished without the use of antifreeze solutions in the chilled water circuit. Lower temperatures make the use of antifreeze solutions mandatory to avoid the danger of freezeup.

Cooling speed control is important during fabrication as it relates to their heating cycle. As an example, a great deal of heat is required in plastics thermoforming operations. The plastic material must be heated to its melting point and sufficiently beyond to ensure that the product is property formed before it assumes its finished shape. Cooling should take place as quickly as possible within this limit for efficient production. During IM, about 80% of the IM cycle consists of mold cooling. The use of mechanical chillers speeds up this cooling process and, in turn, the fabricating machines produce more products per hour.

Plastic materials, during the fabricating process, absorb both sensible and latent heat. The plastic is first heated to its melting point, absorbing sensible heat. It then softens and melts, absorbing latent heat. Further heating past the melting point adds more sensible heat. Table 11.1 shows the heat content of some of the more common plastic materials.

A study of this table will show the abrupt increase in heat content as the latent heat of fusion is added once the melting point is passed. A notable exception is polystyrene (PS), which simply softens and goes through a fluid phase without the addition of latent heat. The temperature heat content relationship is a linear one as shown by a straight line when plotted.

A study of Fig. 11.3 shows that the heat added to various plastics during processing will vary. For example, high-density polypropylene (HDPE) requires more heat than PS to reach a given temperature. Therefore, more cooling will be required after the plastic has been processed to bring it back to its proper temperature.

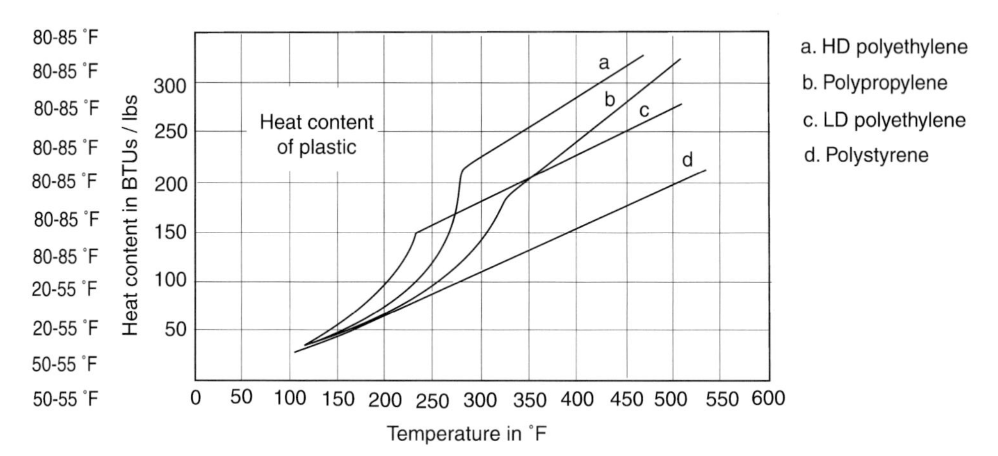

Fig. 11.3 *Heat content of plastics*

It has become common practice to calculate plastics cooling needs in terms of the pounds of a given plastic that can be cooled from processing to handling temperature by 1 ton of refrigeration. One ton of refrigeration is equivalent to a heat transfer rate of 12,000 Btu/h (127.7 J/h), and the plastics cooling rate is also stated in terms of pounds per hour. As an example of this type of calculation, assume that PS is to be processed in an IMM at 500 °F (260 °C) initial temperature and then cooled to about 125 °F (52 °C) final temperature. Figure 11.3 indicates that at 500 °F, PS has a heat content of about 185 Btu/lb (4.3×10^5 J/kg). At the final temperature of 125 °F, the heat content is about 25 Btu/lb (0.6×10^5 J/kg). This indicates a heat removal requirement of 160 Btu/lb (3.7×10^5 J/kg) of the plastic being cooled.

It is common practice to assume that 1 ton of refrigeration will handle the processing of 50 lb/h (22.7 kg/h) of PS. Simple arithmetic indicates that 50 lb/h at 160 Btu/lb (3.7×105 J/kg) account for only 8,000 Btu/h (186.1×105 J/kg). The difference between this figure and the actual 12,000 Btu/h/ton (127.7 J/h/ton) is accounted for by factors other than the heat gained by the plastic. For example, the centrifugal pump circulating the water through the system adds frictional heat. In addition, heat is gained through the chilled water piping and the mold surfaces themselves from the surrounding air. The 4,000 Btu/h difference is accounted for by these external heat gains.

HDPE has considerably more heat added during its processing because of its latent heat requirements. At 440 °F (227 °C), its heat content is about 310 Btu/lb, whereas at 125 °F (52 °C), its heat content is about 35 Btu/lb (0.8×10^5 J/kg) representing a heat gain of 275 Btu/lb (6.4×10^5 J/kg). A rule-of-thumb value often used is 1 ton of cooling for each 30 lb/h (13.6 kg/h). This accounts for 8,250 Btu/h (87.8×10^5 J/kg), with the balance of 3,750 Btu/h (39.9×10^5 J/h) being accounted for by external heat gains.

The heat content of the plastic will vary, with the processing temperature dictated by a particular molding application. The higher the processing temperature, the greater the quantity of heat that must be removed. Following the guidelines established by experience ensures that adequate chiller capacity will be available to meet these varying requirements.

Manufacturers of plastics processing equipment provide data on the number of pounds per hour of a given plastic their machines are capable of handling. For example, an IMM may be rated at 175 lb/h (79.4 kg/h) of PE. If we assume that there are three of these machines in a lineup, a total of 525 lb/h (238 kg/h) of plastic requires cooling. The actual rate is 50% to 60% of 175 lb/h; 175 lb/h requires continuous turning of the screw, which does not happen. Chiller capacity of 1 ton can handle 30 lb/h (13.6 kg/h) of PE. Therefore, 17.5 tons of cooling are required ($525/30 = 17.5$).

Once the load condition has been determined in terms of nominal tons, the supply water temperature required must be evaluated. The nominal ratings of chillers are based on 50 °F (10 °C) supply water temperature. In the case of the job being discussed, a chiller with a nominal capacity of 19 tons would be a good selection if 50 °F water were required.

Chiller capacity decreases with supply water temperature. For the above example, divide the load by 0.8. This establishes the load in terms of 50 °F, and standard rated chiller selections can be made. To determine standard chiller capacities at less than 50 °F, multiply the standard rated capacity times 0.02 for each desired degree. Below 50 °F, a unit would have a 22.08 ton capacity at 48 °F [$100 - 0.04 = 0.96$, 100 being 100% capacity, and at 0.02×2 deg below 50, 0.96 being the correction factor; $23 \times 0.96 = 22.08$, 23 being capacity of a unit at 50 °F or 100% rated capacity, 0.96 the determined correction factor, and 22.08 the capacity of the unit at 48 °F (9 °C). Both chillers selected in this example are of the air-cooled type. In some cases, water-cooled condensers may be used.

The flow through a process and the temperatures entering and leaving must be accurately observed to calculate the cooling requirement. Temperature and flow must be observed simultaneously. Consider carrying three thermometers and submerge them in a glass of water varifying that they agree before making any readings. Be sure to allow enough time for the thermometers to reach an accurate temperature when taking a reading. This may take several minutes if the temperature of the water is far above or below the initial reading on the thermometer. If there are no points in the system where water flow is open to view and accessible for immersing a thermometer bulb, then some means must be found for bleeding water either onto the thermometer bulb or into a container in which the thermometer is immersed.

Flow may be determined in two general ways: from pumping curves or actual timed weight testing.

(a) Flow Determined from Pumping Curve

Pumping curves on all centrifugal pumps has the characteristic shape shown on 2518–3624 series curves (Fig. 11.4). If the pressure difference across the pump is accurately known, it becomes a simple matter of projecting that point over to the proper curve and read flow in gallons per minute (gpm) directly below.

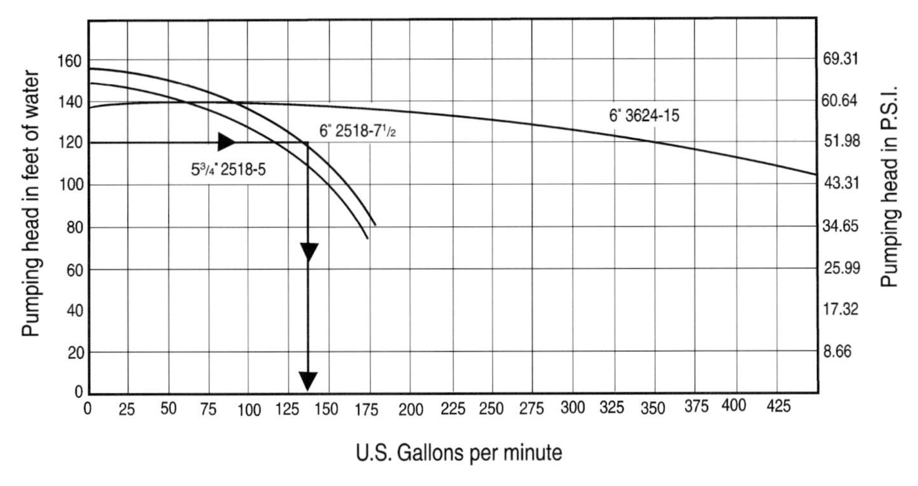

Fig. 11.4 *Flow curves for centrifugal pumps*

Most pressure gauges read psi (kPa) positive and inches (centimeters) of mercury negative. Obtain a good serviceman's test gauge of the compound type. Do not rely on the gauge installed on the job. Readings must be taken at the suction and discharge connections of the pump with the same gauge. The preferred points are the actual test connections furnished on most pumps. Take several readings to be sure of accurate observations. If the suction pressure and discharge pressure are both positive, the pumping head is the difference.

(b) Flow Determined from Measured Water

In determining flow from measured water, the container (such as a 5-gallon (0.02-m^3) bucket) should be weighed empty and then filled. The weight of water will be the difference between the empty weight and the full weight. The accuracy of the quantity held in a 5-gallon container, for example, may be checked. Five gallons of water weigh 8.3×5 or 41.5 lb (18.9 kg).

The difference between empty weight and full weight as observed on a scale should be 41.5 lb if the container holds 5 gallons (0.02 m^3). Record the time required to fill the container with water leaving the process so gpm = lb/(8.3 × min) or gallon/min. Several tests should be run to be certain of the accuracy of observation. The flow must be known accurately.

11.3.2.5 Energy Saver-Heat Pump Chiller

One of the major expenditures in any industrial plant is the cost of the energy required to operate it. Prudent energy management dictates that all means possible be employed to reduce the waste of this precious commodity.

In plastics and similar processing industries, a great deal of heat is required to produce one product. Once used in production, it has been the practice to simply exhaust this heat to the atmosphere. A line of energy-saving process water chillers are available that can recover much of the process heat rejection and put it to practical use.

Heat pump chillers are provided with air-cooled condensers that capture process heat for reuse. During the winter months, this heat is directed into plant or office space, providing supplementary heat. Fuel savings are appreciable when this is done. In summer, the chiller air handling system acts as an efficient exhaust unit to help cool the plant working areas.

11.3.3 Granulator

Scrap, waste, and product rejects in the production of different products can be granulated and recycled. Machines used for size reduction may be broadly classified as granulators, dicers, pelletizers, die face cutters, hammer mills, attrition mills, pin mills, and pulverizers. These machines are universally used to produce granulate, pellet, or powder as required, making it possible for the scrap to be reused or reclaimed into a processible material.

Different designs of granulators with knife blades and sizing screens are available from many different sources. Selection depends on factors such as the type of plastic used, product thickness, hardness, shape, toughness, etc. With heat-sensitive plastics (or even others), the granulator must not cause overheating during the cutting actions. Thick plastics may require a series of different granulators so that incremental reduction occurs, eliminating overheating. Regrind particle size tends to be about 1/16 in. smaller than the screen opening size.

Granulators can be divided into the three categories including energy impacting, energy absorbing, or friable. The energy impact types [for PE, PP, polybutylene (PB), PA-unfilled, thermoplastic rubber (TPR), thermoplastic polyurethane (TPU), etc.] are hard and rigid. They generate fines and require lower rotary speeds. These plastics need to be broken and not cut. The energy absorbing types (for PS, PA-filled, phenolic, etc.) are soft and flexible. They generate angel hairs and require high rotor speeds. A cutting action is required. The friable types are filled plastics [for acrylonitrile–butadiene–styrene (ABS), polymethyl methacryate (PMMA), PVC-rigid, polycarbonate (PC), polyethylene terephthalate (PET), etc.] that generate dust and shatters easily. They use lower rotor speeds.

The selection of a granulator is, for the most part, centered on the plastic material to be granulated (Table 11.4). The first consideration is to examine the fabricating process to determine what type of granulator should be utilized. For an IMM, an under-the-press granulator in the form of an auger granulator may be used. A central granulator or beside-the-press granulator can also be used for this type of process.

Another factor to be considered is the production rate, which is commonly called "throughput." The hourly production rate is determined by the size of the cutting chamber and rotor design. The size of the part will have to be checked to ensure that the hopper and cutting chamber will be able to accept that particular configuration. The basic properties of the plastic materials must be examined to determine if the material is high-impact, energy absorbing, or friable so that the proper knives can be installed in the granulator. The size of the granulate must be determined so that the

Table 11.4 *Example of Specification Listing Standard Equipment and the Optional Equipment Available for Granulators*

Item	Standard	Optional
Hopper	Front feed with top clean-out door	Sheet chute, pipe chute, feed rolls
Cutting chamber	Welded construction, slant-knife seat-safety switch for electrical lockout	Wear-resistant lining
Base	Caster-mounted, bin-type optional front or rear bin removal	Pneu-Vey with 37-in. base for barrel discharge
Rotor	High-shear, two blade slant design, one-piece construction	Three-blade slant design, open-type rotor
Rotor knives	(2) Chrome vanadium alloy steel, (4) for model G-12295M1	(2 + 3) micro-temp, V-7 alloy, V-10 alloy (4 + 6 for model G12295M1)
Bed knives	(2) Chrome vanadium alloy steel, (4) for model G12295M1	Micro-temp, V-7 alloy, V-10 alloy
Screen	Choice of $\frac{1}{4}$-, $\frac{5}{16}$-, or $\frac{3}{8}$-in.-diameter holes	$\frac{3}{16}$- or $\frac{1}{2}$-in.-diameter holes
Motor	3- to 40-hp ODP Lincoln, see model specifications	To customer specifications
Starter	Magnetic across the line in compliance with national electrical code	To customer specifications, 115-V controls
Drive	V-belt	Flywheel

proper screen can be chosen for the machine, as well as the type of material removal to be utilized with the granulator.

Controls should be set for evaluating the performance of granulators or the effect they have when reused. Certain plastics, such as melt heat-sensitive types, can be completely degraded, unless special precautions are taken such as granulating when "frozen" (liquid oxygen, etc.).

The size reduction of plastics can be considered a hazardous operation in the plastics industry. Unfortunately, the lack of knowledge of the machinery can make working with it a dangerous occupation. Rotary shears, densifiers, shredders, and granulators are designed with two main goals: safety and efficiency.

Granulators are equipped with a safety device such as a switch connected to the cutting chamber to terminate operation if it is opened while the motor is running, preventing a person from being injured. At no time should anyone open a machine while it is running. Any adjustment to the knife or drive system should be made with the machine off and power disconnected.

When a serviceman or machine operator has to perform any kind of work on the machine, that person should shut off the machine and disconnect the power at the main power source for that machine. In most cases, this would be the lockout box or disconnect. A padlock should be put on this box, to ensure that the machine does not become energized accidentally. At no time should it be necessary for a person to put any part of his or her body into a granulator, rotary shear, densifier, or any piece of machinery. That includes climbing into a large granulator to free a jam-up or for cleaning purposes. The instructions for the machinery should be read very carefully and the warning label followed in its entirety. The best safety measure is a careful person.

11.3.4 Tooling Dehumidification

Fabricating plastics under variable humidity conditions can cause serious problems during production. As an example, optimum mold cooling temperatures produce condensation on the mold surface that can damage molds and ruin critical products. Cycle times, scrap parts, labor hour, and energy expense must all be analyzed to show how much this condensation problem affects the entire molding operation, all the way from sales through shipment to the bottom line in accounting.

For many years, molders have accepted this problem for processes such as injection blow molding (IBM) and extrusion blow molding (EBM), which require a chilled coolant for their molds and attempted to either compensate for it or eliminate it through various methods.

Until recently, there was actually no set solution to this problem designed for the plastics industry. Specialized processes in the plastics industry had to be analyzed to design a system that would be flexible enough to handle all types of molding and products removal while, above all, maintaining optimum molding conditions.

The main concern in any dehumidification system is dewpoint. A basic definition of dewpoint is the temperature at which moisture in the air will turn into liquid or condense (condensation). Another term that must be defined is relative humidity. This is the amount of moisture in the air as compared to the amount of moisture the air can hold. At 100% relative humidity, air begins to condense and is at its dewpoint. Any percentage less than 100 will have a lower dewpoint, and it is the dewpoint that is the most important factor in dehumidification (Fig. 11.5). To convert pressure dewpoint to atmospheric dewpoint, enter dewpoint at a specific pressure and read down to the corresponding dewpoint.

One basic law of error prevails, that is, warm air can hold more moisture than cold air. As the temperature of warm air is lowered to the dewpoint, moisture in the air condenses and is removed from the air. This can happen as a natural phenomenon or

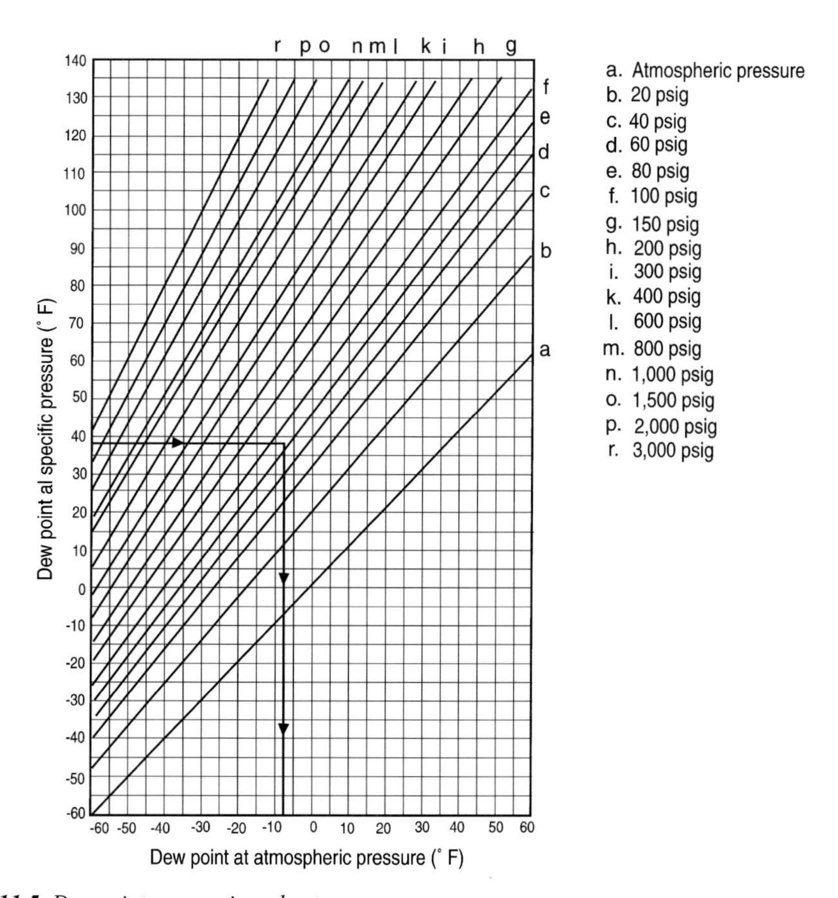

Fig. 11.5 *Dewpoint conversion chart*

by physically lowering the air temperature through some type of mechanical heat exchange.

11.3.5 Feeding/Blending

Equipment manufacturers have increased feeding accuracy using different devices such as microprocessor controllers. Also, materials are being reduced in size with more uniformity to significantly improve uniformity in melt. Processors can use blenders mounted on hoppers that target for precise and even distribution of materials.

Gravimetric blending improves accuracy, better process control, and requires less operator (if any) involvement in calibration, particularly when running processes in which great accuracy is required. Metering by weight eliminates overfeeding expensive additives. The principle of gravimetric feeding with throughput or metered weight control is well established. Equipment can provide an accuracy of at least ±0.25 to 0.50 wt% for ingredient and blend ratios of 2σ (two standard deviations). As an example, with gravimetric metering, coextruders have a simple means of constantly maintaining the average thickness of individual films and overall thickness at better than ±0.5 wt%. By comparison, volumetric and quasi-gravimetric blenders that use batch operation usually have accuracy variations of 2 to 10 wt%.

Feeding and blending can also be performed volumetrically. Several types are used including belt, rotary, slot, vibrator, and single screw feeder. Because they adjust by volume, plastics are not self-adjusting for any variations in bulk density.

Many different blender/mixer designs are used to meet the different requirements of the many compounding materials used in the industry. Because no one type of blender (mixer) is likely to satisfy all demands, purchasing decisions should be determined by such factors as degree of type and form of material components, form of material involved, agitation desired, output rate required, and uniformity of output.

Typical uses of blenders include powder colorants and plastic pellets, reinforcing fillers and plastics, dry blending PVC, blending powdered plastics with additives, and so on. Special considerations have to be understood, such as the amount of work or shear involved, in order to be cost effective in purchasing the mixer and producing the compounds.

11.3.6 Material Handling

Design of the raw material system has a major impact on the plant's manufacturing costs and housekeeping. It is based on the different materials used, annual volume of each material, number of different colors, production run lengths, and so on. Different methods are used that range from manual methods to full automation for either raw material or processed parts. Use includes automatic bulk systems, in-line granulators, parts removal robots, conveyors, stackers or orienters, and so on. The equipment chosen must match the productivity requirements.

When conveying plastics, the shortest distance between two points is a straight line. The maximum conveying distance is usually 800 equivalent ft (244 m). A gradual upward slope is never better than a vertical lift. When the plastic passes through a 45° or 60° elbow, it ricochets back and forth, creating turbulence that destroys its momentum. A properly designed system generates plastic velocities of 5,000 ft/min (1,500 m/min).

Vacuum/pressure conveying provides double the conveying rates of vacuum alone. Plastic lines are not recommended for conveying lines, as static electricity will be generated interesting with the movement of plastics. A rather simple and useful test to determine if material is going to be difficult to convey can be used. Take a handful of the plastic and squeeze it firmly. On opening your hand, if the lines in your palm are filled with fines, it will be difficult.

An important property of bulk storage in regard to the processing of plastics is the bulk density of the particulate material. Bulk density is the weight of a unit volume of the material including the air voids. The actual material density is defined as the weight of unit volume of the plastic, excluding the air voids.

If the bulk density is more than 50% of the actual density, the bulk material likely will be reasonably easy to convey through the plasticator. In this case, the screw channel depth in the feed section does not have to be too large, between 0.1 diameter and 0.2 diameter, and the compression ratio can be at the low end of the range, from about 1 to 3. However, if the bulk density is less than 50% of the actual density, then solids conveying problems are likely to occur. With these materials, the deep feed section has to be rather deep to obtain sufficient solids conveying. As a result, the compression ratio will be at the high end of the range, from about 3 to 5.

When the bulk density becomes less than 30% of the actual density, a conventional plasticator usually cannot handle the bulk material. Such materials may require special feeding devices including crammer feeders or a special extruder design, such as a large-diameter feed section tapering down to a smaller diameter metering section.

Bulk materials handling and conveying methods are based on a combination of theory and experience. Almost any substance can be conveyed pneumatically. Automation of the material handling system increases the advantage of purchase and storage of bulk materials. A pneumatic conveying system can move materials from the storage area to the processing machine, automatically, with little risk of contamination to the materials, which commonly occurs when materials are moved manually.

The major problem with pneumatic conveying of materials is the absence of a standard set of formulas for calculating the equipment sizing and flow characteristics of the material to be conveyed. Standard formulas do not exist because of the vast variety of plastic materials, additives, and their combinations. Most available information on the sizing of material handling systems is based on experiments conducted by equipment manufacturers and the observation of successful conveying systems currently in operation.

Pellets, granules, and powders are pneumatically conveyed through what is classified as a pipeline-conveying system. This type of system transports particles of solid materials through vertical and horizontal pipelines according to the physics principles of kinetic energy, pressure, and aerodynamic lift.

In their basic form, these formulas state that if particles were placed in a pipeline conveying system in a stream of flowing air, at a higher velocity than the terminal velocity of the particles, these particles would move at a velocity equal to the difference between these velocities. The terminal velocity, in this case, would be the least amount of airflow needed to make the particles move.

Velocity is the key to transporting materials pneumatically. Velocity is defined as the rate of motion or speed. Specific terms are used when describing velocity in material-conveying systems. Critical velocity is defined as the minimum superficial air velocity that will convey the material as specified. Dropout velocity is a term commonly used for the minimum amount of velocity needed to move particles through a vertical tube. Settling velocity is the minimum velocity needed to prevent material from falling out of the air stream.

Pipeline conveyors are commonly referred to as dilute-phase systems. Dilute-phase conveying is described as the conveying of a small volume of material by a large volume of air and is expressed as a ratio commonly called the material-to-air ratio and classified in pounds (kg). The counterpart of dilute-phase is dense-phase. Dense-phase conveying is described as a pound of air moving its weight or more of material through the conveying tubes. In other words, material is conveyed in a dense-phase state that will have a low material-to-air ratio. Compactable powders are common materials that are conveyed in a dense state.

The physics principles that play an important role in pneumatic conveying are gravity, pressure differential, inertia, shear, and elasticity. Detailed equations are available in the literature. These formulas explain the flow and work capabilities of air in a pneumatic system. However, these characteristics of air are only one factor in the art of pneumatic conveying. The other factor concerns the actual material that is to be conveyed pneumatically.

Various properties and characteristics of materials used affect the sizing and design of the conveying system. The following explains the characteristics that affect material flow in a system, along with a few tests that can be conducted to determine factors affecting the material flow.

Specific gravity is one of the more important characteristics that pertain to conveying a material in an air stream. It is defined as the ratio of the material's density to that of water. To determine specific gravity, test to see how much water the material displaces when placed in a container holding a specific amount of water.

Powders are commonly used in the plastics industry. The specific gravity test for powders is done by vibrating them in a container until the powders form a densely packed mass. The volume and density are compared to the known density of an equal volume of water.

Particle size is also a consideration in pneumatic conveying systems. The material has to be tested to determine the amount of fines and dust that may be contained in the material. This will help determine the type of airflow in a system, whether it be a vacuum or pressure system, along with the type of filters that will be utilized. Particle size is measured by using sieves that are made to standards set by the American Society of Testing Materials (ASTM) of the US Standards Institute.

A common method of testing for the particle size and range would be to place the material in a sieve with a large mesh opening and shake the material until the particles that are small enough pass through the screen. This procedure is repeated, with screens of a smaller mesh size, until all the particulate is separated.

Tackiness is another characteristic that must be examined. If the material is extremely tacky, it may not be suitable for conveying pneumatically. Tacky material may smear against conveying pipes and cause a buildup of material eventually clogging a line. A simple test for material tackiness would be to take some material into your hand and squeeze it into a compact ball. If the material sticks together, it is classified as a tacky material.

The only way to determine if a tacky material can successfully be conveyed through a pneumatic system is to run the material through the conveying lines and check for proper material flow. The manufacturer can perform this test.

The melting point of a material should be determined. Some plastic materials melt at low temperatures. If these materials are conveyed at a faster rate than necessary, they may slide against the walls of the conveying tubes and heat up by friction, which in turn will cause them to begin melting, producing what is called “angel hair.” This commonly takes place at a bend in a conveying tube due to the centrifugal force that is placed on the pellets, forcing them to slide along the outer periphery of the tube. The melting plastic pellet running along the wall, leaving a thin trail of plastic along the tube wall. This thin plastic will partially peel away from the wall as the pellet moves back toward the center of the air stream, leaving what appears to be a fine hair. If enough of this occurs with other pellets in a particular area, the angel hair will clog the system.

The abrasiveness of a material is another concern in pneumatic conveying. Materials that are abrasive may cause the conveying tubes to wear through quickly. Abrasive materials may have to be conveyed at a lower rate than other materials if at all possible. Other modifications can be made to a system to combat the premature failure of the conveying tubes, such as using wear-resistant material. The only real test that can be performed on a prospective abrasive material is to determine how quickly the material can wear through a conveying tube.

Corrosiveness is a characteristic of powders or other materials that contain acids. Very few of these types of powders are used in the plastics industry. Testing the material for a pH factor can test a material for acid content. A pH of 7 is neutral; any reading below 7 indicates an acid and above 7, that the material is alkaline. Powdered materials with strong acid indications will have to be conveyed through special pneumatic systems to prevent any corrosion from taking place within the system.

Aeration and de-aeration are additional factors to be considered. If a material can be continuously saturated by air in a free flowing state, the material is aerated material. Should a material clump together and block the airflow, the material is de-aerated but can still be conveyed pneumatically. The manufacturer can make recommendations for handling this type of material.

A test can be performed on materials to determine aeration characteristics. The material can be placed in a container with a lid and shaken for a few moments. If the volume of material appears to have increased and takes a long time to settle to the bottom of the container, the material is said to be an aerated material. If the material settles to the bottom of the container quickly, it is said to be a de-aerated material.

Another test for the aeration of material is the angle of repose. The material is put on a horizontal plane, and that plane is lifted at an angle until the material starts to flow (at the angle of repose). If the material starts to flow at a low angle, the material is said to be free-flowing or aerated material. If the material flows at a high angle, the material is said to be de-aerated or hard-flowing material. The angle of repose not only helps with the sizing of pneumatic systems, but also aids in choosing the proper storage system equipment.

Odors and the toxicity of materials should be considered when developing a conveying system. These two related factors are not very common in the plastics industry. However, these characteristics could be a common element when dealing with other chemicals that may be related to the plastics industry. These elements have an effect on the type of conveying and filtration system that should be incorporated in a plant.

Despite the vast amounts of formulas and testing methods, empirical formulas for pneumatic conveying have not been established because of the unlimited supply of materials that can be conveyed by air. However, manufacturers of conveying and storage equipment have established simple sizing charts based on the most commonly used materials in the plastics industry.

11.3.7 Warehousing

In addition to handling plastic products, there is a wide variety of tasks for warehousing such as handling raw materials, additives, auxiliary equipment, spare parts, molds, dies, tools, processed plastic parts, and so on. They require proper handling and storage procedures that are logged economically.The different products to be stored usually have specific requirements such as temperature, height of loads, handling plastic form (pellets, powder), and so on.

Plastic materials can arrive in different size packages or containers depending on requirements. There are drums [from 15 lb (11 kg)], bags [50 lb (23 kg)], gaylords [cardboard box usually lined with plastic sheet holding 1,000 lb (454 kg)], or bulk fabric sack bags [also called super sacks, super bags, or jumbo bags holding 2,000 lb (907 kg)], bulk bin at 5,000 lb, truck at 40,000 lb, to rail car at 180,000 lb. To move materials from these containers, systems include vacuum tube conveyors, dumper and pressure unloader, fork truck hoist, and so on. Use plastic storage box containers rather than bags or drums. Box sizes and weights vary and conform to a standard size pallet on which they are shipped and moved in the plant.

If a plastic is stored in a relatively cold area then brought into the operating plant, it will often become wet (due to moisture condensation) enough to cause processing problems. Different procedures can be used to eliminate this problem, such as moving material to an indoor, closed storage bin to expose the plastic to the same temperature.

All storage and unloading areas should be kept clean and dry to prevent contamination and minimize fire hazard. The store room should be separated from the processing shop by fire-resistant doors. Store materials away from direct sunlight in properly constructed racks, containers, and/or silos. Usually the use of unheated storage areas with natural ventilation is sufficient. Ensure that the plastic does not stagnate in storage by adopting a strict stock control policy. Adopt a first-in, first-out policy (Fi–Fo).

For processors that can make (truck or rail car) bulk purchases, silos with automatic plastic handling systems, although initially costly, will provide economic paybacks. They also provide environmental benefits, save floor space particularly when located outside, reduce handling by people, and leave no mess on shop floors as with sacks, gaylords, or big bags. It takes 40,000 lb/month of a single plastic to justify purchase of a silo.

11.3.8 Part Handling Equipment

Materials and products can be removed during production in many different ways from the primary production equipment as well as upstream and downstream equipment.

Linear guides used in some handling equipment, particularly robots, provide a means of low-friction, precision linear motion through an assortment of rails (round or profile), contact elements (rollers, ball bearings, or full-contact sleeves), and mounting con-figurations. Many types of guides exist, each engineered toward optimized performance in a specific range of applications.

Various application criteria will affect linear guide incorporation. These criteria can be summarized as follows: dynamic load capacity, envelope size, mounting configurations, life, travel accuracy, rigidity, speed/acceleration, cost, and environmental considerations. The priority of these items will determine the appropriate linear guide for the application.

Robots are also utilized for parts handling functions. Slow to fast robots are available. Sturdy/fast robots can be made from lightweight/high strength materials such as carbon fiber/epoxy reinforced plastics that accelerate to at least 25 Gs, and attain linear velocities of at least 60 ft (18 m)/s, all of which permit part retrieval in as little as 300 ms. Their arms can retrieve parts in a definite orientation as fast or faster than they can fall free by gravity. They can be designed to include discrete operations such as parts removal, assembly, decoration, welding, inspection, stacking, wrapping, cartoon filing, palletizing, and lot identification.

11.4 Secondary Equipment

Ideally, fabricating thermoplastic or thermoset products will be completely finished as processed. For example, almost any type of texture or surface finish can be fabricated into the product, as can almost any geometric shape, hole, or projection. There are situations, however, where it is not possible, practical, or economical to have every feature in the finished product. Typical examples where machining might be required are certain undercuts, complicated side coring, flashing, or places where parting line or weld line irregularity is unacceptable. Another common machining/finishing operation with plastics is the removal of the remnant of flash, sprue and/or gate if it is in an appearance area or critical tolerance region of the part.

Many plastic products are decorated to make them multicolored, add distinctive logos, or allow them to imitate wood, metal and other materials. Some plastic products are painted as their as-molded appearance is not satisfactory, as may be the case with reinforced, filled or foamed plastics. Painting or coating also affords product protection. The following section discusses some of the secondary operations frequently used with plastics. As plastics vary widely in their ability to be machined and to accept finishes, this discussion will be general in nature, with details left to other literature dealing with specific plastics.

11.4.1 Machining and Prototyping

Tool shops as well as prototyping shops take full advantage of high-speed machining. To keep pace with shorter product development cycle prototype makers can utilize new classes of high-speed cutting machines to rapidly produce their products. High-speed machining is technically demanding, as users must contend with stringent equipment and processing requirements. They are required to integrate the process effectively within the manufacturing cycle to reap its full potential. Performance and profit gains justify the prerequisite investments in dedicated computer numerical control (CNC) machinery, cutting tools, and integrated computer-aided design and manufacturing software and controls that support high-speed data transmission.

11.4.2 Molded Product Joining and Assembly

Blow molding processes produce individual products. Joining of these products to other products composed of the same or a different plastic material, as well as other materials such as metal, is often necessary. This action is required when:

- the finished assembly is too complex or too large to fabricate in one piece or
- disassembly and reassembly are necessary, for cost reduction, or when different materials must be used within the finished assembly.

The joining of plastic components to each other or to components of other materials is a frequent occurrence in product assembly. The success of a specific technique will depend on whether, as a byproduct of the technique, undesirable sizable stress levels in the plastic product may result. Guarding against potential stresses in the assembly is a very important aspect so proper molding conditions are required to eliminate or reduce stress levels. Many techniques provide assembly of all kinds of products, each having advantages and limitations.

11.4.3 Printing and Decorating

Blow molded products can be decorative as well as functional. Different systems of printing and decorating are shown in Fig. 8.36 and Table 8.7. The various decorating methods are discussed in Chapter 7.

12 Troubleshooting and Maintenance

12.1 Overview

The target in the fabricating plant is to eliminate or at least minimize troubleshooting. Maintenance and specifically preventative maintenance procedures should be used to decrease or eliminate costly downtime and optimize product performance.

Just as during the BM fabricating process, when making changes during maintenance/inspection and troubleshooting, allow enough time to achieve a steady state in the operation prior to collecting data or any output information. It may be important to change one processing operation at a time. For example, with one change, such as an extruder screw speed, temperature zone setting, air pressure, cooling system, or another parameter, allow four time constants to achieve steady state prior to collecting data and evaluating operating performance.

12.2 Maintenance

Periodic checkups and regular maintenance should become a habit (at least use maintenance manuals from the equipment supplier). Include factors such as:

- watch out for plastic, water, and/or oil leaks;
- change oil in gear box and transmission (if oil-driven) at least every 6 months if operating on one shift or 3 to 4 months if operating on more shifts;
- lubricate required parts;
- check electric band heaters (for tightness), thermocouples, pressure transducers, etc.;
- check control circuits (electrical, hydraulic, mechanical, etc.);

- check die opening with a device such as a feeler gauge;
- routinely check the line for alignment, level, and parallelism, particularly windup equipment;
- keep all rolls clean and rust free;
- check safety devices on all equipment;
- keep shop and particularly plastic material clean;
- keep spare parts in stock based on experience and/or equipment suppliers' suggestion;
- set up sessions for personnel to repeat instructions on safety equipment procedures and prevent hazards harmful to personnel, equipment, and product; and
- know the signs of trouble, such as unusual noise and odor, before damaging trouble occurs that shut down the line.

It is important to set up preventive maintenance procedures on all equipment. Processors should make a habit of performing regular checkups and maintenance work based on equipment suppliers' recommendations and one's practical experience.

12.2.1 Tools

Tools are expensive and delicate "instruments." Great care should be taken when in a machine and in the disassembly and cleaning of components. Any protruding parts should be protected against damage in transfer from the fabricating machine.

Disassembly should be done only when the tools have had sufficient time to heat soak or at the end of a run so that they are at operating temperature. Experience has shown a temperature of 450 °F (232 °C) to be adequate for most nondegradable plastics. For degradable plastics, cleanup should begin immediately after shutdown to prevent corrosion on the flow surfaces. While the heat is left on, all the bolts are loosened. The heat should then be turned off and all electrical components and sensors removed carefully. While still hot, the tool is disassembled and thoroughly cleaned with "soft" brass, copper, or aluminum tools.

Their surfaces, especially melt flow die openings and cavities, should be covered with a protective, easy to remove, coating against surface corrosion when not operating as well as during maintenance. Records should be kept to ensure required maintenance is accomplished on a regular time schedule.

12.2.2 Cleaning

Keeping molds or dies clean in order to produce quality products can become a major problem and very costly. The molds and dies have to be handled very carefully. The byproducts that develop during melting of the plastics contain all types of additives that can cause contamination of molds or dies.

Polyethylene, like most plastics, tends to degrade somewhat during processing, especially if the processing temperature is high or if the resin is allowed to remain stagnant in the extruder and die due to interrupted production runs. The degraded products caused by these conditions will coat or “hangup” in the die head, causing die lines to occur in the extruded parison. Die lines, in some cases, can be removed by simply scraping the degraded resin from the annular opening between the mandrel and die with a piece of brass or beryllium–copper shim-stock. However, if the die cannot be cleaned in this manner, the molding process must be stopped and a thorough cleaning operation started.

Die lines are surface imperfections or scratches that are formed on the parison as it is extruded, and are transferred to the blow molded article. The cleanliness of the parison die is important in reducing these lines. Chrome plating of the die or coating with Teflon can be helpful. Some resins, particularly high-density polyethylene (HDPE), will produce die lines even in a clean die. Die lines can be the result of nonuniform melt viscosity. This can be reduced by high-density mixing and use of plastics of uniform molecular weight. The effect of die lines can be minimized by any action that tends to minimize the cooling of the melt until it is pressed against the mold surface. These include

- increasing blowing speed so that localized areas contact the mold before the rest of the piece is filled out;
- increasing air pressure to fill the mold more rapidly and to flatten out the die lines; and
- increasing mold temperature to reduce chilling the parison so that it is still pliable when high pressure develops in the mold.

As the family of commercially important plastics rapidly expands, related processing equipment becomes increasingly sophisticated and much more attention is being directed to the design and metallurgy of ancillary tooling. Among the major considerations are abrasion resistance, corrosion resistance, and the capacity to withstand elevated temperatures and pressures.

A well-designed mold or die with appropriate metallurgical composition and heat treatment should have an excellent service life. The most frequent cause of premature failure is encountered during routine cleaning operations. Because mechanical damage during manual cleaning is the most obvious, the potential problems related to corrosion, distortion, or destruction of original metallurgical properties are frequently overlooked in the selection of a suitable cleaning process.

Different equipment require cleaning on a periodic maintenance time schedule to ensure proper operation. Cleaning devices are available during processing for drool, flash, and so on that operate economically and safely in removing contaminated plastics. The routine techniques used include blow torches, hot plates, hand working, scraping, burnout ovens, vacuum pyrolysis, hot sand, molten salt, dry crystals, high pressure water, ultrasonic chemical baths, heated oil, and lasers.

Commercial cleaning systems include aluminum oxide beds (fluidized beds), salt baths, hot air ovens, and vacuum pyrolysis.

The options for a successful and efficient cleaning system are extremely limited—particularly in those applications in which absolute cleanliness is essential. Current cleaning techniques will now be reviewed.

12.2.2.1 Manual Cleaning

Manual cleaning, whether by wire brush, abrasive buff, torch or a combination of the three, frequently is attempted when hardware configuration is such that it cannot be accommodated with conventional equipment. However, this method is labor intensive and destructive physically and quite often metallurgically.

Before the die or mold is disassembled, it must be heated above the melting point of the resin to ensure that the die can be taken apart and cleaned while the plastic is still molten. It is recommended that electric heaters and temperature controllers be used for heating the die before disassembly. The use of acetylene torches for heating of dies and mandrels can cause local overheating and warpage of the tooling, resulting in problems with part dimensions after reassembly [6].

For safety purposes, the electricity should be turned off and the die heaters removed before dismantling the die. Extreme care should also be exercised during handling of the parts to avoid damage to the die and mandrel.

Materials that will not scratch the steel must be used in cleaning blow molding dies and molds. Typical cleaning supplies, in addition to brass or beryllium–copper knives or scrapers, include brass wool, asbestos gloves, and antiseize compound for threaded parts.

The die and mold may be cleaned in several ways. The most common practice is to remove most of the molten resin from the parts with a brass or beryllium–copper scraper, followed by using brass wool to finish the job. Another method uses a high velocity air jet to remove the molten resin from the die. As the air jet separates the plastic from the metal, the plastic is gently pulled away with a pair of needle-nose pliers. High velocity air can be used to remove most of the plastic from dies used in blow molding HDPE. Generally, it is still necessary to use brass wool to remove the oxidized particles that were not removed with the air jet (never use a screwdriver or other hard metal tools that may scratch the die and/or mold).

12.2.2.2 Solvent Cleaning

Solvent cleaning of hardware can be accomplished with acid or alkaline chemicals, organic agents and organic/inorganic ultrasonic cleaning. Conventional solvent baths require complete breakdown of complex hardware prior to cleaning often resulting in mechanical damage to the hardware. Acid/alkaline cleaning is slow at best and is frequently corrosive to hardware. Although equipment costs are low, chemical costs are high. Sludge removal and disposal are both labor intensive and an increasing environmental problem. Organic solvents can be effective; however, environmental considerations restrict their use and recycling equipment is necessary due to environmental pollution controls.

Solvent cleaning may consist of wiping, immersion in a solvent, spraying, or vapor degreasing. Wiping is the less effective process and may result in distributing contaminate over the surface rather than removing it. Immersion, especially if accompanied by mechanical or ultrasonic scrubbing, is a better process. It is even more effective if followed by either another immersion or a spray rinse.

Vapor degreasing is done by solvent vapors condensing on parts. It is the most effective process because the surfaces do not come in contact with the contaminated solvent bath. Vapor degreasing is carried out in a tank with a solvent reservoir on the bottom. The solvent is heated and vapor condenses on the cool plastic surfaces; the condensate dissolves surface impurities and carries them away. The cleaning action is as fast as a minute.

12.2.2.3 Ultrasonic Solvent

Ultrasonic solvent cleaning will normally improve solvent bath cleaning dramatically. However, corrosion, disposal problems, and equipment costs rise significantly when ultrasonic cleaning is required. It is used, most successfully, for a post-cleaning process for the removal of inorganic residues from a thermal cleaning system.

This method is used for thoroughly cleaning parts, particularly, electrical and mechanical parts. A transducer mounted on the side or bottom of a cleaning tank is excited by a frequency generator to produce high-frequency vibrations in the cleaning medium that dislodge contaminates from crevices and blind holes that normal cleaning methods would not affect.

12.2.2.4 Salt Bath

Molten salt in a container is a very old cleaning method for metal molds, dies, and so on. Cleaning cycles are as short as one minute, depending on tool shape and plastic. Salt baths operate primarily on the basis of thermal oxidative decomposition of the plastic. Chemical corrosion can be associated with this cleaning method and surface defects created by thermal shock during the water rinsing of salt from hardware at an elevated temperature. It is not normally considered for new or replacement installation because of environmental and safety considerations. Other disadvantages include high operating costs, as the system must be left on even when not in use, high replacement costs of salt, and sludging and disposal problems associated with spent salt.

12.2.2.5 Oven

Both vacuum and conventional ovens are still in general use for pyrolysis type cleaning. Generally, the advantages of ovens lie in the relatively low capital equipment cost and convenience in handling large loads. Ovens, however, operate on a very long cleaning cycle and significant post-cleaning operations are required to remove residual carbon, which is normally a manual operation on critical hardware. Glass bead cleaning is used on hardware where precise maintenance or dimensional tolerance is nonessential.

Vacuum ovens tend to require high maintenance, especially on the pumps and seals, and in some cases, special provisions are made when handling certain plastics that require collecting vapors and/or residue released from the molds/dies (different type filters or scrubbers are used).

In both types of ovens, complex assembled hardware is normally broken down prior to cleaning. When the disassembly of parts is not possible, dual cycle cleaning with disassembly between cycles is frequently required. This practice obviously increases both operating costs and turnaround time.

Also, nonuniform heating creates stress and can be preferentially destructive to metallurgical properties induced by prior heat treatment. This consideration becomes more important with the increased use of the precipitation hardening stainless steels, in which dramatic changes in hardness occur over a relatively small temperature differential.

12.2.2.6 Fluidized Bed Cleaning

Fluidized bed cleaning, introduced in the late 1960s, has become the option of choice for most applications. Absolute temperature control and thermal uniformity permit cleaning of the mold. The average cleaning cycle is an order of magnitude shorter than for ovens due to the superior heat transfer characteristics of the fluid bed.

12.2.2.7 Abrasive

Cleaning by abrasion removes surface contamination and increases surface roughness. Removal of surface contamination eliminates a potential weak boundary layer. It has a positive effect on adhesion. Cleaning is usually carried out by several mechanical processes such as dry blasting with nonmetallic grit (flint, silica, aluminum oxide, plastic, walnut shell, etc.), wet abrasive blast (slurry of aluminum oxide), hand or machine sanding, and scouring with tap water and scouring powder.

12.2.2.8 Carbon Dioxide

CO_2 is used for cleaning metals (tools, screws, etc.) to remove plastics by blasting with rice-sized pellets of dry ice (CO_2) at $-104\,^{\circ}$F ($-40\,^{\circ}$C). Residue is removed in a process often compared to sand blasting without the sand. Ice blasting equipment is small and uses shop compressed air. CO_2 ice crystals exert a strong thermomechanical force on impact. Parts can be cleaned while hot and in the machine. With heat, they clean faster. CO_2 is a colorless, odorless gas made, as an example, by passing air over red-hot coke. It is an important intermediate in the production of many plastics and is also used in different fabricating processes in place of air or other gases providing heat, blow molding, and so on resulting in reducing the injection and blow molding cycle time, and so on.

12.2.2.9 Cryogenic Deflashing

Deflashing products, particularly when small and numerous, can be done efficiently using cryogenic tumblers and shot blast. Liquid nitrogen or dry ice is used at −320 °F (−196 °C) for liquid N_2. The parts can be frozen by liquid nitrogen and then blasted usually with a plastic media while tumbling in a basket sealed in an enclosed chamber. The air moving system can be sprayed at about 225 psi (1.6 MPa). Advantages of this procedure include accuracy, repeatable deflashing, and reduced finishing costs.

This system, with or without vibrating devices eliminates tool damage and produces no solvents or other environmental hazardous byproducts. Because the processes operate at relatively low temperatures, no tool dimension changes occur such as warpage and other problems associated with heating.

12.2.2.10 Brass

Brass is a tool used to clear or remove melted plastic during processing that may be trapped in the hopper throat when melt bridges, melt sticking to a tool or screw, and so on. The brass does not damage the metal whereas if steel or other metals were used, the steel would be damaged. Beryllium or aluminum tools are sometimes used but they are harder than brass.

12.2.2.11 Vacuum Pyrolysis

Pyrolytic ovens are a variation to cleaning metal parts. Instead of sealing the chamber and starving oxygen, as burnoff ovens do, they pull a vacuum over parts to remove combustible air. Theoretically, a vacuum oven is the safest method, as it removes all oxygen from the heating chamber. The vacuum pyrolysis cleaner utilizes heat and vacuum to remove the plastic. Most of the plastic is melted and trapped and the remaining plastic is vaporized and appropriately collected in a trap.

12.2.2.12 Assembling Die/Mold

Certain precautions should be taken in reassembling the die and mold components. All mating surfaces should be absolutely free of foreign material to prevent plastic and air leakage. The die bushing should also be free of foreign material, as any contamination could cause misalignment problems.

It is usually recommended that a torque wrench be used when assembling the die and mold onto the extruder or injection head. This practice ensures that the die/mold can be properly centered, and if the parts are properly cleaned prior to assembly, the possibility of leakage around the die is decreased.

12.2.3 PVD Coating

As reviewed, surface treatments are used to protect the processing tool against abrasion and corrosion due to contact with the melt. An example is the physical vapor deposition (PVD) used to optimize the surface properties in a layer up to 10 mm deep that has little effect on the contour of a product. PVD coatings lead to an insignificant and usually discernible roughening of the surface. Tools that have been finished machined can therefore be improved. No expensive post-treatment is necessary. In this vacuum chamber [10^{-2} to 10^{-4} mbar (1 to 0.01 Pa)] process, metals are converted to a gaseous state by the introduction of thermal (electron beam or arc) or kinetic energy (atomization). They condense on the surface to be coated.

12.2.4 Rebuilding and Repair

Retrofit projects should be well planned and evaluated in comparison to buying a die, mold, or other equipment (including new fabricating machines). Even though the initial capital expenditure is much lower than for a new tool or machine, the long-term economical value may be questionable. As an example, machine retrofits can be tailored to meet the customer's performance requirements at 40% to 70% of a new machine's cost. To provide a good basis for a decision, a technical evaluation matrix system using weighted criteria and a time-related method for judging the economical value of an investment are required.

12.2.4.1 Die, Mold, Screw, Barrel

Major rebuilding and repairs involve screws and barrels; also involved are dies and molds. Screws and barrels are expensive and can cause downtime when damaged or worn. It may be practical (cost effective) to repair rather than replace. It is common practice to rebuild a worn screw with hard surfacing materials. Quite often the rebuilt screw will outlast the original screw time in service. The larger the screw, the more economical screw repairing becomes. Usually it does not pay to rebuild 2-in. (50-mm) diameter or smaller screws.

12.2.4.2 Stripping, Polishing, Plating

After a certain period of service, most tools, screws, and such become scratched, carburized, and/or discolored due to the plastic temperature and pressure melting actions. They are difficult to clean and tend to lose their original operating characteristics. It they have been chrome-plated, the chrome may be gone in some places or peeling in others. It is best to refurbish in this condition by stripping the old chrome, polishing, buffing, plating, and buffing again. The part will look much better and will also perform better for little cost and a short delivery time.

12.2.5 Storage

During both short- and long- term storage from hours to months or longer, dies and molds must be protected from water and humidity. Unprotected steel can almost immediately begin to corrode, resulting in damaged tools that will require repolishing, regrinding, and/or repair of at least the surface of the tool. The net result is a cost in both labor and machine downtime. There are excellent rust protectants on the market that operate for different time periods. However, most of these anticorrosion treatments must be completely removed before using the molds. Some may require special cleaners including toxic solvents. Some operations dry off the tool and enclose it in an air evacuated container. For special protection, vacuum containers are used after the mold is properly dried.

12.3 Inspection

Inspection of all equipment and plastic material is required. Equipment manufacturers/material suppliers provide the details on inspection of BM machines, dies and/or molds, plastics, air supply, handling equipment, and all the items reviewed in this book. The following information concerns inspection/maintenance of the plasticator (Chapter 2).

12.3.1 Screw

Screws do not have the same outside continuous diameter. On receiving a machine or just a screw, it is a good idea to check specified dimensions (diameters vs. locations, channel depths, concentricity and straightness, hardness, spline/attach-ment, etc.) and make a proper visual inspection. This information should be recorded so that comparisons can be made following a later inspection. Special equipment should be used other than the usual methods (micrometer, etc.) to ensure that the inspections reproduced accurately. Such equipment are readily available and actually simplify inspections; it also takes less time, particularly for roller and hardness testing. Details on conducting an inspection and conducting important processing behaviors are available from screw and equipment suppliers such as Techware Designs, a subsidiary of Spirex Corp. with a computer software package called the "Extruder's Technician."

Many people hesitate to change the screw to see how a process is running—it seems expensive to take such action. Consequently, many people tend to run a poorly performing screw long after it should have been changed. Converting the cost of a screw into an equivalent volume of plastic or into a profit per day will determine payback. Assume a screw cost $30,000 to $40,000 each with output at 3,000 lb/hr (1,400 kg/h) of a $0.40/lb plastic. If you waste 1,000,000 lb (454,000 kg) of plastic, you justified the cost of the new screw. A new screw would represent 33 h of processing. It pays to replace the screw.

12.3.2 Barrel

To ensure proper performance, different parts of the barrel can be checked to meet tolerance requirements (usually set up by the manufacturer) and determine if any wear has occurred such as inside diameter, straightness and concentricity, and surface condition.

12.3.3 Purging

It has always been a “necessary evil” where at the end of a production run the plasticator has to be cleared of all its plastics in the barrel/screw to eliminate barrel/screw corrosion or contamination; it is also required when a change of material occurs. This action consumes substantial nonproductive amounts of plastics, labor, and machine time. It is sometimes necessary to run hundreds of pounds of plastic to clean out the last traces of a dark color before changing to a lighter one. If a choice exists, process the light color first. Sometimes there is no choice but to pull the screw for a thorough cleaning.

There are a few generally accepted rules on purging agents to use and how to purge. The following tips should be considered:

- try to follow less viscous with more viscous plastics;
- try to follow a lighter color with a darker color plastic;
- maintain equipment by using preventative maintenance;
- keep the materials handling equipment clean; and
- use an intermediate plastic to bridge the temperature gap such as that encountered in going from acetal to nylon.

Ground/cracked cast acrylic and PE-based (typically bottle grade HDPE) materials generally are the main purging agents. Others are used for certain plastics and machines. Cast acrylic, which does not melt completely, is suitable for virtually any plastic. PE-based compounds containing abrasive and release agents have been used to purge the “softer” plastics such as other olefins, styrenes, and certain PVCs. These types of purging agents function by mechanically pushing and scouring residue out of the extruders.

12.4 Troubleshooting

Troubleshooting is one the most important functions in any molding operation. Each problem has its own solution, and each different type of machine, or even each machine of the same type, has its peculiarities. Recognize that no two machines operate in exactly the same manner and also, that plastic materials do not melt as perfect blends, thus operating within permissible limits. The usual problem can be resolved with one or just a few changes during molding. Simplified guides to troubleshooting are given in Tables 12.1 and 12.2. More detailed processors' guides to troubleshooting are given in Tables 12.3, 12.4 and 12.5. Information on troubleshooting guides for granulators, conveying equipment, metering/proportional/feeding equipment, chillers and dehumidifying dryer performance is given in Table 12.6.

A simplified approach to troubleshooting is to develop a check list that reviews basic rules for problem solving.

- Have a plan.
- Molding conditions should be watched.
- Change one condition at a time.
- Allow sufficient time at each change.
- Keep an accurate log of each change.
- Check housekeeping, storage areas, granulators, etc.
- Narrow the areas into which the problem belongs; that is, machine, dies/mold, operating conditions, material, part design and management such as:
  - Change the material. If the problem remains the same, it probably is not the material.
  - Changing the type material may pinpoint the problem.
  - If the trouble occurs at random, it is probably a function of the machine, the temperature control system or the heating bands. Changing the die/mold from one press to the other permits a determination of whether it is in the machine, die/mold and/or resin.
  - If the problem appears, disappears, or changes with the operator, observe the difference in their actions.
  - If the problem appears in about the same position of a single cavity mold, it is probably a function of the flow pattern and system.

— If the problem appears in the same cavity or cavities of a multicavity mold, it is in the cavity.

— If machine operation malfunctions, check hydraulic or electric circuits. As an example, a pump makes oil flow, but there must be resistance to flow to generate pressure. Determine where fluid is going. If actuators fail to move or move slowly, the fluid must be bypassing them or going somewhere else. Trace it by disconnecting lines if necessary. No flow (or less than normal flow) in system will indicate pump or pump drive is at fault. Machine instruction manuals will provide details concerning correcting malfunctions.

— Set up procedure to "break in" new mold:

- Obtain samples and molding cycle information if mold is new to the shop but has been run before.
- Clean mold.
- Visually inspect mold. Obvious corrections, such as improving the polish or removing undercuts, should be done before the mold is put in service.
- Check actions of the mold: try cams, slides, locks, unscrewing devices and other devices on the bench.
- Install safety devices.
- Operate mold in press and move very slowly under low pressure.
- Open mold and inspect it.
- Dry cycle the mold without injecting material. Check knockout stroke, speeds, cushions and low pressure closing.
- After mold is at operating temperature, dry cycle again. Expansion or contraction of the mold parts may affect the fits.
- Take a shot using maximum mold lubrication under conditions least likely to cause mold damage. These are usually low material feed and pressure.
- Build up slowly to operating conditions. Run until stabilized, at least 1 to 2 hours.
- Record operating information.
- Take part to quality control for approval.
- Make required changes.
- Repeat process until approved by quality control and/or customer.

It is important to understand and to develop the proper approach in eliminating molding problems. The following review can assist the molder in finding the cause and probable remedy for problems occurring that result in unsatisfactory molded parts.

Basically, faulty or unacceptable molded parts usually result from problems in one or more of three areas of operation:

- Pre-molding: Material handling and storage
- Molding: Conditions in the processing cycle
- Post-molding: Parts handling and finishing operations

The solution to these problems is usually either quite obvious, or they are very specialized. This review discusses primarily the solution of problems encountered in the molding cycle. Causes of these faults can be grouped as follows:

- Machine
- Dies/molds,
- Operating conditions (time, temperature, pressure),
- Material,
- Part design, and
- Management.

12.5 Detailed Failure Analysis

While millions of parts are produced by the blow molding process each year, it is not a simple processing technique. As a result, there are many types of defects that can appear in molded parts. Although this chapter addresses these problems as they relate to lightweight, thinwalled products such as containers, many of the recommended solutions presented here also apply to heavier, thicker parts. This section provides information on some typical conditions for evaluation.

12.5.1 Rocker Bottoms and Oval Necks

Warpage in blow molded containers is most often seen as rocker bottoms and out-of-round necks. In general, the cause of the problem is the same as found with other molding processes—inadequate or unequal cooling prior to removing from the mold.

There are a number of possible causes to consider. The study should proceed from the simplest to cure to the most complex:

- Insufficient cooling water to the mold. See if an increased flow will stop the warpage.
- Too high a stock temperature. Lower the temperature by small increments to see if it will effect a cure without causing new problems, such as cold spots in the container wall.

Table 12.1 *Simplified Guide to Troubleshooting Polyethylene Bottles*

Difficulty	Suggested solutions
Low gloss	• Increase material temperature • Increase mold temperature • Improve mold venting • Increase air pressure • Increase air blowing rate
Excessive cycle	• Decrease material temperature • Decrease mold temperature • Decrease part wall thickness • Improve mold coring • Increase material density • Increase air pressure • Increase melt index (with decreased stock temperature)
Die lines	• Clean the die • Smooth and polish the die • Increase mold temperature • Increase air pressure • Increase air blowing rate • Increase purge time when changing materials • Improve die streamlining
Low bottle weight	• Increase die to mandrel clearance • Decrease material temperature • Increase extrusion speed • Decrease extrusion die temperature
Surface roughness	• Increase stock temperature • Decrease extrusion speed • Increase die temperature • Improve die streamlining
Weak pinch	• Adjust material temperature • Increase pinch blade land width • Decrease rate of mold closing
Parison curl	• Adjust and center die-parison curls toward thin area • Improve die and head heat uniformity
Wall thickness non-uniform (a) Vertically (sag)	• Decrease material temperature • Increase extrusion speed or accumulator ram pressure • Lower melt index • Increase material density
(b) Circumferentially	• Decrease blowup ratio • Improve head and die heat uniformity • Adjust and center the die
Excessive thinning at parting line	• Decrease stock temperature • Increase mold temperature • Increase air pressure • Increase air blowing rate • Improve mold venting • Improve mold temperature

(continued)

Table 12.1 *(continued)*

Difficulty	Suggested solutions
Excessive shrinkage	• Decrease wall thickness, or make more uniform • Decrease mold temperature • Increase air pressure • Decrease material temperature • Decrease density • Improve mold coring
Excessive parison swell	• Increase material temperature • Increase die temperature • Decrease extrusion speed • Increase melt index • Reduce die size
Doughnut formation	• Wait for mandrel temperature to reach die temperature • Clean the lower surface of the die
Warped top and bottom	• Slow the cycle • Decrease mold temperature • Decrease stock temperature • Decrease part weight • Improve mold coring
Variable bottle weight	• Finer extruder screen pack • Increase screw cooling • Raise rear extruder heats • Decrease extrusion rate

Table 12.2 *Simplified Guide to Troubleshooting PE Dairy Gallon Based on Parison, Structural, and Appearance Problems [3]*

(a)

Parison problems	Causes
Die lines of streaking	• Foreign matter lodged in die gap • Burned-out heater band
Parison pleating or curtaining	• High melt temperature • High weight-swell resin • Misaligned die • Short land between mandrel and bushing
Parison curl	• Cold bushing or mandrel • Misaligned die • Foreign matter on die
Parison stringing	• High melt temperature

(b)

Structural problems	Causes
Thin, weak welds	• High melt temperature • High mold temperature • Short or sharp pinch-off land
Cutting at pinch-off	• Low melt temperature • Fast mold close
Blowouts	• High melt temperature • High pinch-off temperature • Insufficient cooling • Parison stringing • Low clamp pressure • Short parison • High blow-up ratio • Low diameter-swell resin
Webbing in neck or upper handle	• Parison pleating
Warpage	• High melt temperature • Short cooling cycle • Blocked cooling channels • Incorrect parison programming
Tearing of flash	• Long pinch-off land • Low clamp pressure

(c)

Appearance problems	Causes
Foreign matter in container wall	• Contamination resin feed • Poor melt path streamlining • Mismatched head sections • Poor fit between die/adaptor/extruder

(continued)

Table 12.2 *(continued)*

Appearance problems	Causes
Bubbles in container wall	• Low melt temperature • Moisture in resin • Mismatched head sections • Low back pressure
Sharkskin container surface	• Wrong parison drop rate
Cold spots in container wall	• Insufficient working of melt • Low back pressure • Low melt index resin mixed into feed
Poor mold surface reproduction	• Low melt temperature • Low blow pressure • Smooth mold surface • Inadequate venting at mold parting line • Air leak around blow pin
Pitting of container surface	• Moisture in resin • Moisture on mold surface

Table 12.3 Troubleshooting Guide for Extrusion Blow Molding. (a) Part Defects

Cause	Possible remedy
Blowouts	
• Pinch-off too sharp which cuts parison and causes blow-out	• Increase width of pinch-off land
• Pinch-off too wide	• Decrease width of pinch-off land
• Clamp pressure too low	• Adjust clamp pressure
• Pinch-off too hot	• Check cooling water-flow to mold • Check stock temperatures • Check cooling channels for blockage • Check cycle time • Check part design
• Blow pressure too high	• Gradually reduce pressure to avoid stretching the parison at a rate too high
• Parison too short	• Ensure proper parison length
• Blow-up ratio too high	• Check head tooling (change to a larger diameter die may be required)
• Resin swell too low	• Use higher swell resin to ensure proper parison size and to prevent missed handle, accompanied by blow-out
• Moisture in resin	• Check for moisture in resin
• Fines in hopper	• Eliminate hopper fines
• Stock temperature too low	• Increase stock temperature
• Extruder feed zone temperature too low	• Increase extruder feed zone temperature
• Blowing air pressure too high	• Adjust blowing air pressure
Bubbles	
• Moisture in melt (because of condensation appearing on cold resin exposed to warm atmosphere)	• Properly dry hygroscopic resin • Permit resin to warm and evaporate before entering extruder
• Moisture in melt caused by too much water cooling on the throat, producing condensation in the barrel and moisture in melt when running at slow screw speeds	• Adjust throat cooling
• Air inclusion mismatched head sections	• Align head sections and/or increase back pressure
• Contamination	• Check and eliminate contamination source
Cap leakage	
• Worn or damaged blow pin or shear steel	• Replace blow pin and/or shear steel
• Blow pin alignment	• Align blow pin
• Mold alignment (mismatch)	• Replace guide pins and bushings
• Wrong blow pin and shear steel dimensions	• Replace with correct blow pin and shear steel

(continued)

Table 12.3 (continued)

Cause	Possible remedy
• Wrong neck plate dimensions	• Replace neck plates
• Wrong cap dimensions	• Check and ensure correct size caps are being used
• Pleating (vertical web inside neck)	• Adjust preblow system
• Capper tongue too high/too low	• Adjust capper tongue
Cold spots or marbleizing	
• Insufficient back pressure	• (Reciprocating unit) Adjust back pressure gauge • (Continuous extrusion unit) Change screen pack to finer mesh
• Mixture of resin having a lower melt index	• Avoid improper resin mixing
• Melt too cold	• Increase melt temperature
• Cycle too fast	• Increase blow time
Container does not blow or incomplete blowing	
• Blow air restricted	• Check air system
• Pinch-off area to sharp	• Provide broader pinch-off area
Dirty or streaked pieces	
• Contamination	• Inspect storage and heating areas of cleanliness • Purge die and clean hopper
Doughnut forming between top of mold die face during blowing	
• Back pressure on screw set too high	• Slowly decrease screw back pressure
• Die tip temperature too high	• Reduce die tip temperature
• Melt temperature too high	• Slowly decrease melt temperature throughout system
Excessive flash	
• Parison diameter too large	• Change tooling
• Flash pockets too shallow	• Consult mold maker
• Improper mold closure	• Increase locking and check for mold mismatch
Excessive shrinkage	
• Blown part ejected when too hot	• Reduce mold temperature • Reduce melt temperature • Increase blow time
• Inadequate mold cooling	• Clean mold cooling channels
Hole or slit in parting line	
• Melt temperature too low	• Increase stock temperature

(continued)

Table 12.3 (*continued*)

Cause	Possible remedy
• Mold closing too rapidly	• Cushion model closing speed
• Maladjusted trimming equipment	• Adjust trimming equipment
• Excessive pinch-off land length	• Reduce pinch-off land length
Improper part weight	
• Mandrel in up or down position	• Adjust mandrel to increase or decrease part weight
• Continuous extrusion, rpm is too high or too low	• Increase or reduce extruder rpm; note: using a longer die/mandrel also may increase part weight
• Material or tooling too hot	• Reduce stock temperature • Reduce head temperature
• Rupturing blow air	• Check blow pin and shear steel
• Stripper not properly adjusted	• Adjust stripper plate
• Blow pins set too low or too high	• Raise or lower blow pins
Irregular striations and other marks	
• Tube chafing on die	• Align parison to die
Lost handles	
• Misaligned molds	• Realign molds to catch handles
• Parison hooking	• Realign die adjusting ring
• Parison melt stringing	• Reduce melt temperature
• Low flare	• Lower stock temperature • Increase shot pressure
Low environmental stress crack resistance	
• Excessive percentage of regrind	• Reduce regrind rate
• Melt (stock) temperature too low	• Increase melt (stock) temperature
• Melt (stock) temperature too high	• Decrease melt (stock) temperature
• Mismatched molds	• Realign molds
• Worn or damaged trimmer nests	• Replace trimmer nests
• Incorrect resin for application	• Use proper resin
Low impact strength	
• Excessive percentage of regrind	• Reduce regrind rate
• Mismatched molds	• Realign molds
• Worn or damaged trimmer nest	• Replace trimmer nests
Low top load strength	
• Uneven wall distribution	• Reduce parison drop time • Reprogram parison programmer
• Low part weight	• Increase part weight
• Incorrect resin for application	• Use proper resin

(*continued*)

Table 12.3 (continued)

Cause	Possible remedy
Nonuniform wall section circumference	
• Inaccurate die centering	• Adjust die to provide uniform distribution of wall section
• Unsuitable blow head	• Check blow head design
• Head temperature not uniform	• Stagger heater band gaps on head
• Loose die pin	• Tighten die pin
Part too hot	
• Mold temperature too high	• Decrease mold temperature
• Cycle time too short	• Increase blow time to extend cycle time
• Clogged coolant lines/channels	• Clean lines/channels
Part volume too low	
• Part weight too high	• Reduce part weight
• Mold too hot	• Reduce mold temperature
• Cycle time too short	• Increase cycle time
• Mold volume incorrect	• Resize mold
• Volume-reducing inserts used	• Remove inserts for packed parts only
• Blow air pressure too low	• Increase blow air pressure
• Part storage areas too hot	• Reduce storage area temperature or dwell time
• Melt temperature too high	• Reduce melt temperature
Part volume too high	
• Part weight too low	• Increase part weight
• Mold too cold	• Increase mold temperature
• Cycle too long	• Reduce cycle time
• Wrong mold volume	• Resize mold
• Volume-reduction inserts not used	• Install inserts
• Blow air pressure too high	• Reduce blow air pressure
• Part storage area too cold	• Increase storage area temperature or dwell time
Poor definition of detail	
• Blow air pressure too low	• Increase blow air pressure
• Poor mold venting	• Consult mold maker
Poor part surface (roughness, pits, orange peel, etc.)	
• Poor mold surface	• Check mold surface. (It should have a fine matte finish to permit air to vent quickly and parison to conform to mold surface while still hot enough to do so.)
• Inadequate vents which preclude rapid escape of air	• Check vents along mold parting line; minor reworking may be necessary

(continued)

Table 12.3 (*continued*)

Cause	Possible remedy
• Low blow pressure or blow rate	• Increase blow pressure • Ensure that blow pin is large enough to handle required amount of air to fully and rapidly blow part • Check for restrictions or partial plugging of air lines
• Air leak around blow pin	• Check for leakage which prevents sufficient pressure to hold part against the mold for conformation and rapid cooling • Check blow pin and/or shear ring and replace if necessary
• Low stock temperature	• Gradually increase stock temperature but avoid overly high stock temperature
• Condensed water in mold	• Check for low mold temperature, combined with environmental humidity. (If mold temperature cannot be raised, investigate plant humidity control.) • Reduce cooling water flow
• Too much mold lubricant	• Use less lubricant
• Melt temperatures too low	• Increase melt tempreature • (Extruder) if surface roughness appears which may be indicative of melt instability, reduce extrusion pressure until problem disappears • (Reciprocating or accumulator) drop out of melt instability region by further decreasing extrusion rates. (Note: Generally, an unsatisfactory solution because of productivity decrease.) • Increase extrusion rates (decrease drop time) • Increase stock temperature or change to higher melt index resin
Poor weld or pinch-off	
• Thinning of weld	• Check stock and mold temperatures
• Tearing of flash during trimming	• Check pinch-off and length (Recommended land lengths of 0.010 in. to 0.015 in. are common for polyethylene.)
• Cutting at pinch-off	• After checking weld and pinch-off, check for hole or slit along weld which may indicate that stock temperature is too low, thus causing parison to tear as mold closes

(continued)

Table 12.3 *(continued)*

Cause	Possible remedy
• Seams not uniform because of improper line up of mold and parison	• Ensure that blow head is located exactly vertical and improve adjustment of mold to die • Meter tube faster • Adjust mold closure time • Increase stock temperature
• Parison temperature too high	• Decrease parison temperature
• Pinch-off area too hot	• Decrease mold temperature at pinch-off area
Short molding	
• Insufficient hold time on blowing pressure	• Increase blowing time
Sticking in mold	
• Mold temperature too high or premature opening of mold	• Improve coring and heat transfer and slightly reduce mold temperature • Increase cooling time
• Parison hooking	• Adjust parison drop
• Swing arm out of adjustment	• Adjust swing arm
• Molds not opening far enough	• Adjust mold open stop position
• Tail too short	• Increase screw rpm • Increase blow time
• Mold too hot	• Check mold cooling system
• Insufficient exhaust time	• Increase exhaust time
Tail lengths not uniform	
• Manifold chokes out of adjustment	• Adjust manifold chokes making short tails longer first
• Part weights not uniform	• Adjust part weight to ±?g of target weight
• Non uniform head heats	• Check heaters
• Head chokes out of ajustment	• Reset heat chokes
Thin parting line	
• Mold not fully closed	• Clean mold faces • Adjust clamp pressure • Set downward travel of blowpins • Decrease blow air pressure
Tail flash not separating from part	
• Parison too short	• Lengthen parison
• Flash pocket too deep	• Contact mold maker
Top flash not separating from part	
• Dull cutting ring	• Sharpen or replace cutting ring
• Insufficient contact between cutting ring and striker plate	• Adjust down position of blow pin • Increase downward force of blow pin

(continued)

Table 12.3 *(continued)*

Cause	Possible remedy
Unmelted pellets in blown part	
• Melt not homogeneous (insufficient mixing in screw)	• Increase back pressure by choke adjustment • Use finer screw pack
• Melt temperature too low	• Increase heater setting
Vertical lines	
• Foreign matter or degraded resin in head	• Clean die gap • Open to largest die gap and purge • Remove and clean tooling
• Damaged tooling	• Repair or replace tooling as needed
Wall section too thin at die orifice end	
• Excessive draw-down because extrusion rate is too low, melt temperature is too high, or expansion is too late	• Increase extrusion rate • Decrease melt temperature, close mold and expand more rapidly
Wall section too thin in general	
• Improper parison size	• Enlarge parison • Reduce melt temperature • Meter faster
Warpage (rocker bottoms and oval necks)	
• Insufficient mold cooling	• Determine whether an increased water flow will stop warpage • Reduce mold temperature • Increase cycle time to provide longer cooling
• Stock temperature too high	• Drop temperature by small increments to see if it will affect a cure without causing new problems such as cold spots in container wall
• Blocked cooling channels	• Check throughput and clean channels if throughput is significantly less than when mold was new
• Cycle time too short	• Adjust cycle time as needed
• Poorly designed part	• Check material distribution in part for unnecessarily thick or thin sections
Webbing	
• Incorrect mold position relative to parison	• Change position of mold relative to parison
• Incorrect parison diameter	• Change tooling
• Internal parison walls touching	• While mold is closing, add prepinch bars to mold

(continued)

Table 12.3 *Troubleshooting Guide for Extrusion Blow Molding. (b) Parison Defects*

Cause	Possible remedy
Blowout of parison	
• Pinch-off too sharp or hot	• Increase land width of pinch-off • Increase cooling, preferably in pinch-off area
• Blow pressure too high, parison inflation rate too rapid	• Reduce blow pressure and/or inflation rate
• Insufficient mold clamping	• Increase clamp pressure
• Parison too short and not fully captured in pinch-off	• Increase shot • Close mold before full snapback occurs
• Blow-up ratio too high	• Check die head size
Bubbles	
• Small bubbles may be caused by moisture in resin	• Follow correct drying procedures for hygroscopic resins • Make provisions for keeping pellets hot in hopper to prevent pick-up of moisture
• Large bubbles which sometimes burst while leaving die orifice may be caused by trapped air	• Increase back pressure • Raise screw speed if possible
Curtaining and webbing	
• Poor tooling design	• Check land length between bushing and mandrel
• Die misalignment	• Check bushing and mandrel alignment to ensure that different flow rates in different areas are not occurring
• High stock temperature	• Avoid high stock temperatures which will yield a parison that will collapse before it can be blown
• Resin mismatch	• Avoid choosing resin with a weight swell too high
• Parison wall too thin	• Increase die gap
Die lines or streaking in parison	
• Contaminated or degraded resin	• Check for dirt, dust and lint in unprocessed resin • If using regrind, check for foreign matter
• Burned-out heater band (Evidence will appear as rippled streaking in parison causing cold spots)	• Replace non-functioning band
• Foreign matter lodged between die and mandrel	• Increase die gap briefly to purge particles, then reset to original gap

(continued)

Table 12.3 *(continued)*

Cause	Possible remedy
• Carbon or degraded resin in head	• Clean die and mandrel with copper, brass, or material softer than machine parts to prevent marring and scratching
• Melt overheating	• Check blow head, accumulator, and cylinder for dead areas and pockets • Check mandrel for slight gaps with contact surfaces causing hang-up • Check for cylinder hot spots and overheating and reduce temperature or eliminate hot spot
• Damaged tooling	• Repair or replace tooling as needed
Dull, rough, or striated melt surface	
• Insufficient melt or stock temperature	• Increase die temperature and stock temperature to produce melt which is clear and glossy when leaving die orifice
• Incorrect extrusion rate	• Adjust rate
• Dirty die/head tooling	• Clean tooling
• Melt fracture	• Increase melt temperature • Decrease machine rpm • Increase die gap
Excessive tubular melt stretching after emerging from die orifice	
• Melt temperature too high	• Reduce die temperature and/or stock temperature
• Melt extrusion too slow	• Increase extrusion rate
Hole in parison	
• Contamination	• Check for degraded material • Check for hang-up in tooling, head, or screw • Check for foreign material in resin
• Trapped air	• Permit extruder to run for a few minutes. Problem could be caused by letting hopper run out of resin before refilling
• Moisture in resin	• Properly dry hygroscopic resin before using
Horizontal rings	
• Programming changes too drastic	• Decrease flow to programming cylinder • Decrease weight change between steps
Melt fracture (sharkskin)	
• Melt flow is unstable	• Increase or decrease stock temperature • Increase or decrease drop time

(continued)

Table 12.3 (*continued*)

Cause	Possible remedy
	• Check to ensure that head choke and manifold chokes are open as much as possible
	• Clean tooling
Nonuniform parison	
• Improperly designed blow head	• Check blow head design
• Nonuniform heating in blow head	• Check heat distribution in blow head
• Mandrel improperly centered in die ring	• Center die
• Melt flow too rapid	• Reduce extrusion rate
• Melt temperature too high	• Reduce stock temperature
• Die land length too short	• Use longer land die mandrel
Parison curl and doughnutting	
• Cold mandrel or bushing	• Permit sufficient warm up time before attempting production
	• Ensure that die bushing heater is operative, if so equipped
• Misaligned die and mandrel	• Check positioning of die and mandrel. Die bushing may need to be machined to bring mandrel to a flush or lower position with die face
• Foreign matter or degraded resin in die bushing	• Clean die
• Damaged tooling	• Check tooling for damage
• Improper knife cut	• Increase distance between knife and die face
	• Check knife edge for sharpness and cleanliness
Parison deflects without becoming punctured	
• Needle is blunt	• Sharpen needle or increase penetration speed
Parison hooking	
• Nonuniform parison walls	• Center adjusting ring around mandrel to correct
• Die mandrel channel dirty	• Clean channel
• Defective head/die heaters	• Replace heater
• Loose die pin	• Tighten die pin
• Head temperature not uniform	• Stagger heat band gaps on head
• Die gap out of adjustment	• Adjust die around mandrel
• Air flowing on parison	• Shield parisons from moving air
• Warped die or mandrel	• Replace damaged component

(*continued*)

Table 12.3 *(continued)*

Cause	Possible remedy
Parison sag	
• Melt temperature too high	• Decrease melt temperature
• Resin melt index too high	• Use lower melt index resin
• Extrusion rate too slow	• Increase extrusion rate
• Parison drop time too long	• Decrease parison drop/hang time
• Parison weight too heavy	• Reduce parison weight
• Mold closing too slowly	• Increase low pressure close • Lubricate tie bar bushing
Parison stringing	
• Stock temperature too high	• Gradually lower stock temperature
• High fill pressure	• Reduce fill pressure until weeping stops
Pinch-off too thick	
• Pinch-off edges too wide or inadequate mold closure	• Sharpen pinch-off edges • Increase mold closing pressure • Check mold parting line for poor match
Rolling	
• Melt clings to die or mandrel and flows to coolest side	• Adjust die temperature
Smoking	
• Heat controller malfunction	• Check heat controllers
• Contamination in resin	• Check for contamination
• Stock temperature too high	• Decrease stock resin temperature
• Extrusion pressure rate too high	• Decrease extrusion pressure rate
• Ratio of regrind to virgin material too high	• Decrease regrind level

Table 12.3 *Troubleshooting Guide for Extrusion Blow Molding (c) Complete Process Summary*

Problem	Condition	Cause	Solution
Heating interrupted	Temperature falling	(a) Control action to avoid overshoot, time-proportioning or interaction with adjacent zones	Observe if heating resumed in 30 s, otherwise check (b)
		(b) Fail-safe control, indicating broken thermocouple or connection	Check thermocouple continuity
Temperature continues to rise	Temperature above upper control limit	(a) Thermocouple slipped out or not in contact with heater/extruder	Replace, ensuring good thermal contact
		(b) Thermocouple incorrectly connected	Check connections and temperature indication, e.g., in hot water or body heat
Output suddenly falls	Steady screw speed	(a) Feed hopper empty	See barrel and feed unit section in this chapter. Investigate blocked feed line
		(b) Bridging in hopper or into screw	Stop screw and clear with non-metallic rod, e.g. strip of product. Check adequate feed pocket cooling. Check air entrapment esp. with powders
		(c) Polymer melting in feed section	Reduce first barrel zone temperature Increase feed pocket cooling Cool feed end (only) of screw
		(d) Surging	
Polymer leakage from joints or fittings	Excessive pressure	(a) Blocked die or screen	Immediately reduce screw speed.

(continued)

Table 12.3 *(continued)*

Problem	Condition	Cause	Solution
			Check die and adaptor temperature Shut down and clear
		(b) Failed die heater or control	Immediately reduce screw speed Check thermocouple Shut down and replace heater
		(c) Incompletely melted polymer, esp. at start-up or after feed interruption	Stop screw Wait to heat up and restart *slowly*
Screw stops or speed drops	Excessive torque	(a) Low barrel temperature	Reduce speed setting and check heaters
		(b) Excessive pressure	As for 'Polymer leakage' above
		(c) Excess feed, e.g. of strips	Stop or reduce speed setting Increase temperature of first barrel zones
		(d) Excessive speed	Reduce speed and/or increase barrel temperatures
Lumpy or unmelted polymer at die	Incomplete melting, esp. of powders	(a) Insufficient shearing/heating in barrel	
		(b) Excessive screw speed	Reduce screw speed *or* improve melting
		(c) Feed contamination, e.g. with hard-grade polymer	Reduce screw speed Check feed material if necessary, increase barrel temperatures

(continued)

Table 12.3 (*continued*)

Problem	Condition	Cause	Solution
Foaming or bubbles at die (distinguish from later contraction bubbles due to rapid cooling), surface blisters	Steady running or start-up	(a) Moist polymer	Stop feed and restart with dried polymer
		(b) Excess temperature causing decomposition	Stop or reduce screw speed
		(c) Air entrapment in feed	Raise first barrel zone temperature (to reduce unmolten length)
			Fit vent at rear of feed opening

Table 12.4 Troubleshooting Guide for Injection Blow Molding

Cause	Possible remedy
Cocked necks	• Movable bottom plug stuck in • Reset stripper plate • Reduce inject portion of cycle • Increase blow air time/pressure
Color streaks-flow lines	• Raise back pressure • Increase injection pressure • Increase melt temperature • Open nozzle orifice • Change color mix • Change color batch • Add mixing pin to screw • Reduce injection speed • Increase parison mold temperature • Dry material
Contamination (oil/grease on part)	• Wash parison and blow molds with solvent (especially in neck rings) • Wash core rods • Clean air filter • Replace O-rings between molds
Cracked necks	• Increase melt temperature • Increase neck temperature in parison mold • Reduce core rod cooling • Increase neck temperature in blow mold • Reduce retainer grooves in core rods • Movable bottom plugs stuck in (regrind materials only) • Increase injection speed • Balance nozzles for even fill • Check core rod alignment • Check operation of mold temperature controller • Check stripper location and speed (especially in stryrene) • Open nozzle orifice • O-rings in face blocks
Dimensional problems	
• H – height	• Increase by moving parison and/or blow mold out (add shim) • Reduce by moving parison and/or blow mold in (remove shim)
• S – neck finish	• Increase by moving parison mold out (add shim) • Reduce by moving parison and/or blow mold in (remove shim)
• T – average of two dimensions	• Lower parison mold neck temperature

(continued)

Table 12.4 (continued)

Cause	Possible remedy
• To raise T	• Increase injection time • Increase stabilize time • Lower blow mold neck temperature • Increase core rod cooling (internal)
• To lower T	• Increase parison mold neck temperature • Increase blow mold neck temperature • Reduce core rod cooling • Hot line blow mold neck (Note: in some cases the opposite will happen when above is done when the T is being blown out in the blow mold)
• E – taken across major and minor axes	• Refer to T dimension
Distorted shoulder	• Increase blow air pressure • Shorten cycle • Wash out vents in blow mold • Increase parison mold temperature • Clean or replace (plugged) core rods • Increase blow air time • Reset or replace stripper
Engraving	• Increase temperature of body mold temperature controller • Increase melt temperature • Sandblast molds • Adjust engraving depth and width • Adjust vents • Increase blow air pressure • Clean out engraving • Increase blow delay • Balance nozzles for even fill • Clean, check, and set all gaps evenly in core rod valve • Check mold temperature controller operation • Increase venting on blow mold • Adjust parison mold temperature • Adjust blow mold temperature • Increase core pin cooling
Heavy in center	• Increase temperature of gate mold temperature controller (parison mold) • Move nozzles in • Decrease core rod cooling air
Hot spots	• Increase core rod cooling • Reduce injection pressure • Reduce melt temperature

(continued)

Table 12.4 (*continued*)

Cause	Possible remedy
	• Lower parison mold temperature in affected area • Reset stripper for localized external cooling of core rod • Check mold temperature controller operation
Inconsistent shot size	• Increase back pressure • Check hydraulic injection pressures for variations • Increase screw recovery time • Balance nozzles • Check for loose or bad thermocouples • Check for broken element in mixing nozzle • Check mold temperature controller operation • Adjust screw rpm
Nicks	• Replace damaged blow mold • Clean plastic or dirt out of blow mold cavity • Repair or replace damaged parison mold • Replace or repair damaged core rod • Remove strings from nozzles • Remove burrs from parison and blow molds • Repair or replace damaged parison and blow molds • Wash out vents in blow mold • Sandblast blow mold • Reset mold in die set
Nozzle freeze-off	• Remove contaminated material from nozzle • Raise temperature of gate mold temperature controller • Increase manifold temperature • Increase melt temperature • Reduce cycle • Open nozzle orifice • Check manifold heaters, fuses and wiring
Ovality of T & E dimensions (uneven shrinking)	• Increase temperature in shoulder/body area of parison mold • Reduce temperature in neck of parison mold • Lower blow mold neck temperature • Increase melt temperature • Check operation of temperature control unit
Push-up depth	• Increase blow time/pressure • Adjust mold/bottom plug cooling • Lower melt temperature • Clean and check core rod valve; set all gaps evenly • Check movable plug if used • Increase core rod cooling

(*continued*)

Table 12.4 (continued)

Cause	Possible remedy
	• Decrease air pressure • Adjust nozzle • Increase tip cooling
Saddle finish (usually apparent with oval T and E – try to correct oval T & E)	• Parison not packed up tight enough; add inject or screw time and/or pressure • Increase retainer grooves in core rods • Reduce parison mold neck temperature • Increase or decrease cushion • Increase holding pressure • Increase cooling time
Short shots	• Clean nozzle • Open nozzle orifice • Increase high injection pressure • Increase packing pressure • Increase injection time • Increase melt temperature • Raise all parison mold temperatures • Increase back pressure • Increase screw recovery stroke • Lengthen cycle time • Check and clean non-return valve on end of extruder screw • Adjust screw rpm • Increase screw speed • Increase cushion • Clean out hopper and throat
Sticking of parison to core rods (core rod too hot)	• Decrease cushion • Reduce melt temperature • Pack parison harder (increase injection pressure/time) • Reduce screw speed • Add stearate (release agent) • Increase back pressure • Adjust core rod temperature • Check mold temperature controller operation • Check for folds in bottom • Increase internal and external air cooling
Stripping difficulties	• Increase stripping pressure • Check core pins for burrs • Add lubricant to polymer • Adjust stripper bar alignment • V groove too deep

(continued)

Table 12.4 *(continued)*

Cause	Possible remedy
Sunken panels	• Increase blow air time • Reduce blow mold temperature • Shorten inject/transfer portion of cycle • Reduce core rod cooling • Add sink correction to blow mold • Check mold temperature controller operation
Thin parts	• Lower temperature of gate mold temperature controller in parison mold • Check core rod lock-off in parison mold • Check nozzle seats • Reset nozzles • Check parison mold part line • Lower melt temperature • Move nozzles out • Add injection and/or screw time • Replace nozzles • Check to see that mold temperature controller is functioning properly
White or black marks on neck finish caused by gas burning	• Lower melt temperature • Reduce injection speed • Lower temperature in neck of parison mold • Vent (relief) neck ring of parison mold • Open nozzle orifice • Check mold temperature controller operation • Reduce ram speed

Table 12.5 Troubleshooting Guide for Stretched Injection Blow Molding

Cause	Possible remedy
Air bubbles in the preform	
• Air entrapment due to too much decompression in plastifier	• Check dryer settings • Increase back pressure slightly
Bands of thick and thin sections in part wall	
• Improper settings on heat zones (zones colder than others) • Not enough time for equilibration of preform before blowing	• Ensure uniform temperature from capping ring to tip • Equilibration time • Change heat zones
Bands of vertical strips on the preform	
• Too much heat at specific area on preform • Improper rotation for vertical bands	• Reduce heat • Check rotation speed
Blemishes on part	
• Dirt in molds • Water droplets forming in molds or sweating causing condensation	• Blow molds should be cleaned • Increase mold temperature slightly to alleviate condensation
Cloudiness	
• Melt temperature too low • Moisture in injection molds	• Increase melt temperature slightly • Check for condensation on cores or in cavities • Increase water temperature in injection molds
Drag marks on preform	
• Injection mold damaged or scratched	• Polish and possibly rechrome cores and cavities
Fish eyes or zippers	
• Scratch marks on preform surface (normally caused by preforms contacting each other after being ejected from injection mold while still hot)	• Minimize contact of preforms after injection and prior to cooling
Folds in neck area	
• Center rod stretching preform too early	• Synchronize center rod stretch with air delay timers to get blow air to enter at correct time
Heavy material in bottom of part	
• Improper preform design	• Redesign preform

(continued)

Table 12.5 (continued)

Cause	Possible remedy
• Improper cooling in molds causing heavy amount of materials to shrink back	• Improve mold cooling
Knit lines appearing in preform	
• Melt not being injected fast enough or hot enough	• Increase temperature • Raise injection pressure
Long gates	
• Valve gates in mold operating improperly • Incorrect temperature in hot runner system	• Clean mold • Check thermocouples
Mismatch lines on the preform	
• Mold misalignment	• Check cores and cavities for alignment
Off-centered gates	
• Could be caused by not using center rods • Poor concentricity in the preform (preform concentricity should be held to 0.005 in. max.)	• Use center rods • Check injection mold for concentricity of core rod and cavity • Reduce injection speed and injection pressure
Pearlescense (haze in container)	
• Perform stretching too fast for heat in preform	• Increase heat in area showing pearlescense going back to preform
Preform drooling	
• Valve gates are too warm • Not enough packing pressure or time	• Decrease temperature to valve gates • Increase hold time
Radial rings on preform	
• Moisture condensation on core rods	• Increase water temperature in molds
Scratches on part	
• Possible drag marks on preforms from cavity or core of preform	• Polish core and cavity of preform mold
Soft necks or deformed capping rings on finished container	
• Too much heat in top area of preform	• Reduce heating in affected zone. Heat shield may be added to shield capping rings

(continued)

Table 12.5 *(continued)*

Cause	Possible remedy
Undersized parts	
• Not enough high pressure air blow time. (Container not being blown to side wall and held under high pressure to freeze material and set outline of mold.)	• Check mold cooling • Check blow pressures and time • Check vents on mold
Yellowing of preform (indicating oxidization through excessive heating during drying.)	
• Check drying temperature and time	• Adjust drying time as required

Table 12.6 *Troubleshooting Guide for Auxiliary Equipment*

(a) Dryers

Problem	Possible causes	Suggested solutions
Process air temperature too low	• Incorrect temperature selected on control panel	• Dial in correct temperature
	• Controller malfunction	• Check electrical connections, replace controller if necessary
	• Process heating elements	• Check electrical connections, replace elements if necessary
	• Hose connections at wrong location	• Check to make sure delivery hose is entering bottom of hopper
	• Supply voltage different from dryer voltage	• Check supply voltage against name-plate voltage
Process air temperature too high	• Thermocouple not located properly at inlet of hopper	• Secure thermocouple probe into coupling at inlet of dryer
Material not drying	• Process and/or auxiliary filter(s) clogged	• Clean filter
	• Incorrect blower rotation	• Check rotation
	• Regeneration heating elements inoperative	• Check electrical connections; replace elements, if necessary
	• Desiccant assembly not rotating	• Check motor electrical connections. Replace motor, if necessary. Check drive assembly for slippage; adjust
	• Material residence time in hopper too short	• Drying hopper too small for material being processed; replace with larger model
	• Moist room air leaking into dry process air	• Check all hose connections and tighten if required. Check hoses for cracks; replace as necessary. Check filter covers for tightness; secure.
	• Desiccant contamined	• Replace desiccant cartridge

(continued)

Table 12.6 (continued)

Problem	Possible causes	Suggested solutions
(b) Granulators		
Stalled machine	1. Overloading	• Feed material slower
	2. Worn, damaged or improperly set knives	• Readjust or replace as required
	3. Screen and/or blower chute blockage	• Check to see if line is clogged and check rotation of blower
	4. Drive-belt slippage	• Check tensioning
	5. Loss of power	• Check power supply, electrical hook-up and safety switches
	6. Motor running in reverse	• Check directions of rotation and rewire per diagram, if necessary
Material overheating	• See items 1, 2, 3, 6	• Same as above
	• Screen too small	• Change to screen with larger diameter holes
Too many fines in material	• Worn, damaged or improperly set knives	• Readjust or replace as required
Bearing overheating	• Failure to lubricate properly	• Check frequency of lubrication and type of grease used
	• Too much tension on drive belts	• Adjust tensioning
Excessive knife wear	• Highly abrasive material	• Change to higher alloy knives
Knife breakage	• Tramp metal in scrap	• Check scrap material for foreign matter
	• Loose or stretched bolts	• Check bolts and retorque per specification sheet
	• Uneven knife seats	• Inspect and clean seat surfaces
Screen breakage	• Improperly seated	• Check that screen is fitted correctly
Motor won't start	• Power supply failure	• Check main power supply and fuses
	• Overheated motor	• Allow motor to cool, reset starter overloads
	• Starter failure	• Check for burned-out contacts, replace if necessary
	• Inoperative safety switches	• Check that hand guard is secure and all contact points closed. Replace if necessary

(continued)

Table 12.6 *(continued)*

Problem	Possible causes	Suggested solutions
(c) Conveying Equipment		
Spiral conveyors		
• Excessive wear	• System may have kinks, sags, sharp or compound bends or contact with sharp surfaces	• Re-position drive-end tube supports or feed hopped
	• System may exhibit excessive vibration	• Inspect for proper feeding into inlet (consult supplier if material is bridging)
	• Are abrasive materials being conveyed?	• Consult supplier
• Excessive spiral wear or breakage	• There may not be sufficient clearance at inlet end of conveyor for spiral to expand taking into consideration length and inclination of conveyor, and bulk density of material	• Shorten spiral
• Low delivery rate	• Material may not be flowing properly into inlet	• Rotate outer tube so that inlet opening is aligned with hopper feed • Install bin vibrator and/or agitator • Adjust spiral length as per manual instructions • Reverse spiral direction if incorrect • Seal openings in system if material is hygroscopic and system is installed in high-humidity environment
Belt conveyors		
• Slipping clutch	• Oil contamination or excessive lubrication	• Clean system, adjust clutch properly
• Improper belt tracking	• Belt not properly tightened when changed or installed • System damaged in delivery	• Check alignment and tighten

(continued)

Table 12.6 *(continued)*

Problem	Possible causes	Suggested solutions
• Electrical malfunctions	• Motor may be exposed to excessive heat under molding machine	• Install overload-protection temperature limit switches
• Parts jam up or fall off	• Transition points not long enough	• Contact supplier
	• Parts too wide for machine	• Identify to supplier what's being conveyed before purchase
• Belt speed too slow or too fast	• Improper adjustment of variable-speed drive motor	• Change sprockets • Install variable-speed options or speed-adjusting kit
	• Damaged in delivery	• Specify desired speed range to supplier before purchase
Pneumatic parts conveyors		
• Material won't move	• Filters clogged • Filters improperly sized • Blower binded • System improperly sized to suit plant layout	• Inspect filters regularly and frequently. Specify to supplier exactly what types of materials are to be conveyed. Check vacuum pressure gauges. Replace filters. Indicate to supplier anticipated future growth plans, if possible. Check vacuum seats.
• Blower overheating	• Blower binded • Excessive ambient-temperature exposure	• Check filters. Make sure they are properly sized. Install temperature limit switches
• Blower too noisy		• Install muffler. Enclose in a well ventilated sound enclosure. Place blower in a sound-proofed room

(continued)

Table 12.6 *(continued)*

(d) Metering/proportioning/Feeding Equipment		
Problem	Possible causes	Suggested solutions
Dry-solids metering		
• Weight distortion	• Dust or adhesive dry solids accumulation	• Install a dust-exhaust system or hardware to "tare off" dust accumulation in critical areas • Clean belts, trays, augers periodically
• Mechanical and electrical component failure, improper weight signals	• Dust, environmental conditions, materials adhering to underside of belt or other system components	• Evaluate several systems in production trials before purchase • Clean system components periodically
• Belts stretching and mistracking	• Material buildup or proximity of system to moisture	• Inspect regularly. Consult supplier. Install corrective recalibration devices
Liquids metering		
• Materials won't flow	• Improperly sized system—can't handle highly viscous materials	• Install drum pump. Specify to supplier what type of material to be conveyed • Make sure tubing is properly sized
Proportioning loaders		
• Sluggish loading, excessive loading time needed to fill hopper	• Clogged filter in either dust collector or pump	• Clean filter (replace if necessary)
	• Material line clogged	• Clean line
	• Vacuum leak in either material line or vacuum line	• Seal lines
	• Hopper lid or hopper receiver not sealed	• Clamp lid or replace hopper seals if necessary
• Excessive dust carryover into the dust collector	• Valves not sealing	• Check for obstruction and proper air pressure
	• Improper feed tube setting, too much air, not enough material creates high velocities	• Adjust feed tube to give highest obtainable vacuum and smoothest flow
	• Excessive fines and dust in material or improper blending	• Consult material supplier

(continued)

Table 12.6 *(continued)*

(e) Chillers/Mold Temperature Controllers

Problem	Possible causes	Suggested solutions
Cooling water lines frozen	• Thermostat set too low (i.e., below 40 °F without antifreeze)	• Check thermostat and reset if necessary to 40 °F or higher
	• Insufficient antifreeze in process cooling water	• Add antifreeze
System shuts down or cools slowly or poorly while refrigerant pressure is low or drooping; bubbles in refrigerant-level sight glass; oily-looking moisture on coolant-circulating tubes or floor nearby	• Refrigerant leak	• Replace refrigerant, plug leak source, clean condenser
System shuts down while refrigerant pressure is high	• Condenser not getting enough cooling water because constricted or blocked by dirt	• Clean condenser
Low cooling-water pressure	• Leak in process-water circulation lines	• Plug leak
	• Empty water-storage tank	• Fill tank
	• Broken pump motor	• Repair or replace (hermetically sealed motor/pump assemblies must usually be replaced)
	• Broken pump seal	• Replace
Slow or inadequate cooling of molds, high water pressure	• Water flow constricted by dirt or mineral scale	• Clean water circulating system • Treat water • Install intermediate heat exchanger for mold-cooling water (optional)
Slow or inadequate mold cooling, low pressure	• Water-circulating line leak	• Repair leak
	• Pump seal failure	• Replace seal or tighten packing gland
Process water heats slowly or insufficiently	• Heating element encrusted with dirt, mineral scale or, in oil-circulating systems, carbonized oil	• Clean or replace heating element, treat water, replace oil

(continued)

Table 12.6 *(continued)*

(f) Dehumidifying Dryer Performance

System	Possible causes	Cure
Cannot attain desired air inlet temperature	• Heater failure	• Check process air or after heaters—regeneration heaters play no part in this aspect of operation
	• Hose leakages and excessive length air inlet side	• Locate and repair; if the hose is old and brittle, replace • Shorten all hose to minimum lengths
	• Line, hopper, filter blockage	• Check for collapsed or pinched lines, valves that are closed (some makes have air flow valves located on the air inlet side of the hopper). Filters should be changed or cleaned frequently; a good trial period is every four weeks until experience dictates a shorter or longer period
Dew point as measured at air inlet to the hopper is unacceptable	• Loss of regeneration heaters in one or both beds or line fuses	• These can be checked with a volt meter at the control panel
	• Loss of timer or clock motor switching ability from one head to the other, i.e., continuous operation on only one dessicant bed	• Check clock motor for movement by observing either function indicators or valve-shifting mechanisms. Note that loss of regeneration heaters may occur if the clock motor or shifting mechanism malfunctions
	• Desiccant has deteriorated or been contaminated	• Most manufacturers suggest checking the desiccant annually and replacing when it does not meet test criteria. Typically two to three years is a reasonable interval, depending upon the severity of service

(continued)

Table 12.6 *(continued)*

System	Possible causes	Cure
	• Loss of power to one or both desiccant beds	• During regeneration cycle, exterior of the desiccant bed should be hot to touch. Check contacts on relays or printed circuit board for flaws, line fuses if so equipped
Low or nonexistent air flow	• Fan motor burnout	• Replace
	• Loose fan on motor shaft	• Tighten
	• Clogged filter(s)	• Change
	• Restricted or collapsed air lines	• Correct and relieve restrictions
	• Blower motor is reversed	• Use of a pressure gauge or flow meter is suggested. Proper rotation is that in which the highest flow is indicated

- Blocked cooling channels. Check out the throughput and if found appreciably less than when the mold was new, clean the channels.
- If the above conditions have not achieved a cure, consider whether the cycle is too unrealistically short to keep productivity high. If not, proceed to other possibilities.
- Poorly designed cooling channels. If this is determined to be the case, the mold may have to be reworked to increase cooling capability or perhaps to improve the uniformity of cooling (Chapter 11).
- Poorly designed part. Too great a variation in the distribution of material in the part—unnecessarily thick and thin sections—readily cause warping unless that part is thoroughly cooled which may demand an uneconomically long cycle. Depending upon the situation, parison programming or a redesign of the part itself may be the only feasible solutions (Chapter 6).

12.5.2 Defects Within the Blown Wall

Various defects found within the blown wall will be evaluated in this section. They include defects that spoil the appearance of the blown part such as "bubbles" in the wall and "cold spots." In addition, there are 3 distinct types of weld or pinch-off defects including thinning, tearing, and cutting of the pinch-off. Another defect discussed are indented parting lines.

12.5.2.1 Bubbles

These defects are not easily mistaken for foreign matter in the melt for they show up as generally symmetrical areas, clearer than the surrounding area. Moisture in the melt is the most likely cause. It may come from several sources:

- Cold resin that is brought in and exposed to the warm atmosphere of the extruder room will condense moisture on its surface. If the resin is not allowed to warm up and evaporate the moisture before entering the extruder, the moisture will end up as bubbles in the blown part.
- If water cooling is overdone on the throat, moisture may condense in the barrel, and if running at slow screw speeds, the result will be the same as if the resin were wet.

Nonmoisture causes of bubbles include material leaking from mismatched head sections, and insufficient back pressure.

Curing the moisture problem is generally self-evident. Let the resin warm up before using by keeping a few unopened gaylords in a warm room. If drawing directly from hopper cars or trucks, it is advisable to discharge into a surge bin which will protect the resin from moisture condensation and allow it to warm up before use. If condensation is occurring in the barrel, simply ease off on throat cooling.

Further insurance for moisture elimination is to increase barrel temperature in the transition section of the screw, thus flashing off any moisture remaining.

If mismatched head sections are found, the situation should immediately be rectified, for this can also be a source of degraded resin.

Increasing back pressure, if this is suspected as a cause, is easily accomplished on reciprocating units by adjustment of the back pressure gauge. With continuous extruders, a screen pack change to a finer mesh may solve the problem.

12.5.2.2 Cold Spots

These defects show up as non-homogeneous areas in the wall of a blown part. The cause of this defect is too low a back pressure and insufficient working of the melt to ensure that it is completely uniform. Apply the same remedy proposed for bubbles.

Another cause is accidental admixture of some resin having a lower melt index. Extrusion conditions will be inadequate resulting in cold spots from the incompletely melted resin. The obvious cure is to not let such accidents occur.

12.5.2.3 Thinning

A thin, weak weld line is often caused by too high of a stock or mold temperature. As the molds close and the pinch-off is made, the pinch-off lands fail to force sufficient material into the weld line to make a strong weld. Gradually drop the stock temperature to see if this will cure the problem. If it does not, then examine the pinch-offs. Too short or too sharp a pinch-off land length nearly always will result in a thin, weak weld line.

Excessive preblow or high pressure air coming on too early can also cause the pinch-off to thin before it has time to set. Gradually back off the preblow air to a point that does not affect the other areas of the container. If this does not correct the problem then the blow delay time can be increased. Care must also be taken when performing this step because other areas of the container may be affected.

12.5.2.4 Tearing

Weld tearing during trimming is usually a result of too long a pinch-off land length. This may have been selected in the first place to achieve a strong weld by forcing more material into the weld line. It will accomplish the latter, but, if overdone, can prevent the molds from closing completely. This leaves a thick pinch-off line that is apt to tear during trimming, leaving a rough, ragged weld, or the weld may even be torn open. Have the molds reworked to reduce the land length to a point that a good balance is achieved between a strong weld and a trimmable pinch-off. A land length of 0.010 to 0.015 in. (0.254 to 0.381 mm) is most commonly used for polyethylene and most other plastics.

Other areas to investigate when tearing becomes a problem are worn pinch-offs, mismatched molds, damaged molds, and uneven mold clamp pressure. Worn pinch-offs can be temporarily corrected by rolling them back, but since this is only temporary, they will need to be refurbished. Mismatched molds are normally caused by worn locating pins which will need to be replaced. Damaged molds, such as nicks in the pinch-off area, will need to be filled and reground. Molds that appear to have torn pinch-offs in one area and not another are probably not closing and clamping evenly. The clamp will need to be adjusted. If this cannot be accomplished, the molds can be shimmed out at the problem area.

12.5.2.5 Cutting

If the weld and pinch-off both seem to be satisfactory but there is a hole or slit along the weld, the stock temperature may be set too low. If the stock is too cold, the parison may tear as the mold closes. Check a part for cutting as it comes out of the mold with flash attached. If a hole or slit is found, this eliminates the trimming operation as the cause. Examine the mold for defective pinch-offs. If none are found, raise the stock temperature by 5 to 10 °F (3 to 6 °C) increments until the cutting ceases.

Cutting can also result if the molds are allowed to close too fast, slamming or slapping shut. Even more disasterous, this condition may cause the molds to become severely damaged. Ease off on the closing cycle to make better weld and to preserve the molds.

12.5.2.6 Indented Parting Lines

Indented part lines have somewhat of a V shape where they appear to pull into the container. This problem is normally caused by insufficient blow pressure or air entrapment due to poor venting. If raising the blow pressure does not cure the problem, the mold vents need to be cleaned. Should the problem continue to exist, the molds will probably require sand blasting to improve venting.

12.5.3 Poor Bottle Surface

Roughness, pits, orange peel and a variety of other terms describe the poor surfaces frequently found on blown bottles. While an imperfect parison may be a cause, only problems associated with the mold and the blowing process wiil be evaluated.

- Poor mold surface. The mold should have a fine matte finish in order for the air to vent quickly and allow the parison to conform to the mold surface while still hot.
- Plugged or inadequate vents. If the vents along the parting line are dirty or plugged, the part surface will suffer. These vents will need to be cleaned. Vents that are too small or too few in number to permit the rapid escape of air will also cause surface problems. Minor reworking of the mold is the answer.

- Low blow pressure or blow rate. Increase the blow pressure, then check to see that the blow pin is large enough to handle the required amount of air to fully and rapidly blow the part. Check for restrictions or partial plugging of the air lines.
- Air leak around the blow pin. Leakage here prevents sufficient pressure to be maintained to hold the part tightly against the mold for good conformation to shape and rapid cooling.
- Low stock temperature. If the parison is too cold, it will be difficult to make it reproduce the surface of the mold, particularly if there is lettering or fine designs on the surface. Gradually increase stock temperature cautiously. Overdoing this may cause problems that result from overly high stock temperatures.
- Condensed water in the mold. A low mold temperature combined with high humidity in the plant can result in water condensing in the mold between the ejection of a part and the blowing of the next. This can cause bad surface pitting. If the mold temperature cannot be raised, usually not without hurting cycle time, it may be necessary to air condition the area locally or in the entire plant to remove some of the water in the air during periods of excessive humidity.

A particular kind of poor bottle surface results from a rough parison. Such a parison is almost always caused by operating at a shear rate that produced melt instability.

A region of melt instability is a fundamental characteristic of all HDPEs. The region starts at about the upper level of extrusion rates at which continuous extruders are capable, and ends near the lower level of extrusion rates at which reciprocating and accumulator extruders are likely to operate.

Therefore, while running a continuous extruder, if the type of surface roughness indicative of melt instability is encountered, back off on the extrusion rate until the problem disappears.

With either a reciprocating or accumulator type, there are several choices:

- Drop down out of the region of melt instability by further decreasing extrusion rates (increasing drop time), a generally unsatisfactory solution because of the decrease in productivity.
- A better approach is to move up out of the region by increasing extrusion rates (decreasing drop time).
- If it is impractical to increase extrusion rates, try increasing the stock temperature or change to a higher melt index resin.

A second type of rough surface is caused by melt fracture, rarely seen in blow molding, but would result from running at shear rates that only the accumulator or reciprocating type extruders can reach, and only then, if pushed to the limit of their performance. Obviously, the answer is to back off on extrusion pressure until the melt is back in stable territory.

12.5.4 Curtaining and Webbing

Curtaining or folding of the parison as it extrudes from the die is the cause of webbing in necks and handles. If the effect is severe enough, the handle can be partially or completely blocked. Folds in the neck can interfere later with filling operations, or be so restricted as to cause difficultly in reaming and facing, and spoiling the fit of the cap. The several possible causes all imply their own cure, which should be quickly undertaken to eliminate this serious bottle defect.

Causes include:

- Poor tooling design. Too short a land length between the die (bushing) and mandrel will result in poor control of flow through the die gap. There will be areas of high and low flow, and folding will result.
- Die misalignment. Die and mandrel misalignment will also produce differential flow rates in different areas, resulting in folding.
- High stock temperature. Unduly high stock temperature will yield a parison with low melt strength which will collapse before it can be blown, and will fold.
- Resin mismatch. If a resin is chosen with too high a weight swell for the job, the attempt is often made to reduce part weight by decreasing the die gap. The result may be a parison so thin that it quickly collapses and folds. A resin with low diameter swell characteristics may cause lower handle webbing and, in severe cases, can even cause blowouts. This is due to the parison being caught in two places by the inner handle pinch-off. When blown, these areas meet to form a heavy area known as a web.
- Low parison diameter swell. High stock temperature and/or low shot pressure will yield a parison with low diameter swell resulting in lower handle webbing.
- Mold position. Improperly positioned molds that are shifted away from the parison will result in containers having lower handle webbing.
- Hooking parison. A parison hooking away from the handle will also cause the problems stated above.

12.5.5 Blowouts

Blowout of a part can be caused by diverse factors; fortunately the cure in most cases is self-evident. Possible causes are listed below:

- Pinch-off too sharp. This will cut the parison and cause a blowout. Increase the width of the pinch-off land.
- Pinch-off too wide. Too wide a land will cause mold separation and the chance of a blowout.
- Clamp pressure to low. This also can result in mold separation and possibly a blowout.

- Pinch-off too hot. Poor or uneven mold cooling can prevent the pinch-off from cooling sufficiently causing it to cut the parison as if too sharp. Improve the cooling by one of the methods previously decribed under "Warping."
- Blow pressure too high. Stretching the parison at too high a rate can easily lead to blowouts. Back off gradually on the pressure.
- Parison too short. This won't be caught at the bottom pinch-off and guarantees a blowout—the remedy is obvious.
- Blow-up ratio too high. Too great a difference between the diameters of the parison and the part will cause the parison to stretch too much. A thin spot will develop and a blowout will occur. A change of head tooling to a larger diameter die is indicated.
- Resin swell too low. This relates to the previous item, and can be an alternate cause. Too little swell results in a smaller than expected parison, which can cause a missed handle accompanied by a blowout. A higher swell resin is needed.

12.5.6 Parison Curl, Stringing, Hooking, Sag, and Length Consistency

In this section, various problems with parison formation will be evaluated including curl, stringing, hooking, sag and length inconsistency.

12.5.6.1 Parison Curl

Parison curl, sometimes called doughnutting because the parison curls about in a seeming attempt to form itself into a doughnut, usually results from one of three causes:

- Cold mandrel or die
- Misaligned die and mandrel
- Foreign matter or degraded resin in the die (bushing)

If the curl occurs when a machine is started up, then gradually disappears as the unit approaches operating temperature, a cold mandrel or die can be suspected. Allow a longer warmup time before attempting production. Also, check to see that the die heater, equipped with most blow molders, is operative. If it is not, an unnecessarily long warmup time will be needed.

Parison curl on a fully warmed up machine may result from a misalignment of the die and mandrel. The mandrel edge may be recessed within the die so that the parison contacts the die, hangs up on one side and thus curls. Often, the mandrel has deliberately been so positioned in an attempt to blow a lighter-weight bottle than the tooling was designed for. New tooling should be obtained or the die can be machined to bring the mandrel back to a flush or lower position with the die face.

Parison curl on a fully warmed-up machine may also be caused from air leakage around the tooling. This leakage may cause the mandrel or die to cool, resulting in parison curl. The remedy is self-explanatory.

Another possible cause is an uneven buildup of foreign material on the die, which then distorts the parison as it is extruded. This solution is also self-evident: give the die a thorough cleaning.

12.5.6.2 Parison Stringing

Parison stringing is seen as feathery strings of melt attached to the top and bottom flash of the bottle. If stringing is severe, each bottle remains attached to the next. The lower bottle's weight may thin out the parison to the point that blowout can occur, especially in a handled bottle. The stringing is almost always caused by too high a stock temperature, and can be eliminated by gradually lowering the temperature.

Parison stringing can also be caused from too high a fill pressure on reciprocating type extruders, resulting in excessive weeping of polymer from the die. This excessive weeping causes the parison to string during part removal.

In rare cases, when the temperature is lowered enough to eliminate the stringing, other problems, such as cold spots, can arise. In such an event, it may be a matter of the wrong resin being selected for the particular job.

12.5.6.3 Parison Hooking

Parison hooking appears as a parison that will not drop straight, but is hooking from one side or another, causing poor wall distribution, webbing, and/or blowouts in the handle. Parison hooking usually results from any one or a combination of the following causes:

- Nonuniform die temperature
- Warped die or mandrel
- Die gap out of adjustment
- Air flowing on parison

12.5.6.4 Parison Sag

Parison sag or drawdown is seen as a parison that thins closest to the die from its own weight during extrusion. Causes are:

- High stock temperature
- Slow extrusion rate
- Long mold open time
- High die heats
- Resin has too high a melt index

12.5.6.5 Parison Length Inconsistency

Parison length inconsistency is identified as a parison changing from too short, causing blowouts in bottom of container, to a parison that is too long, causing wall distribution problems and tails sticking to the container during unloading. Parison length can change for any of the following reasons:

- Insufficient back pressure in extruder
- Changing regrind virgin mixture
- Malfunctioning temperature controllers
- Bridged or collared screw
- Worn barrel of screw

Parison length inconsistency from head to head is self-explanatory. Causes are as follows:

- Nonuniform head temperature
- Nonuniform bottle weights
- Manifold chocks out of adjustment
- Head chocks out of adjustment

12.5.7 Foreign Matter in the Melt

Foreign matter in the melt—whether from outside contamination of the resin or from particles of degraded resin—discloses its presence in a variety of ways: by die lines or streaking in the parison, by off-colored particles buried in the wall of the molded part, or by holes or "windows" in the blown part. Obviously, the cure is to identify and eliminate the source of the foreign matter.

12.5.7.1 Look for Exterior Sources of Contamination First

Before investigating the more complex problem of degraded resin, first examine closely the material handling procedures and general housekeeping. Are gaylords allowed to stand open to collect dust? Are bags opened and emptied carelessly so as to introduce dirt, lint and dust into the feed? Are parts with foreign matter in them being thrown into the grinder? Is the regrind being protected with the same care as the virgin resin? A contamination problem can be just that simple.

12.5.7.2 Degraded Resin—Many Sources

Tracing the source of degraded resin particles can involve some detective work. Logically approach the task by proceeding with the easiest cured source to the most difficult. Various clues will suggest the source. A very simplified guide to degradation problems is presented that can be integrated with other information presented in this chapter and others.

Cause	*Solution*
• Improper shutdown and/or startup procedures	
• Overheated material	• Reduce barrel temperatures • Reduce residence time, shorten cycle
• Temperature control malfunction	• Calibrate controllers • Check controllers • Check associated relays, etc. • Check thermocouples
• Material hang-up or dead spot in system	• Clean system, purge compound • Disassemble and clean • Improve material flow by streamlining design
• Contaminated material	• Inspect material to be sure it is clean

If a single streak, or perhaps two or three, suddenly appears in the parison, it is likely that some foreign matter has become lodged between the die and mandrel. Just increase the die gap briefly to purge the particles, then reset to the original gap.

If parison streaking or defective bottles (foreign particles or holes) are regularly appearing and, all possibilities of the resin feed being contaminated have been eliminated, then a steady supply of degraded material may be generating somewhere within the head, adaptor or extruder. The cause will probably be a blind spot or non-streamlined area where melt hangs up, degrades, and then breaks away from time to time. A poor fit between head and adaptor, or adaptor and extruder barrel, for example, can produce a sharp break in the melt path and allow resin hangup. Reworking the parts for a smooth, streamlined fit is the solution.

The gradual appearance of many die lines or streaks, after extended periods of operation, points to a definite buildup of carbon or degraded resin in the head. Fortunately, buildup most often occurs on the die mandrel which are readily accessible for cleaning. Extreme caution should be taken to prevent damaging the machine in the process. Use a tool made of copper or brass which is softer than the machine parts and will not mar or scratch their surfaces.

If the dead spots in the head have been eliminated as the real source, there is a need to go deeper into the problem and find the source of degraded resin somewhere back in the extruder.

12.5.7.3 Culprit—Man or Machine?

Degraded material generated in the extruder can result from defective machinery or poor operating practices. Examination of the extruder head may disclose the source: mismatched or nonstreamlined head sections, or inadequate seals that permit material to leak from supposedly matched surfaces.

But more often, the degraded resin will result from some operating error. A poor purge when changing from one resin to another may be the cause. A flood of specks in the bottles upon startup, after a prolonged shutdown, will suggest that the extruder was not allowed to cool down on purge before shut off. The resin in the die lips become oxidized by exposure to the air. This cause can be eliminated, and both shutdown and startup times shortened, by the use of a simple air-lock plate to seal off the die lips from the atmosphere during these periods.

In other cases, the degradation may be occurring back in the extruder barrel. The cause could simply be running at too high a temperature somewhere or along the barrel. The high temperature may be intentional in an effort to run a difficult resin and still maintain a fast cycle. A slower screw speed will probably solve the problem but will also increase the cycle time. Since this is rarely acceptable, the better answer is to switch to a resin that is easier to extrude.

The high temperature may be unintentional. This could result from a broken thermocouple or malfunctioning temperature controller.

In other cases, too much heat will be generated at the wrong point—in the metering zone. Generally, it is best to let this zone hold its contribution to a single function—metering. Try a reverse temperature profile; rather than gradually increasing the temperature of each zone along the extruder, increase it in the transition section, drop it in the metering zone, then let it smoothly increase to the desired final melt temperature through the adaptor and head.

Once a workable temperature profile has been found, a good purge may clear up the source of the degraded resin. But if the bad condition has existed too long, it may be necessary to take the extruder down for a thorough cleaning.

12.5.8 Die Lines or Streaking in the Parison

Parison streaking most often results from particles of foreign matter or degraded resin lodged between the die and mandrel. Since the die gap is no longer of uniform width, the parison is reduced in thickness at each point where a particle is present. This uneven thickness shows up as streaks or die lines in the parison as well as in the blown part.

There is a type of parison streaking that has nothing to do with foreign matter. This shows up as clear or slightly rippling streaks. In such a case before looking for non-existent sources of foreign matter, check the heater bands. A burned-out band will cause a cold spot in the barrel, melt will flow unevenly here, dragging or hanging up and cause the streaking. Replacement of such a non-functioning band will immediately eliminate this type of streaking. Review Section 11.5.7.2 to determine source of degraded resin and solution to problem.

12.5.9 Shrinkage

Shrinkage of a polyethylene (or other plastics) container after molding is a normal occurrence (Chapter 7). However, excessive shrinkage can be caused by insufficient cooling during the molding cycle or heating of the container during storage.

Excessive shrinkage can be caused by:

- Wall thickness or weight too high
- Mold temperature too high
- Stock temperature too high
- Blow pressure too low
- Blow time too short
- Mold volume incorrect
- Bottle storage area too hot
- Resin density too high

12.5.10 Causes of Stress Cracking

Diagnosis of stress crack failures in high density polyethylene bottles is, often, far more complex than just an interaction between the container and its contents. Although specifically for HDPE bottles packaging liquids, many of the analyses can also be applied to similar problems in plastic tote boxes, bowls or even plastic piping.

For blow-molded containers, designed to package a liquid, the ability to withstand the stresses in the presence of a standard liquid surfactant, is generally referred to as its environmental stress-crack resistance (ESCR).The value is reported as F50, hrs; that is, the period of exposure time at which there is a 50% chance that the container will fail.

Actually, four basic types of stress cracking are recognized: environmental, solvent, mechanical and thermal. The last is rarely, if ever, applicable to blow-molded containers. The other three definitely apply and may appear alone or in combination. Although stress cracking is associated with many polymers, researchers were specifically concerned with HDPE, since containers blow-molded from these resins are used for a greater variety of applications and in greater volume than any other polymer.

12.5.10.1 Environmental Stress Cracking

Environmental Stress Cracking (ESC) is the brittle failure of a container, filled with a polar liquid, at a point of stress concentration while under a stress. All of these factors must be operative.

The stress may be the weight of the container contents, or the internal pressure in the container. A point of stress concentration may be a sharp discontinuity in the container design (for example, a sharp change in configuration, or a stress point molded in) or caused by differential rates of cooling of thick and thin sections. The polar liquid is one that does not perceptibly swell, chemically attack or in any other way affect the plastic.

12.5.10.2 Solvent Stress Cracking

This cracking is similar to, and is often confused with, ESC. All factors characteristic of ESC must be present except for liquid content; this one does affect the plastic. It may be absorbed by the polymer, causing it to swell, soften or embrittle, or otherwise exhibit change. There may actually be a chemical attack on the plastic.

12.5.10.3 Mechanical Stress Cracking

This action refers to external stress, typically caused by top loading. It may be this stress rather than an internal stress that contributes to an ESC problem.

When an HDPE container fails in service, it is generally assumed to be an ESC problem. This may be misleading. The resin might not be up to handling the particular product packaged in it. This, however, could have been discovered through a product compatibility test. Such a test should always be made unless the product is essentially identical to others that have been successfully packaged in containers of the same resin.

Where solvents are part of a packaged product, a compatibility test is a must. This need is often overlooked if solvents are only a small part of the formulation, or if certain ingredients are not deemed to be solvents. A petroleum distillate would rarely be overlooked, but others can be, such as alcohols, ketones, aldehydes, chlorinated aromatics and aliphatics, essential oils, etc.

A factor which may set off a futile chase along a false trail is the packaging of the containers or the handling of these packages. By error or by a seemingly harmless change in the specifications for the corrugated shipping container or its divider, the package integrity can be weakened and a greater burden placed on the containers inside. The same is true if the corrugated test strength is decreased or if the height of the case and its dividers are increased.

Similar problems can be caused by cases being stacked higher than recommended or tested. These packaging changes can cause mechanical stress crack failures in the bottles.

Because the causes of stress cracking can be complex, the checklist in Table 12.7 is offered to prevent some apparently minor, but possibly, important factors from being overlooked.

12.5.11 Heat—Too Much or Too Little and What to Do About It

As any blow molder operator knows, temperature problems can occur at any time, affecting quality of the product, bottle weights, cycle times or combination thereof. More often than not, the problem is a matter of too much heat, but cases of too little heat are not rare. Some suggestions for fast, systematic pinpointing of heating problems are discussed below (see also Chapter 6).

Too much heat—consider the following:

- *Wrong set point.* Even if one "knows" this can't happen, check it out first. This only takes seconds and may avoid misspending a lot of energy looking for other possible causes.
- *Insufficient cooling.* If the controller is not calling for heat, yet a zone seems to be overheating, the heating problem may actually be a cooling problem, so check for proper coolant flow.

Table 12.7 *Analyzing Stress Crack Failures in HDPE Bottles*

Category	Specifics
Resin	• Homopolymer or copolymer • Resin supplier designation, lot number
Corrugated appearance	• Package design a. Test weight b. Correct dimensions c. Divider design—correct dimensions • Evidence of damage to package
Bottle history	• Location and date of manufacture • Storage conditions a. Temperature b. Humidities c. Stacking height d. Failure positions within stack e. Date failures discovered f. Date reported g. Percent of total shipment represented by failures
Product compositions	• Surfactants (wetting agents) • Alcohols • Other ingredients with high vapor pressure • Has product compatibility test been run?
Miscellaneous	• Location of failure on bottle a. Bottom chime b. Neck c. Other • Lid seal (tight or venter?)

Source: Norchem/Enron Chemical Co.

- *Shorted thermocouple.* Check the controllers to see if the temperature indication is at or near zero. If it is then the most likely cause of the heating problem is a shorted thermocouple. This will deliver no voltage to the indicator, and the controller will call continuously for heat. A broken thermocouple would produce the same result as a shorted thermocouple in very old instruments not equipped with thermocouple break protection.
- *Poor contact between thermocouple and barrel.* If the thermocouple is not making good contact with the metal of the barrel, it will sense a temperature lower than actual and will cause the zone to overheat.
- *Partial short in heater band.* A partial short will reduce the electrical resistance of the heater band and, as a result, will increase the wattage output. This possibility will take a few minutes to check out. First, note the voltage and wattage values stamped on the heater at the zone believed to be the trouble spot. Then, *with the electrical supply turned off*, disconnect one end of each of the heater bands in the zone and measure the resistance.

Now, using the voltage and wattage values stamped on each heater band, calculate the resistance, using the following formula:

$$\text{Resistance} = \frac{(\text{Voltage})^2}{\text{Watts}}$$

For example, if the voltage reads 220, and the wattage is 2,000, then,

$$\text{Resistance} = \frac{220 \times 220}{2{,}000} = 24.2 \text{ ohms}$$

If the resistance value, as measured, is appreciably lower (10% or more) than the calculated value, then that heater band should be replaced. The actual voltage should also be measured to be certain it is as it should be. If it is not, the problem is probably in the power pack and not the heater.

Too little heat—consider the following:

- *Wrong set point.* Check it out.
- *Broken thermocouple.* When checking the set point on the controller, note if the indication has been driven to the top of the scale. If so, you very likely have a broken thermocouple, and its replacement should cure the problem.
- *Low voltage at heater band.* If a voltage test at a heater band shows low or no voltage reaching the heater, check out the connections back to the controller. A gap in the circuit will, of course, cut all voltage to the heater band. A poor connection, which can put a high resistance in the circuit, will result in low voltage being delivered to the heater.

- *Break or partial break in heater band.* This will increase the resistance of the band and reduce the wattage. Check out the bands with a volt-ohm meter and make the calculation described previously for *Partial short in heater band.*

12.5.11.1 Instrument Recalibration

It's a good idea to check out the instrumentation at least on a semiannual basis. A controller actually controlling 50 °F (28 °C) higher or lower than it indicates can produce some fairly puzzling situations. So, use a pyrometer to test if actual and indicated temperatures agree. If they differ by more than a few degrees, recalibrate.

This test—if not the recalibration—can be done in moments, and probably should be placed high up in priority for checking before going on to tests that take more time and/or require the blow molder to be shut down.

12.5.12 Surging

There are several causes for surging, but the first one to check is the one that will present the most trouble if not quickly dealt with. And that is a bridging or partial blocking of feed in the feed throat or feed zone. A collar or rope of partial melted pellets is formed that impedes the even flow of resin and prevents the screw from maintaining a steady pressure on the melt. This usually results from overheating resin in the feed zone.

To test this possibility, measure the temperature on the rear barrel. If over 350 °F (177 °C), this may be the culprit (for polyethylene resins), so reduce it to a safe level. Also, check the temperature of cooling water at the outlet from the feed throat if the extruder is so equipped. The temperature should not be higher than about 100 to 120 °F (38 to 49 °C).

After the temperatures have been corrected, get rid of the collar by feeding in handle slugs from the trim station or cut up pieces to tail flash.

If this does not work, there is another method. Decrease the feed zone temperature still further, increase the temperature in the transition zone and increase the screw speed. Caution: Do not let the head pressure or the motor load rise too high.

One of these methods usually will eliminate the problem, but if neither does, then the last resort is to shut down, pull the screw and clean it.

If the feed throat and feed zone temperatures are satisfactory, then check for a possible variation in regrind feed rate. This will make waves that can be transmitted all the way to the die.

Still another cause of surging relates to variations in screw speed. Check the rpm meter for a cyclic variation. Variations can result from several causes, some of which can only be discovered while the process is shut down. Included are:

- Voltage fluctuations
- Loose belts
- Broken gear teeth or worn gears
- Misadjusted variable speed drives

A final possibility for the cause of surging, although seldom encountered, is a broken or contaminated screen pack.

12.5.13 Sharkskin

At one time or another, there may be bottles coming off the line with a sharkskin inner surface—wide, alternating bands of thick and thin wall sections. What happened? If running with continuous extrusion, the problem results from extruding too fast. Just back off a few RPM's on the screw speed and the problem will disappear. But, if operating with reciprocating extrusion, increase the pressure for a faster drop time.

The explanation for this odd state of affairs lies in the complex behavior of HDPE melt as it is being pushed through the die. In general as pressure is increased, the extrusion rate increases (or drop time decreases).

However, when real numbers are plotted, instead of a straight, linear relationship, they form a curve like the one in Fig. 12.1. Starting at some low value, the pressure is gradually increased, and for a time, things go as expected—each increase in pressure causes a corresponding increase in extrusion rate. Melt flow is steady, and bottles are smooth. But at some point, the next small increase in pressure results in instant sharkskin. This occurs at the pressure plateau in the curve.

By increasing the extrusion rate to a certain point, the sharkskin vanishes. Actually, neither the parison nor the bottle are as smooth as in the steady-flow region. Close

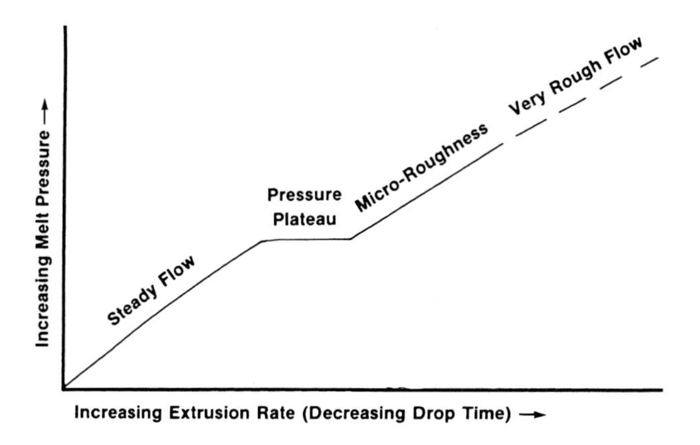

Fig. 12.1 *HDPE melt flow in real life*

examination of the inner bottle surface discloses a micro-fine roughness. However, the bottles are acceptable. Over quite a range of increasing melt pressure, all goes well—but again, at some very high pressure, the system goes completely out of control. The parison will be rough and unmanageable and the bottle unacceptable.

Most *continuous* extrusion takes place in the region of steady flow. Sharkskin will be encountered only if the machine is run at extremely high rates. Most reciprocating extrusion takes place above the pressure plateau and sharkskin appears only if extrusion pressure falls too low.

Strange things happen at the pressure plateau; Fig. 12.2 will help explain them. At some point a very slight increase in pressure causes the extrusion rate to almost double. With this rapid rate increase, the pressure drops slightly, and the extrusion rate falls considerably. Then, with this decrease in extrusion rate, the pressure builds again, and the cycle repeats, each cycle producing a pair of light and heavy sections in the parison, and in the bottle. So, to stay off the pressure plateau, it is only necessary to reduce or increase the melt pressure.

An unexpected feature of the plateau is that the critical pressure corresponding to the upper limit, is basically the same for all high density polyethylenes. There is, however, some control on the *width* of the plateau.

As an example, Enron/Norchem's milk bottle resin 6009, because of its high shear sensitivity, produces a narrower plateau than resins with poor shear sensitivity. The reason is that less pressure is required for a given extrusion rate, and so the onset of the plateau is delayed and its width is less.

Thus, extrusion rates can be increased for better productivity. And for reciprocating extrusion, drop times can be decreased further before the pressure is applied that results in rough parisons and unusable bottles.

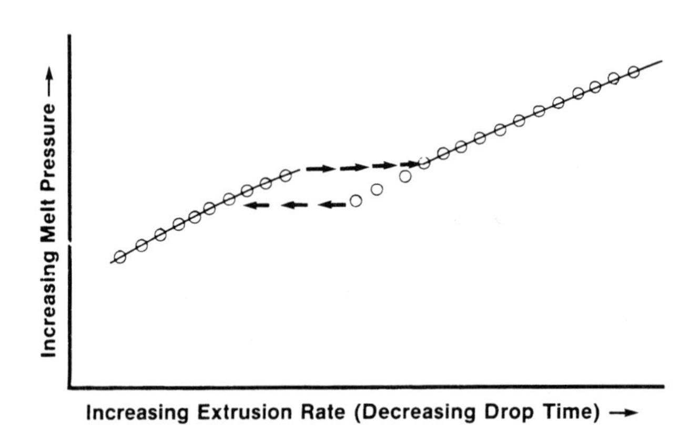

Fig. 12.2 *Pressure and extrusion rate oscillations in the pressure plateau regions produce sharkskin*

12.5.14 Screw Wear Affects Performance

One of the less commonly discussed aspects of maintaining injection and extrusion equipment is the inspection of screws (Fig. 12.3) and barrels. From time to time they should be examined to determine whether they are in condition to render the services expected. Screws do not have a continuous outside diameter. This requires special techniques in manufacturing and inspection. The following methods give reliable results and save time [9].

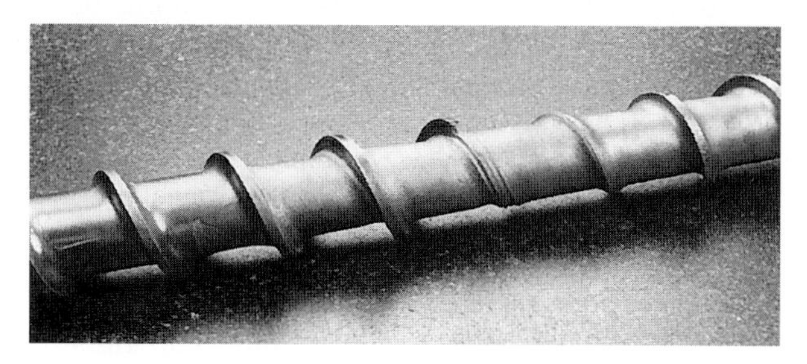

Fig. 12.3 *Abrasive wear on screws (and barrels) is caused by abrasive fillers such as calcium carbonate, talc, glass fibers, titanium dioxide, etc. The mechanism of abrasive wear is much the same as emery cloth on metal such as the wear on these screw slights [41]*

12.5.14.1 Inspection Rollers

Proper visual inspection of a screw or barrel requires that it be turned many times in order to see all sides. Screws and barrels are often heavy and difficult to turn when supported by the usual means. Fig. 12.4 shows a roller support device that uses two

Fig. 12.4 *Inspection rollers make it easy to inspect screws that otherwise would be difficult to handle because of their weight and length*

sets of double-conveyor rollers supported by multi-slotted angle irons. These angle irons are mounted on plain wood blocks that can be spaced to accept the screw.

12.5.14.2 Diameters

The shank and many other diameters are easy to measure by the usual methods. Other diameters such as the root diameter or the outside flighted diameter, require special methods. Measuring the root diameter is not always a reliable way to obtain channel depths. Another problem: if the micrometer sits on the radius on both sides, it can give a false reading. If the outside diameter (OD) is severely worn, this is still the best method to determine the correct channel depths. Pitches less than square or very deep channels make the problem worse. The best way to obtain root diameter is to find the OD, then subtract the channel depths.

The OD is measured with the assistance of a "mike" bar spanning two flights (Fig. 12.5). The thickness of the bar is subtracted from the measurement obtained. The usual technique is to place the bar on top and hold the anvil of the micrometer against the slight at the bottom with the left hand. The right hand adjusts the micrometer while making rocking motions along the screw. The bar will rock with the micrometer as the final setting is reached. Of course, it is essential that the bar be straight and of uniform thickness. It is best to check the OD at 90° from the original set of measurements because screws can be manufactured egg-shaped or become worn that way.

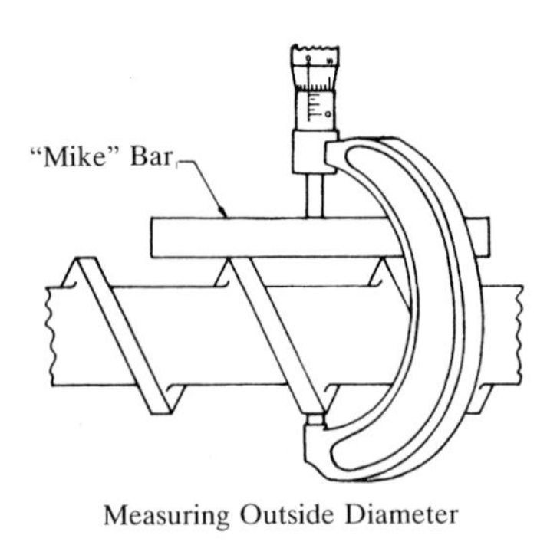

Fig. 12.5 O.D. measurement

12.5.14.3 Depths

The channel depth of small screws can be easily checked with a standard depth micrometer. With larger screws, the depth micrometer will not span from flight to flight. If checking many screws, it is best to make a screw-depth indicator or buy one. The screw-depth indicator consists of a wide angle "V" block with a dial indicator mounted on top and the probe extending down through the center of the V. The indicator is placed with the V resting on top of the screw and the probe on top of the flight. The gauge is then adjusted to zero. When the tip moves down into the root, the dial gives an accurate indication of channel depth. It is also very fast, allowing many measurements to be made very rapidly, and it can give continuous readings as the screw is rotated.

This last feature is also helpful in locating the starting and ending points of the feed, transition, and metering sections of a screw. This is not easy to do without the inspection rollers. The original channel depths of a severely worn screw are difficult to determine by this method. In the case of a severely worn screw, it is best to use the root-diameter method. Sometimes deep-jawed calipers can help if micrometers are running on the radii. A Spirex channel-depth gauge is shown in Fig. 12.6.

12.5.14.4 Concentricity and Straightness

Checking for straightness is difficult for the average plastics processor. If a good, long granite inspection table can be found, it will be helpful in checking concentricity and straightness. A preliminary check for straightness is possible just by rolling the screw on the table. If the screw is not straight, it will roll unevenly and show light under the flights in the low areas. This technique is appropriate only if the screw is not worn. The approximate amount that the screw is bent can be determined by feeler gauges. This technique is not completely accurate because the weight of the screw will tend to straighten it against the table. Most injection screws can be mounted between centers on a lathe and checked with an indicator while they are rotated. This requires an accurate center on both ends.

Extrusion screws usually do not have a center on the discharge end, requiring that a center be installed, then rewelded after testing and possible straightening. Checking the runout on the flighted portion is done with a "T" bar that spans at least three flights. The bar leans against the flights and an indicator measures the movement of the bar.

Straightness and concentricity can be determined on screws that don't have centers by rotating them in V blocks on an accurate inspection table. This is done with the help of a height gauge. This method is particularly useful in inspecting the pilot at the discharge end of injection screws. A sensitive dial indicator is best used as shown in Fig. 12.7.

Fig. 12.6 Channel-depth gauge

Fig. 12.7 *To obtain a pilot runout measurement at the discharge end of a screw, a sensitive dial indicator as shown here is recommended*

12.5.14.5 Hardness

The hardness of most portions of a screw is difficult to measure, because the screw is usually too large to test in a Rockwell-type tester. Also, the curved surfaces of the screw present a problem, and it is undesirable to make penetration marks on the surface. A satisfactory method for checking screw hardness is the impact or falling-ball type of tester such as a Shore scleroscope (Fig. 12.8). It is portable, works on curved surfaces, and can be reliable if checked against calibrated reference samples.

12.5.14.6 Finish and Coating Thickness

Finishes can be verified by a number of profilometers. A portable thickness tester can be used to test for chrome-plating thickness (Fig. 12.9), or, with experience, nickel and other nonmagnetic coatings. A surface profilometer is shown in Fig. 12.10.

12.5.14.7 Screw Manufacturing Tolerances

For reference, standard screw manufacturing tolerances are published by suppliers and made available on request. These tolerances have been established according to practical application requirements and reasonable ease of machining. Some tolerances can be held closer than indicated on the published list if necessary for a specific application.

12.5.15 Barrel Inspection

12.5.15.1 Inside Diameters

The very front of a barrel can be measured with inside micrometers; however, this is a limited measurement, and will not show far enough into the barrel. An inexpensive cylinder gauge can be rigged up with a long handle to slide the full length. Such a device is accurate enough to determine the need for a replacement or repairs. It is not

Fig. 12.8 *A Shore scleroscope checks screw hardness using the impact or falling ball technique. The Shore device is portable and works on curved surfaces*

accurate enough for machining purposes. There is a problem with light and the return glare off the gauge from a flashlight. There are cylinder and bore gauges that measure accurately at great depths and have the indicator outside the barrel for easy reading (Figs. 12.11 and Fig. 12.12).

Fig. 12.9 *This portable thickness tester can be used to test chrome thickness or, in the hands of an expert user, check nickel or other coatings*

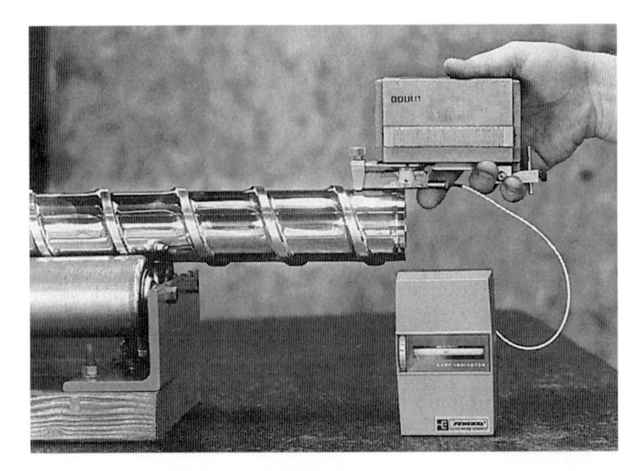

Fig. 12.10 Finishes can be verified by a profilometer

Fig. 12.11 The Starrett cylinder gauge, available in various sizes, measures accurately at great depths and has an indicator outside the barrel for easy reading

12.5.15.2 Straightness and Concentricity

Barrel straightness is difficult to determine by conventional methods. The internal diameter (ID) and OD are not always exactly concentric. The first method to measure straightness is to set the barrel on precision rollers at each end. The ID is then indicated for runout with a dial indicator. This is limited in depth. Tolerances allow straightness deviations to accumulate to a total allowable indicated runout.

The second method uses a test bar usually about 70% of the barrel length. The test bar is a slotted and chrome-plated bar that is precision-ground to approximately

Fig. 12.12 A boregauge, such as this Sunnen Products Co. gauge, measures accurately at great depths and permits easy reading

mid-range of the normal screw size tolerances. In theory, if the test bar slides easily through the barrel, the screw will also. In addition, this procedure will catch rapid changes or kinks that would, otherwise, be allowed under the accumulation of tolerances. A different-size bar is needed to test each ID. Because these bars are quite expensive, their use is impractical for most organizations except for the barrel manufacturers.

12.5.15.3 Barrel Hardness

Obviously, the standard hardness testers are not able to get inside a barrel to measure hardness. The instrument most commonly used for this purpose is the internal mobile hardness tester. When testing and comparing the hardness of bimetallic lines, one must remember that absolute hardness, as measured, is not necessarily in direct proportion to wear resistance. Sometimes one may be measuring the softer matrix rather than the wear-resistant carbides. The degree of lubricity or how well the screw and barrel slide against each other is very important. This is not always related to hardness, but is very critical to barrel and screw wear.

12.5.15.4 Barrel Specifications

Chemistry and hardness information for various types of domestic barrels will be supplied for the processor's reference on request. Chemical information supplied is for the "as cast" condition. The actual chemistry may vary widely after final machining is complete. It is important to note that the chemistry and hardness are not necessarily indicative of wear resistance. Also important is how materials (such as tungsten and carbide) are combined and how near to the bore they are located.

Putting into practice these inspection tips to detect problems and taking steps to solve those problems will enhance the operation of any processor using barrels and screws.

12.5.16 Detecting Container Leakage

It is often thought that a container produced as fast and inexpensively as a blow molded product does not warrant leak testing. Unless a major problem occurs, the incidence of leakage is very small. However, because of the speed at which the containers are produced and filled, such a small incidence of leakers can rapidly multiply into a substantial loss. It is worth noting that a machine producing 1,800 1-gallon milk containers per hour, and operating at 99% efficiency (one leaker in 100 containers) will cost the dairy some $200,000.00 per year in lost milk alone.

From a leak detection aspect, it is vitally necessary to establish a definition of terms. For instance, what is leakage? It is generally understood that the term implies the passage of commodity through a barrier which is meant to retain it: however, when the search for leakage was started it was discovered that the commodity could penetrate the barrier in several different ways, and required different techniques to find it.

Therefore the air monitor grouped types of leakage into four categories, the first being termed an "open" leak, which means that the barrier is thin and presents no restriction to a commodity passing through an orifice in it. This means that the commodity would experience an immediate pressure drop from one side of the barrier to the other.

The same size of orifice, occurring in a thicker barrier, would present a restriction to the flow of the commodity, resulting in a much slower pressure drop.

This restriction would be even more apparent if the passage had a bend or turn in it. To distinguish this type of leak from the "open" leak, it is referred to as a "channel." The two types present entirely different problems when detecting the leakage.

When there are a number of orifices in a thin barrier or, in other words, more than one open leak, it is referred to as a porous condition. This is not the general concept of porosity, but to distinguish between this and a number of "channel" orifices, the term "seepage" is used. The reason for this distinction becomes apparent when considering the methods available for detecting such leaks on a production basis.

The methods for detecting leaks of blow molded containers, fall into three basic categories:

- *The Achieved Pressure Test* The Achieved Pressure Test consists of filling a container with pressure at a controlled rate of flow and registering, by means of a pressure sensor, that the required pressure has been achieved in the container.

  Leakage from the container will prevent pressure building up to the required level, and therefore, the signal to accept the container will not be initiated.

  It will be seen that the system would be infallible if the orifice through the flow control could be the same size as the smallest leak to be detected. It would only allow pressure to flow into the container at the same rate as the leak would release it, and therefore, no pressure buildup would occur. Unfortunately, feeding the pressure through such a small orifice would take a long time to fill a container, far too long for production purposes. The process is speeded up by using a larger input flow orifice and relying upon the ability of the pressure sensor to distinguish the difference in the pressure buildup between leaking and nonleaking containers. A time limitation for the buildup can also be imposed, often by using the duration of the machine cycle for the test.

Because the size of the feed orifice is directly proportional to the pressure differential caused by leakage, the volume of a container tested by this method is limited by the time allowed for the test. This method can only be for "open" type leakage.

- *The Pressure Drop Test* The second method of leak detection consists of filling a container to a prescribed pressure, then shutting off the flow and determining that the container retains this pressure.

  This test is considerably more definitive than the achieved pressure test and can be used for any type of leakage. It will take longer to accomplish than the achieved pressure test when small volumes are involved, as the pressure must be allowed to "settle" before measurement is commenced. This is necessary as the inrush of air will cause an initial surge of pressure, which will drop substantially immediately when the flow is shut off. Such pressure drop will indicate leakage.

  In the case of larger containers the settling time may well be less than the time taken to fill the container for an achieved pressure test.

- *The Displacement Test* The displacement test can be applied as an achieved pressure or pressure drop test, the difference being that a definite quality of air is released into a container to achieve the test pressure, required. This is accomplished by filling a known volume to a prescribed pressure and then releasing the pressure/volume combination into the container being tested. The main advantage of this method is that it controls the inrush pressure surge and permits a faster test. In some cases, the displacement is actually used to generate the test pressure by using the volume chamber as a pump, which is actuated by contacting the container.

  This test is extremely susceptible to variations in volume and is very useful where such changes in volume are the only indication of leakage. It can, for instance, be used to determine that a cap is sealing correctly by covering the cap area, or the entire container if necessary, and releasing the known pressure/volume combination into the cavity created by the sealing action.

  Should some of the quantity of air introduced into the cavity escape into the container, via a leaking cap, the pressure build-up will not be achieved, or hold, as the case might be. This is true even though the cavity left in the container, when filled with a product, is minute.

These three methods do not, by any means, represent the only methods of leak testing available, but they are the methods most used in the testing of blow molded products.

Combination of the methods can be employed, such as controlling the flow into a container for a pressure drop test. This helps to prevent the inrush surge, and automatically includes an achieved pressure test. By using a timer to shut off the input flow, the combination of controlled pressure and time becomes a form of displacement test.

References

1. Rosato, D. V., Plastics Research and Development, Kluwer, 2001.
2. Rosato, D. V., PIA Plastics Engineering Manufacturing & Data Handbook, Kluwer, 2001.
3. Rosato, D. V., Concise Encyclopedia of Plastics (25,000 entries), Kluwer, 2000.
4. Rosato, D. V., Rosato's Plastics Encyclopedia and Dictionary, Hanser, 1993.
5. Rosato, D. V., Extruding Plastics: A Practical Processing Handbook, Kluwer, 1998.
6. Rosato, D. V., Injection Molding Handbook, 3rd edit., Kluwer, 2000.
7. Rosato, D. V., Injection Molding, in R. F. Jones, Guide to Short Fiber RP, Hanser, 1998.
8. Rosato, D. V., Plastics Processing Data Handbook, 2nd edit., Kluwer, 1997.
9. Rosato, D. V., Plastics Design Handbook, Kluwer, 2001
10. Rosato, D. V., Designing with Plastics and Composites: A Handbook, Kluwer, 1991.
11. Rosato, D. V., Designing with Plastics, Rhode Island School of Design, Lectures 1987–1990.
12. Rosato, D. V., Design Features Influence Performance: Detractors/Constraints, SPE-ANTEC, May 1991.
13. Rosato, D. V., Follow All Opportunities, SPE-IMD Newsletter, Issue 58, 2001.
14. Rosato, D. V., Statistical Process Control, SPE-IMD Newsletter Issues 56-58, 2001.
15. Rosato, D. V., Parts Removal with On-Robot Secondary Operations, Tool & Manufacturing Engineers Handbook, Vol. 9, Materials & Parts Handling, SME, 1998.
16. Rosato, D. V., Current and Future Trends in the Use of Plastics for Blow Molding, SME, 1990.
17. Rosato, D. V., Choosing Blow Molding Machines to Meet Product Demand, Plastics Today, Jan. 1989.
18. Rosato, D. V., Blow Molding Handbook, Hanser, 1989.
19. Rosato, D. V., Blow Molding Expanding Technology & Market, SPE-ANTEC, May 1987.
20. Rosato, D. V., Extrusion: Technology, Markets, Economics Seminar, University of Lowell, 1987.
21. Rosato, D. V., Plastics and Solid Waste, RISD, Oct. 1989.
22. Rosato, D V., et al., Plastics Industry Safety Handbook, Cahners, 1973.
23. Rosato, D. V., et al., Markets For Plastics, Kluwer, 1969.
24. Abbott, W. H., Statistics Can Be Fun, A. Abbott, Chesterland, OH 44026.
25. Arnold-Feret, B., Rapid Tooling & Plastics—where the RT Industry Stands in 2001 on Better Alternative Tooling Methods, SPE-ANTEC, May 2001.
26. ASTM Book of Standards, Section 8: Plastics, Three Volumes, Annual Issues.
27. Avallone E., et al., Mark's Standard Handbook of Mechanical Engineers, 10th Edit., McGraw Hill, 1996.
28. Bass, M., Handbook of Optics, 2nd edit., McGraw-Hill, 1995.
29. Beal, G., Textured Finishes, IM, Jun. 2000.
30. Beaumont, J. P., Mold Filling Imbalance in Geometrically Balanced Runner Systems, SPE-IMD Newsletter 49, Fall 1998.

31. Beck, N., Plastics Product Design, Kluwer, 1980.
32. Bernhardt, A., Computer Modeling Predicts the Dimensional Tolerances of Molded Parts, SPE-IMD Newsletter No. 55, Fall, 1990.
33. Bernhardt, A., et al., Rationalization of Molding Machine Intelligent Setting & Control, SPE-IMD Newsletter, No. 54, Summer 2000.
34. Biodegradable Polymers, Chemical Industry Newsletter, SRI Consulting, Issue 1, 2000.
35. Brassine, C., Adhesion between PP-Based Elastomer & PVDF in Layered Structures by Interleafing a Grafted Copolymer, SPE-ANTEC, May 2001.
36. Busch, J., & F. Field, IBIS Associates, Wellesley, MA (tel. 781-239-0666), Communication 2000.
37. Cappelletti, M., In Defense of Plastics, World Plastics Technology, 2000.
38. Carvill, J., Mechanical Engineer's Data Handbook, CRC Press, 1993.
39. Cerwin, N., Your Guide to Mold Steels, Plastics Auxiliaries, Mar. 1999.
40. Choffel, J. E., et al., Verifying the Accuracy of Internal Stresses from BM Simulation Software Used for FEA Stress Analysis, SPE-ANTEC, May 2001.
41. Colby, P. N., Plasticating Components Technology: Screw & Barrel Technology, Spirex Bulletin, Spirex Corp., 2000.
42. Coleman, B. D., Thermodynamics of Materials with Memory Treatise, Arch. Rat. Mech. Anal., 1964.
43. Computer More than Child's Play, The Inside Line, Mar, 1999.
44. Consumers Computers Intolerable, The Inside Line, Mar. 1999.
45. Daido Steel, Tool Steel Selection Software Targets Mold Making Startups, Modern Mold, 2000.
46. Defosse, M., SBM Machine Makers Eye More Turnkey Lines, MP, Apr. 2001.
47. Dealy, J. M., in Rosato et al. (eds.) Blow Molding Handbook, Chapter 17, Hanser, 1989.
48. Dowler, B. L., et al., Re-Inventing the Wheel: Breaking Ground with Rotary Shuttle Technology, SPE-ANTEC, May 2001.
49. Facts & Figures of the U.S. Plastics Industry, Annual updates.
50. Ferris, R. M., >35,000 Plastic Materials (via IDES), PE, Jul. 2000.
51. Glanville, A. B., Plastics Engineers Data Book, Machine Publ., 1971.
52. Gross, H. G., New Technology to Vary the Thickness of the Parison in Circumferential Direction During the BM Process, SPE-ANTEC, May 2001.
53. Gust, P., et al., The Early Use of Process Simulation to Optimize the Wall Thickness of Blow Molded Plastic Parts, SPE-ANTEC, May 2001.
54. Hannagan, T., The Use and Misuse of Statistics, Harvard Management Update, May 2000.
55. Jivaj, N., et al., Large Part BM of HDPE: Parison Behavior, SPE-ANTEC, May 2001.
56. Kirkland, C., Manufacturing Challenges: Time, Money, & Resistance to Change, IM, Jun. 2000.
57. Knights, M., Plastics Plants Can Be Safe Even for Blind Workers, PT, Jul. 2000.
58. Lee, N. C., Blow Molding Design Guide, Hanser, 1998.
59. Lowen, R., et al., Fuzzy Logic, Kluwer, 1993.
60. Luker, K., Surge Suppression—A New Means to Limit Surging, SPE-ANTEC, May 2001.
61. McCullough, M. D., et al., Verification of Parison Sag & Swell with CAE Simulation Software, SPE-ANTEC, May 2001.
62. Benedikt, G. M., Metallocene Technology and Modern Catalytic Methods in Commercial Applications, PDL, 1999.
63. Missing Link in Computer Integrated Manufacturing, IM, May 2000.
64. Ng, W., et al., Suppliers Unveil New Approaches to Produce Controlled Rheology PP, MP, Apr. 2000.

65. Oberg, E., et al., Machinery's Handbook, 25th edit., Industrial Press, 1996.
66. Oliver, D., Developing Design Control Strategies to Meet Technology Advances, MD&DI, Sep. 2000.
67. Plastic Beer Bottles on Trial, Plastics Solutions International, 2000.
68. Rejon, A. G., et al., Injection/Stretch Blow Molding of Polymer/Clay Nanocomposites, SPE-ANTEC, May 2001.
69. Rao, N., Design Data for Plastics Engineers, Hanser, 1998.
70. Rao, N., Design Formulas for Plastics Engineers, Hanser, 1991.
71. Rauwendaal, C., Statistical Process Control in Extrusion, Hanser, 1993.
72. Rauwendaal, C., Statistical Process Control in Injection Molding and Extrusion, Hanser, 2000
73. Rees, H., Understanding Injection Mold Design, Hanser, 2001.
74. Report Says Recycling No Panacea, MP, Apr. 2000.
75. Rios, J. B., et al., Die Swell Estimation for HDPE BM Grade Resins Using the Wagner Model Constitutive Equation, SPE-ANTEC, May 2001.
76. Rohsenow, W. M., et al., Handbook of Heat Transfer, 3rd edit., McGraw-Hill, 1998.
77. Schmidt, L. R., et al., Biaxial Stretching of Heat-Softened Plastic Sheets, Pergamon Press, Int. J. Engng Sci., Vol. 13, pp. 563–578, 1975.
78. Shah, V., Handbook of Plastics Testing Technology, 2nd edit., Wiley, 1998.
79. Sherratt. D., Taking a Risk-Based Approach to Medical Device Design, MD&DI, Sep. 1999.
80. Sims, F., Engineering Formulas, Industrial Press, 1999.
81. Stoeckhert, K., Mold Making Handbook, 2nd edit., Hanser, 1998.
82. Straff, R., et al., Commercialization of Microcellular Blow Molding, SPE-ANTEC, May 2001.
83. Thielen, M., et al., 3D Blow Molding Today, An Overview About Different Systems Which Are Established in the Market Today, SPE-ANTEC, May 2001.
84. Verbruggen, H. B., et al., Fuzzy Logarithms for Control, Kluwer, 1999.
85. Wigotsky, V., Marketing and Management—A Wider Focus, PE, May 1996.
86. Wenskus, J. J., Transfer Point Control Comparison Between Mold Parting Line and the Standard Strategies, SPE-IMD Newsletter 22, Fall 1989.
87. Wood, K., Tooling Challenges, IM, May 2000.
88. Woodward, A. E., Understanding Polymer Morphology, Hanser, 1995.
89. Wright, D. C., Environmental Stress Cracking of Plastics, PDL, 1996.
90. Wypych, G., Weathering of Plastics, PDL, 2000.
91. Young, W. C., Roark's Formulas for Stress and Strain, 6th edit., McGraw-Hill, 1998.
92. Zahnel, K., et al., Introduction to Microcontrollers, Butterworth-Heinemann, 1997.

Index

F

G

H

T

U

V

W

Z

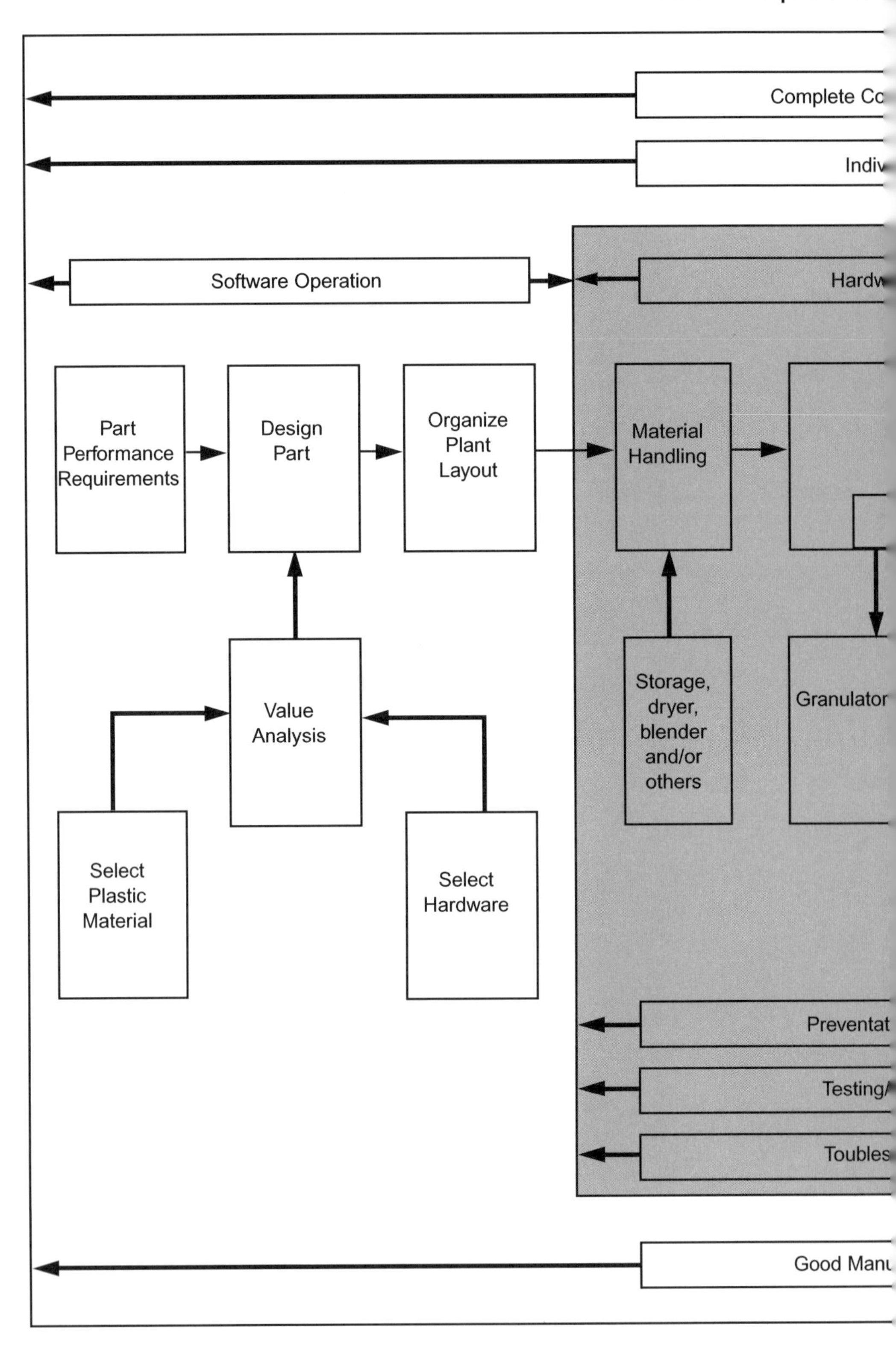

Software Operation
Part Performance Requirements
Design Part
Organize Plant Layout
Material Handling
Value Analysis
Storage, dryer, blender and/or others
Granulator
Select Plastic Material
Select Hardware